D1189657

ORGANIC CHEMISTRY AS A SECOND LANGUAGE, 3e

ORGANIC CHEMISTRY AS A SECOND LANGUAGE, 3e

Second Semester Topics

DAVID KLEIN
Johns Hopkins University

JOHN WILEY & SONS, INC.

ASSOCIATE PUBLISHER	Petra Recter
EDITORIAL ASSISTANT	Lauren Stauber
MARKETING MANAGER	Kristine Ruff
SENIOR PRODUCTION EDITOR	Sujin Hong; Production Management Services provided by Prepare, Inc.
CREATIVE DIRECTOR	Harry Nolan
SENIOR COVER DESIGNER	Wendy Lai
COVER CREDITS	Background: © William Hopkins/iStockphoto
	Test tube: Untitled X-Ray/Nick Veasey/ Getty Images, Inc.
	Bicycle: Igor Shikov/Shutterstock

This book was set in 9/11 Times Roman by Prepare, Inc. and printed and bound by Courier Westford. The cover was printed by Courier Westford.

This book is printed on acid-free paper. ∞

Founded in 1807, John Wiley & Sons, Inc. has been a valued source of knowledge and understanding for more than 200 years, helping people around the world meet their needs and fulfill their aspirations. Our company is built on a foundation of principles that include responsibility to the communities we serve and where we live and work. In 2008, we launched a Corporate Citizenship Initiative, a global effort to address the environmental, social, economic, and ethical challenges we face in our business. Among the issues we are addressing are carbon impact, paper specifications and procurement, ethical conduct within our business and among our vendors, and community and charitable support. For more information, please visit our website: www.wiley.com/go/citizenship.

Evaluation copies are provided to qualified academics and professionals for review purposes only, for use in their courses during the next academic year. These copies are licensed and may not be sold or transferred to a third party. Upon completion of the review period, please return the evaluation copy to Wiley. Return instructions and a free of charge return mailing label are available at www.wiley.com/go/returnlabel. If you have chosen to adopt this textbook for use in your course, please accept this book as your complimentary desk copy. Outside of the United States, please contact your local sales representative.

ISBN 978-1-118-14434-3

Printed in the United States of America

10 9 8 7 6

CONTENTS

CHAPTER 8 *AMINES* **281**

CHAPTER 1

IR SPECTROSCOPY

Did you ever wonder how chemists are able to determine whether or not a reaction has produced the desired products? In your textbook, you will learn about many, many reactions. And an obvious question should be: "how do chemists *know* that those are the products of the reactions?

Until about 50 years ago, it was actually VERY difficult to determine the structures of the products of a reaction. In fact, chemists would often spend many months, or even years to elucidate the structure of a single compound. But things got a lot simpler with the advent of spectroscopy. These days, the structure of a compound can be determined in minutes. Spectroscopy is, without a doubt, one of the most important tools available for determining the structure of a compound. Many Nobel prizes have been awarded over the last few decades to chemists who pioneered applications of spectroscopy.

The basic idea behind all forms of spectroscopy is that electromagnetic radiation (light) can interact with matter in predictable ways. Consider the following simple analogy: imagine that you have 10 friends, and you know what kind of bakery items they each like to eat every morning. John always has a brownie, Peter always has a French roll, Mary always has a blueberry muffin, etc. Now imagine that you walk into the bakery just after it opens, and you are told that some of your friends have already visited the bakery. By looking at what is missing from the bakery, you could figure out which of your friends had just been there. If you see that there is a brownie missing, then you deduce that John was in the bakery before you.

This simple analogy breaks down when you really get into the details of spectroscopy, but the basic idea is a good starting point. When electromagnetic radiation interacts with matter, certain frequencies are absorbed while other frequencies are not. By analyzing which frequencies were absorbed (which frequencies are missing once the light passes through a solution containing the unknown compound), we can glean useful information about the structure of the compound.

You may recall from your high school science classes that the range of all possible frequencies (of electromagnetic radiation) is known as the electromagnetic spectrum, which is divided into several regions (including X-rays, UV light, visible light, infrared radiation, microwaves, and radio waves). Different regions of the electromagnetic spectrum are used to probe different aspects of molecular structure, as seen in the table below:

Type of Spectroscopy	Region of Electromagnetic Spectrum	Information Obtained
NMR Spectroscopy	Radio Waves	The specific arrangement of all carbon and hydrogen atoms in the compound
IR Spectroscopy	Infrared	The functional groups present in the compound
UV-Vis Spectroscopy	Visible and Ultraviolet	Any conjugated π system present in the compound

We will not cover UV-Vis spectroscopy in this book. Your textbook will have a short section on that form of spectroscopy. In this chapter, we will focus on the information that can be obtained with IR spectroscopy. Chapter 2 will cover NMR spectroscopy.

1.1 VIBRATIONAL EXCITATION

Molecules can store energy in a variety of ways. They rotate in space, their bonds vibrate like springs, their electrons can occupy a number of possible molecular orbitals, etc. According to the principles of quantum mechanics, each of these forms of energy is quantized. For example, a bond in a molecule can only vibrate at specific energy levels:

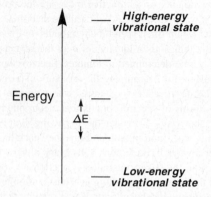

The horizontal lines in this diagram represent allowed vibrational energy levels for a particular bond. The bond is restricted to these energy levels, and cannot vibrate with an energy that is in between the allowed levels. The difference in energy (ΔE) between allowed energy levels is determined by the nature of the bond. If a photon of light possesses exactly this amount of energy, the bond (which was already vibrating) can absorb the photon to promote a *vibrational excitation*. That is, the bond will now vibrate more energetically (a larger amplitude). The energy of the photon is temporarily stored as vibrational energy, until that energy is released back into the environment, usually in the form of heat.

Bonds can store vibrational energy in a number of ways. They can *stretch,* very much the way a spring stretches, or they can *bend* in a number of ways. Your textbook will likely have images that illustrate these different kinds of vibrational excitation. In this chapter, we will devote most of our attention to stretching vibrations (as opposed to bending vibrations) because stretching vibrations generally provide the most useful information.

For each and every bond in a molecule, the energy gap between vibrational states is very much dependent on the nature of the bond. For example, the energy gap for a C—H bond is much larger than the energy gap for a C—O bond:

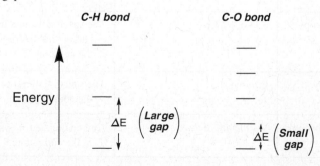

Both bonds will absorb IR radiation, but the C—H bond will absorb a higher energy photon. A similar analysis can be performed for other types of bonds as well, and we find that each type of bond will absorb a characteristic frequency, allowing us to determine which types of bonds are present in a compound. For example, a compound containing an O—H bond will absorb a frequency of IR radiation characteristic of O—H bonds. In this way, *IR spectroscopy can be used to identify the presence of functional groups in a compound.* It is important to realize that IR spectroscopy does NOT reveal the entire structure of a compound. It can indicate that an unknown compound is an alcohol, but to determine the entire structure of the compound, we will need NMR spectroscopy (covered in the next chapter). For now, we are simply focusing on identifying which functional groups are present in an unknown compound. To get this information, we simply irradiate the compound with all frequencies of IR radiation, and then detect which frequencies were absorbed. This can be achieved with an IR spectrometer, which measures absorption as a function of frequency. The resulting plot is called an IR absorption spectrum (or IR spectrum, for short).

1.2 IR SPECTRA

An example of an IR spectrum is shown below:

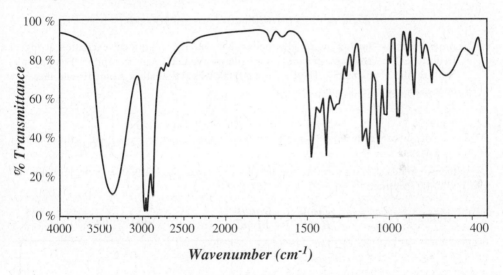

Wavenumber (cm⁻¹)

Notice that all signals point down in an IR spectrum. The location of each signal on the spectrum is reported in terms of a frequency-related unit, called wavenumber ($\widetilde{\nu}$). The wavenumber is simply the frequency of light (ν) divided by a constant (the speed of light, c):

$$\widetilde{\nu} = \frac{\nu}{c}$$

The units of wavenumber are inverse centimeters (cm^{-1}), and the values range from 400 cm^{-1} to 4000 cm^{-1}. Don't confuse the terms wavenumber and wavelength. Wavenumber is proportional to frequency, and therefore, a larger wavenumber represents higher energy. Signals that appear on the left side of the spectrum correspond with higher energy radiation, while signals on the right side of the spectrum correspond with lower energy radiation.

Every signal in an IR spectrum has the following three characteristics:

1. the *wavenumber* at which the signal appears
2. the *intensity* of the signal (strong vs. weak)
3. the *shape* of the signal (broad vs. narrow)

We will now explore each of these three characteristics, starting with wavenumber.

1.3 WAVENUMBER

For any bond, the wavenumber of absorption associated with bond stretching is dependent on two factors:

1) *Bond strength* – Stronger bonds will undergo vibrational excitation at higher frequencies, thereby corresponding to a higher wavenumber of absorption. For example, compare the bonds below. The C≡N bond is the strongest of the three bonds and therefore appears at the highest wavenumber:

$$C\equiv N \qquad\qquad C=N \qquad\qquad C-N$$

$$\sim 2200\ cm^{-1} \qquad \sim 1600\ cm^{-1} \qquad \sim 1100\ cm^{-1}$$

2) *Atomic mass* – Smaller atoms give bonds that undergo vibrational excitation at higher frequencies, thereby corresponding to a higher wavenumber of absorption. For example, compare the bonds below. The C—H bond involves the smallest atom (H) and therefore appears at the highest wavenumber.

$$C-H \qquad\qquad C-D \qquad\qquad C-O \qquad\qquad C-Cl$$

$$\sim 3000\ cm^{-1} \qquad \sim 2200\ cm^{-1} \qquad \sim 1100\ cm^{-1} \qquad \sim 700\ cm^{-1}$$

Using the two trends shown above, we see that different types of bonds will appear in different regions of an IR spectrum:

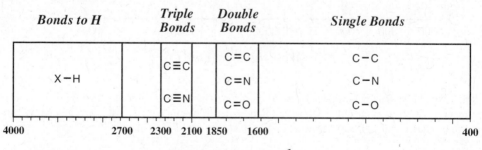

$$\textbf{\textit{Wavenumber (cm}}^{-1}\textbf{\textit{)}}$$

Single bonds appear on the right side of the spectrum, because single bonds are generally the weakest bonds. Double bonds appear at higher wavenumber (1600–1850 cm^{-1}) because they are stronger than single bonds, while triple bonds appear at even higher wavenumber (2100–2300 cm^{-1}) because they are even stronger than double bonds. And finally, the left side of the spectrum contains signals produced by X—H bonds (such as C—H, O—H, or N—H), all of which stretch at a high wavenumber because hydrogen has the smallest mass.

IR spectra can be divided into two main regions, called the diagnostic region and the finger-print region:

DIAGNOSTIC REGION			FINGERPRINT REGION
Bonds to H	*Triple Bonds*	*Double Bonds*	*Single Bonds*

4000   3500   3000   2500   2000   **1500**   1000   400

Wavenumber (cm⁻¹)

The diagnostic region generally has fewer peaks and provides the most information. This region contains all signals that arise from the stretching of double bonds, triple bonds, and X—H bonds. The fingerprint region contains mostly bending vibrations, as well as stretching vibrations of most single bonds. This region generally contains many signals, and is more difficult to analyze. What appears like a C—C stretch might in fact be another bond that is bending. This region is called the fingerprint region because each compound has a unique pattern of signals in this region, much the way each person has a unique fingerprint. For example, IR spectra of ethanol and propanol will look extremely similar in their diagnostic regions, but their fingerprint regions will look different. For the remainder of this chapter, we will focus exclusively on the signals that appear in the diagnostic region, and we will ignore signals in the fingerprint region. You should check your lecture notes and textbook to see if you are responsible for any characteristic signals that appear in the fingerprint region.

PROBLEM 1.1   For the following compound, rank the highlighted bonds in order of increasing wavenumber.

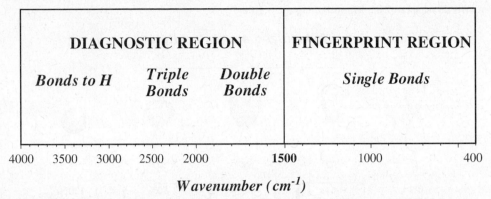

Now let's continue exploring factors that affect the strength of a bond (which therefore affects the wavenumber of absorption). We have seen that bonds to hydrogen (such as C—H bonds) appear on the left side of an IR spectrum (high wavenumber). We will now compare various kinds of C—H bonds. The wavenumber of absorption for a C—H bond is very much dependent on the hybridization state of the carbon atom. Compare the following three C—H bonds:

sp^3 sp^2 sp

$>$C—H $=$C—H $\equiv$C—H

~ 2900 cm⁻¹   ~ 3100 cm⁻¹   ~ 3300 cm⁻¹

Of the three bonds shown, the C_{sp}—H bond produces the highest energy signal (~3300 cm^{-1}), while a C_{sp^3}—H bond produces the lowest energy signal (~2900 cm^{-1}). To understand this trend, we must revisit the shapes of the hybridized atomic orbitals:

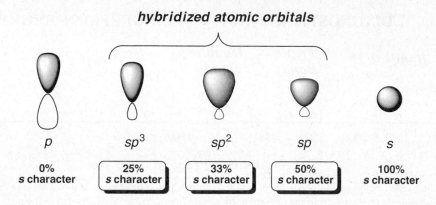

hybridized atomic orbitals

p	sp^3	sp^2	sp	s
0% s character	**25%** s character	**33%** s character	**50%** s character	**100%** s character

As illustrated, sp orbitals have more s character than the other hybridized atomic orbitals, and therefore, sp orbitals more closely resemble s orbitals. Compare the shapes of the hybridized atomic orbitals, and note that the electron density of an sp orbital is closest to the nucleus (much like an s orbital). As a result, a C_{sp}—H bond will be shorter than other C—H bonds. Since it has the shortest bond length, it will therefore be the strongest bond. In contrast, the C_{sp^3}—H bond has the longest bond length, and is therefore the weakest bond. Compare the spectra of an alkane, an alkene, and an alkyne:

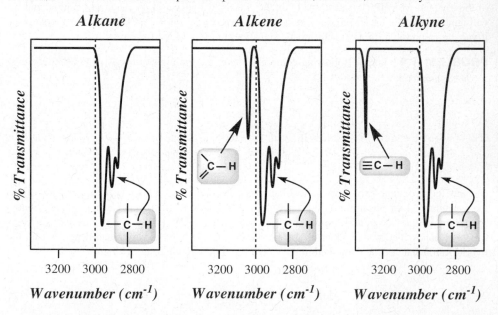

In each case, we draw a line at 3000 cm^{-1}. All three spectra have signals to the right of the line, resulting from C_{sp^3}—H bonds. The key is to look for any signals to the left of the line. An alkane does not have a signal to the left of 3000 cm^{-1}. An alkene has a signal at 3100 cm^{-1}, and an alkyne has a signal at 3300 cm^{-1}. But be careful—the absence of a signal to the left of 3000 cm^{-1} does *not* necessarily indicate the absence of a double bond or triple bond in the compound. Tetrasubstituted double bonds do not possess any C_{sp^2}—H bonds, and internal triple bonds also do not possess any C_{sp}—H bonds.

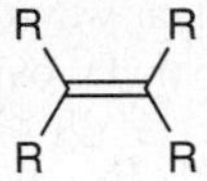

no signal at 3100 cm^{-1}

$\left(\text{no } C_{sp^2}\text{−}H\right)$

no signal at 3300 cm^{-1}

$\left(\text{no } C_{sp}\text{−}H\right)$

PROBLEMS For each of the following compounds, determine whether or not you would expect its IR spectrum to exhibit a signal to the left of 3000 cm^{-1}

1.2 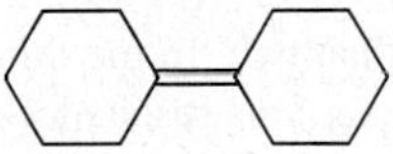   **1.3** 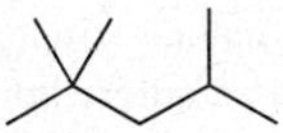   **1.4** 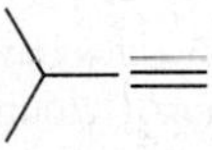   **1.5**

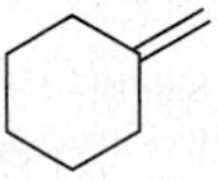

Now let's explore the effects of resonance on bond strength. As an illustration, compare the carbonyl groups (C=O bonds) in the following two compounds:

A ketone

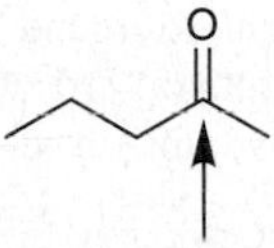

A conjugated ketone

1720 cm^{-1}

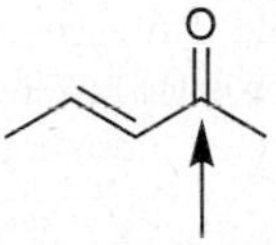

1680 cm^{-1}

The second compound is called an unsaturated, conjugated ketone. It is *unsaturated* because of the presence of a C=C bond, and it is conjugated because the π bonds are separated from each other by exactly one single bond. Your textbook will explore conjugated π systems in more detail. For now, we will just analyze the effect of conjugation on the IR absorption of the carbonyl group. As shown, the carbonyl group of an unsaturated, conjugated ketone produces a signal at lower wavenumber (1680 cm^{-1}) than the carbonyl group of a saturated ketone (1720 cm^{-1}). In order to understand why, we must draw resonance structures for each compound. Let's begin with the ketone.

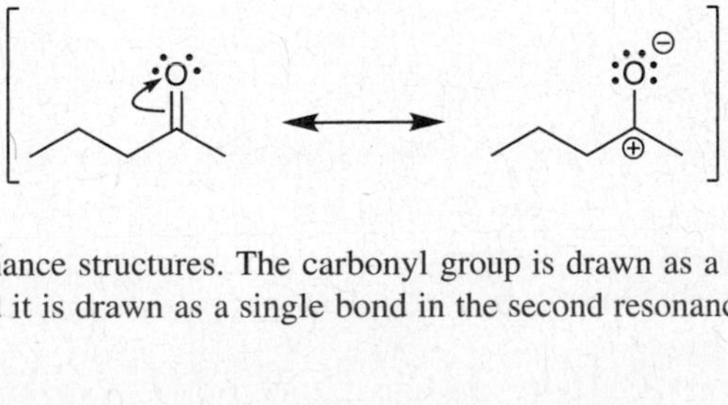

Ketones have two resonance structures. The carbonyl group is drawn as a double bond in the first resonance structure, and it is drawn as a single bond in the second resonance structure. This means that the carbonyl group has some double-bond character and some single-bond character. In order to determine the nature of this bond, we must consider the contribution from each resonance structure. In other words, does the carbonyl group have more double-bond character or more single-bond character? The second resonance structure exhibits charge separation, as well as a carbon atom (C^+) that has less than an octet of electrons. Both of these reasons explain why the second resonance structure contributes only slightly to the overall character of the carbonyl group. Therefore, the carbonyl group of a ketone has mostly double-bond character.

Now consider the resonance structures for a conjugated, unsaturated ketone.

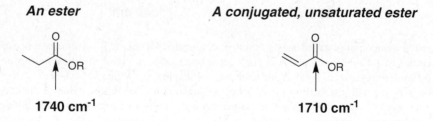

one additional
resonance structure

Conjugated, unsaturated ketones have three resonance structures rather than two. In the third resonance structure, the carbonyl group is drawn as a single bond. Once again, this resonance structure exhibits charge separation as well as a carbon atom (C^+) with less than an octet of electrons. As a result, this resonance structure also contributes only slightly to the overall character of the compound. Nevertheless, this third resonance structure does contribute some character, giving this carbonyl group slightly more single-bond character than the carbonyl group of a saturated ketone. With more single-bond character, it is a slightly weaker bond, and therefore produces a signal at a lower wavenumber (1680 cm^{-1} rather than 1720 cm^{-1}).

Esters exhibit a similar trend. An ester typically produces a signal at around 1740 cm^{-1}, but conjugated, unsaturated esters produce lower energy signals, usually around 1710 cm^{-1}. Once again, the carbonyl group of a conjugated, unsaturated ester is a weaker bond, due to resonance.

An ester

1740 cm^{-1}

A conjugated, unsaturated ester

1710 cm^{-1}

PROBLEM 1.6 The following compound has three carbonyl groups. Rank them in order of increasing wavenumber in an IR spectrum:

1.4 SIGNAL INTENSITY

In an IR spectrum, some signals will be very strong in comparison with other signals on the same spectrum:

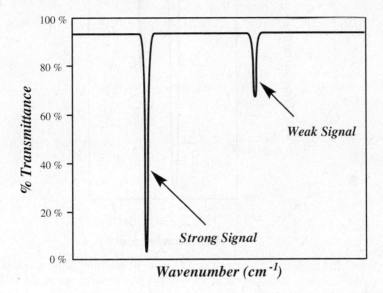

That is, some bonds absorb IR radiation very efficiently, while other bonds are less efficient at absorbing IR radiation. The efficiency of a bond at absorbing IR radiation depends on the strength of the dipole moment for that bond. For example, compare the following two high-lighted bonds:

Each of these bonds has a measurable dipole moment, but they differ significantly in strength. Let's first analyze the carbonyl group (C=O bond). Due to resonance and induction, the carbon atom bears a large partial positive charge, and the oxygen atom bears a large partial negative charge. The carbonyl group therefore has a large dipole moment. Now let's analyze the C=C bond. One vinylic position is connected to electron-donating alkyl groups, while the other vinylic position is connected to hydrogen atoms. As a result, the vinylic position bearing two alkyl groups is slightly more electron-rich than the other vinylic position, producing a small dipole moment.

Since the carbonyl group has a larger dipole moment, the carbonyl group is more efficient at absorbing IR radiation, producing a stronger signal:

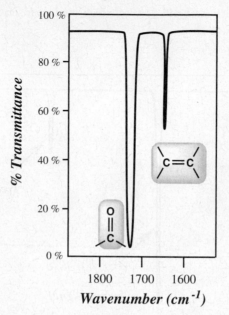

Carbonyl groups often produce the strongest signals in an IR spectrum, while C=C bonds often produce fairly weak signals. In fact, some alkenes do not even produce any signal at all. For example, consider the IR spectrum of 2,3-dimethyl-2-butene:

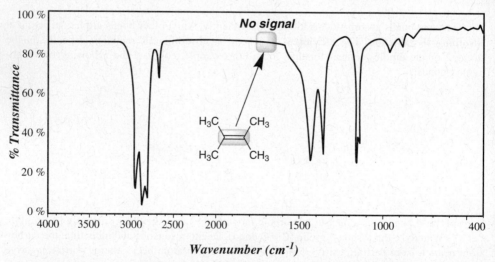

This alkene is symmetrical. That is, both vinylic positions are electronically identical, and the bond has no dipole moment at all. As such, this C=C bond is completely inefficient at absorbing IR radiation, and no signal is observed. The same is true for symmetrical C≡C bonds.

There is one other factor that can contribute significantly to the intensity of signals in an IR spectrum. Consider the group of signals appearing just below 3000 cm^{-1} in the previous spectrum. These signals are associated with the stretching of the C—H bonds in the compound. The intensity of these signals derives from the number of C—H bonds giving rise to the signals. In fact, the signals just below 3000 cm^{-1} are typically among the strongest signals in an IR spectrum.

PROBLEMS

1.7 Predict which of the following C=C bonds will produce the strongest signal in an IR spectrum:

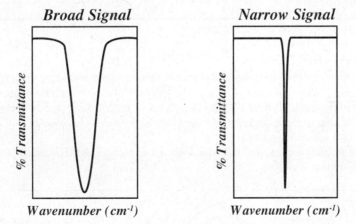

1.8 The C=C bond in the following compound produces an unusually strong signal. Explain using resonance structures:

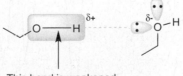

1.5 SIGNAL SHAPE

In this section, we will explore some of the factors that affect the shape of a signal. Some signals in an IR spectrum might be very broad while other signals can be very narrow:

Broad Signal *Narrow Signal*

% Transmittance

Wavenumber (cm^{-1})

% Transmittance

Wavenumber (cm^{-1})

Concentrated alcohols commonly exhibit broad O—H signals, as a result of hydrogen bonding, which weakens the O—H bonds.

This bond is weakened
as a result of H-bonding

At any given moment in time, the O—H bond in each molecule is weakened to a different extent. As a result, all of the O—H bonds do not have a uniform bond strength, but rather, there is a *distribution* of bond strength. That is, some molecules are barely participating in H-bonding, while others are participating in H-bonding to varying degrees. The result is a broad signal.

The shape of an O—H signal is different when the alcohol is diluted in a solvent that cannot form hydrogen bonds with the alcohol. In such an environment, it is likely that the O—H bonds will not participate in an H-bonding interaction. The result is a narrow signal. When the solution is neither very concentrated nor very dilute, two signals are observed. The molecules that are not participating in H-bonding will give rise to a narrow signal, while the molecules participating in H-bonding will give rise to a broad signal. As an example, consider the following spectrum of 2-butanol, in which both signals can be observed:

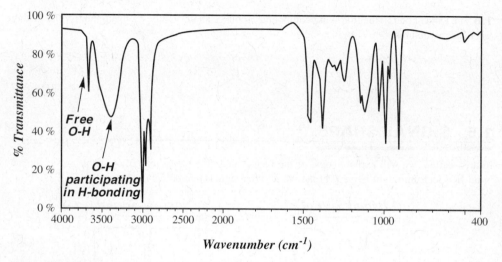

Wavenumber (cm^{-1})

When O—H bonds do not participate in H-bonding, they generally produce a signal at approximately 3600 cm^{-1}. That signal can be seen in the spectrum above. When O—H bonds participate in H bonding, they generally produce a broad signal between 3200 cm^{-1} and 3600 cm^{-1}. That signal can also be seen in the spectrum. Depending on the conditions, an alcohol will either give a broad signal, or a narrow signal, or both.

Carboxylic acids exhibit similar behavior; only more pronounced. For example, consider the following spectrum of a carboxylic acid:

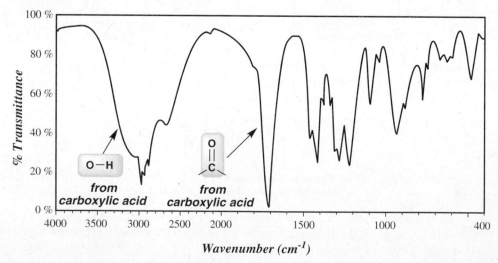

Wavenumber (cm^{-1})

Notice the very broad signal on the left side of the spectrum, extending from 2200 cm^{-1} to 3600 cm^{-1}. This signal is so broad that it extends over the usual C—H signals that appear around 2900 cm^{-1}. This very broad signal, characteristic of carboxylic acids, is a result of H bonding. The effect is more pronounced than alcohols, because molecules of the carboxylic acid can form two hydrogen-bonding interactions, resulting in a dimer.

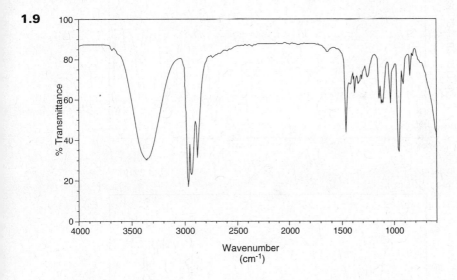

The IR spectrum of a carboxylic acid is easy to recognize, because of the characteristic broad signal that covers nearly one third of the spectrum. This broad signal is also accompanied by a broad C=O signal just above 1700 cm^{-1}.

PROBLEMS   For each IR spectrum below, identify whether it is consistent with the structure of an alcohol, a carboxylic acid, or neither.

1.9

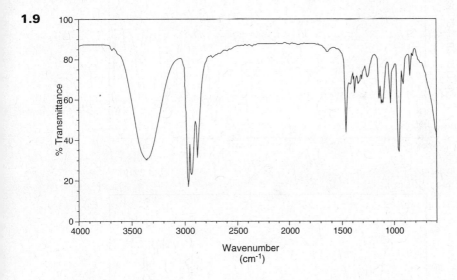

1.10

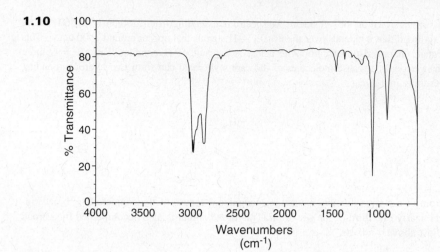

1.11

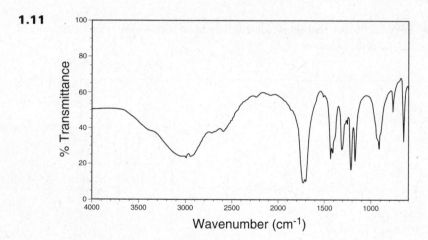

1.12

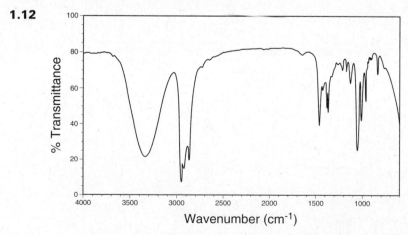

1.13

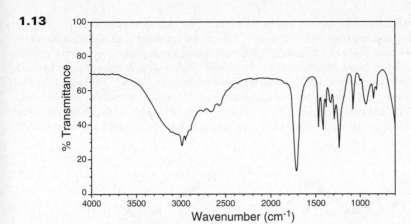

1.14

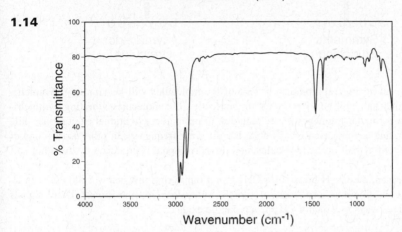

There is another important factor in addition to H bonding that affects the shape of a signal. Consider the difference in shape of the N—H signals for primary and secondary amines:

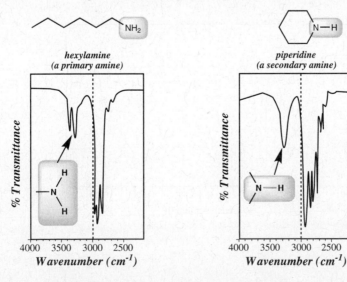

hexylamine
(a primary amine)

piperidine
(a secondary amine)

The primary amine exhibits two signals; one at 3350 cm^{-1} and the other at 3450 cm^{-1}. In contrast, the secondary amine exhibits only one signal. It might be tempting to explain this by arguing that each N—H bond gives rise to a signal, and therefore a primary amine gives two signals because it has two N—H bonds. Unfortunately, that simple explanation is not accurate. In fact, both N—H bonds of a single molecule will together produce only one signal. The reason for the appearance of two signals is more accurately explained by considering the two possible ways in which the entire NH$_2$ group can vibrate. The N—H bonds can be stretching in phase with each other, called *symmetric stretching,* or they can be stretching out of phase with each other, called *asymmetric stretching:*

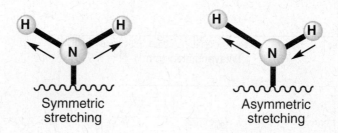

Symmetric
stretching

Asymmetric
stretching

At any given moment in time, approximately half of the molecules will be vibrating symmetrically, while the other half will be vibrating asymmetrically. The molecules vibrating symmetrically will absorb a particular frequency of IR radiation to promote a vibrational excitation, while the molecules vibrating asymmetrically will absorb a different frequency. In other words, one of the signals is produced by half of the molecules, and the other signal is produced by the other half of the molecules.

For a similar reason, the C—H bonds of a CH$_3$ group (appearing just below 3000 cm^{-1} in an IR spectrum) generally give rise to a series of signals, rather than just one signal. These signals arise from the various ways in which a CH$_3$ group can be excited.

PROBLEMS   For each IR spectrum below, determine whether it is consistent with the structure of a ketone, an alcohol, a carboxylic acid, a primary amine, or a secondary amine.

1.15

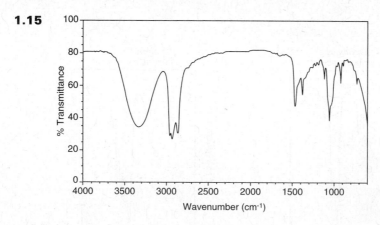

1.16

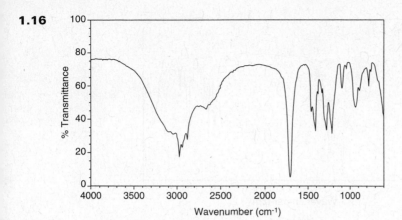

1.17

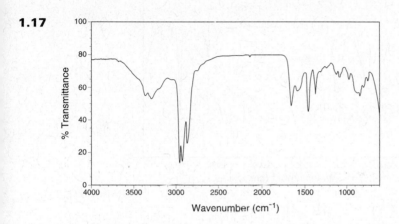

1.18

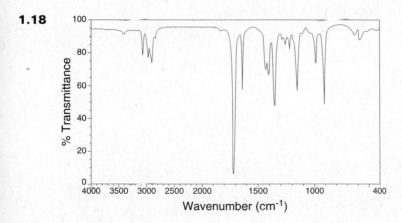

1.19

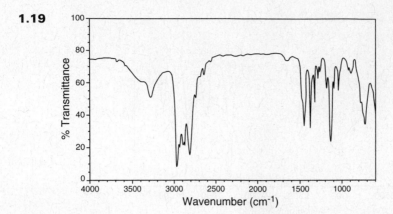

1.20

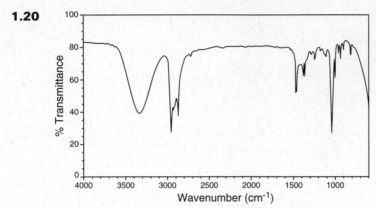

1.6 ANALYZING AN IR SPECTRUM

The following table is a summary of useful signals in the diagnostic region of an IR spectrum:

Useful Signals in the Diagnostic Region			
Structural Unit	**Wavenumber (cm⁻¹)**	**Structural Unit**	**Wavenumber (cm⁻¹)**
Single Bonds (X-H)		*Double Bonds*	
—O—H	3200 - 3600	Cl—C=O	1750 - 1850
(O)—O—H	2200 - 3600	RO—C(R)=O	1700 - 1750
N—H	3350 - 3500	HO—C(R)=O	1700 - 1750
≡C—H	~ 3300	R—C(R)=O	1680 - 1750
C—H	3000 - 3100	H₂N—C(R)=O	1650 - 1700
—C—H	2850 - 3000	C=C	1600 - 1700
(O)C—H	2750 - 2850	(benzene ring)	1450 - 1600 1650 - 2000
Triple Bonds			
—C≡C—	2100 - 2200		
—C≡N	2200 - 2300		

When analyzing an IR spectrum, the first step is to draw a line at 1500 cm⁻¹. Focus on any signals to the left of this line (the diagnostic region). In doing so, it will be extremely helpful if you can identify the following regions:

Double bonds: 1600–1850 cm⁻¹
Triple bonds: 2100–2300 cm⁻¹
X—H bonds: 2700–4000 cm⁻¹

Remember that each signal appearing in the diagnostic region will have three characteristics (wavenumber, intensity, and shape). Make sure to analyze all three characteristics.

When looking for X—H bonds, draw a line at 3000 cm^{-1} and look for signals that appear to the left of the line:

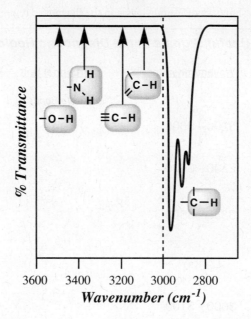

EXERCISE 1.21 A compound with molecular formula $C_6H_{10}O$ gives the following IR spectrum:

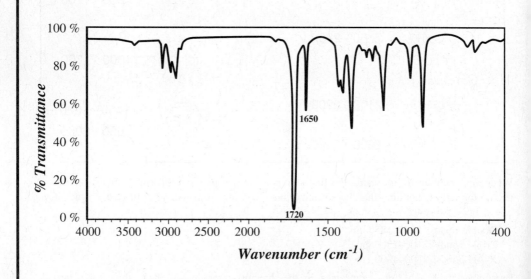

Identify the structure below that is most consistent with the spectrum:

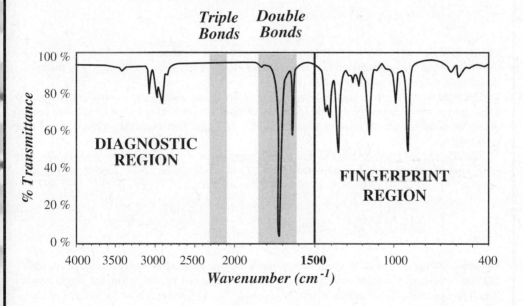

Solution Draw a line at 1500 cm^{-1}, and focus on the diagnostic region (to the left of the line). Start by looking at the double-bond region and the triple-bond region:

There are no signals in the triple-bond region, but there are two signals in the double-bond region. The signal at 1650 cm^{-1} is narrow and weak, consistent with a C≡C bond. The signal at 1720 cm^{-1} is strong, consistent with a C=O bond.

Next, look for X—H bonds. Draw a line at 3000 cm^{-1}, and identify if there are any signals to the left of this line.

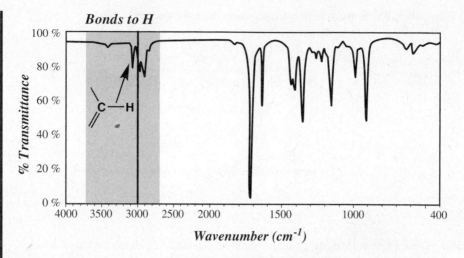

This spectrum exhibits one signal just above 3000 cm^{-1}, indicating a vinylic C—H bond.

The identification of a vinylic C—H bond is consistent with the observed C=C signal present in the double-bond region (1650 cm^{-1}). There are no other signals above 3000 cm^{-1}, so the compound does not possess any OH or NH bonds.

The little bump between 3400 and 3500 cm^{-1} is not strong enough to be considered a signal. These bumps are often observed in the spectra of compounds containing a C=O bond. The bump occurs at exactly twice the wavenumber of the C=O signal, and is called an overtone of the C=O signal.

The diagnostic region provides the information necessary to solve this problem. Specifically, the compound must have the following bonds: C=C, C=O, and vinylic C—H. Among the possible choices, there are only two compounds that have these features:

To distinguish between these two possibilities, notice that the second compound is conjugated, while the first compound is *not* conjugated (the π bonds are separated by more than one single bond). Recall that ketones produce signals at approximately 1720 cm^{-1}, while conjugated ketones produce signals at approximately 1680 cm^{-1}.

1720 cm^{-1}   **1680 cm^{-1}**

In the spectrum provided, the C=O signal appears at 1720 cm^{-1}, indicating that it is not conjugated. The spectrum is therefore consistent with the following compound:

PROBLEM 1.22   Match each compound with the appropriate spectrum

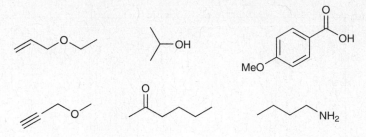

Spectrum A

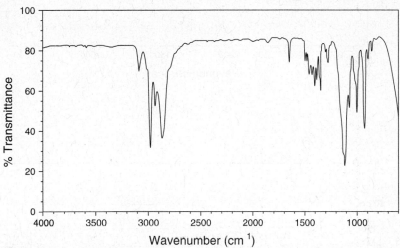

Spectrum B

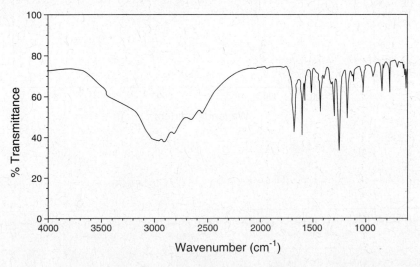

Spectrum C

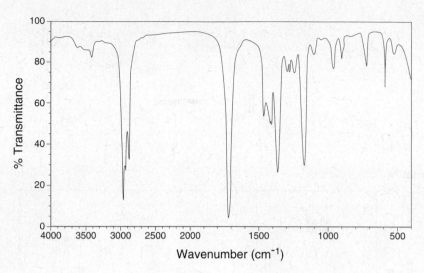

Spectrum D

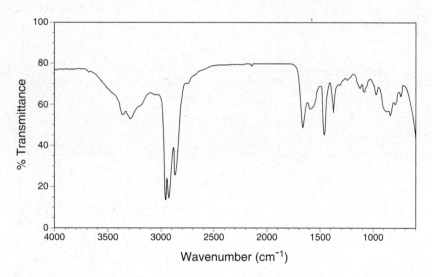

Spectrum E

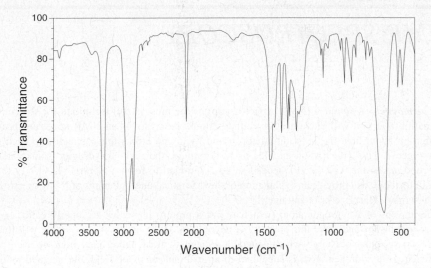

Spectrum F

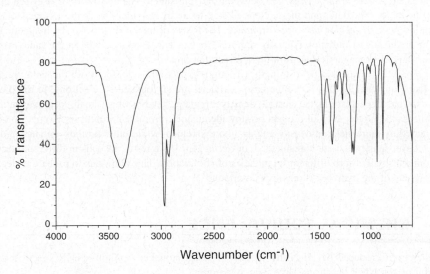

CHAPTER 2

NMR SPECTROSCOPY

Nuclear magnetic resonance (NMR) spectroscopy is the most useful technique for structure determination that you will encounter in your textbook. Analysis of an NMR spectrum provides information about how the individual carbon and hydrogen atoms are connected to each other in a molecule. This information enables us to determine the carbon-hydrogen framework of a compound, much the way puzzle pieces can be assembled to form a picture.

Your textbook will provide an explanation of the theoretical underpinnings of NMR spectroscopy (how it works). Here is a brief summary:

The nucleus of a hydrogen atom (which is just a proton) possesses a property called *nuclear spin*. A true understanding of this property is beyond the scope of our course, so we will think of it as a rotating sphere of charge, which generates a magnetic field, called a *magnetic moment*. When the nucleus of a hydrogen atom is subjected to an external magnetic field, the magnetic moment can align either with the field (called the α spin state) or against the field (called the β spin state). There is a difference in energy (ΔE) between these two spin states. If a proton occupying the α spin state is subjected to electromagnetic radiation, an absorption can take place IF the energy of the photon is equivalent to the energy gap between the spin states. The absorption causes the nucleus to *flip* to the β spin state, and the nucleus is said to be in *resonance* with the external magnetic field; thus the term *nuclear magnetic resonance*. NMR spectrometers employ strong magnetic fields, and the frequency of radiation typically required for nuclear resonance falls in the radio wave region of the electromagnetic spectrum (called rf radiation).

At a particular magnetic field strength, we might expect all nuclei to absorb the same frequency of rf radiation. Luckily, this is not the case, as nuclei are surrounded by electrons. In the presence of an external magnetic field, the electron density circulates, establishing a small, local magnetic field that *shields* the proton. Not all protons occupy identical electronic environments. Some protons are surrounded by more electron density and are more shielded, while other protons are surrounded by less electron density and are less shielded, or *deshielded*. As a result, protons in different electronic environments will absorb different frequencies of rf radiation. This allows us to probe the electronic environment of the hydrogen atoms in a compound.

2.1 CHEMICAL EQUIVALENCE

The spectrum produced by ^1H NMR spectroscopy (pronounced "proton" NMR spectroscopy) is called a proton NMR spectrum. Here is an example:

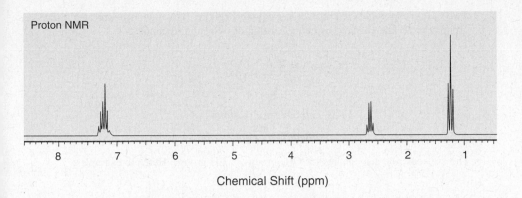

Chemical Shift (ppm)

The first valuable piece of information is the number of signals (the spectrum above appears to have three different signals). In addition, each signal has the following important characteristics:

1. The *location* of each signal indicates the electronic environment of the protons giving rise to the signal.

2. The *area* under each signal indicates the number of protons giving rise to the signal.

3. The *shape* of the signal indicates the number of neighboring protons.

We will discuss these characteristics in the upcoming sections. First let's explore the information that is revealed by counting the number of signals in a spectrum.

The number of signals in a proton NMR spectrum indicates the number of different kinds of protons (protons in different electronic environments). Protons that occupy identical electronic environments are called *chemically equivalent,* and they will produce only one signal.

Two protons are chemically equivalent if they can be interchanged by a symmetry operation. Your textbook will likely provide a detailed explanation, with examples. For our purposes, the following simple rules can guide you in most cases.

- The two protons of a CH_2 group will generally be chemically equivalent if the compound lacks stereocenters. But if the compound has a stereocenter, then the protons of a CH_2 group will generally not be chemically equivalent:

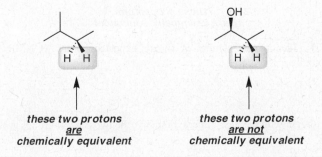

these two protons
are
chemically equivalent

these two protons
are not
chemically equivalent

- Two CH_2 groups will be equivalent to each other (giving four equivalent protons) if the CH_2 groups can be interchanged by either rotation or reflection. Example:

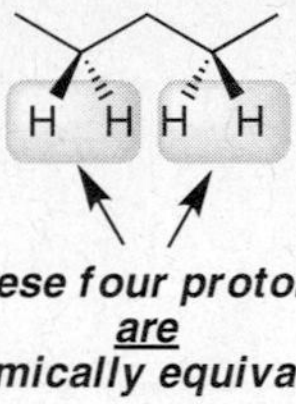

these four protons
are
chemically equivalent

- The previous rule also applies for CH_3 groups or CH groups. Here are examples:

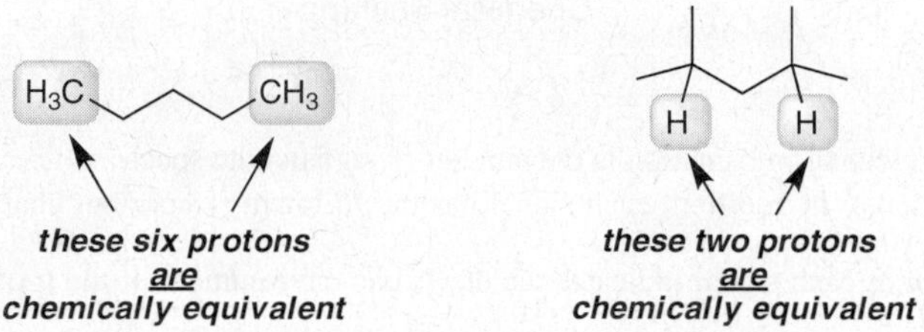

these six protons   *these two protons*
are   *are*
chemically equivalent   *chemically equivalent*

- The three protons of a CH_3 group are always chemically equivalent, even if there are stereocenters in the compound:

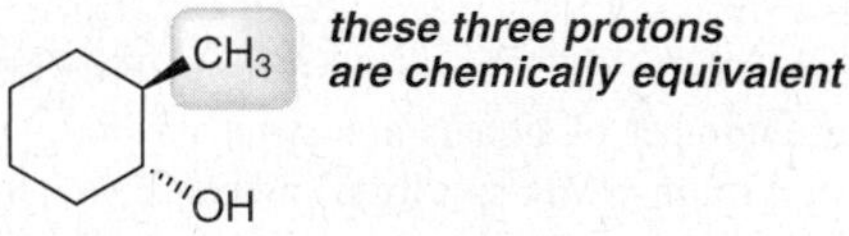

these three protons
are chemically equivalent

- For aromatic compounds, it will be less confusing if you draw a circle in the ring, rather than drawing alternating π bonds. For example, the following two methyl groups are equivalent, which can be easily seen when drawn in the following way:

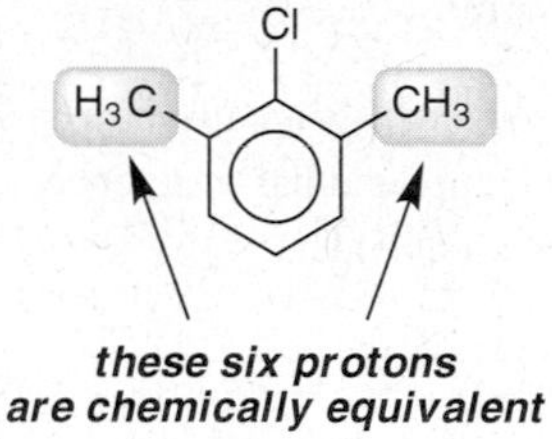

these six protons
are chemically equivalent

EXERCISE 2.1   Identify the number of signals expected in the ^1H NMR spectrum of the following compound:

Answer   Let's begin with the *gem*-dimethyl moiety (the two methyl groups at the center of this compound):

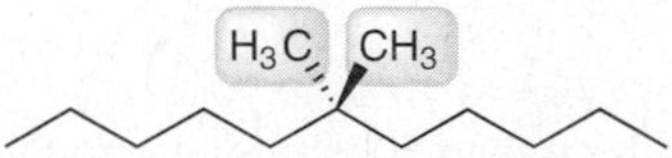

These two methyl groups are equivalent to each other, because they can be interchanged by either rotation or reflection (only one type of symmetry is necessary, but in this case we have both, reflection and rotation, so these six protons are certainly chemically equivalent). We therefore expect one signal for all six protons.

Now let's consider the following methylene (CH_2) groups:

For each of these methylene groups, the two protons are chemically equivalent because there are no stereocenters. In addition, these two methylene groups will be equivalent to each other, since they can be interchanged by rotation or reflection. We therefore expect one signal for all four protons.

The same argument applies for the following two methylene groups:

These four protons are chemically equivalent. But they are different from the other methylene groups in the compound, because they cannot be interchanged with any of the other methylene groups via rotation or reflection.

In a similar way, the following four protons are chemically equivalent:

And the following four protons are also chemically equivalent:

And finally, the following six protons are also chemically equivalent:

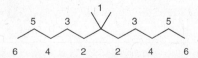

Note that these six protons are different from the six protons of the *gem*-dimethyl group in the center of the compound, because the first set of six protons cannot be interchanged with the other set of six protons via either rotation or reflection.

In total, there are six different types of protons:

So we would expect the proton NMR spectrum of this compound to have six signals.

PROBLEMS Identify the number of signals expected in the proton NMR spectrum of each of the following compounds.

2.2

2.3

2.4

2.5

2.6 OH

2.7 HO OH

2.8 OH

2.9 OCH$_3$

2.10

2.11 If you look at your answers to the previous problems, you should find that one of the structures was expected to produce an NMR spectrum with only one signal. In that structure (problem 2.4), all six methyl groups were chemically equivalent. Using that example as guidance, propose two possible structures for a compound with molecular formula C_9H_{18} that exhibits an NMR spectrum with only one signal.

2.2 CHEMICAL SHIFT (BENCHMARK VALUES)

We will now begin exploring the three characteristics of every signal in an NMR spectrum. The first characteristic is the location of the signal, called its *chemical shift* (δ), which is defined relative to the frequency of absorption of a reference compound, tetramethylsilane (TMS):

$$H_3C - \underset{\underset{CH_3}{|}}{\overset{\overset{CH_3}{|}}{Si}} - CH_3$$

Tetramethylsilane (TMS)

Your textbook will go into greater depth in explaining chemical shift and why it is a unitless number that is reported in parts per million (ppm). For now, we will simply point out that for most organic compounds, the signals produced will fall in a range between 0 and 10 ppm. In rare cases, it is possible to observe a signal occurring at a chemical shift below 0 ppm, which results from a proton that absorbs a lower frequency than the protons in TMS. Most protons in organic compounds absorb a higher frequency than TMS, so most chemical shifts that we encounter will be positive numbers.

The left side of an NMR spectrum is described as *downfield*, and the right side of the spectrum is described as *upfield*:

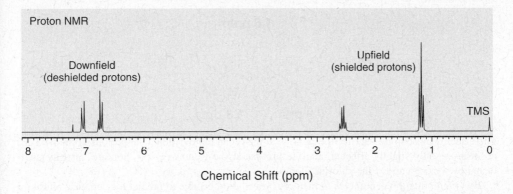

But keep in mind that these terms are used in a relative way. For example, we would say that the signal at 2.5 ppm (in the spectrum above) is downfield from the signal at 1.2 ppm. Similarly, the signal at 6.8 ppm is upfield from the signal at 7.1 ppm.

The protons of alkanes typically produce signals between 1 and 2 ppm. We will now explore some of the effects that can push a signal downfield (relative to the protons of an alkane). Recall that electronegative atoms, such as halogens, withdraw electron density from neighboring atoms:

$$H_3C \longrightarrow X$$

(X = F, Cl, Br, or I)

This inductive effect causes the neighboring protons to be deshielded (surrounded by less electron density), and as a result, the signal produced by these protons is shifted downfield—that is, the signal appears at a higher chemical shift than the protons of an alkane. The strength of this effect depends on the electronegativity of the halogen. Compare the chemical shifts of the protons in the following compounds:

1.0 ppm   **2.2 ppm**   **2.7 ppm**   **3.1 ppm**   **4.3 ppm**

Fluorine is the most electronegative element, and therefore produces the strongest effect. When multiple halogens are present, the effect is generally additive, as can be seen when comparing the following compounds:

1.0 ppm   **3.1 ppm**   **5.5 ppm**   **7.3 ppm**

Each chlorine atom adds approximately 2 ppm to the chemical shift of the signal. An important aspect of inductive effects is the fact that they taper off drastically with distance, as can be seen by comparing the chemical shifts of the protons in the following compound:

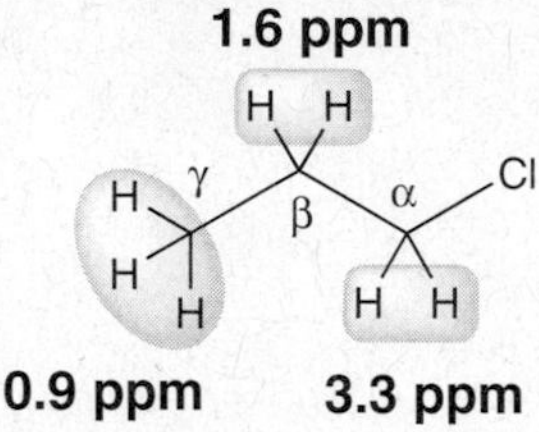

1.6 ppm

0.9 ppm   **3.3 ppm**

The effect is most significant for the protons at the alpha (α) position. The protons at the beta (β) position are only slightly affected, and the protons at the gamma (γ) position are virtually unaffected by the presence of the chlorine atom.

By committing a few numbers to memory, you should be able to predict the chemical shifts for the protons in a wide variety of compounds, including alcohols, ethers, ketones, esters, and carboxylic acids. The following numbers can be used as benchmark values:

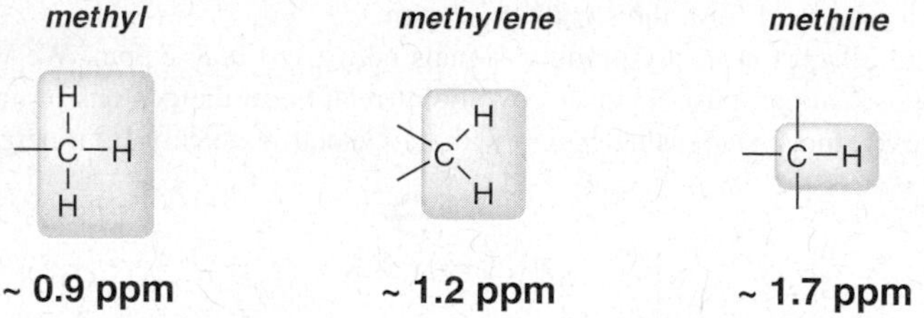

methyl   *methylene*   *methine*

~ **0.9 ppm**   ~ **1.2 ppm**   ~ **1.7 ppm**

In the absence of inductive effects, a methyl group (CH_3) will generally produce a signal near 0.9 ppm, a methylene group (CH_2) will produce a signal near 1.2 ppm, and a methine group (CH) will produce a signal near 1.7 ppm. These benchmark values are then modified by the presence of neighboring functional groups, in the following way:

Functional group	*Effect on α protons*	*Example*	
Oxygen of an alcohol or ether	+ 2.5		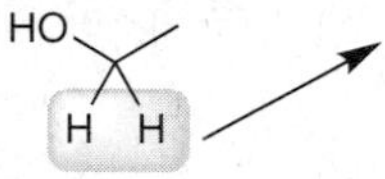 methylene group (CH_2) = 1.2 ppm next to oxygen = +2.5 ppm **3.7 ppm**
Oxygen of an ester	+3		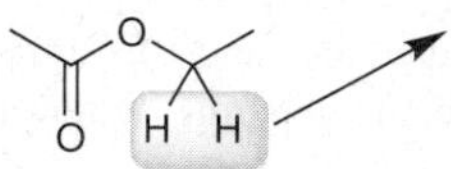 methylene group (CH_2) = 1.2 ppm next to oxygen = + 3.0 ppm **4.2 ppm**
Carbonyl group (C=O) All carbonyl groups, including ketones, aldehydes, esters, etc.	+1		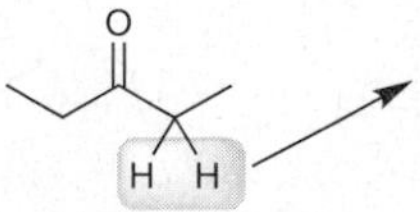 methylene group (CH_2) = 1.2 ppm next to carbonyl = + 1.0 ppm **2.2 ppm**

The values above represent the effect of a few functional groups on the chemical shifts of alpha protons. The effect on beta protons is generally about one-fifth of the effect on the alpha protons. For example, in an alcohol, the presence of an oxygen atom adds +2.5 ppm to the chemical shift of the alpha protons, but adds only +0.5 ppm to the beta protons. Similarly, a carbonyl group adds +1 ppm to the chemical shift of the alpha protons, but only +0.2 to the beta protons.

The values above (together with the benchmark values for methyl, methylene, and methine groups), enable us to predict the chemical shifts for the protons in a wide variety of compounds. Let's see an example.

EXERCISE 2.12   Predict the chemical shifts for the signals in the proton NMR spectrum of the following compound:

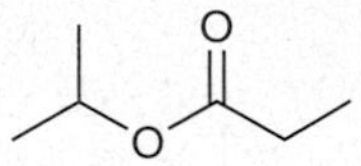

Solution   First determine the total number of expected signals. In this compound, there are four different kinds of protons, giving rise to four distinct signals. For each type of signal, identify whether it represents methyl (0.9 ppm), methylene (1.2 ppm), or methine (1.7 ppm) groups:

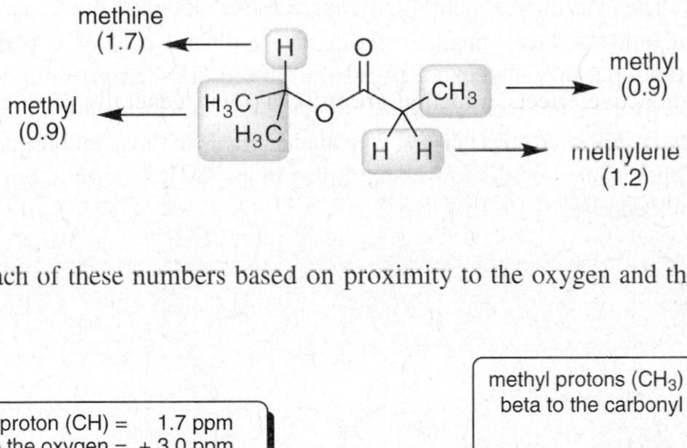

Finally, modify each of these numbers based on proximity to the oxygen and the carbonyl group:

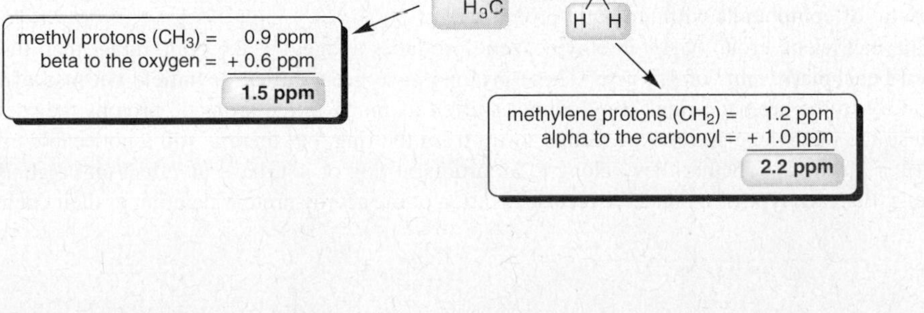

methine proton (CH) =   1.7 ppm
alpha to the oxygen = + 3.0 ppm
4.7 ppm

methyl protons (CH₃) =   0.9 ppm
beta to the carbonyl = + 0.2 ppm
1.1 ppm

methyl protons (CH₃) =   0.9 ppm
beta to the oxygen = + 0.6 ppm
1.5 ppm

methylene protons (CH₂) =   1.2 ppm
alpha to the carbonyl = + 1.0 ppm
2.2 ppm

These values are only estimates, and the actual chemical shifts might differ slightly from the predicted values. Note that for the methine proton, we did not count the distant C=O bond and add +0.2, because the ester moiety is considered as one group, which has the effect of adding +3.0 to the chemical shift of the methine proton. Similarly, note that for the methylene protons, we did not add +0.5 for a distant oxygen. Once again, the ester moiety is considered as one group, which has the effect of adding +1.0 to the chemical shift of the methylene protons.

PROBLEMS   Predict the chemical shifts for the signals in the proton NMR spectrum of each of the following compounds:

2.13   **2.14**   **2.15**

2.16   **2.17**   **2.18**

The chemical shift of a proton is also sensitive to the presence of nearby π electrons. This effect is particularly strong for aromatic protons (protons connected directly to an aromatic ring). In your textbook, you will find a diagram showing how (and why) the aromatic protons are affected. Here is a brief summary. The external magnetic field causes the π electrons to circulate, and this flow of electrons causes an induced, local magnetic field. The aromatic protons experience not only the external magnetic field, but they also experience the induced, local magnetic field. As a result, aromatic protons feel a stronger net magnetic field, which causes their signals to be shifted downfield, significantly. In fact, aromatic protons generally produce signals in the neighborhood of 7 ppm (sometimes as high as 8 ppm, sometimes as low as 6.5 ppm) in an NMR spectrum. For example, consider the structure of ethylbenzene:

Ethylbenzene has three different kinds of aromatic protons (make sure you can identify them), producing three overlapping signals just above 7 ppm. A complex signal around 7 ppm is characteristic of compounds with aromatic protons.

The methylene group (CH$_2$) in ethylbenzene produces a signal at 2.6 ppm, rather than the expected benchmark value of 1.2 ppm. These protons have been shifted downfield because of their proximity to the aromatic ring. They are not shifted as much as the aromatic protons themselves, because the methylene protons are farther away from the ring, but there is still a noticeable effect.

All π electrons, whether they belong to an aromatic ring or not, have an effect on neighboring protons. For each type of π bond, the precise location of the nearby protons determines their chemical shift. For example, aldehydic protons produce characteristic signals at approximately 10 ppm. Below are some important chemical shifts. It would be wise to become familiar with these numbers, as they will be required in order to interpret proton NMR spectra:

Type of proton	Chemical Shift (δ)	Type of proton	Chemical Shift (δ)
methyl R—CH$_3$	~0.9	alkyl halide  R—C—X (H, R)	2 – 4
methylene 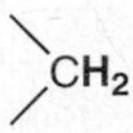 CH$_2$	~1.2	alcohol R—O—H	2 – 5
methine 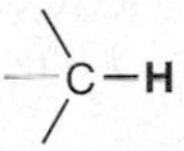 C—H	~1.7	vinylic 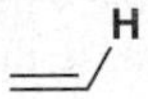 H	4.5 – 6.5
allylic 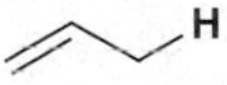 H	~2	aryl H 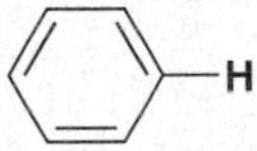	6.5 – 8
alkynyl R—≡—H	~2.5	aldehyde 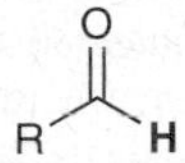 R(C=O)H	~10
aromatic methyl 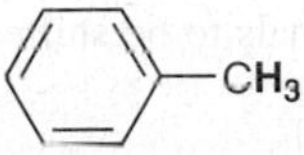 CH$_3$	~2.5	carboxylic acid 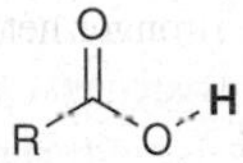 R(C=O)O—H	~12

PROBLEM 2.19 Predict the expected chemical shift for each type of proton in the following compound:

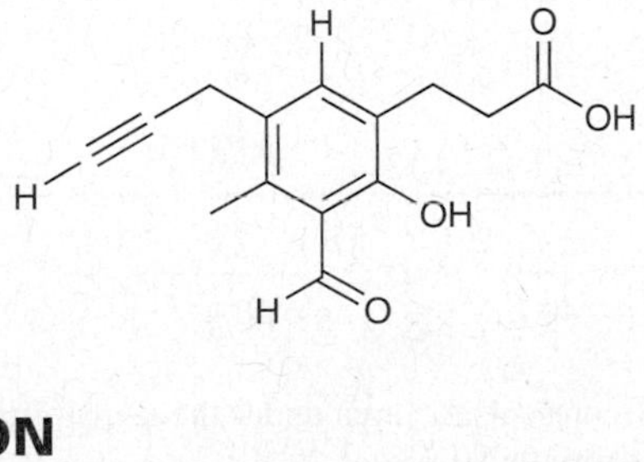

2.3 INTEGRATION

In the previous section, we learned about the first characteristic of every signal, chemical shift. In this section, we will explore the second characteristic, *integration,* which is the area under each signal. This value indicates the number of protons giving rise to the signal. After acquiring a spectrum, a computer calculates the area under each signal, and then displays this area as a numerical value placed under the signal:

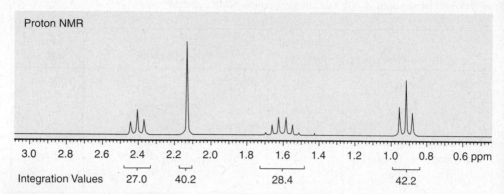

These numbers only have meaning when compared to each other. In order to convert these numbers into useful information, choose the smallest number (27.0 in this case), and then divide all integration values by this number:

$$\frac{27.0}{27.0} = 1 \qquad \frac{40.2}{27.0} = 1.49 \qquad \frac{28.4}{27.0} = 1.05 \qquad \frac{42.2}{27.0} = 1.56$$

These numbers provide the *relative number,* or ratio, of protons giving rise to each signal. This means that a signal with an integration of 1.5 involves one and a half times as many protons as a signal with an integration of 1. In order to arrive at whole numbers (there is no such thing as half a proton), multiply all the numbers above by two, giving the same ratio now expressed in whole numbers, 2 : 3 : 2 : 3. In other words, the signal at 2.4 ppm represents 2 equivalent protons, and the signal at 2.1 ppm represents 3 equivalent protons.

Integration values are often represented by *step-curves,* for example:

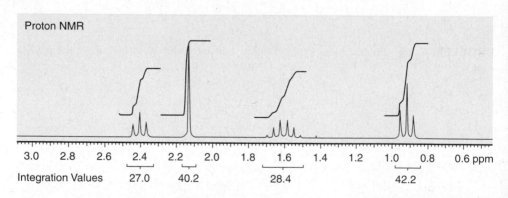

The height of each step curve represents the area under the signal. In this case, a comparison of the heights of the four step curves reveals a ratio of 2 : 3 : 2 : 3.

When interpreting integration values, don't forget that the numbers are only relative. For example, consider the structure of butane:

Butane has two kinds of protons, and will therefore produce two signals in its proton NMR spectrum. The methyl groups give rise to one signal and the methylene groups give another signal. A computer analyzes the area under each signal and provides numbers that allow us to calculate a ratio of 2 : 3. This ratio only indicates the relative number of protons giving rise to each signal, not the exact number of protons. In this case, the exact numbers are 4 (for the methylene groups) and 6 (for the methyl groups). When analyzing the NMR spectrum of an unknown compound, the molecular formula provides extremely useful information because it enables us to determine the exact number of protons giving rise to each signal. If we were analyzing the spectrum of butane, the molecular formula (C_4H_{10}) would indicate that the compound has a total of 10 protons. This information then allows us to determine that the ratio of 2 : 3 must correspond with 4 protons and 6 protons, in order to give a total of 10 protons.

The previous example illustrated the important role that symmetry can play on integration values. Here is another example:

This compound has only two kinds of protons, because the two methylene groups are equivalent to each other, and the two methyl groups are equivalent to each other. The proton NMR spectrum is therefore expected to exhibit only two signals, with relative integration values of 2 : 3. But once again, the values 2 and 3 are just relative numbers. They actually represent 4 protons and 6 protons. This can be determined by inspecting the molecular formula ($C_4H_{10}O$) which indicates a total of 10 protons in the compound. Since the ratio of protons is 2 : 3, this ratio must represent 4 and 6 protons, in order for the total number of protons to be 10. This analysis indicates that the molecule possesses symmetry.

EXERCISE 2.20   A compound with molecular formula $C_5H_{10}O_2$ has the following NMR spectrum. Determine the number of protons giving rise to each signal:

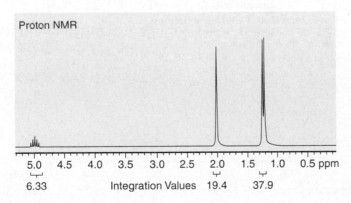

Solution   The spectrum has three signals. Begin by comparing the relative integration values: 6.33, 19.4, and 37.9. Divide each of these three numbers by the smallest number (6.33):

$$\frac{6.33}{6.33} = 1 \qquad \frac{19.4}{6.33} = 3.06 \qquad \frac{37.9}{6.33} = 5.99$$

This gives a ratio of 1 : 3 : 6, but these are just relative numbers. To determine the exact number of protons giving rise to each signal, look at the molecular formula, which indicates a total of 10 protons in the compound. Therefore, the numbers 1 : 3 : 6 are not only relative values, but they are also the exact values. Exact integration values are sometimes expressed in the following way:

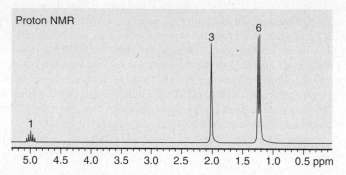

PROBLEMS

2.21 A compound with molecular formula $C_8H_{10}O$ has the following proton NMR spectrum. Determine the number of protons giving rise to each signal.

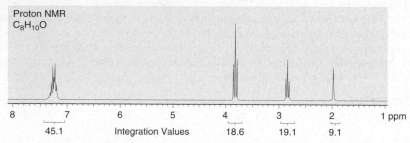

2.22 A compound with molecular formula $C_7H_{14}O$ has the following proton NMR spectrum. Determine the number of protons giving rise to each signal.

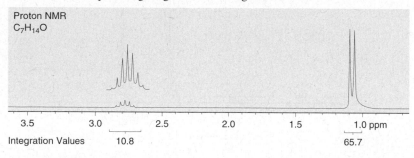

2.23 A compound with molecular formula $C_4H_6O_2$ has the following proton NMR spectrum. Determine the number of protons giving rise to each signal.

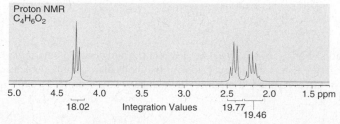

2.4 MULTIPLICITY

The third, and final, characteristic of each signal is its *multiplicity,* which refers to the number of peaks in the signal. A *singlet* has one peak, a *doublet* has two peaks, a *triplet* has three peaks, a *quartet* has four peaks, a *quintet* has five peaks, etc:

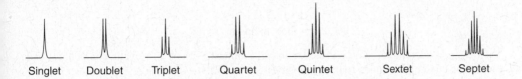

| Singlet | Doublet | Triplet | Quartet | Quintet | Sextet | Septet |

The multiplicity of a signal is the result of the magnetic effects of neighboring protons, and therefore indicates the number of neighboring protons. Your textbook will explain the cause for this effect in detail. The net effect can be summarized with the $n + 1$ *rule*, which states the following: if n is the number of neighboring protons, then the multiplicity will be $n + 1$. For example, a proton with three neighbors ($n = 3$) will be split into a quarter ($n + 1 = 3 + 1 = 4$ peaks).

It is important to realize that nearby protons do not always split each other. There are two major factors that determine whether or not splitting occurs:

1. **Equivalent protons do not split each other.** Consider the two methylene groups in the following compound:

H H
\ /
Cl—C—C—Cl
/ \
H H

Four equivalent protons
no splitting

All four protons are chemically equivalent, and therefore, they do not split each other. Instead, they produce one signal that has no neighboring protons ($n = 0$), so the signal is a singlet ($n + 1$). In order for splitting to occur, the neighboring protons must be different than the protons producing the signal.

2. **Splitting is most commonly observed when protons are separated by either two or three sigma bonds:**

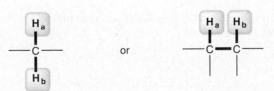

| non-equivalent protons separated by two sigma bonds | or | non-equivalent protons separated by three sigma bonds |

However, when two protons are separated by more than three sigma bonds, splitting is generally not observed:

$$\begin{array}{c} H_a \qquad\quad H_b \\ \Big| \qquad \Big| \qquad \Big| \\ -C-C-C- \\ \Big| \qquad \Big| \qquad \Big| \end{array}$$

too far apart

Such long-range splitting is only observed in rigid molecules, such as bicyclic compounds, or in molecules that contain rigid structural moieties, such as allylic systems. For purposes of this introductory treatment of NMR spectroscopy, we will avoid examples that exhibit substantial long-range coupling. Make sure to look through your lecture notes to see if you covered any examples of long-range coupling.

EXERCISE 2.24   Determine the multiplicity of each signal in the expected proton NMR spectrum of the following compound:

Solution   Begin by identifying the different kinds of protons. That is, determine the number of expected signals.

three equivalent methyl groups

methyl group

methylene group

This compound is expected to produce three signals in its proton NMR spectrum. Now let's analyze each signal, using the $n + 1$ rule:

0 neighbors
$n+1 = 1$
singlet

two neighbors
$n+1 = 3$
triplet

three neighbors
$n+1 = 4$
quartet

Notice that the *tert*-butyl group (on the left side of the molecule) appears as a singlet, because the following carbon atom has no protons:

no protons

This quaternary carbon atom is directly connected to each of the three neighboring methyl groups, and as a result, each of the three methyl groups has no neighboring protons. This is characteristic of *tert*-butyl groups.

PROBLEMS Predict the multiplicity of each signal in the expected proton NMR spectrum of each of the following compounds:

2.25   **2.26**   **2.27**

2.28   **2.29**   **2.30**

When signal splitting occurs, the distance between the individual peaks of a signal is called the *coupling constant,* or *J value,* and is measured in Hz. Neighboring protons always split each other with equivalent *J values.* For example, consider the two kinds of protons in an ethyl group:

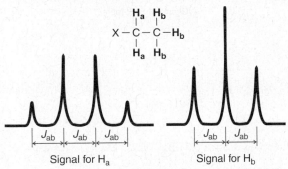

Signal for H_a Signal for H_b

The H_a signal is split into a quartet under the influence of its three neighbors, while the H_b signal is split into a triplet under the influence of its two neighbors. H_a and H_b are said to be coupled to each other. The coupling constant, J_{ab}, is the same in both signals. *J* values can range anywhere from 0 to 20 Hz, depending on the type of protons involved.

2.5 PATTERN RECOGNITION

There are a few splitting patterns that are commonly seen in proton NMR spectra, and you will save yourself time on an exam if you can recognize these patterns:

Ethyl   **Isopropyl**   ***tert*-butyl**

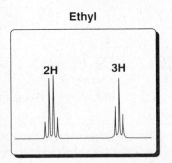   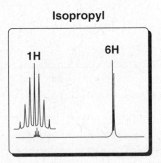   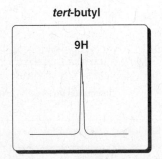

An ethyl group is characterized by a triplet with an integration of 3 and a quartet with an integration of 2. An isopropyl group is characterized by a doublet with an integration of 6 and a quartet with an integration of 1. A *tert*-butyl group is characterized by a singlet with an integration of 9.

Let's get some practice recognizing these patterns.

PROBLEMS Below are NMR spectra of several compounds. Identify whether these compounds are likely to contain ethyl, isopropyl, and/or *tert*-butyl groups:

2.31

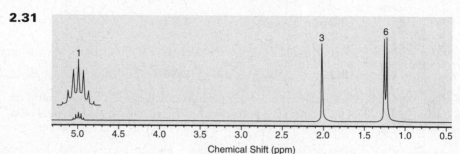

Chemical Shift (ppm)

2.32

Proton NMR
$C_9H_{10}O$

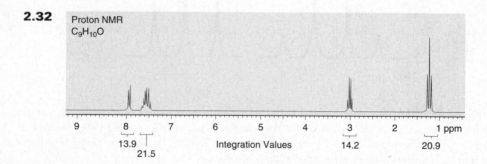

2.33

Proton NMR
$C_7H_{14}O$

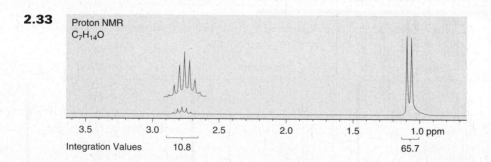

2.34

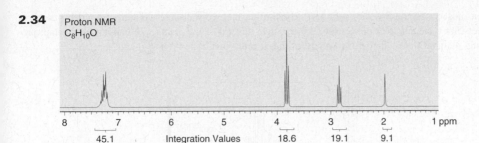

Proton NMR
$C_8H_{10}O$

Integration Values

45.1   18.6   19.1   9.1

2.6   COMPLEX SPLITTING

Complex splitting occurs when a proton has two different kinds of neighboring protons. For example, consider the splitting pattern that you might expect for the protons labeled H_b in the following compound:

$$H_a-C-C-C-X$$

with H_a, H_a on the first carbon, H_b, H_b on the second carbon, and H_c, H_c on the third carbon.

The signal for H_b is being split into a quartet because of the nearby H_a protons, AND it is being split into a triplet because of the nearby H_c protons. The signal is therefore expected to have twelve peaks (4×3). The appearance of the signal will depend greatly on the J values. If J_{ab} is much greater than J_{bc}, then the signal will appear as a quartet of triplets, as shown in the following splitting tree:

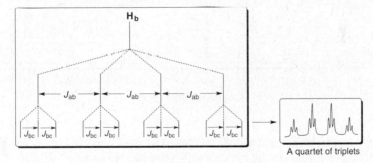

A quartet of triplets

If, however, J_{bc} is much greater than J_{ab}, then the signal will appear as a triplet of quartets:

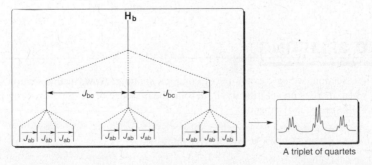

A triplet of quartets

In most cases, the *J* values will be fairly similar, and we will observe neither a clean quartet of triplets nor a clean triplet of quartets. More often, several of the peaks will happen to overlap, producing a signal that is difficult to analyze and is often just called a *multiplet*.

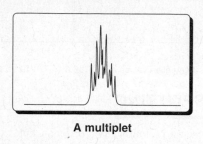

A multiplet

In some cases, J_{ab} and J_{bc} will be almost identical. For example, consider the proton NMR spectrum of 1-nitropropane:

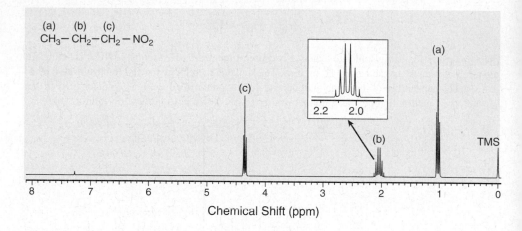

Look carefully at the splitting pattern of the H_b protons (at approximately 2 ppm). This signal looks like a sextet, because the J_{ab} and J_{bc} are so close in value. In such a case, it is "as if" there are five equivalent neighbors, even though all five protons are not equivalent.

2.7   NO SPLITTING

In the previous section, we saw examples of complex splitting. Now, in this section, we will explore cases where there is no splitting at all, despite the presence of neighboring protons. Consider the proton NMR spectrum of ethanol:

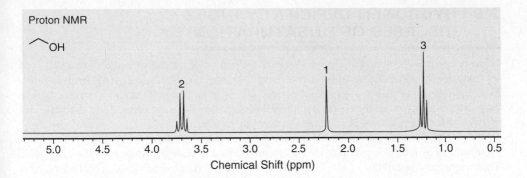

As expected, the spectrum exhibits the characteristic signals of an ethyl group (a quartet with an integration of 2 and a triplet with an integration of 3). In addition, there is another signal at 2.2 ppm, representing the hydroxyl proton (OH). Hydroxyl protons typically produce a signal between 2 and 5 ppm, and it is often difficult to predict exactly where that signal will appear. In the proton NMR spectrum above, notice that the hydroxyl proton is not split into a triplet from the neighboring methylene group. Generally, no splitting is observed across the oxygen of an alcohol, because proton exchange is a very rapid process that is catalyzed by trace amounts of acid or base:

Hydroxyl protons are said to be **labile**, because of the rapid rate at which they are exchanged. This proton transfer process occurs at a faster rate than the timescale of an NMR spectrometer, producing a blurring effect that averages out any possible splitting effect. It is possible to slow down the rate of proton transfer by scrupulously removing the trace amounts of acid and base dissolved in ethanol. Such purified ethanol does in fact exhibit splitting across the oxygen atom, and the signal at 2.2 ppm is observed to be a triplet.

There is one other common example of neighboring protons that often do not produce observable splitting. Aldehydic protons, which generally produce signals near 10 ppm, will often couple only weakly with their neighbors (i.e., a very small *J* value):

~ 10 ppm

This J value is often very small

Depending on the size of the *J* value, splitting may or may not be noticeable. If the *J* value is too small, then the signal near 10 ppm will appear to be a singlet, despite the presence of neighboring protons.

2.8 HYDROGEN DEFICIENCY INDEX (DEGREES OF UNSATURATION)

In the previous sections of this chapter, we have learned all of the individual tools that you need for analyzing a proton NMR spectrum (considering the number of signals, analyzing chemical shifts, assessing integration values, interpreting the multiplicity of each signal, pattern recognition, etc.). Now, we are just about ready to put all of these tools together. But there is one more important tool that you will need, and we will cover that tool in this section.

Imagine that you have an unknown compound with a molecular formula of $C_6H_{12}O$. The molecular formula by itself does not provide enough information to draw the structure of the compound. There are many constitutional isomers of $C_6H_{12}O$. Nevertheless, a careful analysis of the molecular formula can often provide helpful clues about the structure of the compound. To see how this works, let's begin by analyzing the molecular formula of several alkanes.

Compare the structures of the following alkanes, paying special attention to the number of hydrogen atoms attached to each carbon atom.

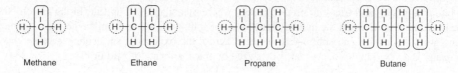

| Methane | Ethane | Propane | Butane |

In each case there are two hydrogen atoms on the ends of the structures (circled), and there are two hydrogen atoms on every carbon atom. This can be summarized like this:

$$H–(CH_2)_n–H$$

where n is the number of carbon atoms in the compound. Accordingly, the number of hydrogen atoms will be $2n + 2$. In other words, all of the compounds above have a molecular formula of C_nH_{2n+2}. This is true even for compounds that are branched rather than having a straight chain.

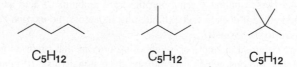

$$C_5H_{12} \qquad C_5H_{12} \qquad C_5H_{12}$$

The compounds above are said to be *saturated*—that is, they possess the maximum number of hydrogen atoms possible relative to the number of carbon atoms present.

A compound with a π bond (a double or triple bond) will have fewer than the maximum number of hydrogen atoms. Such compounds are said to be *unsaturated*.

$$C_5H_{10} \qquad C_5H_8$$

A compound containing a ring will also have fewer than the maximum number of hydrogen atoms, just like a compound with a double bond. For example, compare the structures of 1-hexene and cyclohexane:

C_6H_{12} C_6H_{12}

Both compounds have molecular formula (C_6H_{12}) because both are "missing" two hydrogen atoms [6 carbon atoms can accommodate (2×6) + 2 = 14 hydrogen atoms]. Each of these compounds is said to have one *degree of unsaturation*. The *hydrogen deficiency index (HDI)* is a measure of the number of degrees of unsaturation. A compound is said to have one degree of unsaturation for every two hydrogen atoms that are missing. For example, a compound with molecular formula C_4H_6 is missing four hydrogen atoms (if saturated, it would be C_4H_{10}), so it has two degrees of unsaturation (HDI − 2). There are several ways for a compound to possess two degrees of unsaturation: two double bonds, or two rings, or one double bond and one ring, or one triple bond. Let's explore all of these possibilities for C_4H_6:

Two double bonds	One triple bond	Two rings	One ring and one double bond

These are all of the possible constitutional isomers for C_4H_6. With this in mind, let's expand our skills set. Let's explore how to calculate the HDI when other elements are present in the molecular formula.

Halogens: Compare the following two compounds:

ethane   chloroethane

Notice that chlorine takes the place of a hydrogen atom. Therefore, for purposes of calculating the HDI, treat a halogen as if it were a hydrogen atom. For example, C_4H_9Cl should have the same HDI as C_4H_{10}.

Oxygen: Compare the following two compounds:

ethane   ethanol

Notice that the presence of the oxygen atom does not affect the expected number of hydrogen atoms. Therefore, whenever an oxygen atom appears in the molecular formula, it should be ignored for purposes of calculating the HDI. For example, C_4H_8O should have the same HDI as C_4H_8.

Nitrogen: Compare the following two compounds:

$$
\begin{array}{ccc}
& H\ \ H & \\
& |\ \ \ | & \\
H-&C-C&-H \\
& |\ \ \ | & \\
& H\ \ H &
\end{array}
\qquad\qquad
\begin{array}{cccc}
& H\ \ H & & H \\
& |\ \ \ | & & \diagup \\
H-&C-C&-N & \\
& |\ \ \ | & & \diagdown \\
& H\ \ H & & H
\end{array}
$$

ethane   ethyl amine

Notice that the presence of a nitrogen atom changes the number of expected hydrogen atoms. It gives one more hydrogen atom than would be expected. Therefore, whenever a nitrogen atom appears in the molecular formula, one hydrogen atom must be subtracted from the molecular formula. For example, C_4H_9N should have the same HDI as C_4H_8.

In summary:

- Halogens: *Add* one H for each halogen
- Oxygen: *Ignore*
- Nitrogen: *Subtract* one H for each N

These rules will enable you to determine the HDI for most simple compounds. Alternatively, the following formula can be used:

$$HDI = (2C + 2 + N - H - X)/2$$

C is the number of carbon atoms, N is the number of nitrogen atoms, H is the number of hydrogen atoms, and X is the number of halogens. This formula will work for all compounds containing C, H, N, O, and X.

Calculating the HDI is particularly helpful, because it provides clues about the structural features of the compound. For example, an HDI of zero indicates that the compound cannot have any rings or π bonds. That is extremely useful information when trying to determine the structure of a compound, and it is information that is easily obtained by simply analyzing the molecular formula. Similarly, an HDI of one indicates that the compound must have either one double bond *or* one ring (but not both). If the HDI is two, then there are a few possibilities: two rings, or two double bonds, or one ring and one double bond, or one triple bond. Analysis of the HDI for an unknown compound can often be a useful tool, but only when the molecular formula is known with certainty.

We will use this technique in the next section. The following exercises are designed to develop the skill of calculating and interpreting the HDI of an unknown compound whose molecular formula is known.

EXERCISE 2.35

Calculate the HDI for a compound with molecular formula $C_5H_8Br_2O_2$, and identify the structural information provided by the HDI.

Answer Use the following calculation:

# of H's:	8
Add 1 for each Br:	+2
Ignore each O:	0
Subtract 1 for each N:	0
Total =	10

This compound will have the same HDI as a compound with molecular formula C_5H_{10}. To be fully saturated, 5 carbon atoms would require $(5 \times 2) + 2 = 12$ H's. According to our calculation, two hydrogen atoms are missing, and, therefore, this compound has one degree of unsaturation. HDI = 1.

Alternatively, the following formula can be used:

$$HDI = (2C + 2 + N - H - X)/2 = (10 + 2 + 0 - 8 - 2)/2 = 2/2 = 1$$

With one degree of unsaturation, the compound must contain either one ring or one double bond, but not both. The compound cannot have a triple bond, as this would require two degrees of unsaturation.

PROBLEMS   Calculate the degree of unsaturation for each of the following molecular formulas:

2.36 $C_6H_{10}O_4$ **2.37** $C_7H_{11}N$ **2.38** $C_8H_{14}O_2$

2.39 $C_5H_{12}O_2$ **2.40** $C_6H_{15}N$ **2.41** $C_8H_{10}O$

2.9   ANALYZING A PROTON NMR SPECTRUM

In this section, we will practice analyzing and interpreting NMR spectra, a process that involves four discrete steps:

1. Always begin by inspecting the molecular formula (if it is given), as it provides useful information. Specifically, calculating the hydrogen deficiency index (HDI) can provide important clues about the structure of the compound. An HDI of zero indicates that the compound does not possess any rings or π bonds. An IIDI of 1 indicates that the compound has either one ring or one π bond. An HDI of four should indicate the likely presence of an aromatic ring:

four degrees of unsaturation
(HDI = 4)

2. Consider the number of signals and integration of each signal (this gives clues about the symmetry of the compound).

3. Analyze each signal (chemical shift, integration, and multiplicity), and then draw fragments consistent with each signal. These fragments become our puzzle pieces that must be assembled to produce a molecular structure.

4. Assemble the fragments into a molecular structure.

The following exercise illustrates how this is done.

EXERCISE 2.42

Identify the structure of a compound with molecular formula $C_9H_{10}O$ that exhibits the following proton NMR spectrum:

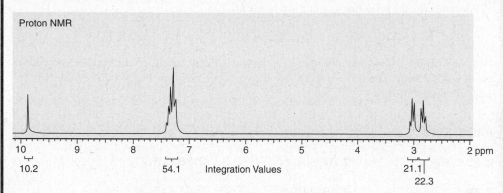

Proton NMR

10.2   54.1   Integration Values   21.1
22.3

Answer The first step is to calculate the HDI. The molecular formula indicates 9 carbon atoms, which would require 20 hydrogen atoms in order to be fully saturated. There are only 10 hydrogen atoms, which means that 10 hydrogen atoms are missing, and therefore, the HDI is 5. This is a large number, and it would not be efficient to think about all the possible ways there are to have five degrees of unsaturation. However, anytime we encounter an HDI of 4 or more, we should be on the lookout for an aromatic ring. We must keep this in mind when analyzing the spectrum. We should expect an aromatic ring (HDI = 4) plus one other degree of unsaturation (either a ring or a double bond).

The second step is to consider the number of signals and the integration value for each signal. Any signals with large integration values would suggest the presence of symmetry elements. For example, a signal with an integration of 4 would suggest two equivalent CH_2 groups. In this spectrum, we see four signals. In order to analyze the integration of each signal, we must first divide by the lowest number (10.2):

$$\frac{10.2}{10.2} = 1 \qquad \frac{54.1}{10.2} = 5.30 \qquad \frac{21.1}{10.2} = 2.07 \qquad \frac{22.3}{10.2} = 2.19$$

The ratio is 1 : 5 : 2 : 2. Now look at the molecular formula. There are 10 protons in the compound, so the relative integration values represent the actual number of protons giving rise to each signal:

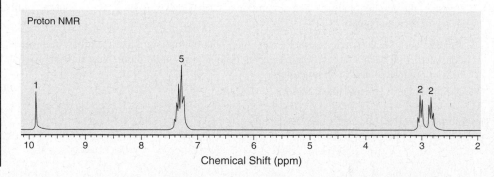

Proton NMR

Chemical Shift (ppm)

The next step is to analyze each signal. Starting upfield, there are two triplets, each with an integration of 2. This suggests that there are two adjacent methylene groups:

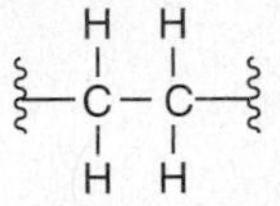

These signals do not appear at 1.2 where methylene groups are expected, so one or more factors is shifting these signals downfield. Our proposed structure must take that into account.

Moving downfield through the spectrum, the next signal appears just above 7 ppm, characteristic of aromatic protons (just as we suspected after analyzing the HDI). The multiplicity of aromatic protons only rarely gives useful information. More often, a messy multiplet of overlapping signals is observed. But the integration value gives important information. Specifically, there are five aromatic protons, which means that the aromatic ring is monosubstituted.

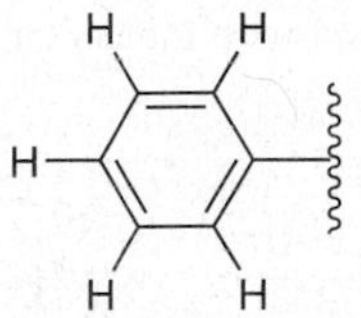

Five aromatic protons

Moving on to the last signal, we see a singlet at 10 ppm with an integration of 1. This is suggestive of an aldehydic proton.

In summary, our analysis has produced the following fragments:

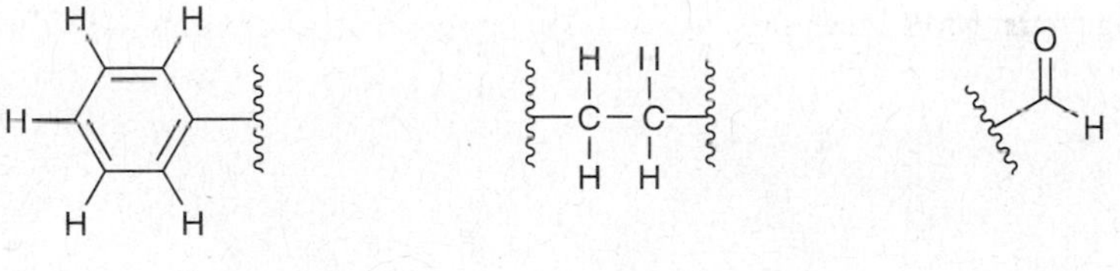

The final step is to assemble these fragments. Fortunately, there is only one way to assemble these three puzzle pieces.

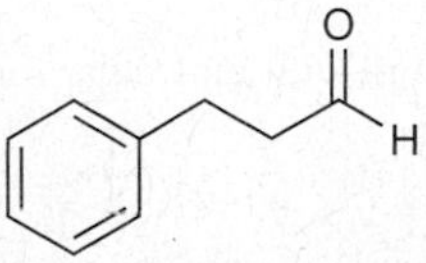

We mentioned before that each methylene group is being shifted downfield by one or more factors. Our proposed structure explains the observed chemical shifts. In particular, one methylene group is shifted significantly by the carbonyl group and slightly by the aromatic ring. The other methylene group is being shifted significantly by the aromatic ring and slightly by the carbonyl group.

PROBLEMS

2.43 Propose a structure for a compound with molecular formula $C_5H_{10}O$ that is consistent with the following proton NMR spectrum.

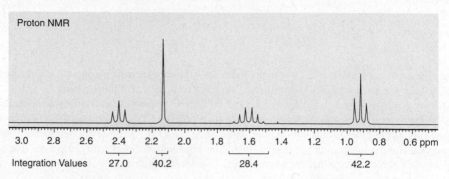

2.44 Propose a structure for a compound with molecular formula $C_5H_{10}O_2$ that is consistent with the following proton NMR spectrum.

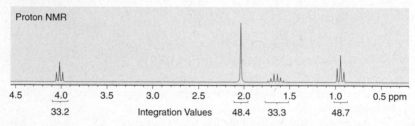

2.45 Propose a structure for a compound with molecular formula $C_4H_6O_2$ that is consistent with the following proton NMR spectrum.

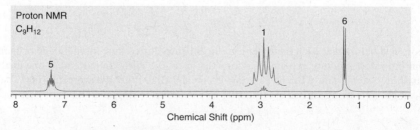

2.46 Propose a structure for a compound with molecular formula C_9H_{12} that is consistent with the following proton NMR spectrum.

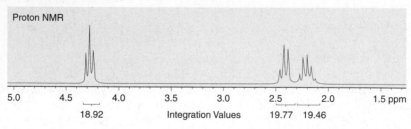

2.47 Propose a structure for a compound with molecular formula $C_6H_{12}O_2$ that is consistent with the following proton NMR spectrum.

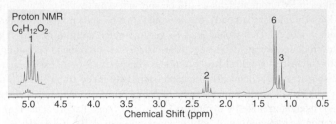

2.48 Propose a structure for a compound with molecular formula $C_8H_{10}O$ that is consistent with the following proton NMR spectrum.

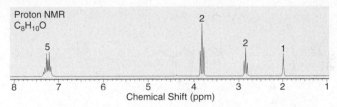

Your textbook has many more spectroscopy problems, including problems in which you are given both IR and NMR spectra. I recommend that you do ALL of those problems. The skills in this chapter were meant to prepare you for those problems.

2.10   ¹³C NMR SPECTROSCOPY

Many of the principles that apply to ¹H NMR spectroscopy also apply to ¹³C NMR spectroscopy, but there are a few major differences, and we will focus on those. For example, ¹H is the most abundant isotope of hydrogen, but ¹³C is only a minor isotope of carbon, representing about 1.1% of all carbon atoms found in nature. As a result, only one in every hundred carbon atoms will resonate, which demands the use of a sensitive receiver coil for ¹³C NMR.

In ¹H NMR spectroscopy, we saw that each signal has three characteristics (chemical shift, integration, and multiplicity). In ¹³C NMR spectroscopy, only the chemical shift is important. The integration and multiplicity of ¹³C signals are not reported, which greatly simplifies the interpretation of ¹³C NMR spectra. Integration values are not routinely calculated in ¹³C NMR spectroscopy because the pulse technique employed by NMR spectrometers has the undesired effect of distorting the integration values, rendering them useless in most cases. Multiplicity is also not a common characteristic of ¹³C NMR signals.

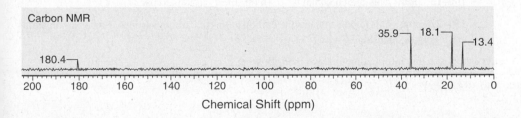

Notice that all of the signals are recorded as singlets. This is a result of a special technique, called broadband decoupling, that suppresses all ^{13}C–^1H splitting. If we did not use this technique, then the signal of each ^{13}C atom nucleus would be split not only by the protons directly connected to it (separated by only one sigma bond), but it would also be split by the protons that are two or three sigma bonds removed. This would lead to very complex splitting patterns, and the signals would overlap to produce an unreadable spectrum. The use of broadband decoupling causes all of the ^{13}C signals to collapse to singlets, which renders the spectrum more easily interpreted.

In ^{13}C NMR spectroscopy, chemical shift values typically range from 0 to 220 ppm. The number of signals in a ^{13}C NMR spectrum represents the number of carbon atoms in different electronic environments (not interchangeable by symmetry). Carbon atoms that are interchangeable by symmetry (either rotation or reflection) will only produce one signal. To illustrate this point, consider the compounds below. Each compound has eight carbon atoms, but does not produce eight signals. The unique carbon atoms in each compound are highlighted.

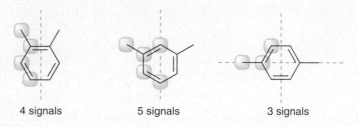

| 4 signals | 5 signals | 3 signals |

Each carbon atom that is not highlighted is equivalent to one of the highlighted carbon atoms.

The location of each signal is dependent on shielding and deshielding effects, just as we saw in proton NMR spectroscopy. Below are chemical shifts of several important types of carbon atoms.

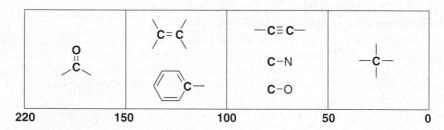

- **0–50 ppm:** This region contains signals from sp^3–hybridized carbon atoms (methyl, methylene, and methine groups).

- **50–100 ppm:** This region contains sp^3–hybridized carbon atoms that are deshielded by electronegative atoms, as well as sp–hybridized carbon atoms.

- **100–150 ppm:** This region contains sp^2–hybridized carbon atoms.

- **150–220 ppm:** This region contains the carbon atoms of carbonyl groups. These carbon atoms are highly deshielded.

Let's now use this information to solve the following exercise.

EXERCISE 2.49   Predict the number of signals and the location of each signal in the expected ¹³C NMR spectrum of the following compound:

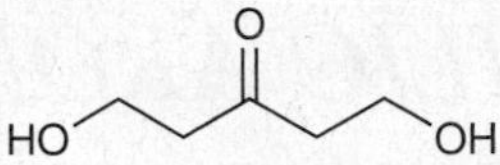

Answer   Begin by determining the number of expected signals. The compound has five carbon atoms, but we must look to see if any of these carbon atoms are interchangeable by symmetry. In this case, there is symmetry, and we expect only three signals in the ¹³C NMR spectrum.

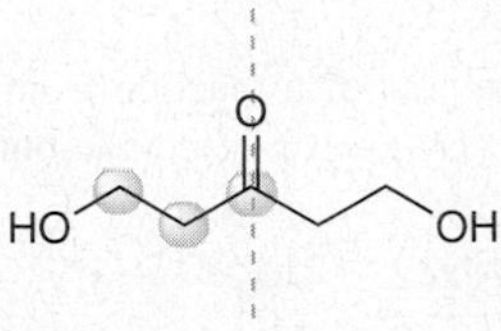

The expected chemical shifts are shown below, categorized according to the region of the spectrum in which each signal is expected to appear:

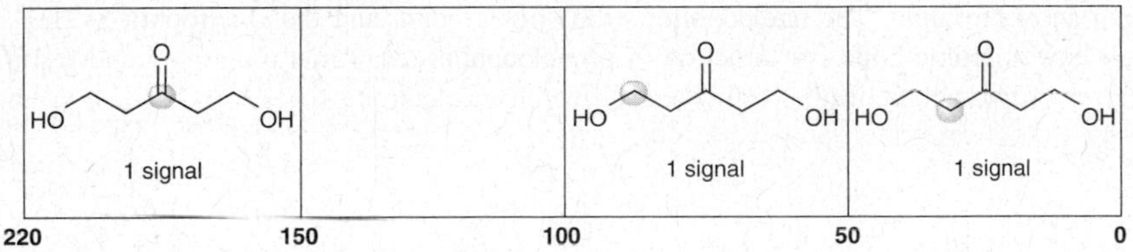

PROBLEMS   For each compound below, predict the number of signals and the location of each signal in the expected ¹³C NMR spectrum.

2.50

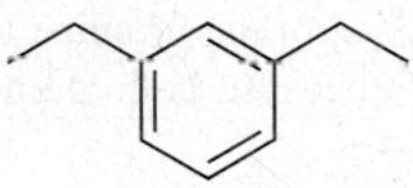

2.51

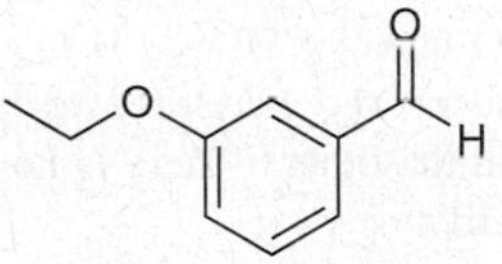

2.52

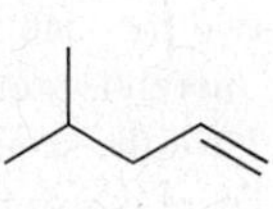

2.53

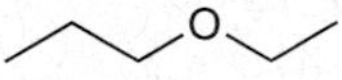

2.54

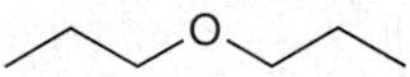

2.55

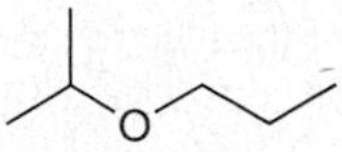

ELECTROPHILIC AROMATIC SUBSTITUTION

We must begin this chapter with a review of a reaction from the first semester of organic chemistry. Recall the addition of bromine (Br_2) across a double bond:

When we learned this reaction in the first semester, we saw that this reaction involves a nucleophile attacking an electrophile. The nucleophile is the double bond, and the electrophile is Br_2. To understand how a double bond can function as a nucleophile, recall that a double bond results from the overlap of two neighboring p orbitals, each with one electron:

Therefore, a double bond represents a region in space of high-electron density. Even though there is no full negative charge anywhere, the double bond can function as a nucleophile and can attack an electrophile. But the obvious question is: why is Br_2 an electrophile? After all, the bond between the two bromine atoms is covalent, and therefore, we cannot say that one of the bromine atoms has any more electron density than the other. (There is no induction here because both atoms, Br and Br, have the same electronegativity.)

There is a simple reason why bromine can act as an electrophile here. We need to consider what happens when a molecule of Br_2 approaches an alkene. To help us see this, think of Br_2 in terms of the electron cloud surrounding it:

$$Br \text{———} Br$$

As a molecule of Br_2 gets close to the pi bond of an alkene, the electron density of the pi bond begins to repel the electron cloud around Br_2. This effect gives the Br_2 molecule an induced dipole moment (this is a temporary interaction—it only happens while the bromine molecule is near the alkene):

$$\overset{\delta+}{Br} \text{———} \overset{\delta-}{Br}$$

So, we have an electron-rich alkene, which can attack the nearby, electron-poor bromine atom. This gives the reaction that we saw in the first semester of organic chemistry:

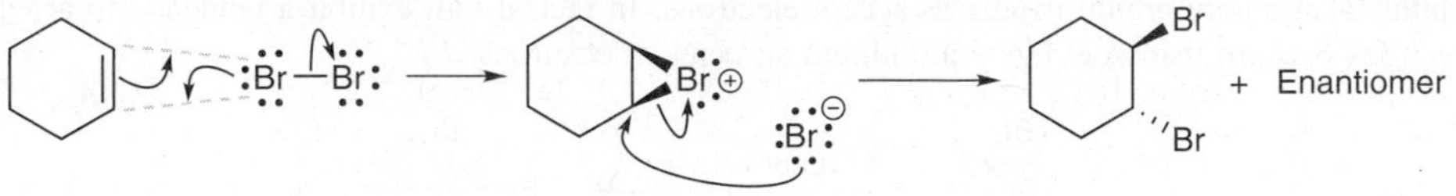

Now let's consider what happens if we try to do this exact same reaction with benzene as our nucleophile. So we are trying to perform this reaction:

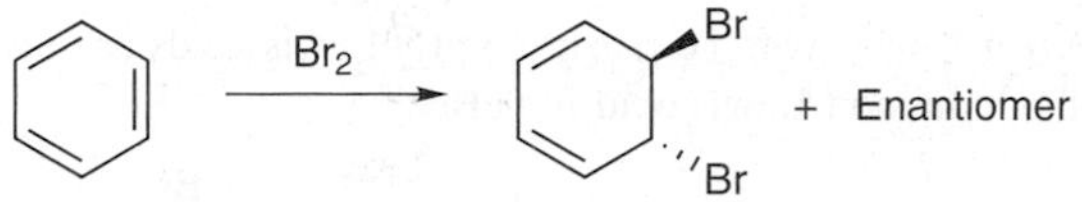

But when benzene is heated in the presence of Br_2, no reaction is observed.

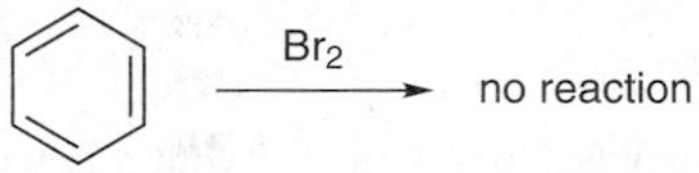

We can understand this because benzene is an aromatic compound. It has a special stability due to its aromaticity. If we add Br_2 across benzene, aromaticity will be lost. And that is why the reaction does not take place—it would be going "uphill" in energy. But is it possible to try to "force" the reaction to happen?

This brings us to a simple and important concept in organic chemistry. One of the driving forces for any reaction between a nucleophile and an electrophile is the difference in the electron density between the two compounds. The nucleophile is electron-rich, and the electrophile is electron-poor. Therefore, they are attracted to each other in space (opposite charges attract). So, if the reaction is not proceeding, we can try to force it along by making the attraction even stronger between the nucleophile and the electrophile. We can accomplish this in one of two ways. We can either make the nucleophile even more electron-rich (more nucleophilic), or we can make the electrophile even more electron-poor (more electrophilic).

In this chapter, we will explore both of these scenarios. For now, let us start by trying to make the electrophile a better electrophile. How do we make Br_2 a better electrophile? Let's remember why Br_2 is an electrophile in the first place. Just a few moments ago, we saw that an induced dipole moment is formed when Br_2 gets close to an alkene. This creates a partial positive charge on one of the bromine atoms. Clearly, if we had Br^+ instead of Br_2, then that would be an even better electrophile. We would not have to wait around for Br_2 to become slightly polarized.

But how do we form Br^+? That is where Lewis acids come into the picture.

3.1 HALOGENATION AND THE ROLE OF LEWIS ACIDS

Consider the compound $AlBr_3$. The central atom in this structure is aluminum. Aluminum is in Column 3A of the periodic table, and therefore, it has three valence electrons. It uses each of these electrons to form a bond, which is why we see three bonds to the aluminum atom in $AlBr_3$:

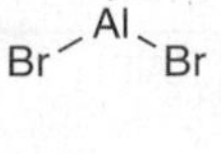

But you should notice that the aluminum atom does not have an octet. If you count the electrons around the aluminum atom, there are only six electrons. That means that aluminum has one empty orbital. That empty orbital is able to accept electrons. In fact, it will exhibit a tendency to accept electrons because that would give aluminum an octet of electrons:

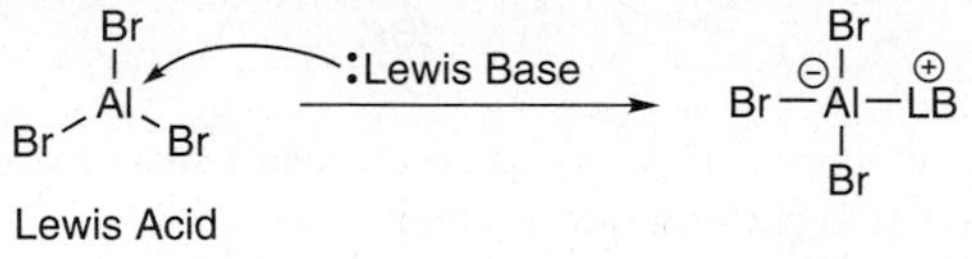

Therefore, we call AlBr$_3$ a *Lewis acid*. To put it simply, Lewis acids are just compounds that can *accept electrons*. Another common Lewis acid is FeBr$_3$:

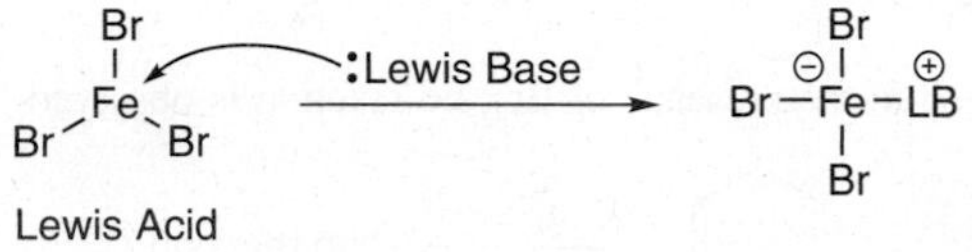

Now let's consider what happens when Br$_2$ is treated with a Lewis acid. The Lewis acid can accept electrons from Br$_2$:

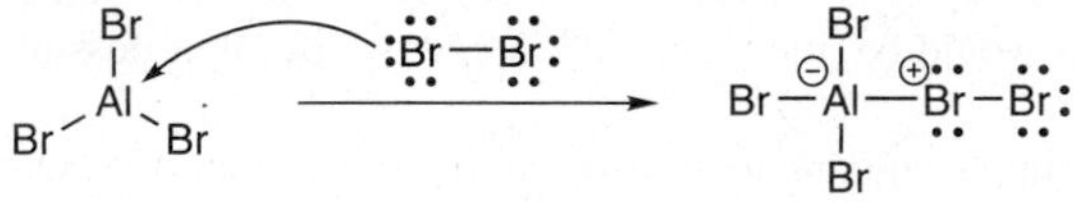

The resulting complex can then serve as a source of Br$^+$, like this:

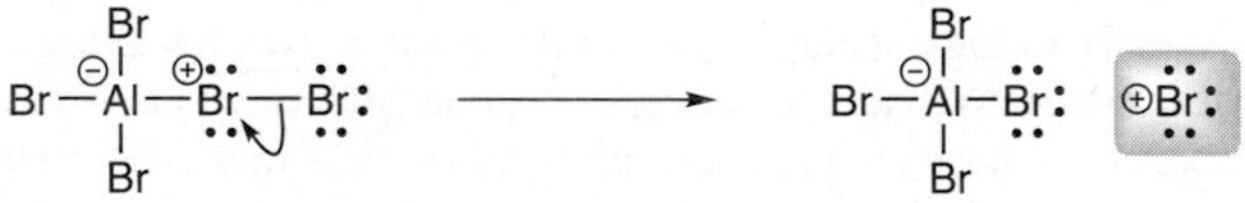

It is probably not accurate to think of this as a free Br$^+$ that can exist in solution by itself. Rather, the complex can *transfer* Br$^+$ to an attacking nucleophile:

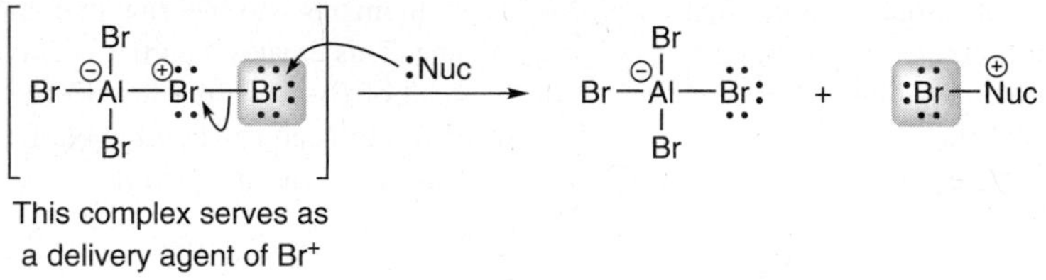

This complex serves as
a delivery agent of Br$^+$

The important point is that this complex can function as a delivery agent of Br$^+$, and that is what we needed in order to force a reaction between benzene and bromine. So, now let's try our reaction again. When we treat benzene with bromine in the presence of a Lewis acid, such as AlBr$_3$, a reaction is indeed observed. BUT it is not the reaction that we expected. Look closely at the product:

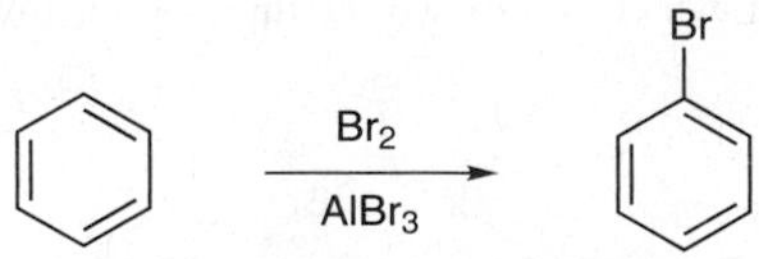

This is NOT an addition reaction. Rather, it is a *substitution* reaction. One of the aromatic protons was replaced with bromine. Since the ring is being treated with an electrophile (Br^+), we call this reaction an *electrophilic aromatic substitution*.

To see how this reaction occurs, let's take a close look at the accepted mechanism. It is absolutely critical that you fully understand this mechanism, because we will soon see that ALL electrophilic aromatic substitution reactions follow a similar mechanism. The first step shows the ring acting as a nucleophile to attack the complex, thereby transferring Br^+ to the aromatic ring:

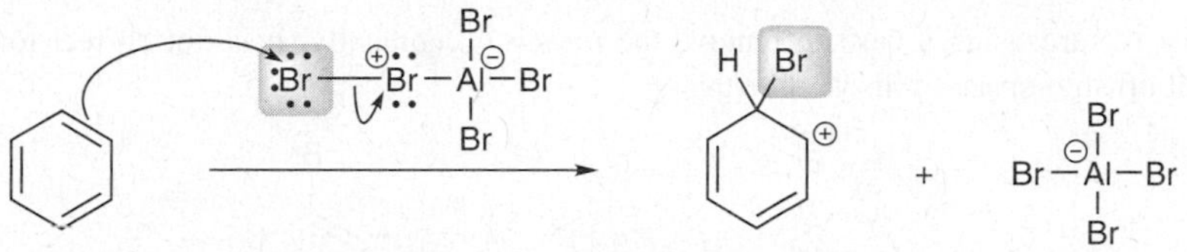

This step generates an intermediate that is not aromatic. It is true that aromaticity has been *temporarily* destroyed, but it will soon be reestablished in the second step (final step) of the mechanism. In this first step of the mechanism (shown above), the ring attacks Br^+ to form an intermediate that has three important resonance structures:

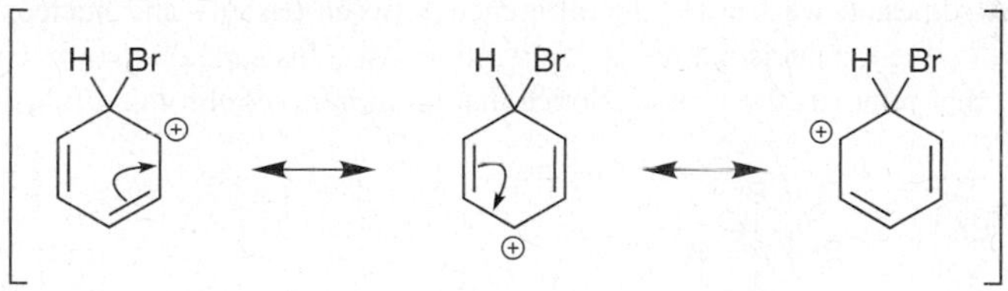

It is important to remember what resonance structures represent. Recall from the first semester that resonance is NOT a molecule flipping back and forth between different states. Rather, resonance is the way we deal with the inadequacy of our drawings. There is no one single drawing that adequately captures the essence of the intermediate, so we draw three drawings, and we meld them all together in our minds to gain a better understanding of the intermediate.

Attempts have been made to draw a single drawing for this intermediate:

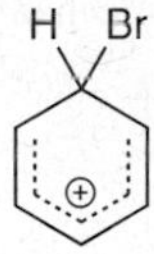

You might even see this in your textbook. I usually try to avoid using this drawing because it could easily lead you to think, erroneously, that the positive charge is spread over five atoms in the ring. This is not the case. The majority of the positive charge is actually only spread over three atoms of the ring (which we can clearly see when we look at all three resonance structures above).

This intermediate has some special names. We often call it a sigma complex, or sometimes, we call it an Arrhenium ion. These are just two different names for the same intermediate. From now on, in this book, we will call it a sigma complex:

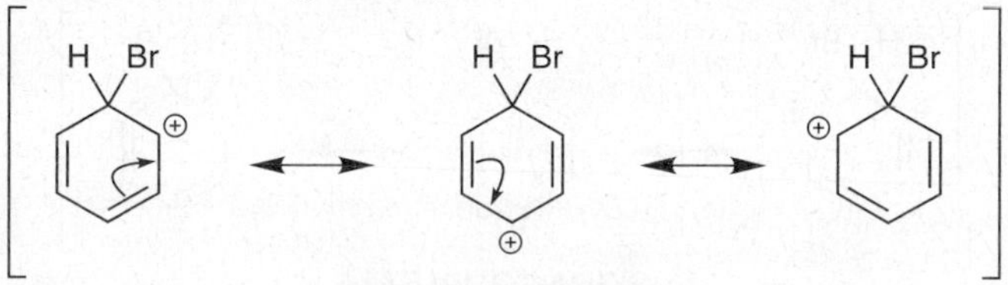

SIGMA COMPLEX

The second step (last step) of the mechanism involves transfer of a proton to re-form aromaticity:

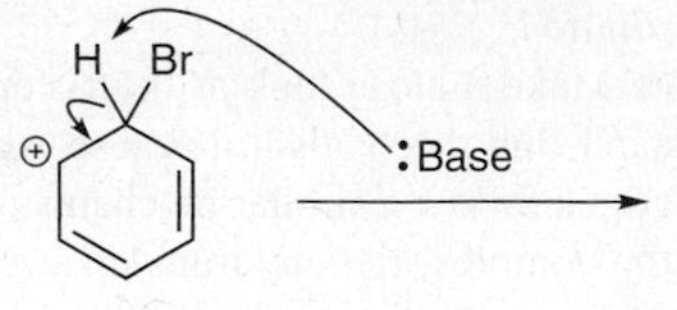

Notice that we are using a base to remove the proton. Technically, it is not correct to just let a proton fall off into space by itself, like this:

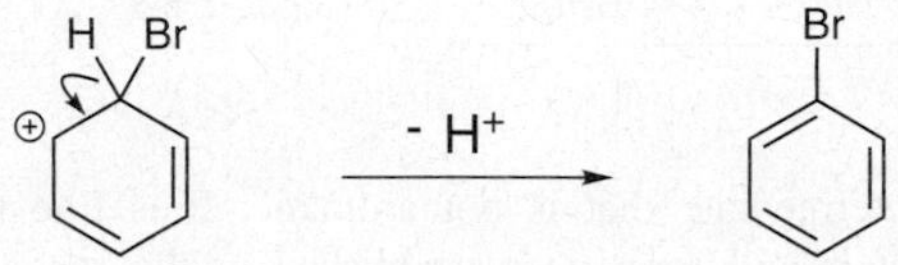

Whenever you are drawing a proton transfer, you should show the base that is removing the proton. In this particular case, it might be tempting to use Br⁻ to remove the proton. But Br⁻ is not a good base. (In the first semester, we learned the difference between basicity and nucleophilicity, and we saw that Br⁻ is a very good nucleophile but a very poor base.) Instead, aluminum tetrabromide functions as the base that removes the proton. Notice that aluminum tetrabromide functions as a "delivery agent" of Br⁻.

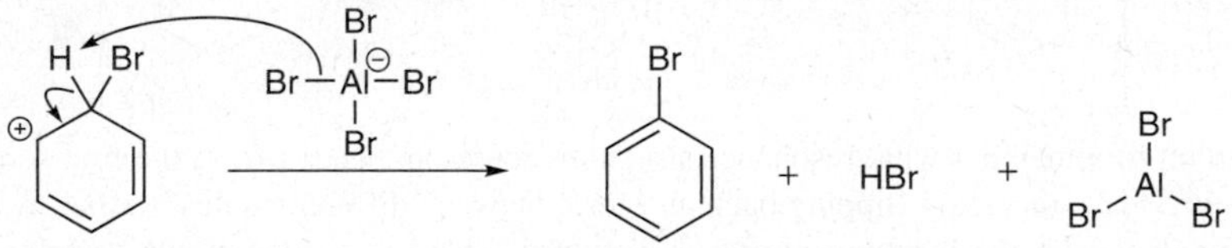

Notice that, in the end, the Lewis acid (AlBr$_3$) is regenerated. So the Lewis acid is actually not being consumed by the reaction. It is only there to help the reaction along, which is why we call it a *catalyst* in this case. That is why the presence of even a small quantity of the Lewis acid will suffice.

Now that we have seen both steps of the mechanism, let's take a close look at the mechanism all at once:

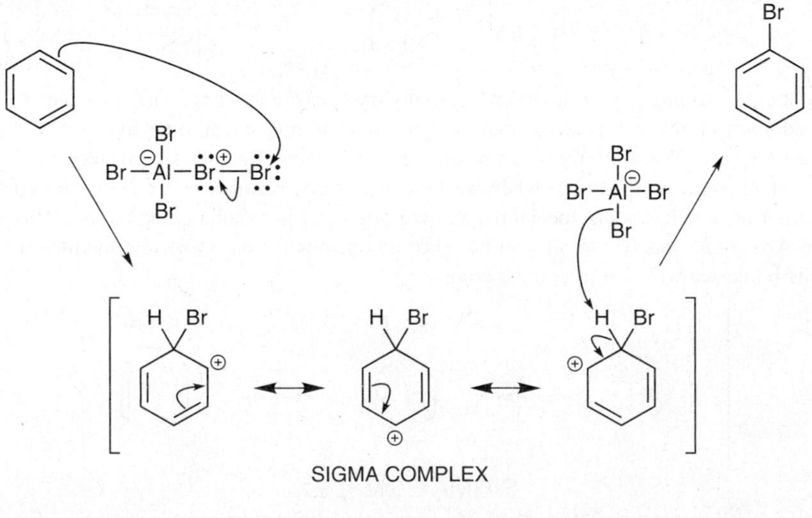

SIGMA COMPLEX

On the surface, it might seem like many steps. However, remember that resonance structures are not actually steps. Those three resonance structures (in the center of the mechanism) are necessary so that we can understand the nature of the one and only intermediate (the sigma complex). Indeed, the mechanism has only two steps. In the first step, benzene acts as a nucleophile attacking Br^+ to form the sigma complex, and in the second step, a proton is removed from the ring to reestablish aromaticity. In summary, the two steps are: attack, then deprotonate. To put it in other terms: Br^+ comes on, and then H^+ comes off. That's all there is to it.

PROBLEM 3.1 Without looking at the mechanism above, try to redraw the entire mechanism on a separate sheet of paper. Don't look above—you can figure it out. Just remember that there are two steps: E^+ on and then H^+ off. Don't forget to draw all three resonance structures of the intermediate sigma complex.

PROBLEM 3.2 Consider the following reaction, in which an aromatic ring undergoes chlorination, rather than bromination:

The mechanism is very similar to the mechanism for bromination. First, Cl_2 reacts with $AlCl_3$ to generate a complex that can serve as a source of Cl^+. Draw a mechanism for formation of this complex.

PROBLEM 3.3 Draw a mechanism for the electrophilic aromatic substitution reaction that occurs when benzene is treated with the complex from problem 3.2. The mechanism is exactly the same as the mechanism for installing a Br on the ring. But PLEASE, do not look back at that mechanism to copy it. Try to do it *without* looking back. Then, when you are finished, compare your answer to the answer in the back of the book (and compare every arrow to make sure that all of your arrows were drawn correctly):

PROBLEM 3.4 Aromatic rings will also undergo iodination when treated with a suitable source of I^+. There are many ways to form I^+; you should look in your textbook and in your lecture notes to see if you are responsible for knowing how to iodinate benzene. If so, be aware that the mechanism is exactly the same as what we have seen. The only difference will be in the mechanism of how I^+ is formed. Draw a mechanism for the reaction between benzene and I^+ to form iodobenzene. In the first step of your mechanism, simply draw I^+ as the electrophile (rather than a complex which delivers I^+), and make sure to draw all resonance structures of the resulting sigma complex. Then, in the last step of your mechanism, use H_2O as the base that removes the proton to restore aromaticity (H_2O will be present for many of the methods that are used to prepare a source of I^+).

3.2 NITRATION

In the previous section, we saw the mechanism of an electrophilic aromatic substitution reaction. We saw that the mechanism is the same, whether you are installing Br^+, Cl^+, or I^+ on the ring. We also said that this same mechanism explains how any electrophile (E^+) can be installed on an aromatic ring. For example, let's say we wanted to convert benzene into nitrobenzene:

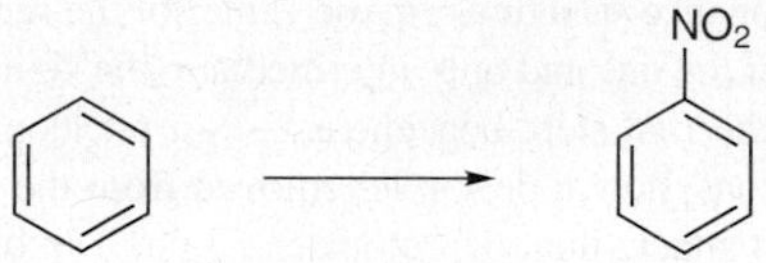

In order to form nitrobenzene, we will need NO_2^+ as our electrophile. But how do we make NO_2^+? If we look at how we made Br^+ or Cl^+ in the previous section, we might be tempted to use NO_2Br and $AlBr_3$, to get the following complex:

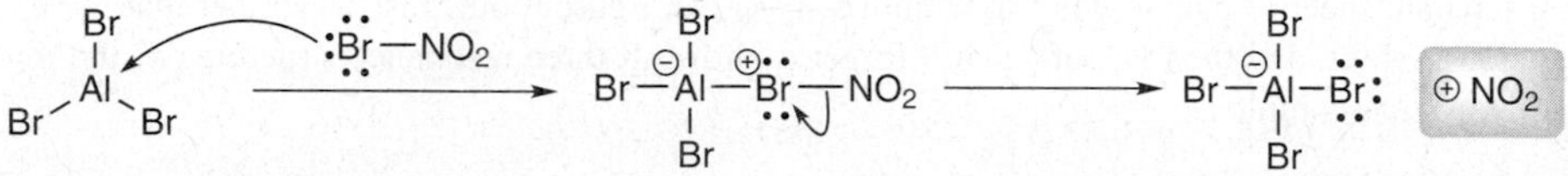

This complex could then serve as a source of NO_2^+. The problem is that NO_2Br is nasty stuff, and you probably would not want to work with it in a lab, especially since there is a much simpler way to make NO_2^+. We can form NO_2^+ by mixing sulfuric acid with nitric acid:

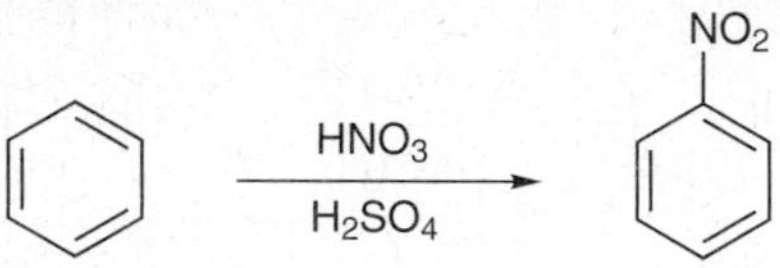

We need to take a close look at how NO_2^+ is formed under these conditions. Let's begin by drawing the structures of sulfuric acid and nitric acid:

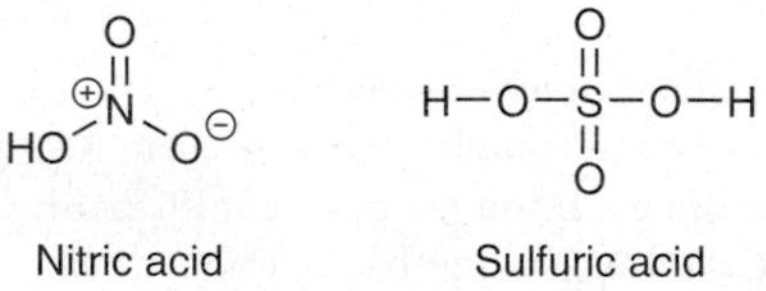

Nitric acid   Sulfuric acid

Notice that nitric acid exhibits charge separation. It might be tempting to remove the charges and draw it like this:

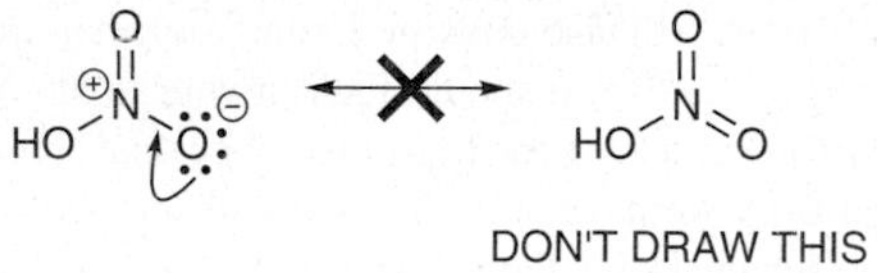

DON'T DRAW THIS

But you cannot do that because it would give five bonds to the central nitrogen atom. Nitrogen cannot EVER have five bonds because it only has four orbitals that it can use to form bonds. So nitric acid must be drawn with charge separation.

Now that we have seen the structures of both nitric acid and sulfuric acid, we must remember that the term *acidic* is a relative term. It is true that nitric acid is acidic, and it is also true that sulfuric acid is acidic. But between the two of them, sulfuric acid is a *stronger* acid. In fact, it is so much stronger as an acid that it is able to give a proton to nitric acid:

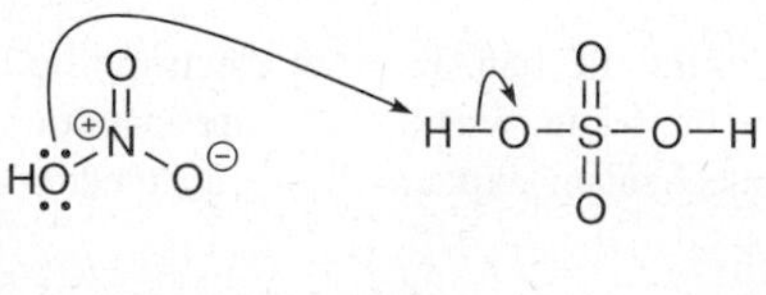

That's right—it might seem weird because nitric acid is essentially functioning *as a base* to remove a proton from sulfuric acid. And it might make us uncomfortable to use nitric acid as a base, but that is exactly what is happening. Why? Because *relative* to sulfuric acid, nitric acid is a base. It's all relative.

OK, so nitric acid removes a proton from sulfuric acid. But the obvious question is: why does the uncharged oxygen remove the proton? Wouldn't it make more sense for the negatively charged oxygen to remove the proton? Like this:

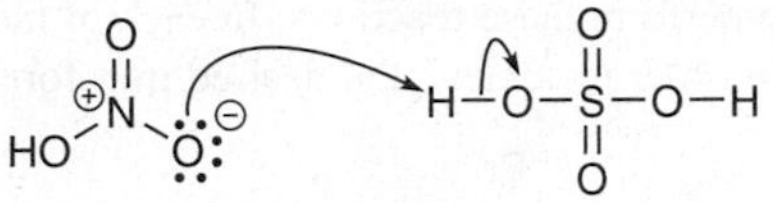

The answer is: yes, this probably would make more sense. And it probably happens like this a lot more often. The oxygen with the negative charge probably removes the proton much more readily than the uncharged oxygen atom does. However, proton transfers are reversible. Protons are being transferred back and forth all of the time. And all of this is happening very quickly. So, it is true that the negatively charged oxygen atom removes the proton more often—but when that happens, the only thing that can happen next is for the proton to be given right back to re-form nitric acid.

Every once in a while, however, the uncharged oxygen atom can remove the proton. And when that happens, something new can happen next: water can leave:

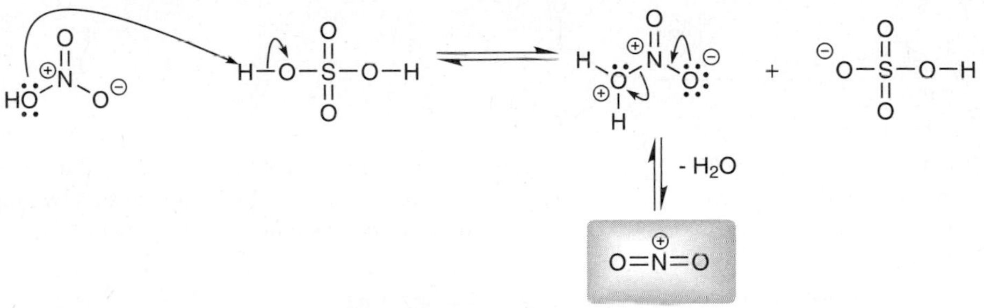

$$- H_2O$$

And when this happens, NO_2^+ is generated. So when we mix sulfuric acid and nitric acid, we get a little bit of NO_2^+ in the equilibrium mixture, and that NO_2^+ functions as the electrophile we need in order to install a nitro group on a benzene ring.

Once again, the following mechanism is essentially the same mechanism that we saw in the previous section: NO_2^+ on and then H^+ off.

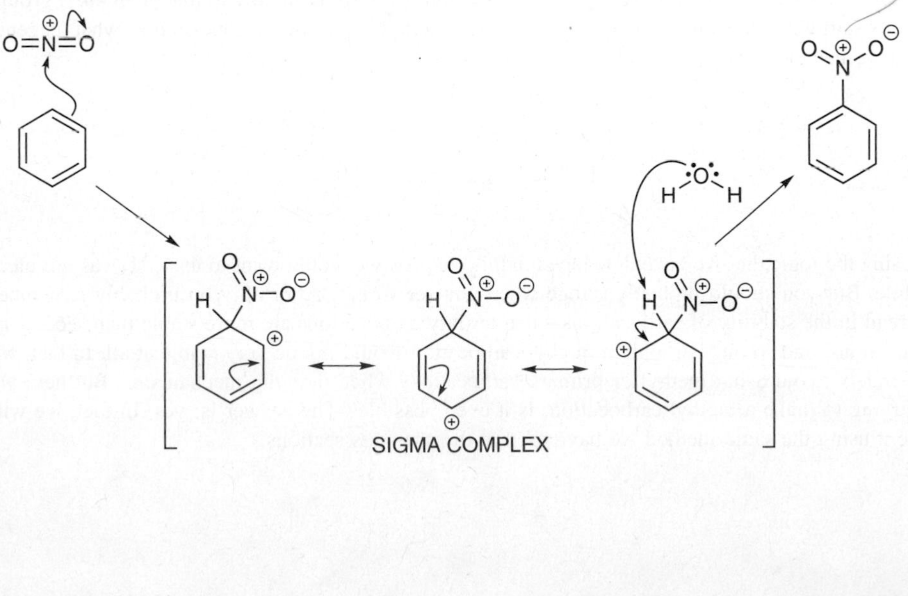

SIGMA COMPLEX

In this case, we are using water to remove the proton (instead of $AlBr_4^-$), which should make sense because we don't have any $AlBr_4^-$ in this reaction. There is plenty of water, because nitric and sulfuric acids are both aqueous solutions. Notice that the mechanism is very similar to what we have already seen in the previous reactions.

So far, we have seen how to install a halogen (Cl, Br, or I) on an aromatic ring, and we have seen how to install a nitro group. Before we move on, let's just make sure that you are familiar with the reagents necessary to perform these reactions. In each of the following cases, identify the reagents that you would use in order to achieve the desired transformation.

3.5

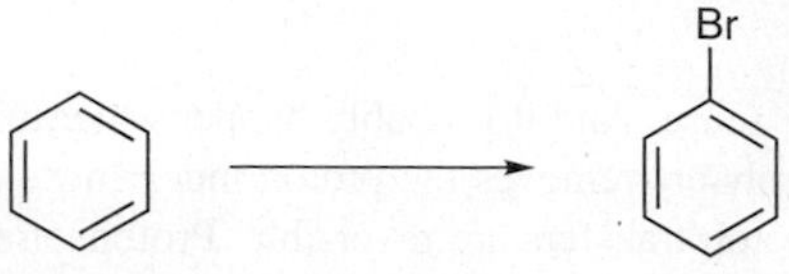

3.6

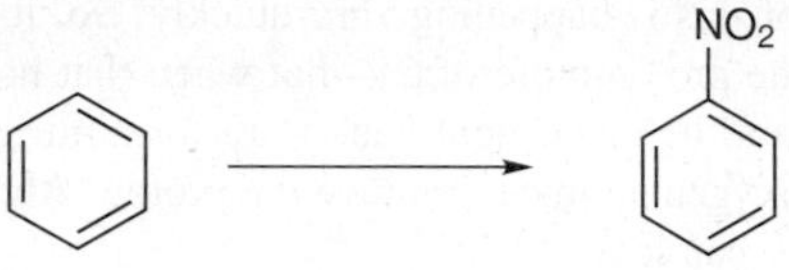

3.7

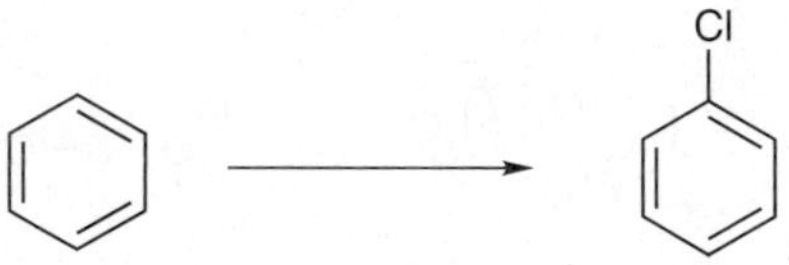

3.8 *Without looking back* at the previous section, try to draw the mechanism for the nitration of benzene. You will need a separate piece of paper to record your answer. Make sure to start by drawing the mechanism for formation of NO_2^+, and then show the reaction of benzene with NO_2^+.

3.3 FRIEDEL-CRAFTS ALKYLATION AND ACYLATION

In the previous sections, we saw how to install several different groups (Br, Cl, I, or NO_2) on a benzene ring, using an electrophilic aromatic substitution. In each case, the mechanism was the same: E^+ *on* the ring, and then H^+ *off*. In this section, we will learn how to install an alkyl group.

Let's start with the simplest of all alkyl groups: a methyl group. So, the question is: what reagents would we need to achieve the following transformation:

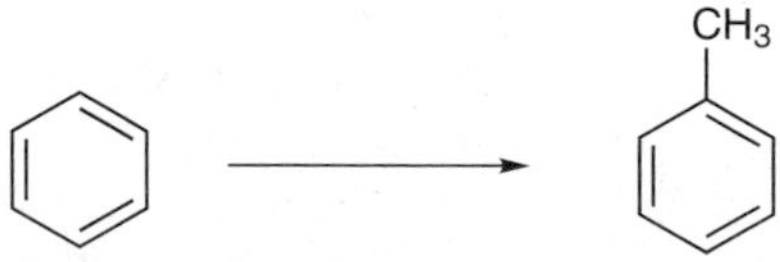

Using the logic that we have developed in this chapter, we would want to use CH_3^+ as our electrophile. But you should probably cringe when you see CH_3^+. After all, you probably remember the trend in the stability of carbocations—that tertiary carbocations are more stable than secondary carbocations, and so on. Certainly a methyl carbocation would not be very stable at all. In fact, we deliberately avoid using methyl or primary carbocations when drawing mechanisms. But here we are, trying to make a methyl carbocation. Is it even possible? The answer is: yes. In fact, we will make it using the same method we have used in the previous sections.

If we take methyl chloride and we mix it with a pinch of $AlCl_3$, we will have a source of CH_3^+:

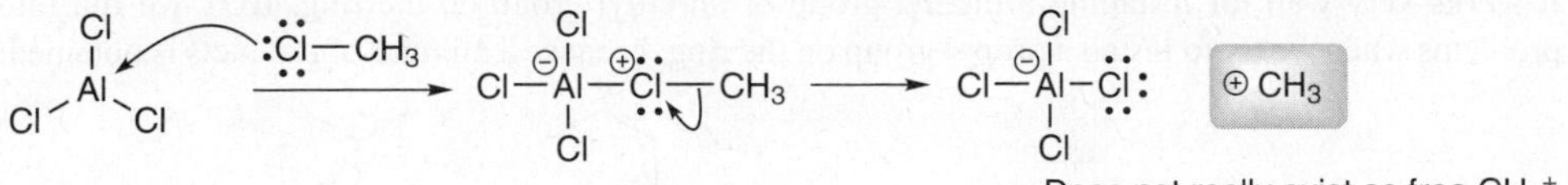

Does not really exist as free CH_3^+

The truth is that we are NOT really forming a free methyl carbocation that can float off into solution. A free CH_3^+ would be too unstable to form. So, instead, we must view this as a complex that can serve as a "source" of CH_3^+.

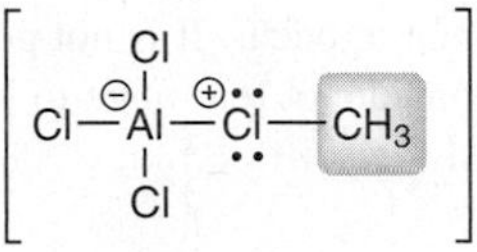

Source of CH_3^+

This provides us with a method for methylating a benzene ring:

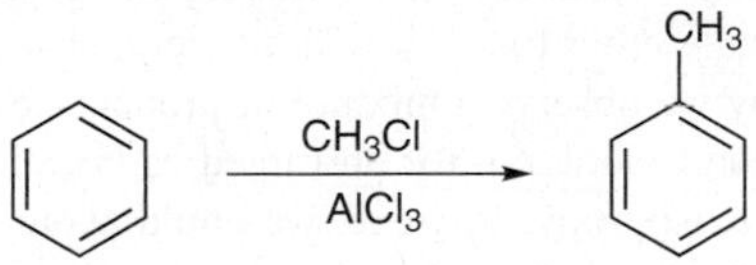

$$\xrightarrow[AlCl_3]{CH_3Cl}$$

And the mechanism is, once again, the same mechanism that we have seen over and over again. It is an electrophilic aromatic substitution: CH_3^+ *on* the ring and then H^+ *off*:

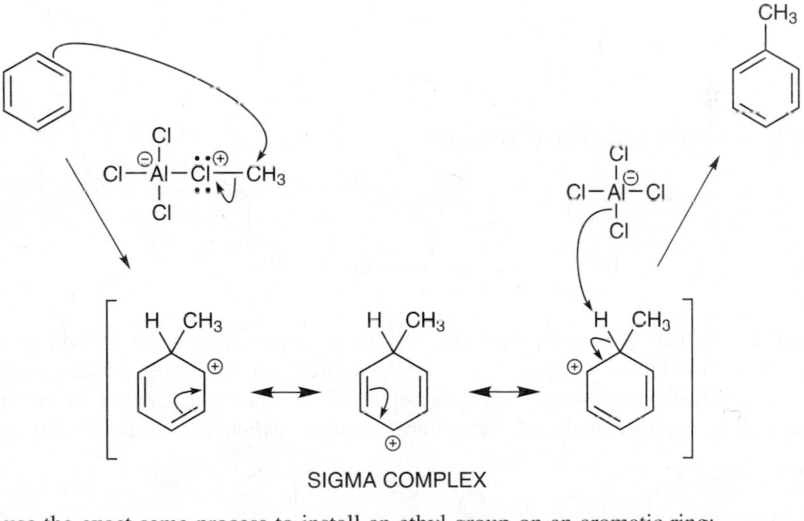

SIGMA COMPLEX

We can use the exact same process to install an ethyl group on an aromatic ring:

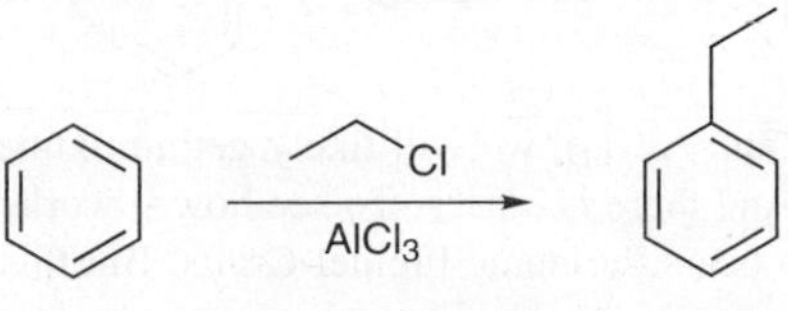

$$\xrightarrow[AlCl_3]{}$$

This process (installing an alkyl group onto an aromatic ring) is called a Friedel-Crafts alkylation. It works very well for installing a methyl group or an ethyl group on the ring. BUT we run into problems when we try to install a propyl group on the ring, because a mixture of products is obtained:

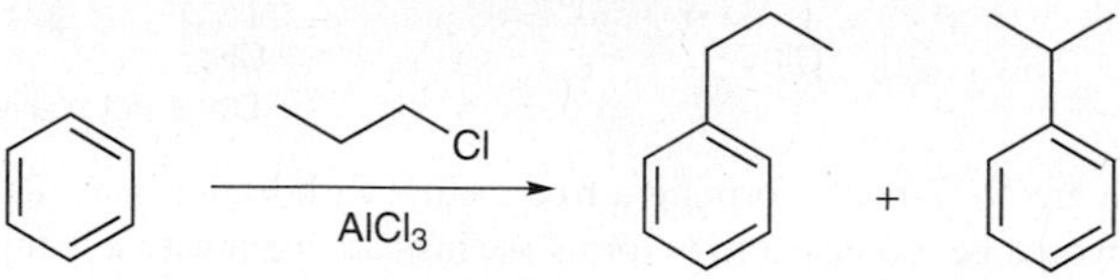

The reason for this is simple. Since we are forming a complex with carbocationic character, it is possible for a carbocation rearrangement to occur. It is not possible for a methyl carbocation to rearrange. Similarly, an ethyl carbocation cannot rearrange to become any more stable. But a propyl carbocation CAN rearrange (via a hydride shift):

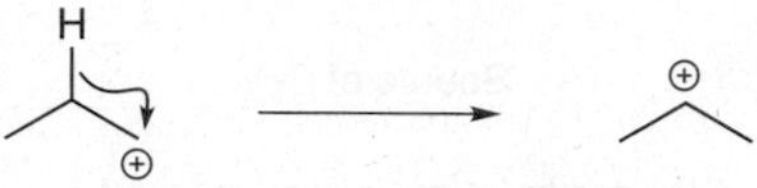

And since we are forming a propyl carbocation, we can expect that sometimes it will rearrange before reacting with benzene (while other times, it will not get a chance to rearrange before it reacts with benzene). And that is why we observe a mixture of products. So you need to be careful when using a Friedel-Crafts alkylation to look out for unwanted carbocation rearrangements.

Now, if we wanted to make isopropyl benzene, we could avoid this whole issue by just using isopropyl chloride as our reagent:

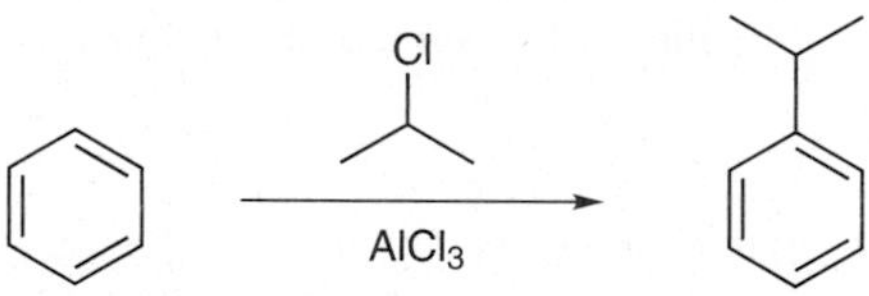

But what if we want to make propyl benzene?

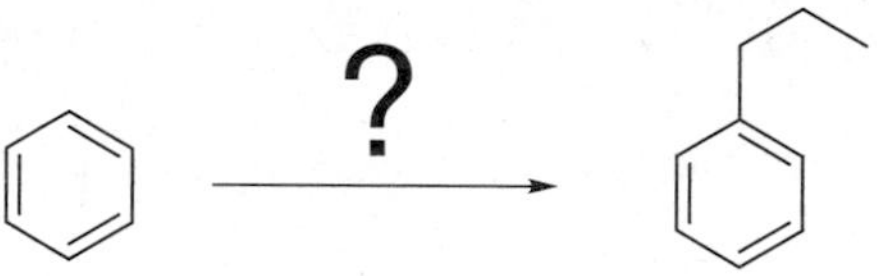

How would we do that? If we use propyl chloride, we have already seen that we will get some rearrangement, and we will not get a good yield of the desired product. In fact, we can generalize the question like this: how do we install *any* alkyl group and avoid a potential carbocation rearrangement. For example, how could we do the following transformation, ***without*** a carbocation rearrangement?

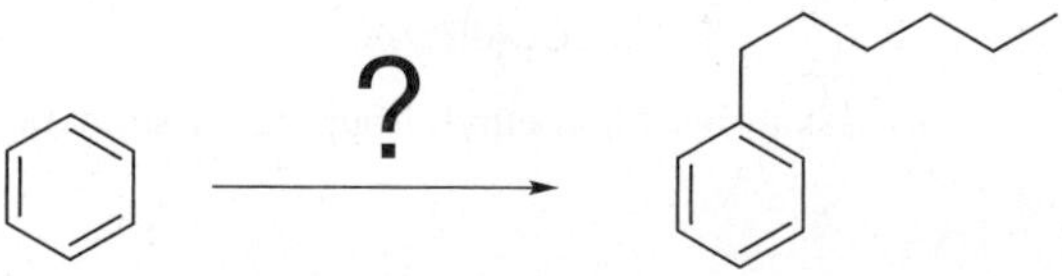

If we just use chlorohexane (and AlCl$_3$), we will likely get a mixture of products.

Clearly, we need a trick. And there is a trick. To see how it works, we need to take a close look at a similar reaction that also bears the name Freidel-Crafts. But this reaction is not an ***alkyl***ation.

Rather, it is called an *acyl*ation. To see the difference, let's quickly compare an alkyl group with an acyl group.

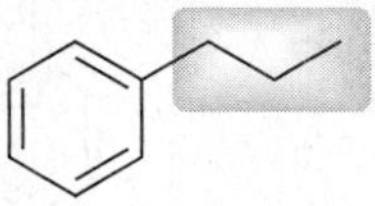

Alkyl group 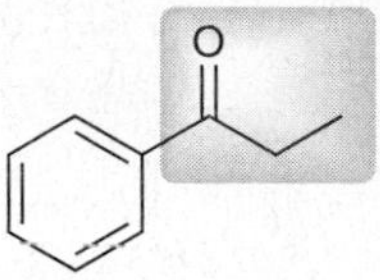   *Acyl* group

We can install an *acyl* group on a benzene ring in exactly the same way that we installed an alkyl group on the ring. We simply use the following reagents:

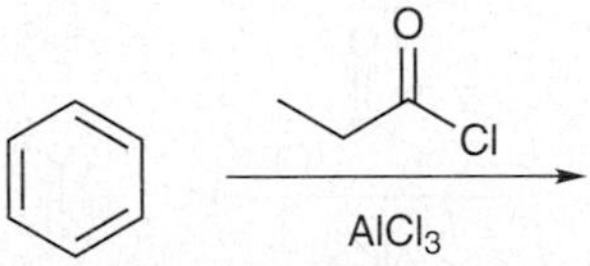

The first reagent is called an acyl chloride (or acid chloride), and we are already familiar with the role of $AlCl_3$ (the Lewis acid). The Lewis acid interacts with the Cl atom of the acyl chloride, generating a resonance-stabilized acylium ion:

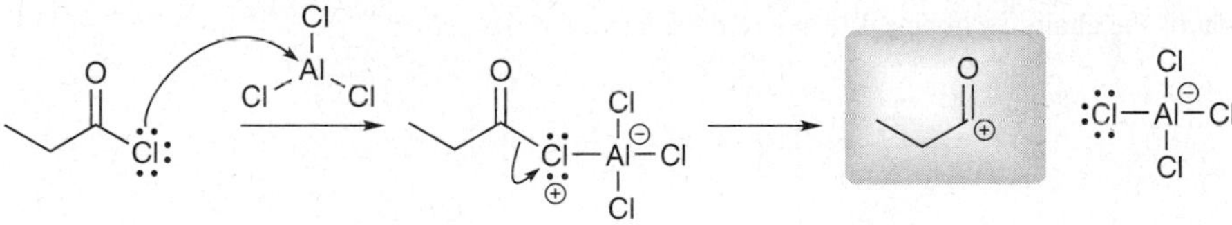

The term *acylium* ion should make sense—"*acyl*" because we can see that this electrophile has an acyl group; and "*ium*" because there is a positive charge. This electrophile actually has an important resonance structure:

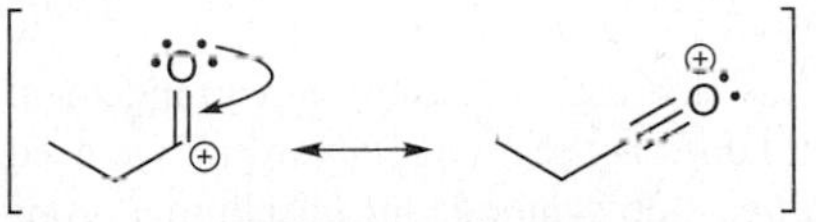

ACYLIUM ION

These resonance structures are important because they indicate that an acylium ion is *stabilized* by resonance, and therefore, it will NOT undergo a carbocation rearrangement. (If it did, it would lose this resonance stabilization.) Compare the following two cases:

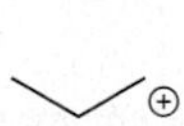

CAN REARRANGE 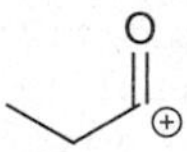   CANNOT REARRANGE

Using a Friedel-Crafts *acyl*ation, we can cleanly install an acyl group on a benzene ring (without any rearrangements):

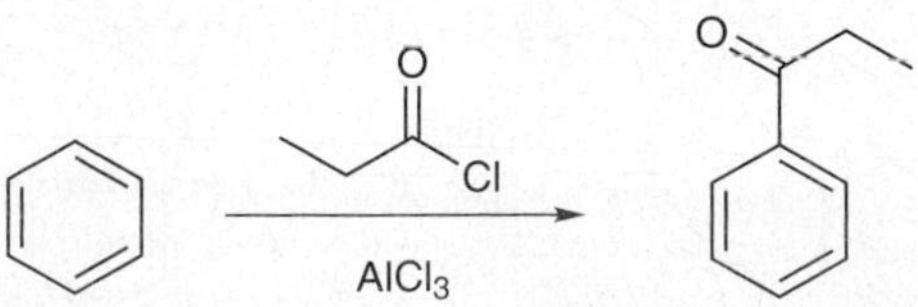

We don't observe any side products that would result from a carbocation rearrangement. Once again, compare a Friedel-Crafts *alkyl*ation with a Friedel-Crafts *acyl*ation:

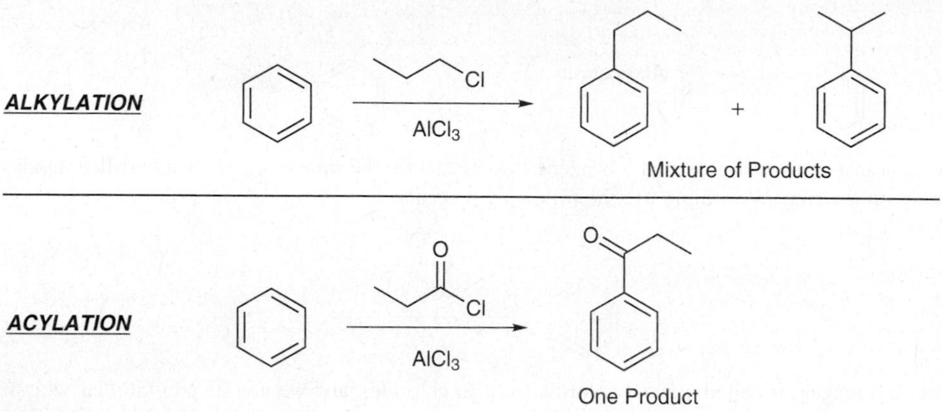

ALKYLATION

Mixture of Products

ACYLATION

One Product

Now take a close look at the acylation above, and let's point out a very important feature. Notice that we have installed a three-carbon chain on the ring, with the chain being attached by the *first* carbon of the chain, as opposed to the middle carbon of the chain:

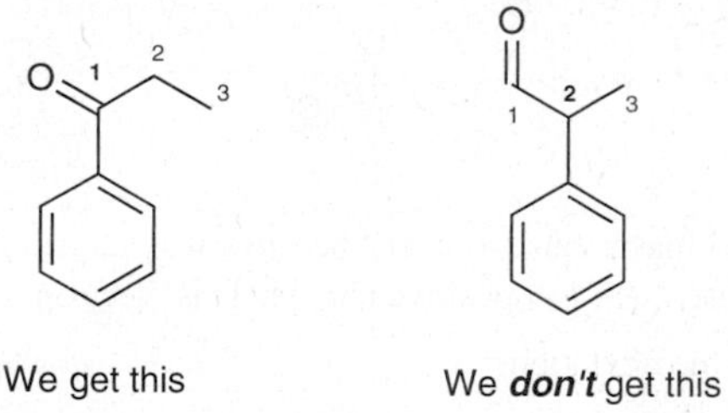

We get this   We *don't* get this

Once again, it is attached by the first carbon because a rearrangement does not occur (the acylium ion is resonance stabilized, and does not rearrange). Now, all we need is a way to remove the oxygen, and then we will have a two-step synthesis for installing a propyl group on a benzene ring:

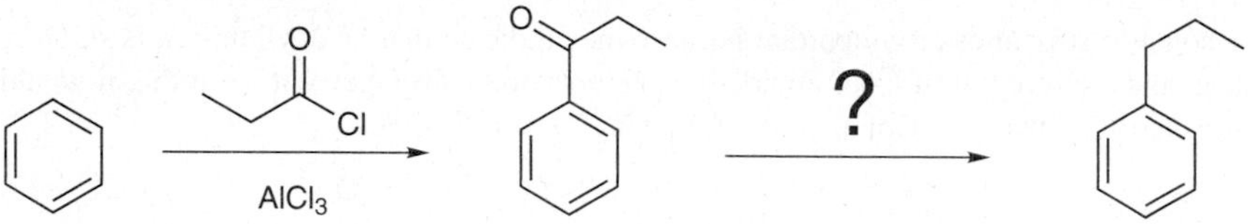

And luckily, there is a simple way to remove the oxygen; in fact, there are *three* very common ways to remove the oxygen. We will just focus on one method right now (which uses acidic conditions), but keep in mind that we will see two other methods of doing this in the upcoming chapters (one method uses basic conditions and the other method uses neutral conditions). When we reduce a ketone under acidic conditions, we call the reaction a Clemmensen reduction:

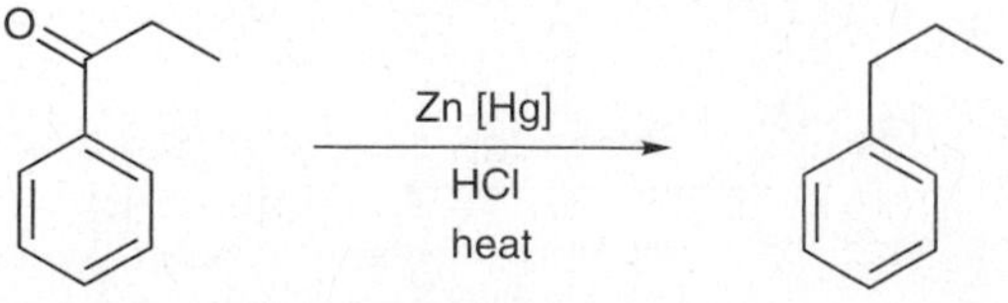

In the presence of a zinc amalgam and HCl, the C=O bond is completely reduced and replaced with two C—H bonds. In this way, a Clemmensen reduction can be used as the second step of a two-step synthesis that installs an alkyl group on an aromatic ring *without* any rearrangement taking place:

Before we move on, there is one subtle point to mention about a Friedel-Crafts acylation (step 1 of the synthesis above). Remember that this reaction takes place in the presence of a Lewis acid (AlCl₃), which is constantly on the lookout for electrons with which it can interact. Well, the product of the acylation step is a ketone. And a ketone has electrons that the Lewis acid is seeking:

Therefore, whenever we perform an acylation reaction, we need to pull the Lewis acid off our product in the end. And there is a simple way to do this. We just give the Lewis acid some other source of electrons to interact with instead. The easiest thing to use is water (H₂O). So, whenever you use a Friedel-Crafts acylation in a synthesis problem, it is appropriate to include water in your list of reagents (immediately after the acylation):

To summarize, we have seen that a Friedel-Crafts acylation can be followed up by a Clemmensen reduction, as a clever way of installing an alkyl group on an aromatic ring (without rearrangements). But there are actually times when you will want to install an acyl group on the ring, and you won't want to do a Clemmensen reduction afterward. For example:

To achieve this transformation, you would just use a Friedel-Crafts acylation, and that's it. There is no need for a Clemmensen reduction, because in this case, we don't want to reduce the C=O bond.

EXERCISE 3.9 Show the reagents you would use to achieve the following synthesis:

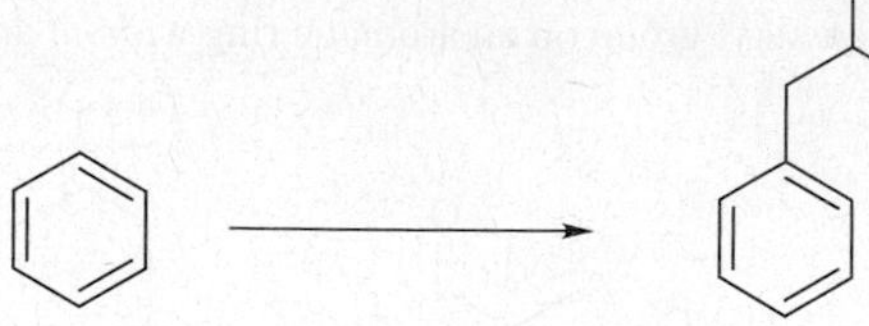

Answer In this problem, we need to install an alkyl group on a benzene ring. So, we first look to see if we could do this one step, using a Friedel-Crafts alkylation. In this case, we cannot do it in one step because we have to worry about a carbocation rearrangement. If we think about the electrophile that we would need to make, we will see that it could rearrange:

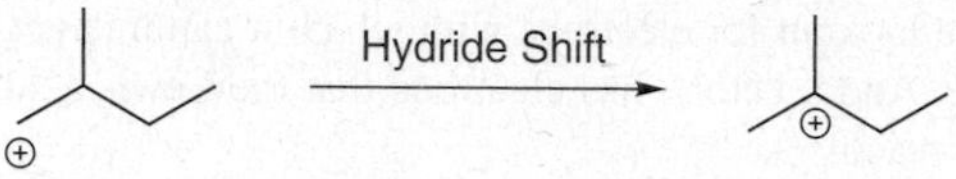
Hydride Shift

And this would give us a mixture of products:

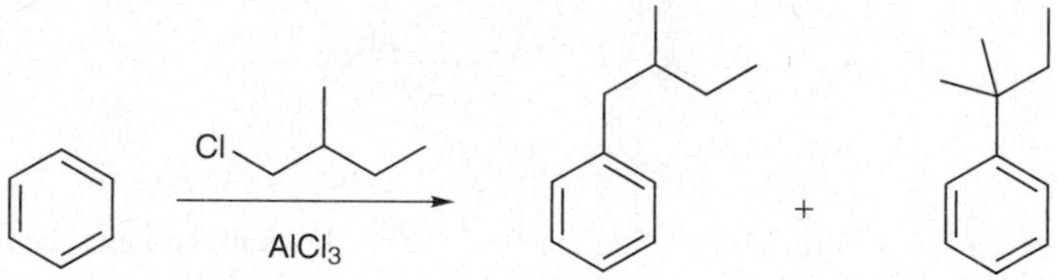
Cl
$AlCl_3$
+

So, instead we will have to use a Friedel-Crafts acylation followed by a Clemmensen reduction:

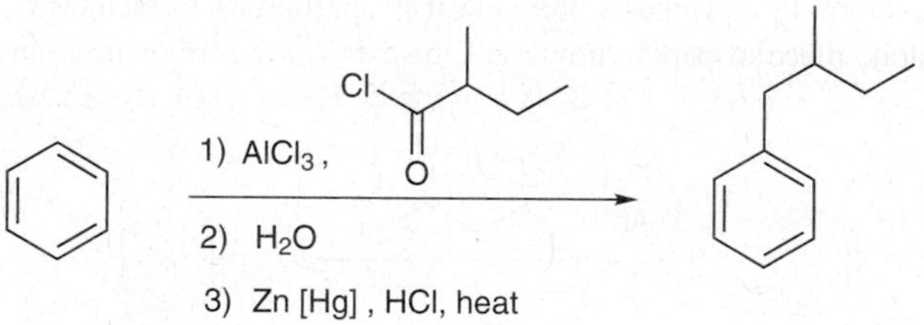
Cl
O
1) $AlCl_3$,
2) H_2O
3) Zn [Hg] , HCl, heat

For each of the following problems, show what reagents you would use to accomplish the transformation. In some situations, you will want to use a Friedel-Crafts alkylation, while in other situations, you will want to use a Friedel-Crafts acylation.

3.10

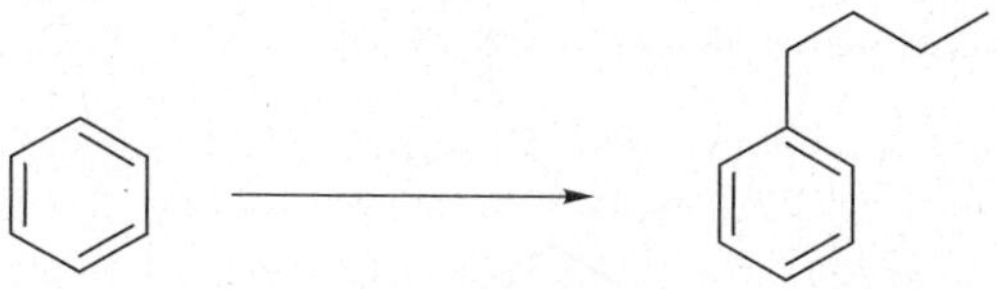

3.11

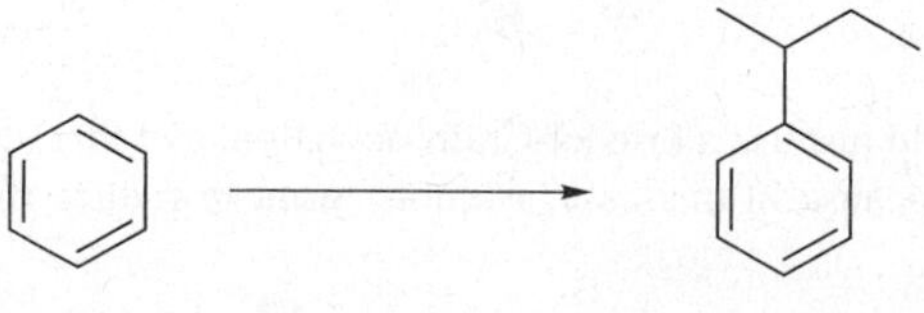

3.12

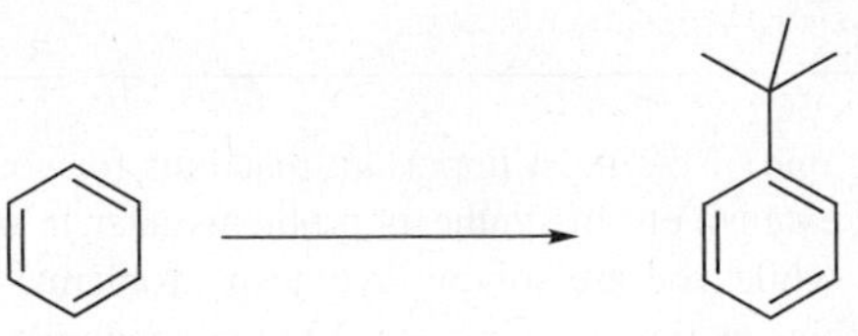

3.13

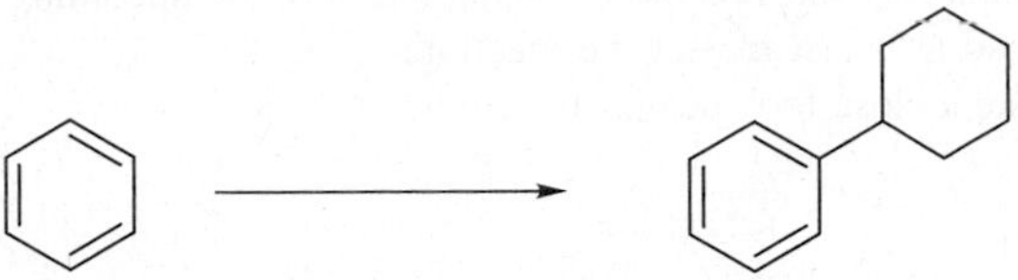

3.14

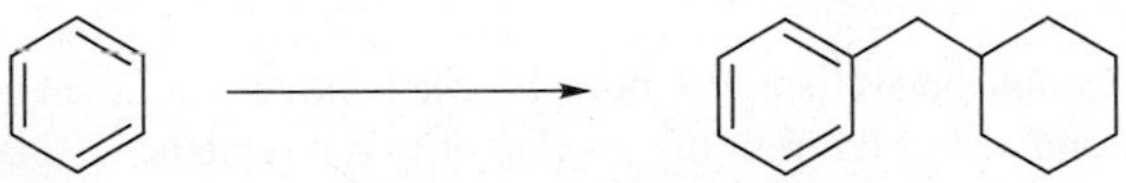

3.15 Predict the products of the following reaction.

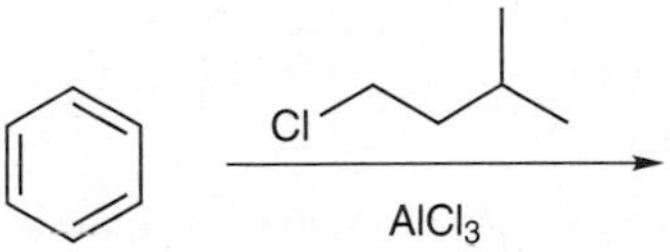

(*Hint*: There should be a mixture of *multiple* products in this case. Be sure to consider all of the possible rearrangements that can take place. If you are rusty on carbocation rearrangements, then you should go back and review them now.)

3.16 On a separate piece of paper, draw the mechanism of formation for each one of the three products from the previous problem.

3.17 On a separate piece of paper, draw a mechanism for the following transformation. Make sure to show the mechanism of formation of the acylium ion that reacts with the ring:

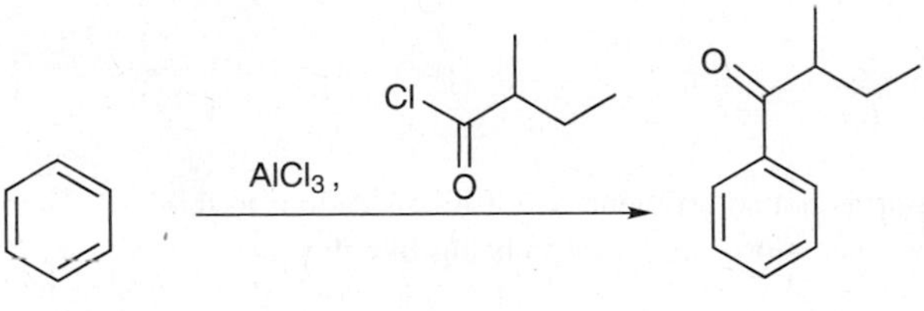

Friedel-Crafts reactions have a few limitations. You should take a moment to read about them in your textbook. The two most important limitations are as follows:

1. When performing a Friedel-Crafts *alkylation*, it is often difficult to install just one alkyl group. Each alkyl group makes the ring *more* reactive toward a subsequent attack on the same ring.

2. When performing a Friedel-Crafts *acylation*, it is generally not possible to install more than one acyl group. The presence of one acyl group makes the ring *less* reactive toward a second acylation.

We need to understand WHY an alkyl group makes the ring more reactive, and WHY an acyl group makes the ring less reactive. We will explain this in greater detail during the upcoming sections. But first, we have one more electrophile to discuss.

3.4 SULFONATION

The reaction we will discuss now is probably one of the most important reactions for you to have at your fingertips. This reaction will be used extensively in synthesis problems later in this chapter. If you do not keep this reaction in mind while you are solving synthesis problems, then you will be at a severe loss. We will explain why this reaction is so important in the upcoming sections. For now, just take my word for it, and let's just master the reaction.

The electrophile is SO_3. Let's take a close look at the structure:

Notice that there are three $S=O$ double bonds here. But these double bonds are not such great double bonds. Recall that a double bond is formed from the overlap of two *p* orbitals:

When we are talking about a carbon-carbon double bond, the overlap of the *p* orbitals is efficient because the *p* orbitals are the same size. But what happens when you try to overlap the *p* orbital of an oxygen atom with the *p* orbital of a sulfur atom? The *p* orbitals are different sizes (oxygen is in the second row of the periodic table, which means that it is using a *p* orbital from the second energy level; but sulfur is in the third row of the periodic table, so it is using a *p* orbital from the third energy level):

Therefore, the overlap is not so efficient, and it is misleading to think of $S=O$ as being a double bond. It is probably much closer in nature to being like this:

$$\overset{\oplus}{S}-\overset{\ominus}{O} \quad \text{rather than} \quad S=O$$

When we do this analysis for each of the three double bonds in SO_3, we begin to see that the sulfur atom is VERY electron-poor:

In fact, the sulfur atom is so electron-poor that it is an *excellent* electrophile, even though the compound is overall neutral (no net charge). Now we are going to see a reaction that uses SO_3 as an electrophile. But first, let's see where SO_3 comes from.

Sulfuric acid is constantly in equilibrium with SO_3 and water:

$$H_2SO_4 \quad \rightleftharpoons \quad SO_3 \ + \ H_2O$$

That means that any bottle of sulfuric acid will have some SO_3 in it. At room temperature, SO_3 is a gas, and it is possible to add extra SO_3 gas to sulfuric acid (which shifts the equilibrium). When we do this, we call the mixture *fuming* sulfuric acid. So, from now on, whenever you see concentrated, fuming sulfuric acid, you should realize that we are talking about SO_3 as the reagent.

And here is the reaction:

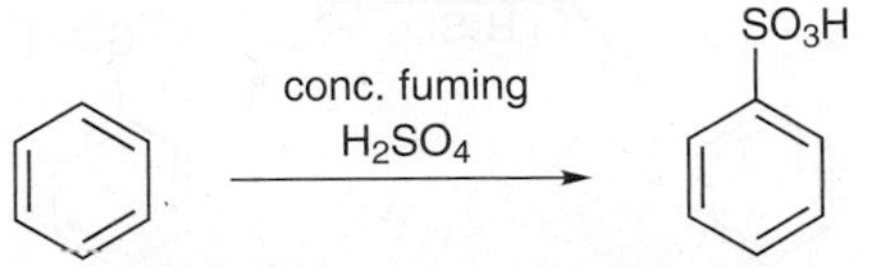

Notice that, in the end, we have installed the SO_3H group on the ring. The obvious question is: why is the H attached to the SO_3 in the end? To see why, let's take a closer look at the mechanism. Remember our two steps for any electrophilic aromatic substitution: E^+ goes on the ring, and then H^+ comes off. But wait a second. .... In this case, we are not using an electrophile with a net positive charge. The electrophile in this case has no net charge. In all of the reactions we have seen so far, we put something positively charged onto the ring, and then we took something positively charged off of the ring. So in the end, our ring never gained or lost any charges. But in this case, we are putting something neutral (SO_3) onto the ring, and then we are removing something positively charged (H^+). That should leave our product with an overall negative charge, which is exactly what happens:

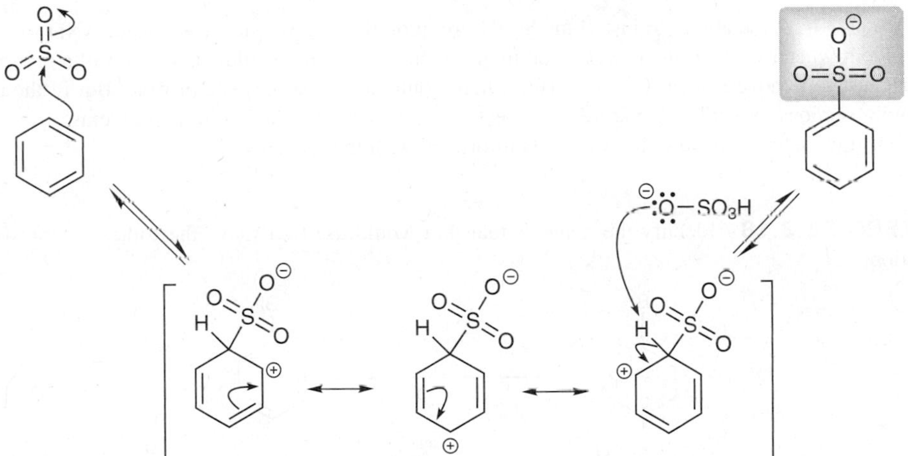

SIGMA COMPLEX

So, we need to add one more step to our mechanism. The negatively charged oxygen atom removes a proton from sulfuric acid:

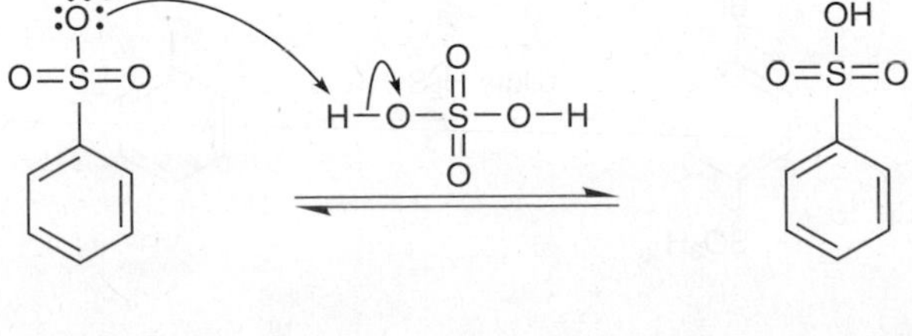

Although this reaction has this one extra step at the end of the mechanism, keep in mind that this extra step is just a proton transfer. The core reaction is still the same as we have seen in all of the previous reactions: the electrophile comes on the ring, and then H$^+$ comes off of the ring.

An important feature of this reaction (and this is the feature that will make this reaction so important for synthesis problems) is how easily the reaction can be reversed. The amount of product is equilibrium controlled, and it is very sensitive to the conditions. So, if you use dilute sulfuric acid instead, the equilibrium leans the other way:

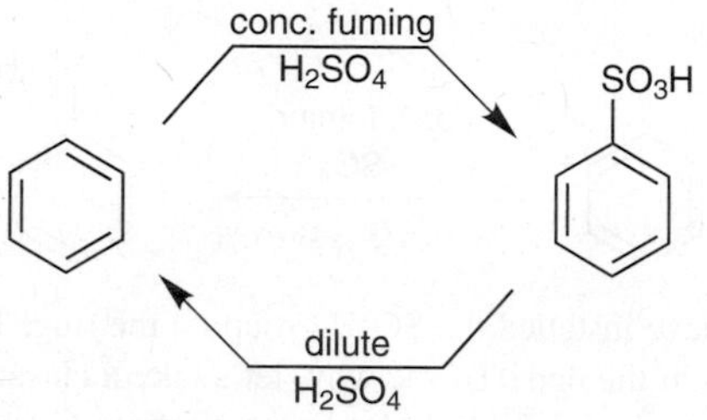

We can use this to our advantage because this provides a way to remove the SO$_3$H group whenever we want. We would just use dilute sulfuric acid to remove it:

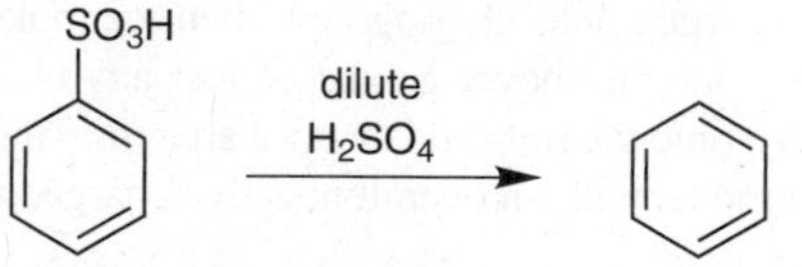

So we now have the ability to install the SO$_3$H group on the ring whenever we want, AND we can remove it whenever we want as well. You might wonder why you would want to install a group in order just to remove it later. On the surface, that would seem like a waste of time. But in the upcoming sections, we will see that this will become very important in synthesis problems.

For now, let's make sure that we are comfortable with the reagents.

EXERCISE 3.18 Identify the reagents that you would use to achieve the following transformation:

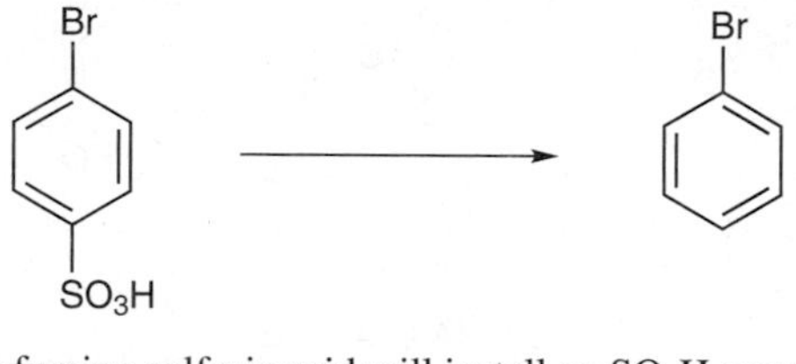

Answer We know that fuming sulfuric acid will install an SO$_3$H group on an aromatic ring, and dilute sulfuric acid will remove the group. In this case, we are removing the group, so we need to use dilute sulfuric acid:

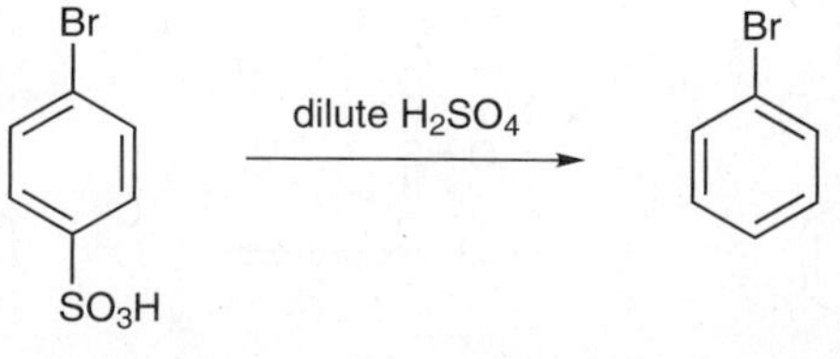

Identify the reagents you would use to achieve each of the following transformations:

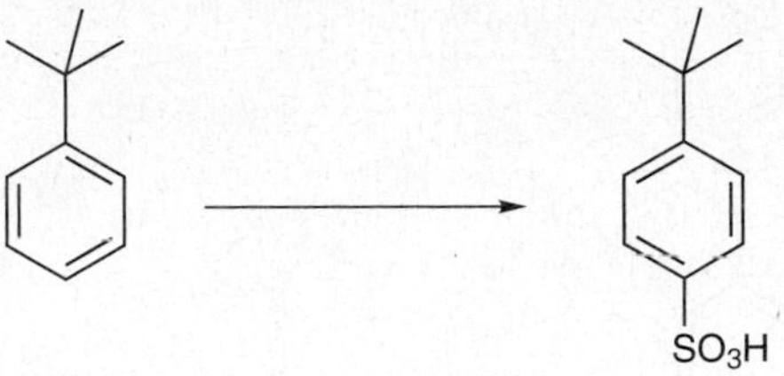

3.19

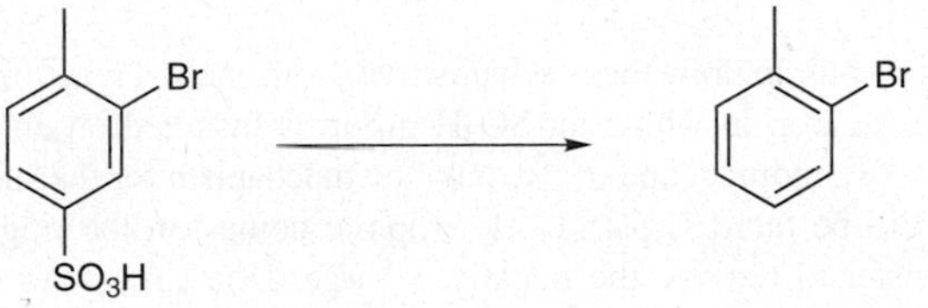

3.20

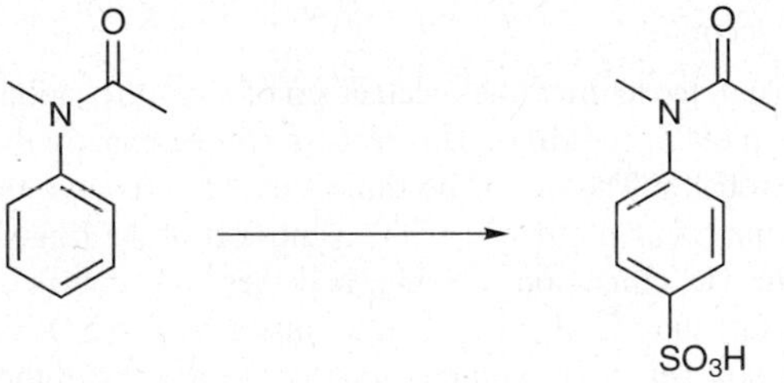

3.21

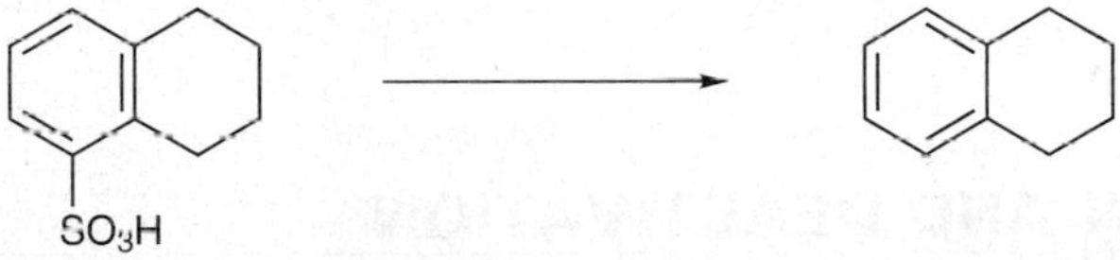

3.22

And to make sure that you are not getting rusty on the other reactions we have learned in this chapter so far, fill in the reagents you would use for the following transformations:

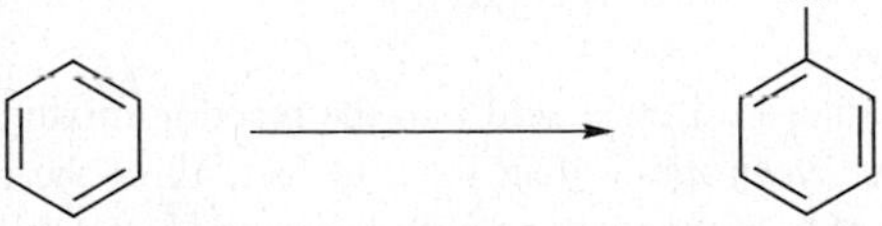

3.23

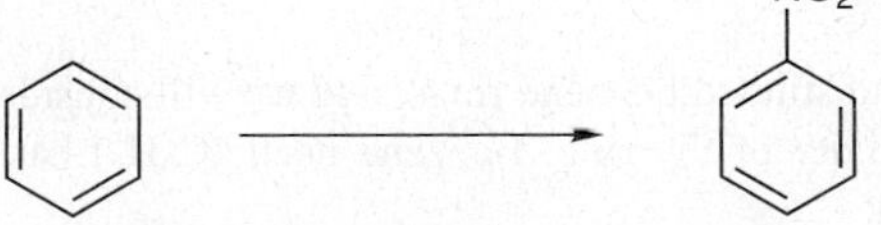

3.24

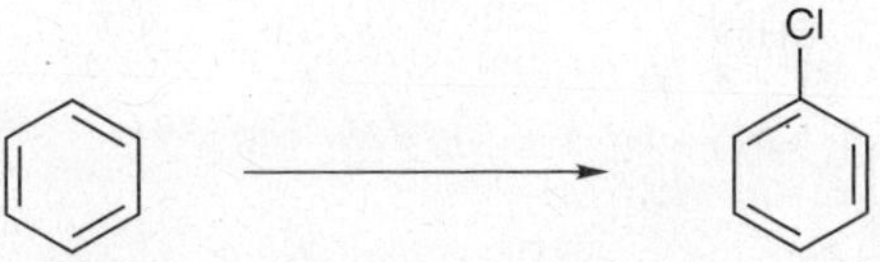

3.25

3.26

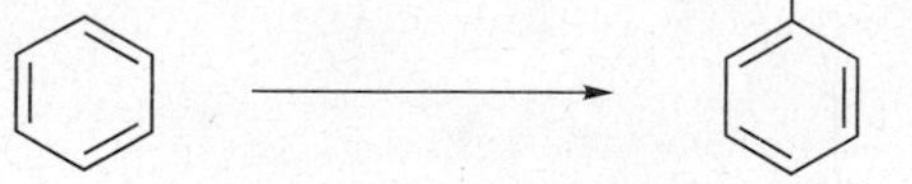

3.27

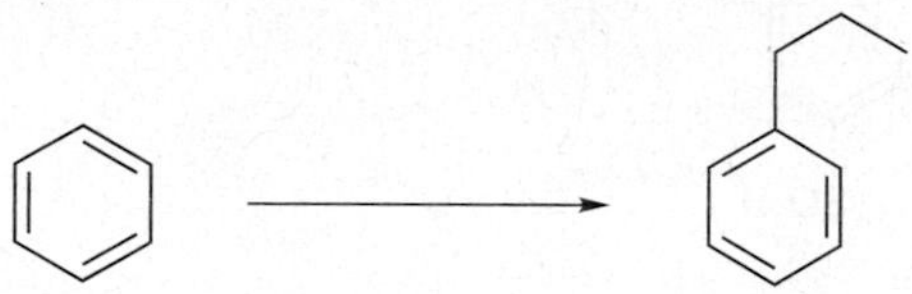

3.28   Now, let's just make sure that you can draw the mechanism of a sulfonation reaction. (That's just the fancy name we give to the reaction in which an SO$_3$H group is installed on an aromatic ring.) On a separate piece of paper, take a moment and try to draw the mechanism for the sulfonation of benzene. Remember that there will be three steps: (1) electrophile comes on the ring, (2) H$^+$ comes off, and then (3) proton transfer to remove the negative charge. Don't forget to draw the resonance structures of the intermediate sigma complex. Try to draw the mechanism without looking back to where it is shown earlier in this section.

3.29   And now, for a challenging problem, try to draw the mechanism of a desulfonation reaction (a reaction where we take the SO$_3$H group off of the ring). The process will be exactly the reverse of what you just drew in the previous problem. There will be three steps: (1) remove the proton from the SO$_3$H group, (2) H$^+$ comes on the ring, and then (3) SO$_3$ comes off of the ring. The truth is that there are only two core steps here: H$^+$ comes on the ring, and then SO$_3$ comes off of the ring. You can actually pull the proton off of the SO$_3$H group at the same time that SO$_3$ comes off of the ring. Try to do it yourself, and if you get stuck, you can look at the answer in the back of the book. Make sure that your mechanism involves an intermediate sigma complex (with a resonance-stabilized positive charge).

3.5   ACTIVATION AND DEACTIVATION

On the first page of this chapter, we saw that benzene is unreactive toward bromine:

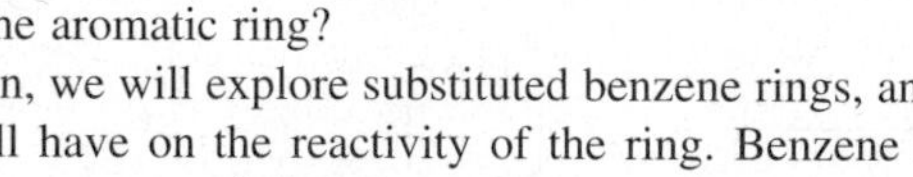

In order to force a reaction to occur, we introduced a Lewis acid into the reaction mixture, which generated a better electrophile (Br$^+$ is a better electrophile than Br$_2$). In fact, all of the reactions we have explored thus far have been examples of benzene reacting with powerful electrophiles (Cl$^+$, NO$_2^+$, alkyl$^+$, acyl$^+$, and SO$_3$). Now, we will turn our attention to the nucleophile—how can we modify the reactivity of the aromatic ring?

To answer this question, we will explore substituted benzene rings, and we will consider the effect that a substituent will have on the reactivity of the ring. Benzene itself (C$_6$H$_6$) has no substituents. But consider the structure of phenol:

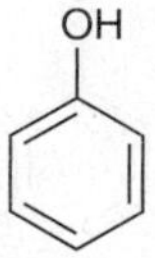

In this compound, the aromatic ring has one substituent: an OH group. What effect does this substituent have on the nucleophilicity of the aromatic ring? Is this compound a better nucleophile than benzene?

To answer this question, we must explore the effect of an OH group on the electron density of the aromatic ring. There are two factors to consider: *induction* and *resonance*. Let's begin with induction. Recall from the first semester that inductive effects can be evaluated by comparing the relative electronegativity of the atoms. In our example, we are looking specifically at the C—O bond connecting the OH group to the ring. Oxygen is more electronegative than carbon, so there is an inductive effect, shown by the following arrow:

The oxygen atom is withdrawing electron density from the ring. Remember that the aromatic ring is only a nucleophile in the first place because it is electron-rich (from all of those π electrons), so withdrawing electron density from the ring (via induction) should render the ring *less* nucleophilic. But we're not done yet. We need to consider one other factor: resonance.

Below are resonance structures of phenol:

Notice that there is a negative charge spread throughout the ring. When we meld all of these resonance structures together in our minds, we obtain the following information:

The $\delta-$ shows that there is a partial negative charge spread throughout the ring. Therefore, the effect of resonance is to *donate* electron density to the ring. So, now we have a competition. By induction, the OH group is *electron-withdrawing*, which makes the ring *less* nucleophilic. But by resonance, the OH group is *electron-donating*, which makes the ring *more* nucleophilic. Which effect is stronger? Resonance or induction? This is a common scenario in organic chemistry (where induction and resonance are in competition), and the general rule is: ***resonance is usually a stronger factor than induction***. There are some important exceptions. In fact, we will soon see one of these exceptions, but, in general, resonance wins.

Now let's apply this general rule to our case of phenol. If we say that resonance is stronger than induction, then the net effect of the OH group is to *donate* electron-density to the ring. And therefore, the net effect of the OH group is to make the ring *more nucleophilic* (as compared with benzene).

Indeed, experiments reveal that phenol is significantly more nucleophilic than benzene. The effect of the OH group on the nucleophilicity of the aromatic ring is called "activation." The OH group is said to be activating the ring (making it more nucleophilic). So, the OH group is called an *activator*. Alkyl groups (such as methyl or ethyl) are also activators (recall from the first semester that alkyl groups are electron donating because of an effect called hyperconjugation). There are some groups, though, that actually *withdraw* electron density from the ring, and we call those groups *deactivators*, because they deactivate the ring (make the ring *less* nucleophilic). An excellent example is the nitro group. Consider the structure of nitrobenzene:

Once again, the effect of the nitro group can be evaluated by exploring two factors: induction and resonance. Induction is simple. The nitrogen atom is more electronegative than the carbon atom to which it is connected (especially since the nitrogen atom bears a positive charge). So the nitro group withdraws electron density from the ring via induction, which should render the ring less nucleophilic. But we're not done yet. We still have to consider resonance. Below are resonance structures of nitrobenzene:

Notice that there is now a positive charge spread throughout the ring (rather than a negative charge, as we saw in the case of phenol). When we meld all of these resonance structures together in our minds, we obtain the following information:

The $\delta+$ shows that there is a partial positive charge spread throughout the ring. Therefore, the effect of resonance is to *withdraw* electron density from the ring.

In summary, the nitro group is electron-withdrawing via resonance *and* via induction. In other words, there is no competition between resonance and induction. Both factors dictate that a nitro group should *deactivate* the ring. And that is indeed what we observe in the laboratory.

3.6 DIRECTING EFFECTS

Now let's consider electrophilic aromatic substitution reactions with substituted benzene rings. In order to explore this topic, we must first review important terminology that we will use frequently throughout the remainder of this chapter. The various positions on a monosubstituted benzene ring are referred to in the following way:

The two positions that are closest to the substituent (R) are called *ortho* positions. Then we have the *meta* positions. And finally, the farthest position from the substituent is called the *para* position. With this terminology in mind, let's consider the products obtained when toluene or nitrobenzene undergo bromination. Both reactions are shown below:

The first reaction (with toluene) is certainly faster, because the methyl group activates the ring toward electrophilic aromatic substitution, while the nitro group deactivates the aromatic ring. But consider the difference in regiochemical outcome. Specifically, notice that the methyl group directs the reaction to occur at the *ortho* and *para* positions, while the nitro group directs the reaction to occur at the *meta* position. Your textbook will give an explanation for this observation, and you should read that explanation, but here is the bottom line:

- Activators are *ortho-para* directors

- Deactivators are *meta* directors

We will not encounter any exceptions to the first rule (all activators that we encounter will be *ortho-para* directors). But there is one important exception to the second rule above. Halogens (F, Br, Cl, or I) are deactivators, so we might expect them to be *meta* directors. But instead, they are actually *ortho-para* directors. Let's try to understand why halogens are the exception.

In the previous section, we analyzed the effect of an OH group on an aromatic ring, and we saw that there were two competing effects: induction and resonance. We saw that induction *withdraws* electron density from the ring, but resonance *donates* electron density to the ring. In order to know which factor dominates, we gave a general rule: **resonance is usually a stronger factor than induction**. We also said that there was an important exception to this general rule that we would see later. Well, it is now later. Halogens are the exception. Let's take a closer look. As an example, consider the structure of chlorobenzene:

The substituent in this case (Cl) is an *ortho-para* director for the same reason that an OH group is an *ortho-para* director (the explanation for this directing effect can be found in your textbook). But, unlike OH (which is an activator), Cl is a deactivator. To explain why, we must explore the effect of a halogen (such as Cl) on the electron density of the aromatic ring. As we have seen in the previous section, our analysis must focus on two factors: induction and resonance. Let's start with induction. Just like an OH group, a halogen is also electron-withdrawing by induction:

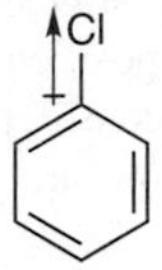

However, we also need to consider resonance effects. So, we draw the resonance structures:

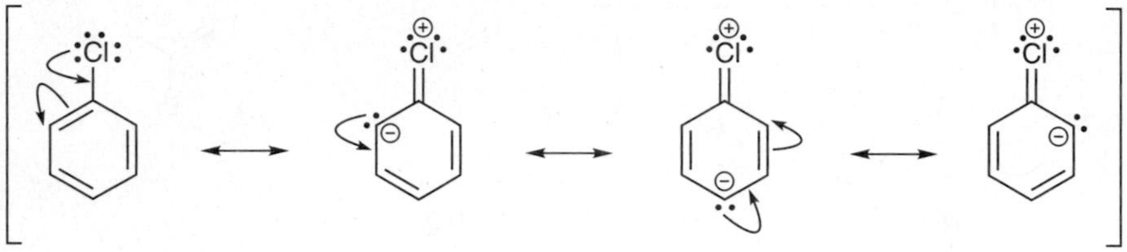

And once again, we see that a halogen is very similar to an OH group. It is donating electron density by resonance:

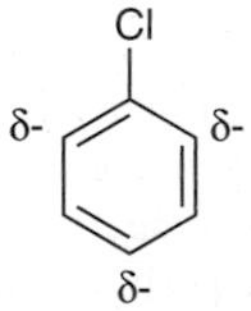

Now let's consider the net effect of a halogen. Just as we saw with the OH group, there is a competition between resonance and induction. And just as we saw with the OH group, a halogen will *withdraw* electron density by induction, and it will *donate* electron density by resonance. But, in the case of the OH group, we used the argument that resonance beats induction (we said that was a general rule that holds true most of the time). Therefore, the net effect of the OH group was to donate electron density to the ring (thus, the OH group is an activator). But with a halogen, resonance does NOT beat induction. This is one of the rare cases where induction actually beats resonance. Why is resonance not the predominant factor in this case? To answer this question, notice that the resonance structures of chlorobenzene exhibit a positive charge on Cl. Halogens do not easily bear positive charges. So, these resonance structures do not contribute very much character to the overall structure of the compound. Resonance is a weak effect in this case, so induction actually beats resonance. Therefore, the net effect of a halogen substituent is to *withdraw* electron density from the aromatic ring, rendering the ring less nucleophilic (deactivation).

Now we are ready to modify the rules we gave earlier when we said that all activators are *ortho-para* directors and all deactivators are *meta* directors. Here is our new-and-improved formula:

- All activators are *ortho-para* directors.

- All deactivators are *meta* directors, ***except for halogens (which are deactivators, but nevertheless, they are <u>ortho-para</u> directors)***.

With that in mind, let's try to predict some directing effects.

EXERCISE 3.30   Look closely at the following monosubstituted benzene ring.

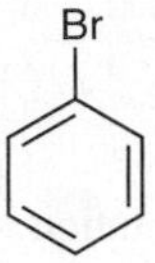

If this compound were to undergo an electrophilic aromatic substitution reaction, predict where the incoming substituent would be installed.

Answer   Br is a halogen (remember that the halogens are F, Cl, Br, and I). We have seen that halogens are the one exception to the general rules. That is, they are deactivators but they are not *meta* directors, like most other deactivators. Rather, they are *ortho-para* directors. Therefore, if we use this compound in an electrophilic aromatic substitution, we expect substitution to take place at the *ortho* and *para* positions:

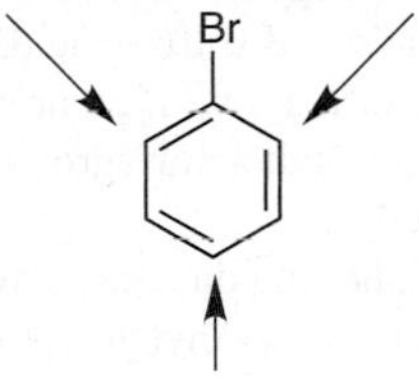

Identify the expected directing effects that would be observed if each of the following compounds were to undergo an electrophilic aromatic substitution reaction.

3.31

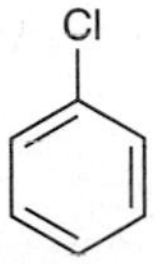

3.32

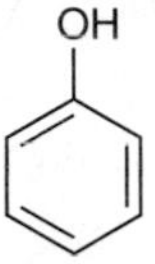

3.33

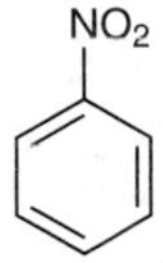

3.34 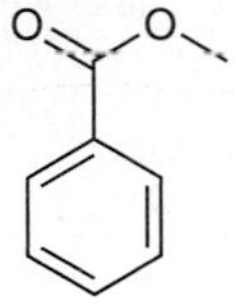   This group is a deactivator.

3.35 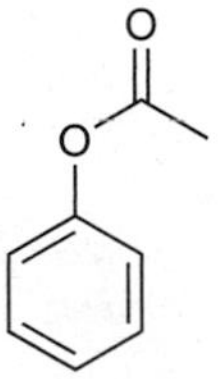   This group is an activator.

3.36 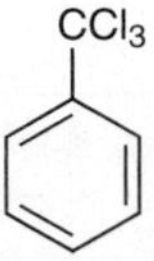   This group is a deactivator.

3.37 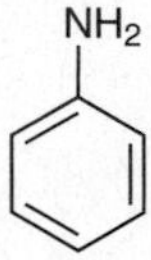 This group is an activator.

Clearly, you can only predict where the substitution will take place if you know whether the group is an activator or a deactivator. In the next section, we will learn how to predict whether a group is an activator or a deactivator. But for now, let's get some practice predicting products.

EXERCISE 3.38 Predict the products of the following reaction:

$$\text{(benzene)} \xrightarrow[\text{H}_2\text{SO}_4]{\text{HNO}_3}$$

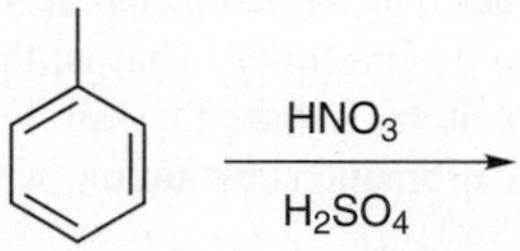

Answer We begin by looking at the reagents, so that we can determine what kind of reaction is taking place. The reagents are nitric acid and sulfuric acid. We have seen that these reagents generate NO_2^+ as an electrophile, which can react with an aromatic ring in an electrophilic aromatic substitution reaction. The end result is to install a nitro group on an aromatic ring. So, now the question is: where is the nitro group installed?

To answer this question, we must predict the directing effects of the group already present on the ring (before the reaction takes place). There is a methyl group on the ring, and we have seen that methyl groups are activators. Therefore, we predict that the reaction will take place at the *ortho* and *para* positions, relative to the methyl group:

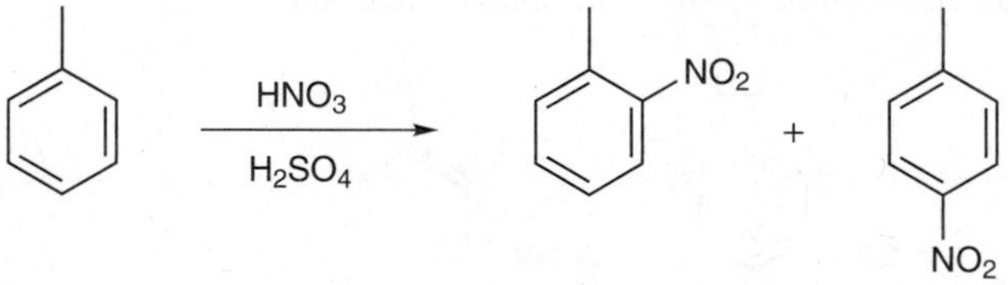

$$\text{(toluene)} \xrightarrow[\text{H}_2\text{SO}_4]{\text{HNO}_3} \text{(o-nitrotoluene)} + \text{(p-nitrotoluene)}$$

Notice that in this case, we don't draw a substitution at both *ortho* positions because we get the same product either way:

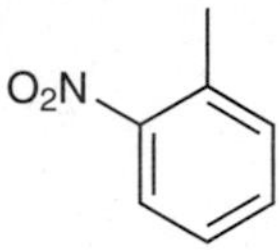

$$\text{(2-nitrotoluene, drawn left)} \quad \text{is the same as} \quad \text{(2-nitrotoluene, drawn right)}$$

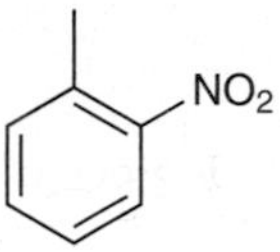

Predict the products for each of the following reactions:

3.39
$$\text{(bromobenzene)} \xrightarrow[\text{H}_2\text{SO}_4]{\text{HNO}_3}$$

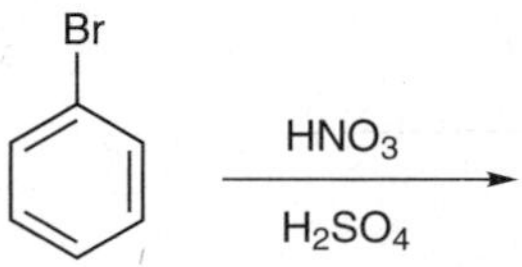

3.40
$$\text{(toluene)} \xrightarrow[\text{AlCl}_3]{\text{CH}_3\text{Cl}}$$

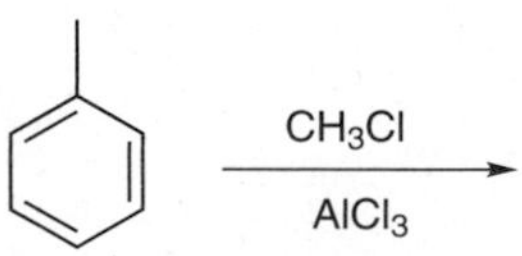

3.41

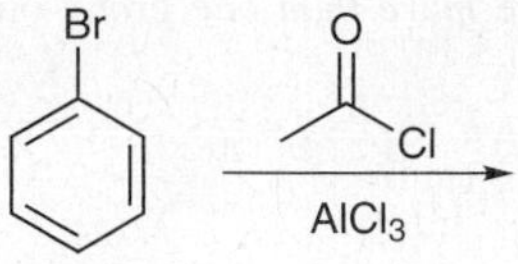

3.42

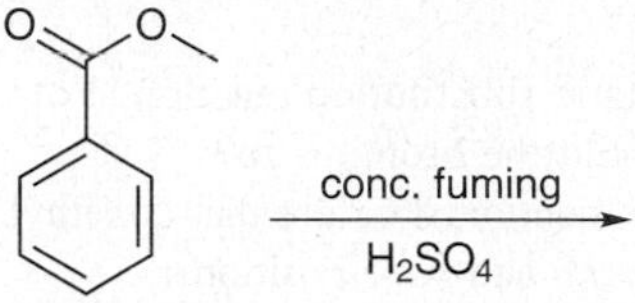

Hint: The group on the ring is a deactivator.

3.43

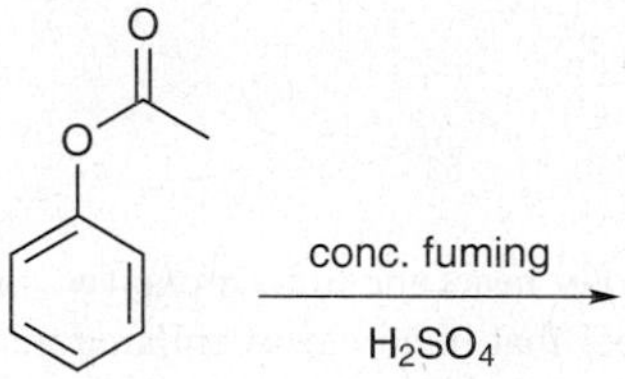

Hint: The group on the ring is an activator.

3.44

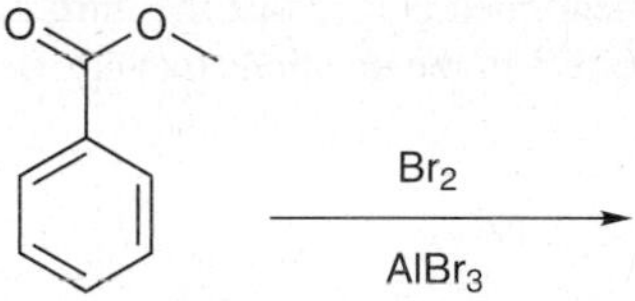

Hint: The group on the ring is a deactivator.

3.45

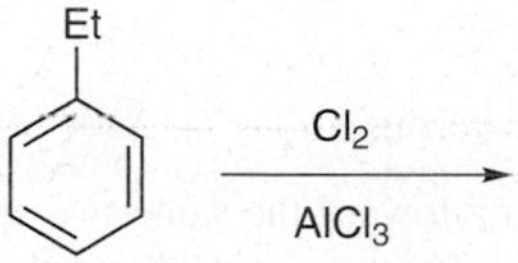

Hint: The group on the ring is an activator.

So far, we have focused on the directing effects when you have *only one group* on a ring. And we have seen that activators direct toward the *ortho* and *para* positions, whereas deactivators direct toward the *meta* positions:

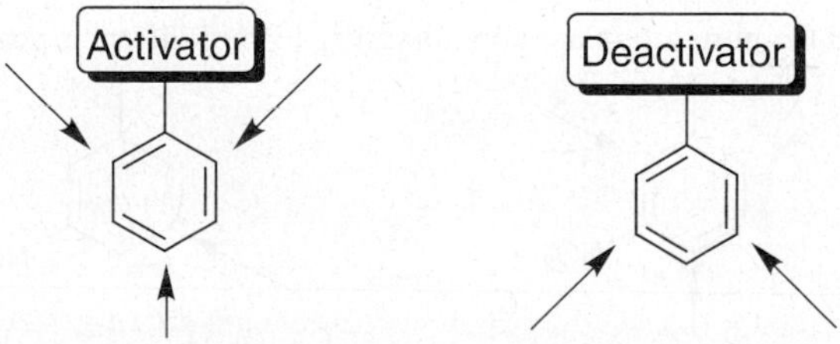

and we saw only one exception (the halogens).

But how do you predict the directing effects when you have *more than one group* on a ring? For example, consider the following compound:

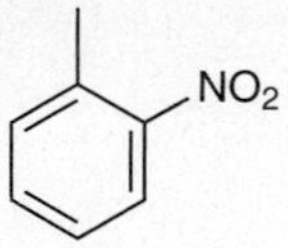

What if we used this compound in an electrophilic aromatic substitution reaction? For example, let's say we try to brominate this compound. Where would the bromine go?

Let's first consider the effect of the methyl group. We mentioned before that a methyl group is an activator, so we predict that it will direct toward the *ortho* and *para* positions:

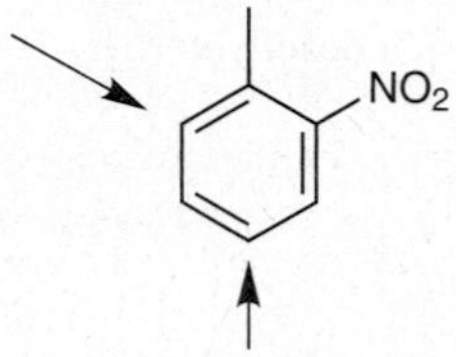

Notice that we do **not** point to the *ortho* position that already bears the nitro group (we only look at positions where there are currently no groups—remember that in an electrophilic aromatic substitution, E+ comes on the ring and **H+** comes off). So, the methyl group is directing toward *two* spots, as shown above.

Now let's consider the effect of the nitro group. We mentioned before that the nitro group is a powerful deactivator. Therefore, we predict that it should direct to the positions that are *meta* **to the nitro group**:

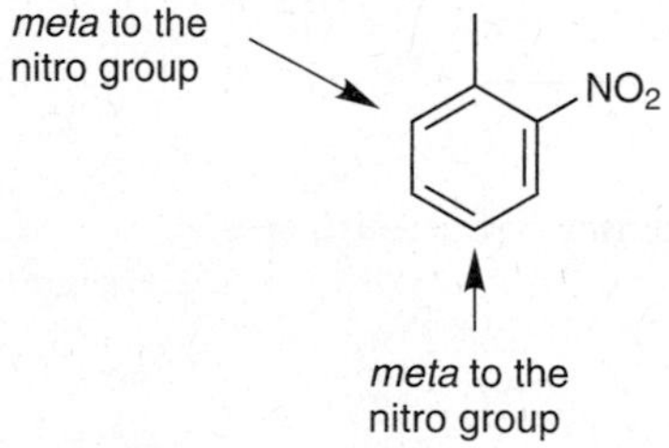

meta to the
nitro group

meta to the
nitro group

So, we see that the nitro group and the methyl group are directing toward the same two spots. So, in this case there is no conflict between the directing effects of the nitro group and the methyl group.

But consider this case:

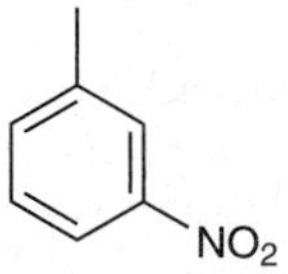

The methyl group and the nitro group are now directing toward different positions:

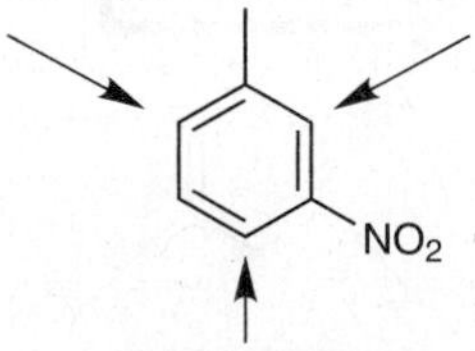

Directing effects
of the methyl group
(*ortho-para* director)

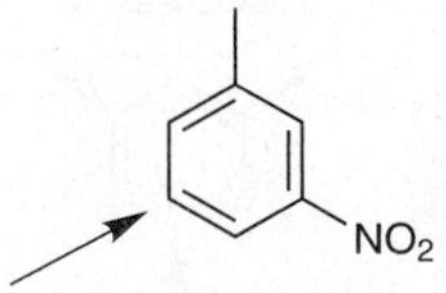

Directing effects
of the nitro group
(*meta* director)

So, the big question is: which group wins? It turns out that the directing effects of the methyl group trump the directing effects of the nitro group. So, if we brominate this ring, we will get the following products (where the Br goes *ortho* or *para* to the methyl group):

This product is not obtained in significant yield, for reasons that we will soon see

It is common to see a situation where the directing effects of two groups are competing with each other (like the methyl and the nitro groups in the above example). So we clearly need rules for determining which group wins. It turns out that you need to know just two simple rules in order to determine which group will dominate the directing effects:

1. *Ortho-para directors always beat meta directors.* The example we just saw is a perfect illustration of this rule. The methyl group is an activator (an *ortho-para* director), and the nitro group is a deactivator (a *meta* director), so the methyl group wins.

2. *Strong activators always beat weak activators.* For example, consider the following case:

The OH group is a *strong* activator, and the methyl group is a *weak* activator. (We will learn in the next section how to predict which groups are strong and which are weak—for now, just take my word for it.) So, the OH group will win, and the directing effects are as follows:

So, we have seen two rules:

- *ortho-para directors always beat meta directors.*
- *Strong activators always beat weak activators.*

Keep in mind that the first rule always trumps the second rule. So if you have a weak activator against a strong deactivator, the weak activator wins. Even though the activator is weak, it still beats a strong deactivator because activators (*ortho-para* directors) always beat deactivators (*meta* directors). This rule was already seen in one of our previous examples:

The methyl group is a weak activator, and the nitro group is a strong deactivator. So, in this case, the methyl group wins (and the directing effects are *ortho* and *para* to the methyl group; and **not** *meta* to the nitro group):

EXERCISE 3.46 Predict the directing effects in the following scenario.

For this problem, you should assume that the deactivator is *not* a halogen.

Answer We have two groups. The activator will direct toward the positions that are *ortho* or *para* to itself, and the deactivator will direct toward the positions that are *meta* to itself:

Directing effects of the activator

Directing effects of the deactivator

So, there is a competition in the directing effects. Between the two groups, the strong activator beats the strong deactivator because the strong activator is an *ortho-para* director. So the directing effects are:

If we performed an electrophilic aromatic substitution on a compound of this type, we might expect three products (because the directing effects are toward three positions, shown above). Here is a specific example of a reaction like this.

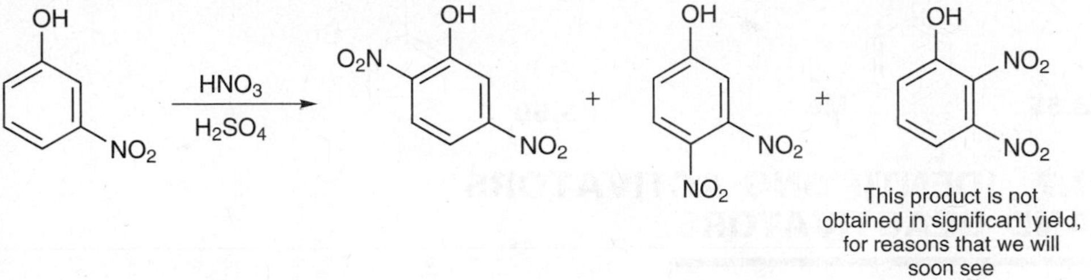

This product is not obtained in significant yield, for reasons that we will soon see

The starting aromatic ring has two substituents. The OH group is a strong activator, and the nitro group is a strong deactivator.

Predict the directing effects in each of the following scenarios. Unless otherwise indicated, assume that anything labeled as a deactivator is not a halogen (unless it is specifically indicated as a halogen).

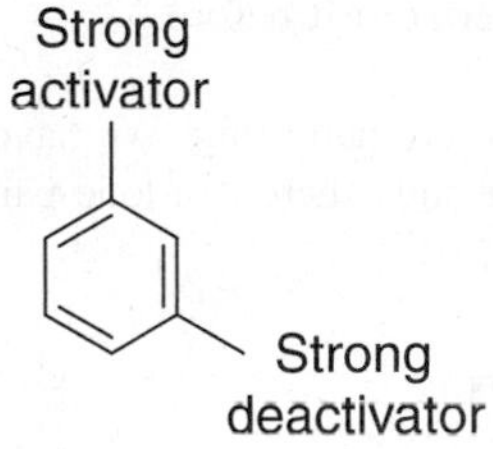

3.47

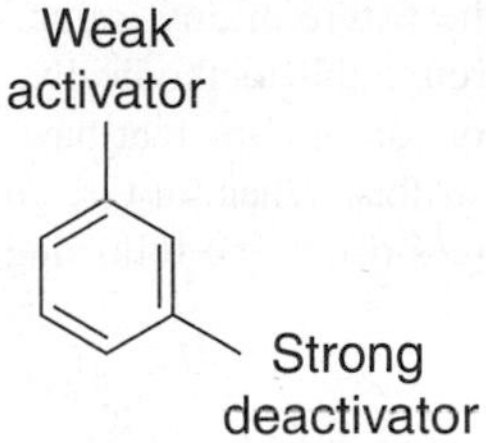

3.48

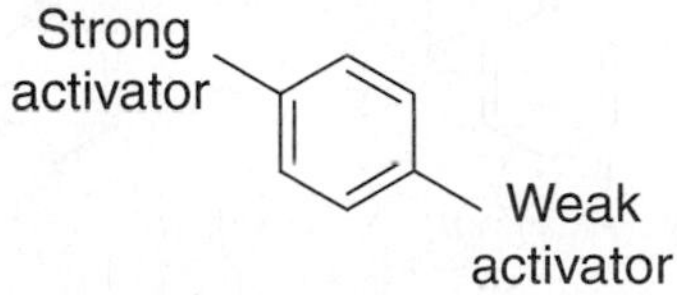

3.49

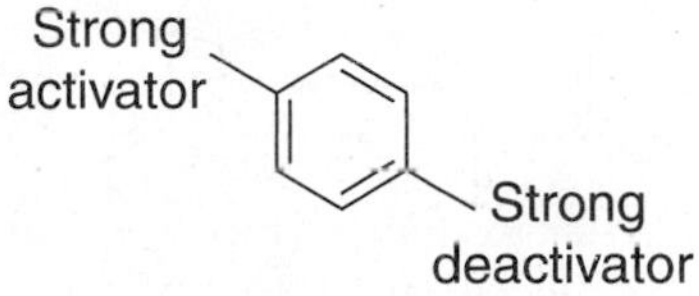

3.50

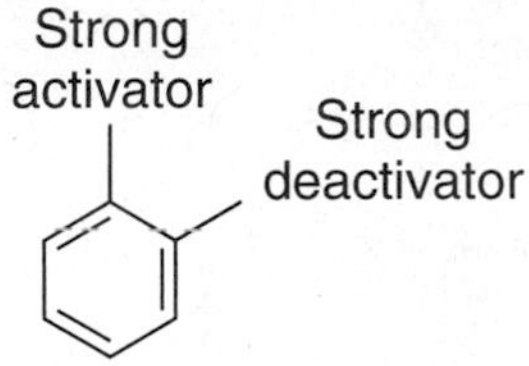

3.51

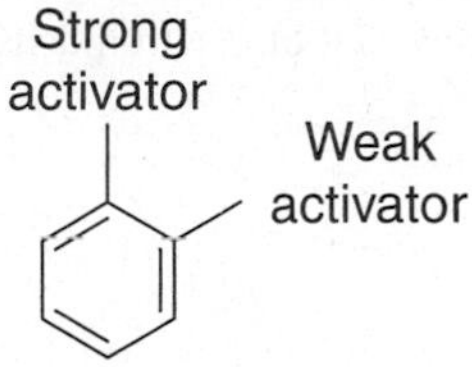

3.52

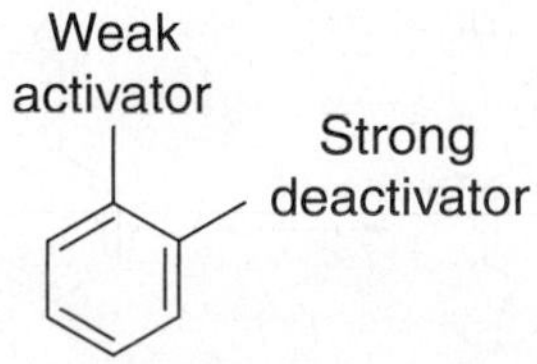

3.53

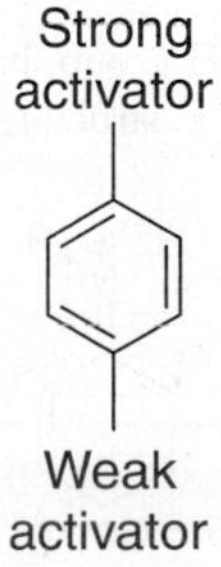

3.54

Strong
deactivator

3.55

Me

Strong
activator

Br

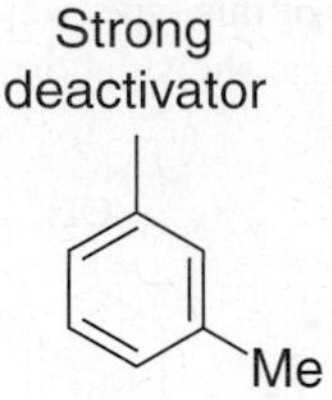

3.56

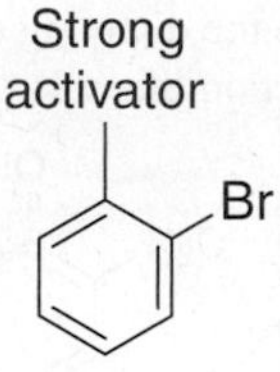

3.7 IDENTIFYING ACTIVATORS AND DEACTIVATORS

In the previous section, we learned how to predict the directing effects in a situation where you have more than one group on the ring. But in all of the cases in the previous section, I had to tell you whether each group was an activator or a deactivator and whether it was strong or weak. In this section, we will learn how to predict this, so that you won't have to memorize the characteristics of every possible group. In fact, very little memorization is actually involved here. We will see a few concepts that should make sense. And with those concepts, you should be able to identify the nature of any group, even if you have never seen it before.

We will go through this methodically, starting with strong activators.

Strong activators are groups that have a lone pair next to the aromatic ring. We have already seen an example of this. When an OH group is connected to the ring, there is a lone pair next to the ring, which gives rise to the following resonance structures:

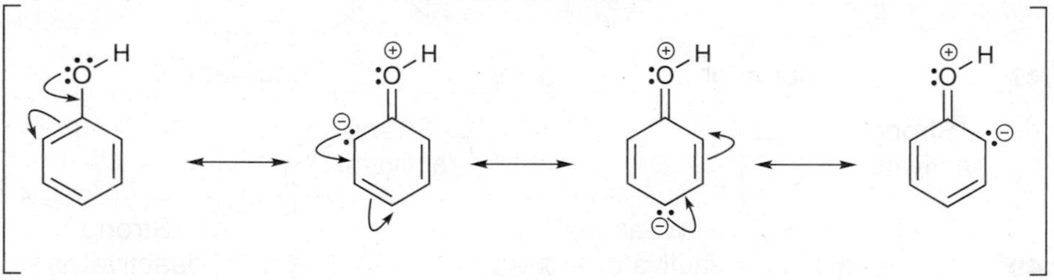

We concluded in the previous section that this resonance effect is very strong and that the OH group is therefore donating a lot of electron density to the ring:

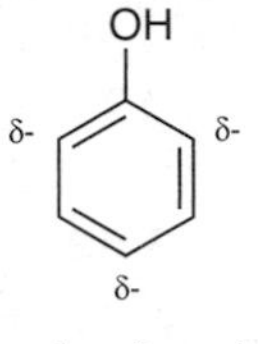

This is true, not only for the OH group, but also for other groups that have a lone pair next to the ring. The same kind of resonance structures can be drawn for an amino group connected to a ring:

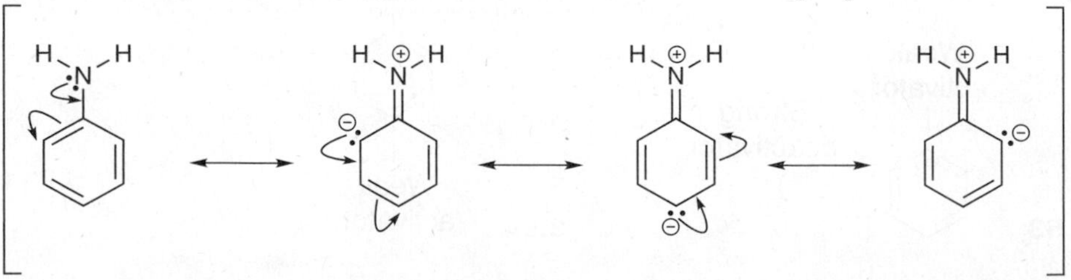

Here are several examples of strong activators. Make sure that you can easily see the common feature (the lone pair next to the ring):

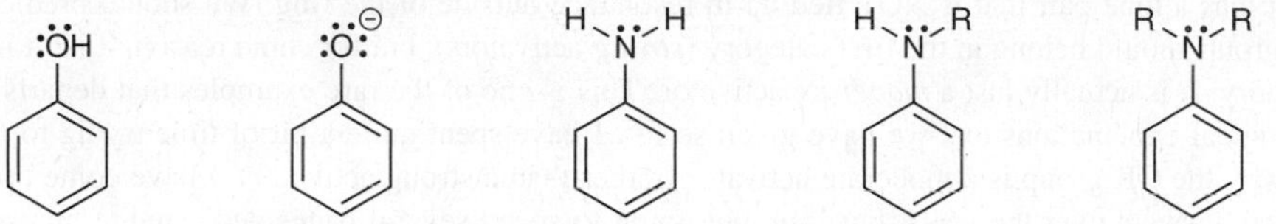

Next, we move on to the *moderate* activators. Moderate activators are groups that have a lone pair next to the ring, BUT that lone pair is already partially tied up in resonance. For example, consider the following group:

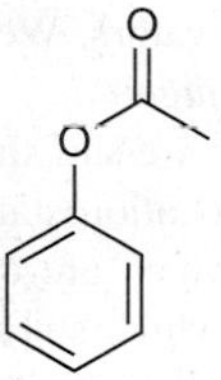

This compound has all of the resonance structures that place electron density into the ring (just like an OH group does):

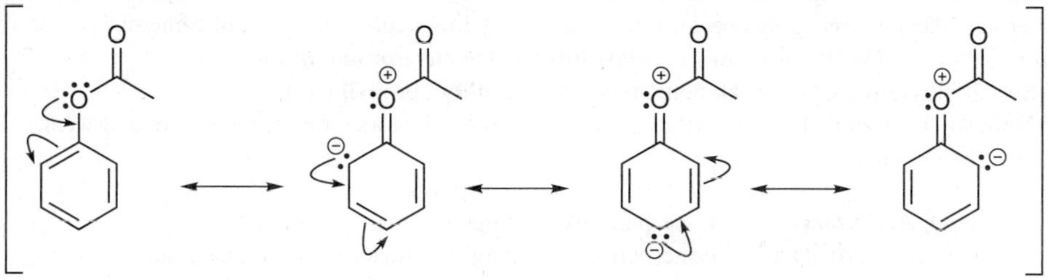

BUT there is an additional resonance structure, which has the electron density *outside* of the ring:

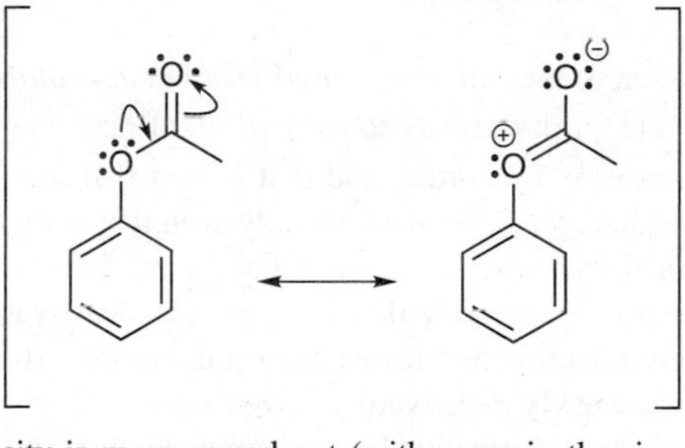

Therefore, the electron density is more spread out (with some in the ring and some out of the ring). This group is therefore not a *strong* activator. Rather, we call it a *moderate* activator. (Some textbooks do not point out this subtle distinction between strong activators and moderate activators.) Here are several examples of moderate activators:

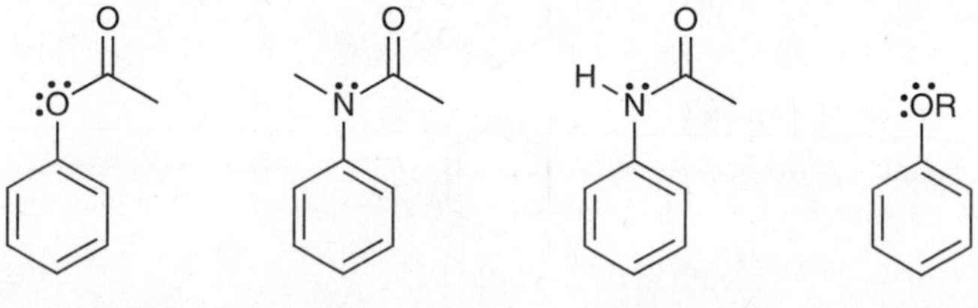

Look closely at the examples above. They all have a lone pair that is tied up in resonance outside of the ring. BUT WAIT A SECOND. What about the last group on this list (the OR group)? This group has a lone pair that is NOT tied up in resonance outside of the ring. We should predict that this group should belong in the first category (*strong* activators), but for some reason, it isn't in that category. It is actually just a *moderate* activator. This is one of the rare examples that departs from the logical explanations that we have given so far. I have spent quite a bit of time trying to figure out why the OR group is a moderate activator (rather than a strong activator). I have come up with several answers over the years, but I am not going to spend several pages dedicated to an esoteric topic that you will certainly not need for your exams (perhaps you might think of it as a brain teaser—something to think about . . .). For now, you will just have to remember that the OR group doesn't follow the trends we have seen. It is a moderate activator.

Now let's turn our attention to *weak* activators. Weak activators donate electron density to the ring through a weak effect, called *hyperconjugation*.

In the first semester of organic chemistry, we saw that *alkyl groups are electron-donating*. That was important when we learned about carbocation stability (we saw that tertiary carbocations are more stable than secondary carbocations, which are more stable than primary carbocations—because *alkyl groups are electron-donating*, which stabilizes a carbocation). There is a simple reason why alkyl groups are electron-donating. It is due to a phenomenon called hyperconjugation. If you do not remember this term from the first semester, you can go back and review it if you like. Whether or not you go back, make sure that you remember that *alkyl groups are electron-donating*. I keep stressing this because there are many more concepts in organic chemistry that you can understand only if you know that *alkyl groups are electron-donating*.

So, all alkyl groups are weak activators (methyl, ethyl, propyl, etc.)

Now we have seen all of the different categories of activators (strong, moderate, and weak). To review, this is what we saw:

Strong Activators	Lone pair next to ring
Moderate Activators	Lone pair next to ring, but tied up in resonance outside of ring as well
Weak Activators	Alkyl groups

Now, we will turn our attention to the different categories of *deactivators*. This time, we will begin with the *weak* deactivators and work our way toward *strong* deactivators (rather than starting with strong). There is a reason for using this order, and that reason will soon become clear.

Weak deactivators are the halogens. We have already seen that halogens are the one case where induction beats resonance (and therefore, we argued that the net effect of a halogen is to *withdraw* electron density from the ring). So, we saw that halogens are deactivators. But you should know that the competition between induction and resonance (in the case of the halogens) is a close competition, so halogens are only weakly deactivating.

We will summarize all of this information in one complete chart, but for now let's move on to moderate deactivators.

Groups that withdraw electron density from the ring via resonance are moderate activators. For example, consider the following group:

This group does *not* have a lone pair next to the ring (so it is *not* an activator). But it does have a pi bond next to the ring, giving rise to the following resonance structures:

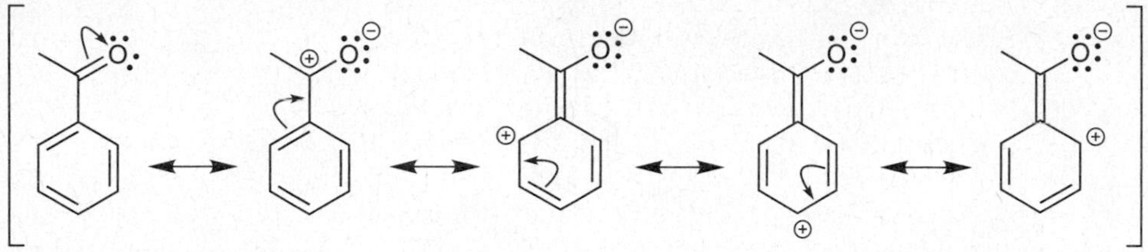

When we look closely at these resonance structures, we can see that the substituent is *withdrawing* electron density from the ring:

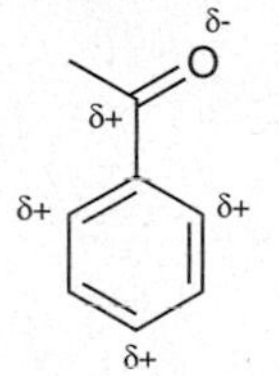

Therefore, this group is a *moderate deactivator*. Numerous other similar groups can also withdraw electron density from the ring. Here is a list of many examples:

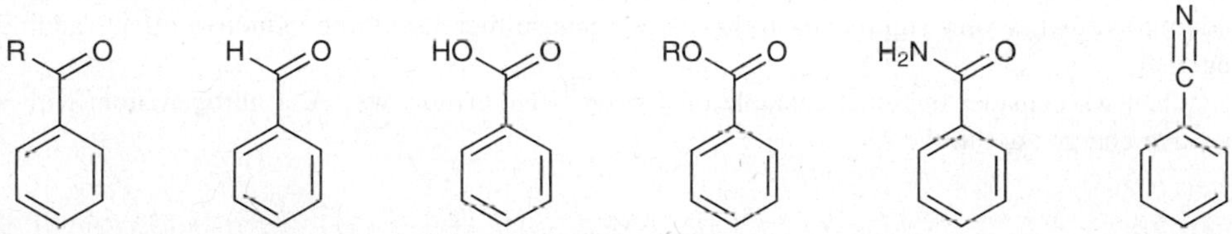

All of these substituents are withdrawing electron density from the ring via resonance. They all have one feature in common: a pi bond to an electronegative atom. Take a close look at the last example. A cyano group is a pi bond (a triple bond) to an electronegative atom (nitrogen). So we see that a triple bond can also be included in this category.

And now for the last category: *strong* deactivators. There are a few common functional groups that fall into this category:

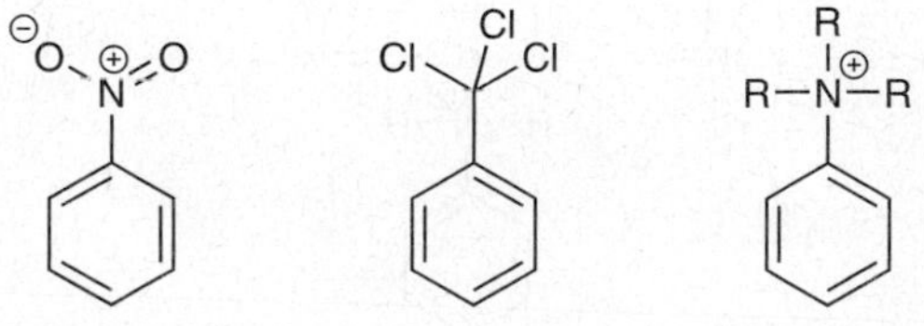

We have already explained why the nitro group is so powerfully electron-withdrawing. The nitro group is electron-withdrawing by resonance *and* induction.

To understand why the second group above (the trichloromethyl group) is a strong deactivator, we need to focus on the collective inductive effects of all of the chlorine atoms:

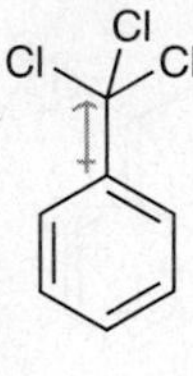

The inductive effects of each chlorine atom add together to give one very powerful deactivating group. Be careful not to confuse this group with a halogen on a ring:

When a halogen is connected directly to the ring (above right), then there are lone pairs next to the ring, so there are resonance effects to consider. (We spent a lot of time talking about the competition between resonance and induction in the case of halogens.) The group we are talking about now (above left) does not have any resonance effects to consider because the lone pairs are *not* directly next to the ring. So, there is only an inductive effect to consider, and this inductive effect is very significant in this case (because there are three inductive effects adding together).

When we consider our final example of a *strong deactivator*, we see a nitrogen atom with a positive charge next to the ring:

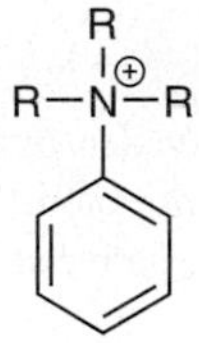

The nitrogen atom is so poor in electron density that it is practically sucking electron density out of the ring like a vacuum cleaner:

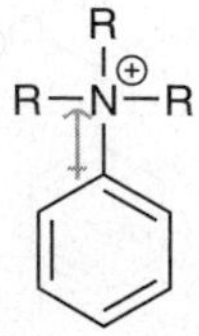

Now we are ready to summarize everything we have seen into one chart:

Common Feature		*Some Examples*

ACTIVATORS

Strong   Lone pair next to ring

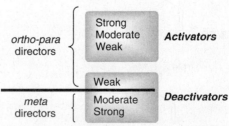

$:\ddot{O}H$ $:\ddot{O}:^{\ominus}$ $H-\ddot{N}-H$

Moderate   Lone pair next to ring
But tied up in resonance
outside of ring as well

$:\ddot{O}$... $\overset{|}{N}$... $:\ddot{O}R$

Does not fit
the pattern

Weak   Alkyl groups

Me   Et   Pr

DEACTIVATORS

Weak   Halogens

Cl   Br   I

Moderate   Pi bond to an electronegative
atom (next to ring)

Strong   Very powerfully electron-
withdrawing

Take a close look at this chart and make sure that
every category makes sense to you. As you look
over the chart, you should be able to remember the
arguments that we gave for each category. If you
have trouble with this, you might want to review
the last few pages of explanation.

And now we can understand why we looked at
weak deactivators first (before strong deactivators). When we organize it like this, we can clearly
organize the directing effects in our minds:

ortho-para
directors {
Strong
Moderate
Weak
} ***Activators***

{
Weak
}

meta
directors {
Moderate
Strong
} ***Deactivators***

This chart shows that all activators are *ortho-para* directors and all deactivators are *meta* directors, with the exception of the weak deactivators (halogens).

EXERCISE 3.57   Look closely at the following substituent:

OH
|
O=S=O

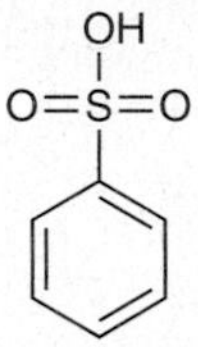

Try to predict what kind of group it is (a strong activator, a moderate activator, a weak activator, a weak deactivator, a moderate deactivator, or a strong deactivator).

   Use this information to predict the directing effects.

Answer   This group does not have a lone pair next to the ring, and it is not an alkyl group. Therefore, it is not an activator. This group has a pi bond to an oxygen atom (next to the ring), and therefore, it is a moderate deactivator.

   Because all deactivators are *meta* directors (except for weak deactivators—halogens), we predict the following directing effects:

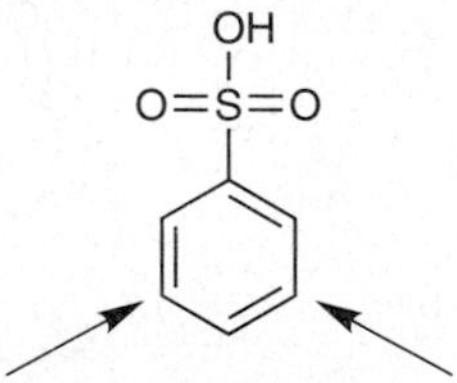

OH
|
O=S=O

   For each of the following substituents, determine what kind of group it is (a strong activator, a moderate activator, a weak activator, a weak deactivator, a moderate deactivator, or a strong deactivator). Place your answer on the space provided. Try to do this **without** looking at the chart that we constructed. You won't have access to this chart on an exam. Try to remember and apply the explanations that we used.

   Then, use that information to predict the directing effects. Indicate the directing effects using arrows for pointing to the positions where you would expect an electrophilic aromatic substitution to occur:

3.58 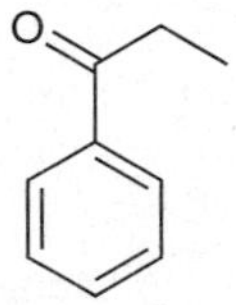   Answer: _____   **3.59** 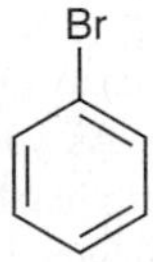   Answer: _____

3.60 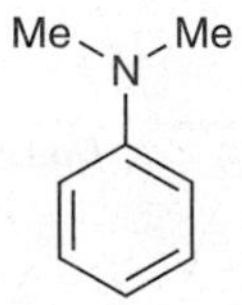   Answer: _____   **3.61** 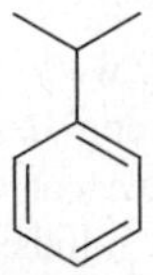   Answer: _____

3.62 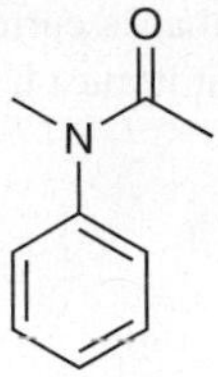   Answer: ———

3.63 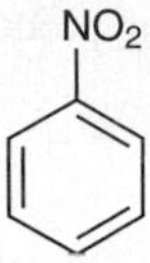   Answer: ———

3.64 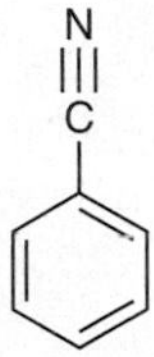   Answer: ———

3.65 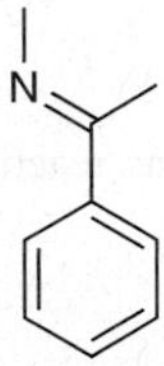   Answer: ———

3.66 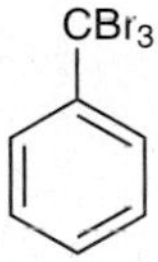   Answer: ———

3.67 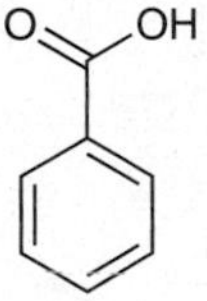   Answer: ———

3.68   Can you explain why the following group is a strong activator:

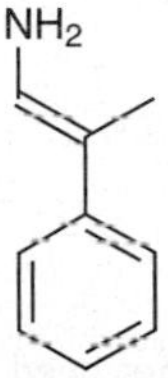

(*Hint*: Think about what strong activators have in common.)

Now we can use the skills that we developed in this section to predict the products of a reaction. Let's see an example:

EXERCISE 3.69   Predict the product of the following reaction:

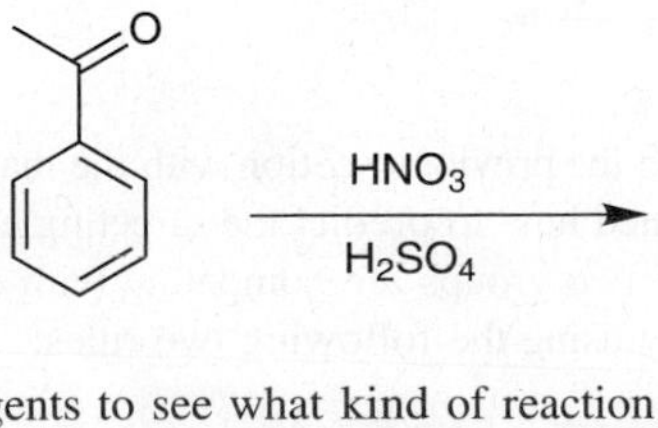

$$\xrightarrow[\text{H}_2\text{SO}_4]{\text{HNO}_3}$$

Answer   We look at the reagents to see what kind of reaction we expect. The reagents are nitric acid and sulfuric acid. These reagents generate NO_2^+, which is an excellent electrophile. So, we know that the reaction will install a nitro group on the ring. But the question is: where?

To answer this question, we must predict the directing effects of the group that is currently on the ring. We recognize that this group is a moderate deactivator, which means that it must be a *meta* director. So, we predict the following product:

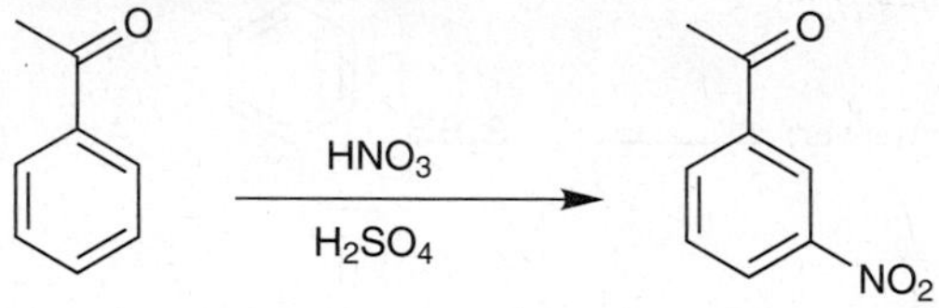

Predict the products of the following reactions:

3.70

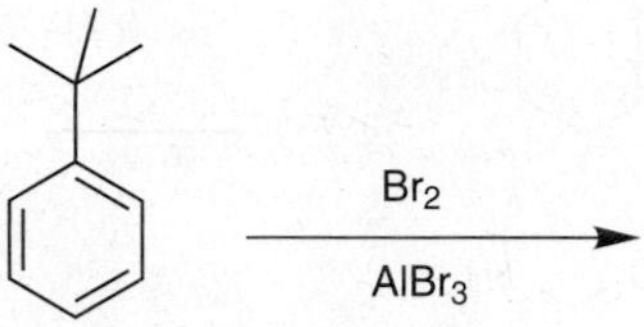

$$\xrightarrow[\text{AlBr}_3]{\text{Br}_2}$$

3.71

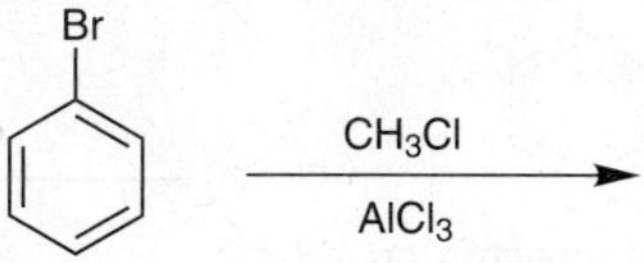

$$\xrightarrow[\text{AlCl}_3]{\text{CH}_3\text{Cl}}$$

3.72

$$\xrightarrow{\text{Br}_2}$$

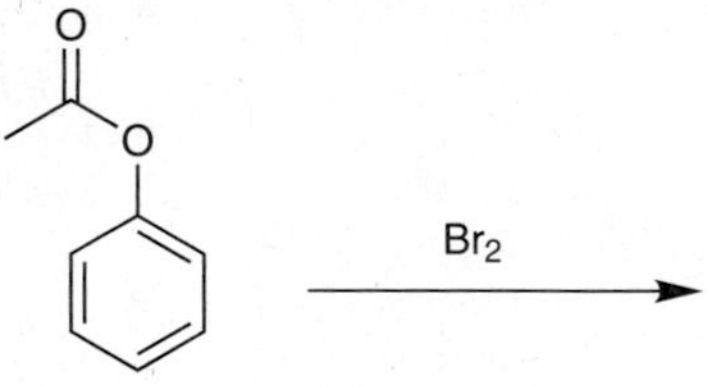

Notice that in this reaction, we do not need a Lewis acid. Can you explain why not?

3.73

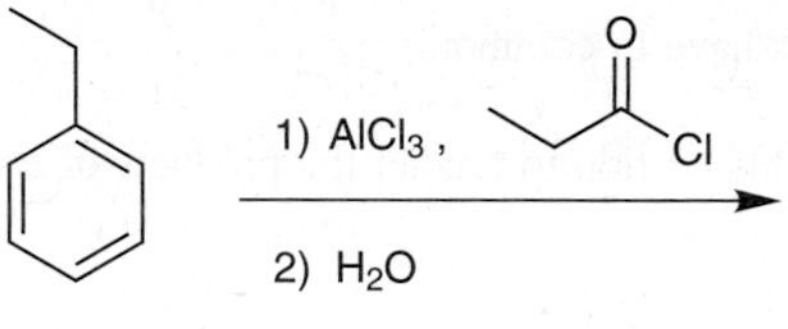

1) AlCl$_3$,

2) H$_2$O

3.74

$$\xrightarrow[\text{H}_2\text{SO}_4]{\text{HNO}_3}$$

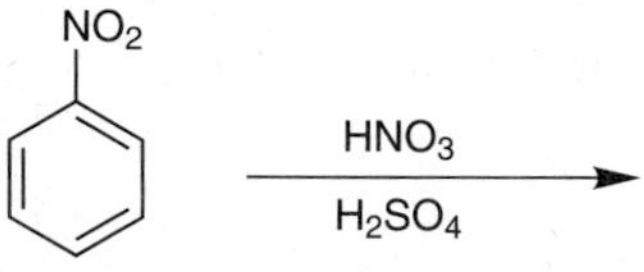

Now let's combine what we did in the previous section with the material in this section. Recall from the previous section that we learned how to predict the directing effects when you have more than one group on the ring. When the two groups are competing with each other, we saw that you can determine the directing effects by using the following two rules:

- *ortho-para directors always beat meta directors.*
- **Strong** *activators always beat* **weak** *activators.*

Now that we have learned how to categorize the various kinds of groups, let's get practice using our skills to predict products:

EXERCISE 3.75 Predict the product of the following reaction:

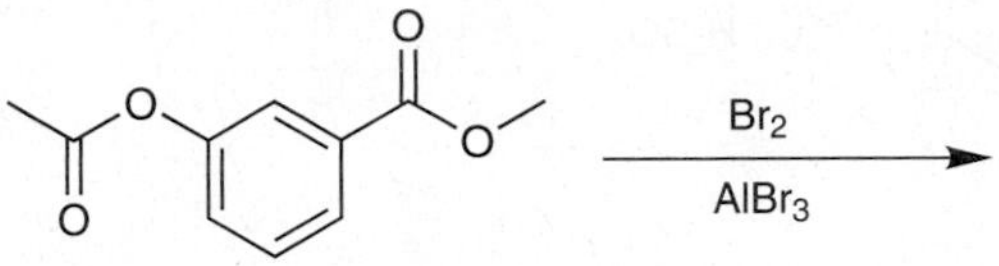

Answer We look at the reagents to see what kind of reaction we expect. The reagents are bromine and aluminum tribromide. These reagents generate Br^+, which is an excellent electrophile. So, we know that we will be installing a Br on the ring. But the question is: where?

To answer this question, we must predict the directing effects of the two groups that are currently on the ring. The group on the left is a moderate activator (make sure that you know why), and therefore, it directs *ortho-para* to itself:

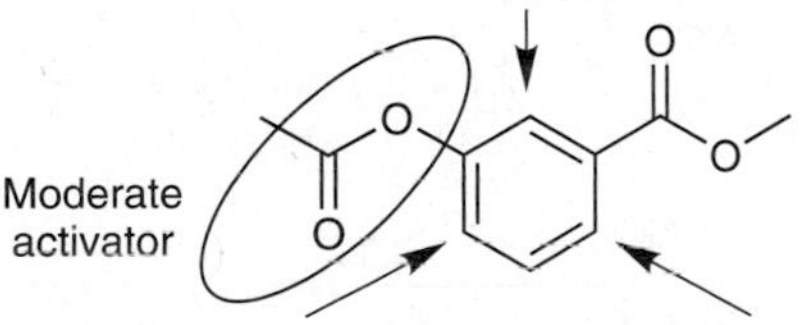

The group on the right is a moderate deactivator (make sure you know why), so it directs *meta* to itself:

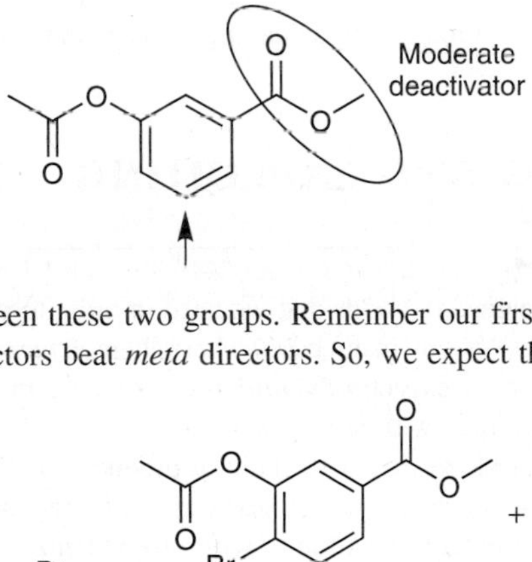

There is a competition between these two groups. Remember our first rule for determining which group wins: *ortho-para* directors beat *meta* directors. So, we expect the following products:

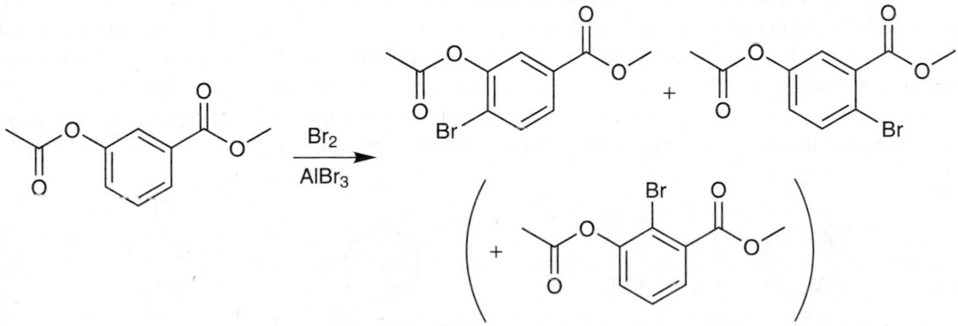

Notice that I placed the last product in parentheses. This product is actually a very minor product. We will see why in the next section. For now, we will just write that we expect three products. Then, we will fine-tune this prediction in the next section.

PROBLEMS Predict the products of the following reactions:

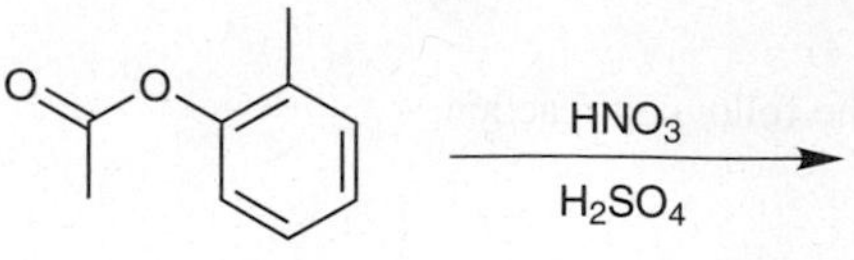

3.76

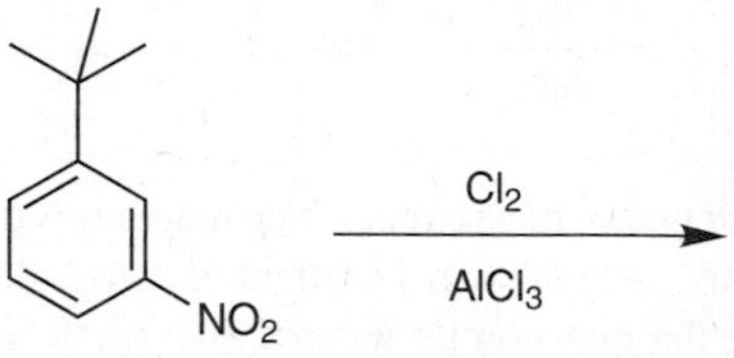

$$\xrightarrow[\text{H}_2\text{SO}_4]{\text{HNO}_3}$$

3.77

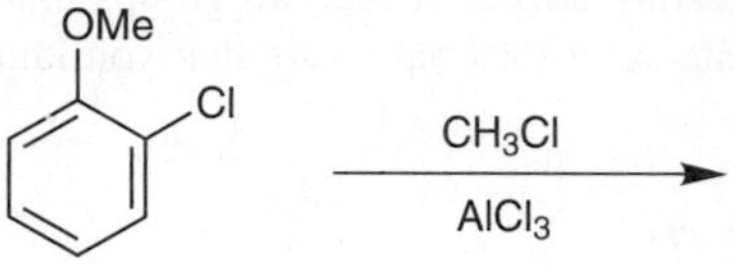

$$\xrightarrow[\text{AlCl}_3]{\text{Cl}_2}$$

3.78

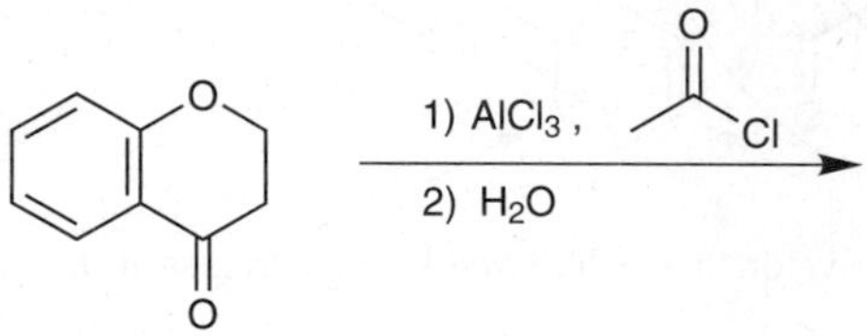

$$\xrightarrow[\text{AlCl}_3]{\text{CH}_3\text{Cl}}$$

3.79

$$\xrightarrow[\text{2) H}_2\text{O}]{\text{1) AlCl}_3,}$$

(*Hint*: Consider this aromatic ring as having two separate substitutents, and analyze each separately.)

3.8 PREDICTING AND EXPLOITING STERIC EFFECTS

In the previous sections, we learned the skills that we need in order to predict the products of an electrophilic aromatic substitution. We saw many cases where there is *more* than one product. For example, if the ring is activated, then we expect *ortho and para* products. In this section, we will see that it is possible to predict which product will be the major product and which will be the minor product (*ortho* vs. *para*). It is even possible *to control* the ratio of products (*ortho* vs. *para*). This is VERY important for synthesis problems, which will be the next (and final) section of this chapter.

Consider an electrophilic aromatic substitution with propyl benzene. The propyl group is a weak activator, and therefore, we expect the directing effects to be *ortho-para*:

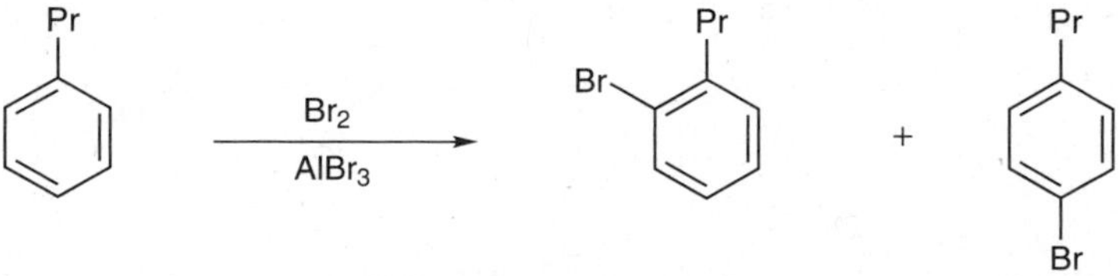

$$\xrightarrow[\text{AlBr}_3]{\text{Br}_2}$$

There are two products here. But let's try to figure out which one of these products is the major product? *ortho* or *para*? At first, we might be tempted to say that the *ortho* product should be major.

Let's see why. The propyl group is an *ortho-para* director, so there should be a total of three positions that can get attacked (two *ortho* positions and one *para* position):

Therefore, the chances of attacking an *ortho* position should be two-thirds (or 67%), while the chances of attacking the *para* position should be one-third (33%). From a purely statistical point of view, we should therefore expect our product distribution to be 67% *ortho* and 33% *para*. But the product ratio is different from what we might expect, because of steric considerations. Specifically, the propyl group is fairly large, and it partially "blocks" the *ortho* positions. We do still observe *ortho* products, but less than 67%. In fact, the *para* product is the major product in this case:

This is usually the case (that *para* is the major product). A notable exception is toluene (methylbenzene), for which the ratio of *ortho* and *para* products is sensitive to the conditions employed, such as the choice of solvent. In some cases, the *para* product is favored; in other cases, the *ortho* product is favored. Therefore, it is generally not wise to use the directing effects of a methyl group to favor a reaction at the *para* position over the *ortho* position.

But this is only the case with a methyl group. With just about any other group, we should expect that the *para* product will be the major product. Keep that in mind because it is very important— *para* is usually the major product.

With that in mind, imagine that I asked you to propose an efficient synthesis for the following transformation:

This is simple to do. The *tert*-butyl group is so large that we expect the *para* product to be the major product. So we just use Br_2 and $AlBr_3$, and we should obtain the desired product.

But suppose we wanted substitution to occur at the *ortho* position:

How would we do this? When confronted with this problem, students often suggest using the same reaction as before, with the understanding that the *ortho* product will be a minor product (so *some* ortho product will definitely be formed). But you can't do that. Whenever you have a synthesis problem, you must choose reagents that give you the desired compound as the MAJOR product. If you propose a synthesis that would produce the desired product as a MINOR product, then your synthesis is not efficient. So we have a problem here. How do we run the reaction so that the *ortho* product will be the major product?

The answer is: we cannot do it in one step. There is no way to "turn off" steric effects. However, there is a way to exploit them. At the start of this chapter, we learned about sulfonation (using fuming sulfuric acid to install an SO_3H group on the ring). We saw that this group can be installed on the ring, **and** it can be removed from the ring very easily. We said that this feature (reversibility) would be VERY important in synthesis problems. Now we are ready to see why.

If we perform a sulfonation reaction first, we will expect the SO_3H group to go predominantly in the *para* position (the major product will be from *para* substitution):

Now think about what we have done. We have "blocked" the *para* position. Now if we brominate, the incoming Br will be installed in the *ortho* position (because the *para* position is already taken). So, the reaction places the Br in the desired location:

Finally, we can perform a desulfonation to remove the SO_3H group. To do so, remember that we need to use dilute sulfuric acid:

And this is the desired product. In summary, here is our entire synthesis:

Notice that it took three steps (where the first step was used to block the *para* position and the third step was used to unblock the *para* position). Three steps might seem inefficient, BUT we did not need to rely on isolating minor products. At each step of the way, we were using the major product to move on to the next step.

If you think about what we have done, you should realize that this trick is really very clever. We recognized that we cannot just "turn off" the steric effects. So, instead, we developed a strategy that *uses* the steric effects. Notice that the SO_3H group is not in our final product at all. It was just used temporarily, as a "blocking group." This type of concept is very important in organic chemistry. As you move through the course, you will see a few other examples of blocking groups (in reactions that have nothing to do with electrophilic aromatic substitution). The basic strategy is applicable elsewhere. By temporarily blocking the position where the reaction would primarily occur (and then unblocking after you perform the desired reaction), it is possible to form a product that would otherwise be the minor product.

Now let's make sure that we know how to use this technique:

EXERCISE 3.80   Propose an efficient synthesis for the following transformation:

Answer   We see that we need to install an acyl group in the *ortho* position. If we just perform a Friedel-Crafts acylation, we would expect the *para* product to be the major product (because of steric effects). So, we must perform a sulfonation reaction to block the *para* position. Our answer is:

Propose an efficient synthesis for each of the following transformations. Make sure that sulfonation is necessary (I am purposefully giving you at least one problem that does not require sulfonation—to make sure that you understand *when* to use this blocking technique):

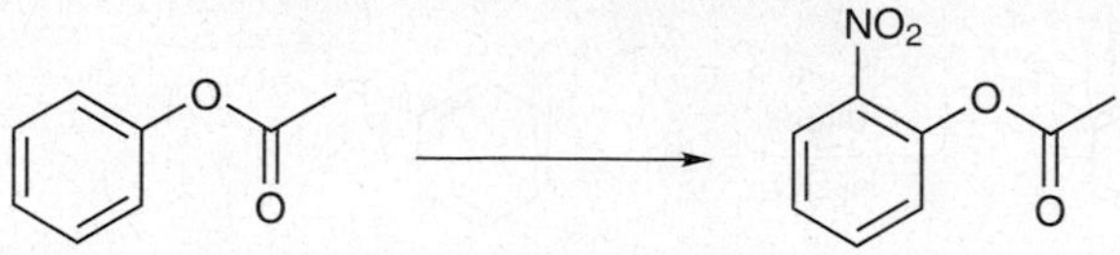

3.81

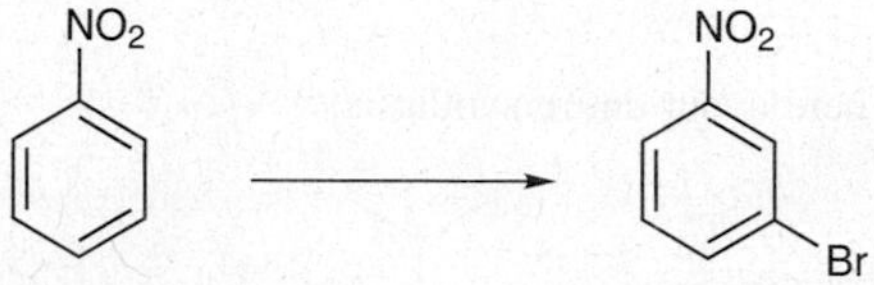

3.82

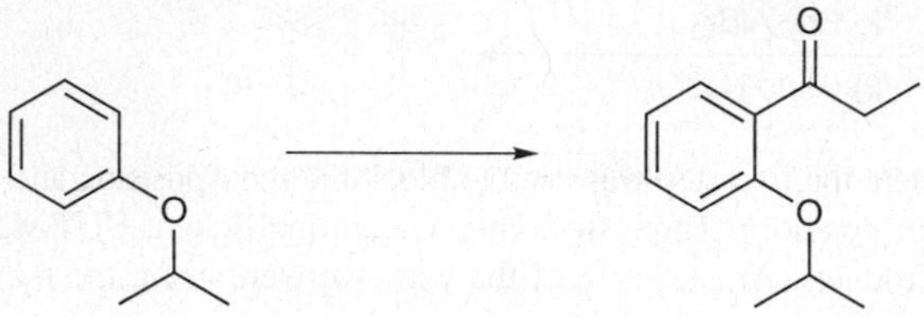

3.83

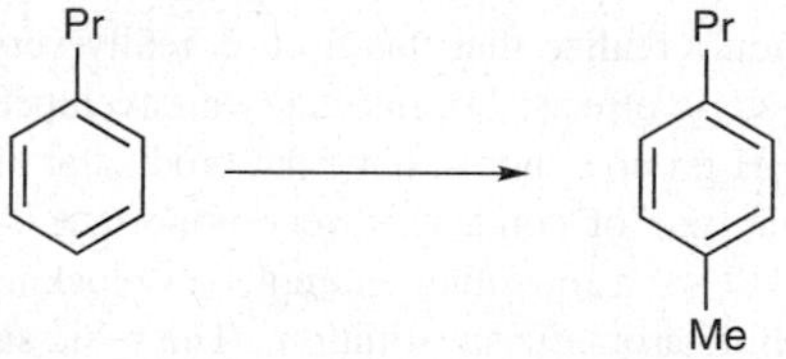

3.84

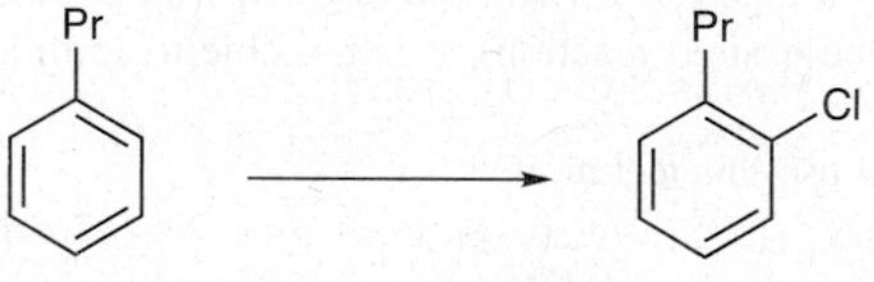

3.85

Before we move on to the final section of this chapter, you should be familiar with a few other steric effects. So far, we have seen the steric effects of ONE group on a ring. But what happens when we have two groups on a ring. For example, consider the directing effects of *meta*-xylene:

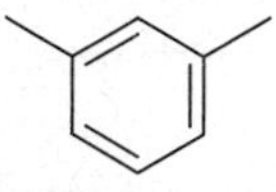

This compound has *two* methyl groups on the ring. Both methyl groups are directing to the same three positions:

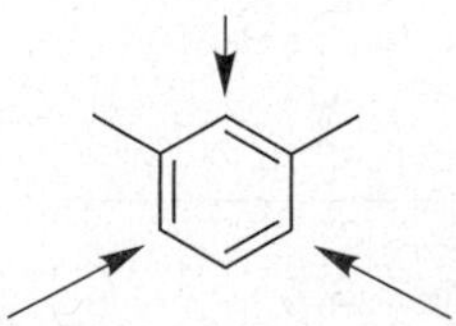

Two of these positions are essentially the same because of symmetry:

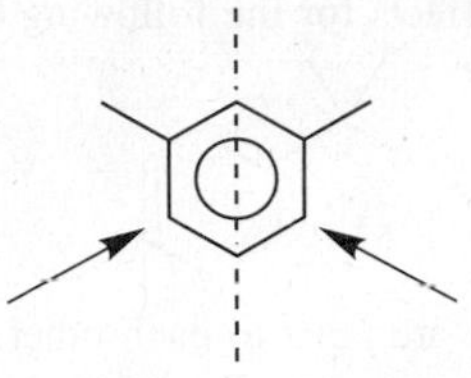
Attacking either of these two positions
would yield the same product

So, if we brominate this compound, we will expect to get only two products (rather than three):

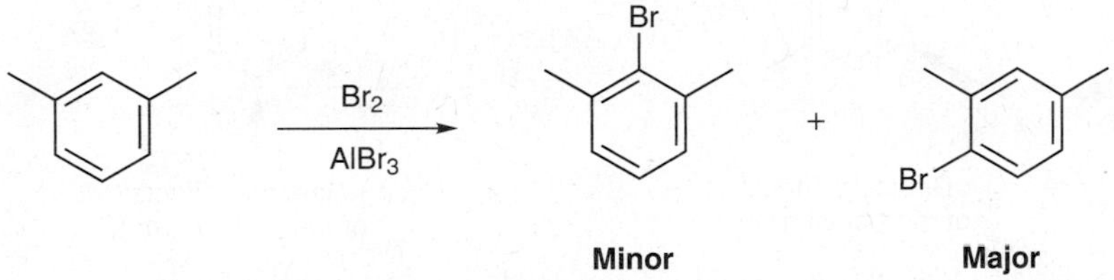
Br
Br_2
$AlBr_3$
+
Br
Minor
Major

Notice that we have indicated that one of the products is major. To understand why, we must consider steric effects. The position in between the two methyl groups is more sterically hindered than the other positions. Therefore, we primarily get just one product.

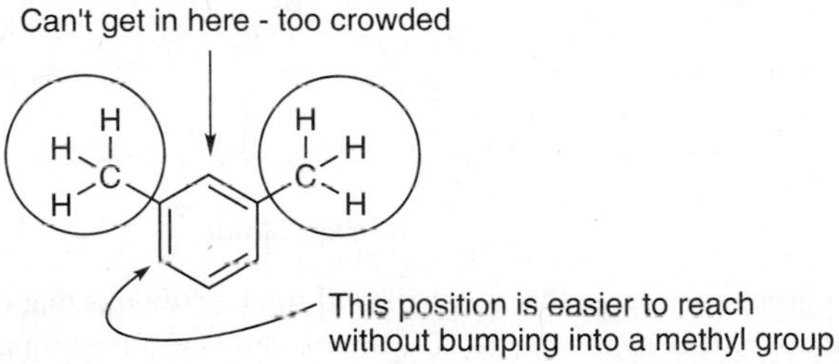
Can't get in here - too crowded
This position is easier to reach
without bumping into a methyl group

This type of argument can be used in a variety of similar situations. For example, you might remember that we saw the following reaction earlier in this chapter:

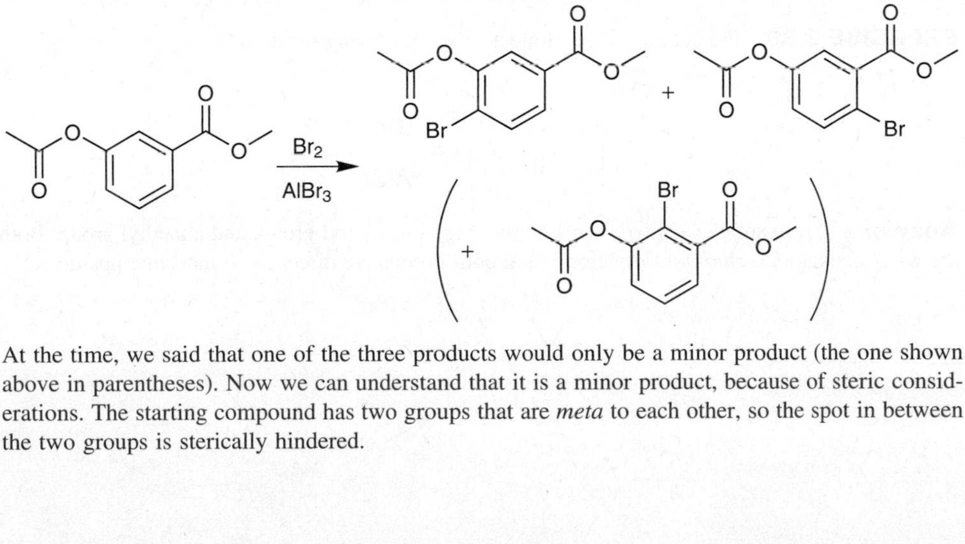
Br_2
$AlBr_3$
Br
Br
Br

At the time, we said that one of the three products would only be a minor product (the one shown above in parentheses). Now we can understand that it is a minor product, because of steric considerations. The starting compound has two groups that are *meta* to each other, so the spot in between the two groups is sterically hindered.

But suppose you have a disubstituted benzene ring where the two groups are *para* to each other. For example, consider the directing effects for the following compound:

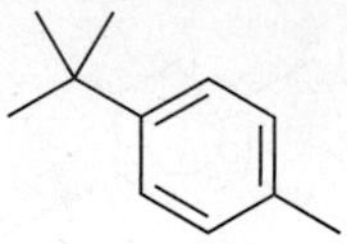

In this case, we have two groups that are *para* to each other. Here is a summary of the directing effects of each group:

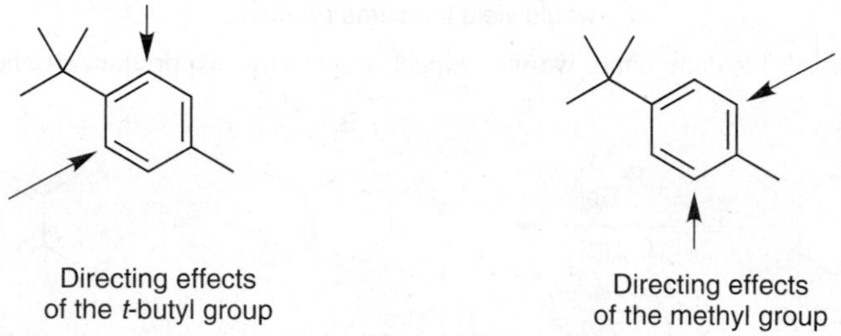

Directing effects
of the *t*-butyl group

Directing effects
of the methyl group

So, these two groups are directing to all four potential spots. Both groups are weak activators (alkyl groups). So, when we consider electronic factors, we don't really see any preference among the possible spots. However, when we consider steric factors, we notice that the *tert*-butyl group is very large compared to the methyl group. As a result, we see the following results:

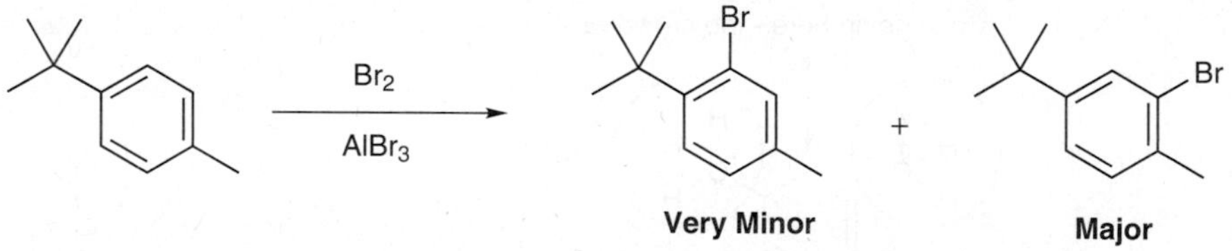

Very Minor   **Major**

In fact, the *tert*-butyl group is so large that you will find some textbooks that do not even show the minor product above at all. It is so minor that it is almost not worth mentioning.

In this section, we have seen many examples where steric effects play a significant role in determining the product distribution. Now let's get some practice using these principles.

EXERCISE 3.86 Predict the major product of the following reaction:

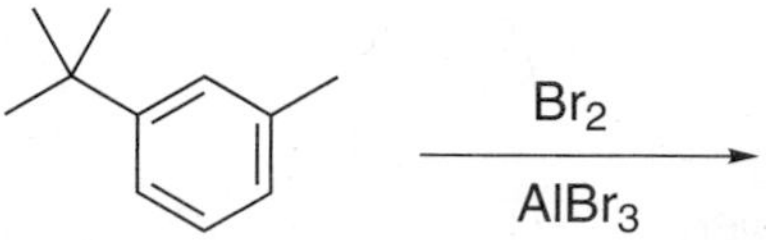

$$\xrightarrow[\text{AlBr}_3]{\text{Br}_2}$$

Answer This example has two groups on the ring: a *tert*-butyl group, and a methyl group. Both are weak activators (*ortho-para* directors), and both groups are directing to the same positions:

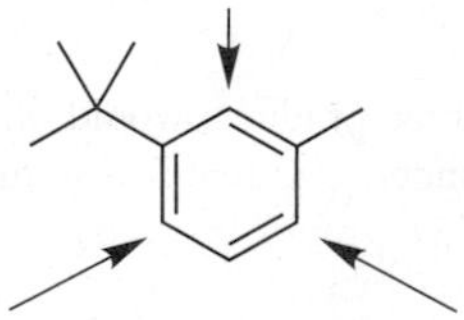

Of these three positions, the one in between the two groups is the most sterically hindered. We won't expect the reaction to take place at that spot very often. Also, the position next to the *tert*-butyl group is fairly hindered, so we won't expect the reaction to take place there either. Thus, we expect the reaction to take place most often at the position next to the methyl group:

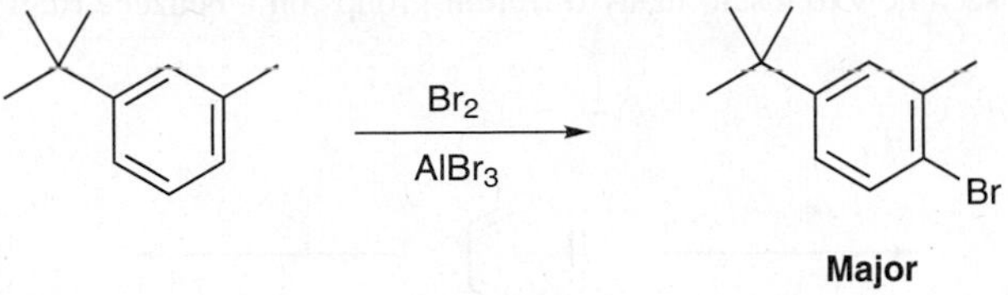

$$\xrightarrow[\text{AlBr}_3]{\text{Br}_2}$$

Major

Predict the MAJOR product of each of the following reactions (you do NOT need to show the minor products in these problems):

3.87

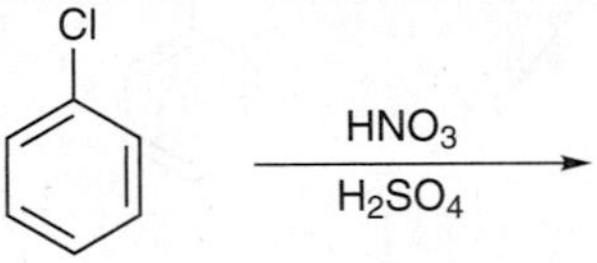

$$\xrightarrow[\text{H}_2\text{SO}_4]{\text{HNO}_3}$$

3.88

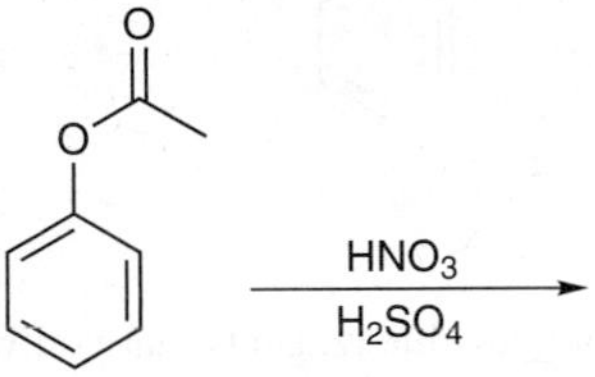

$$\xrightarrow[\text{H}_2\text{SO}_4]{\text{HNO}_3}$$

3.89

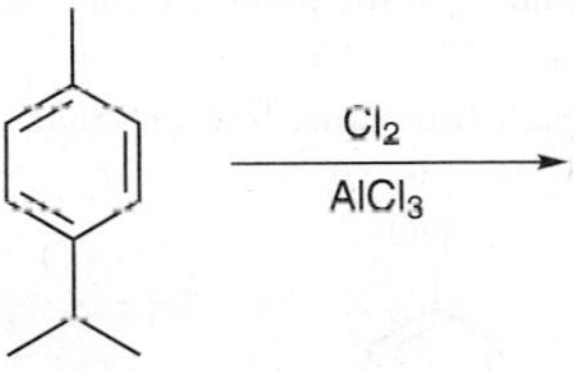

$$\xrightarrow[\text{AlCl}_3]{\text{Cl}_2}$$

3.90

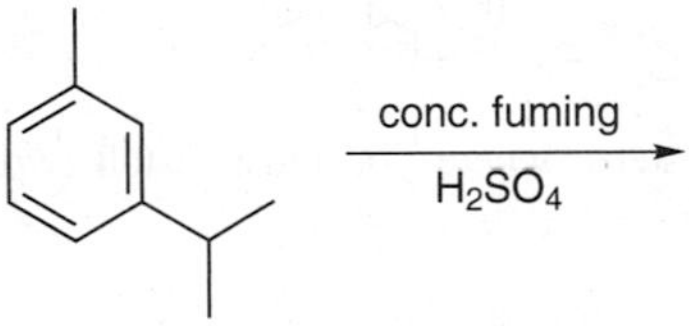

$$\xrightarrow[\text{H}_2\text{SO}_4]{\text{conc. fuming}}$$

3.91

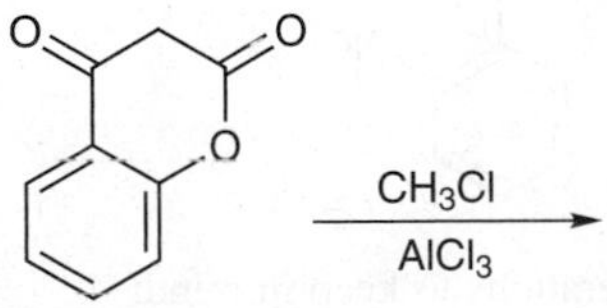

$$\xrightarrow[\text{AlCl}_3]{\text{CH}_3\text{Cl}}$$

3.92

$$\xrightarrow[\text{AlBr}_3]{\text{Br}_2}$$

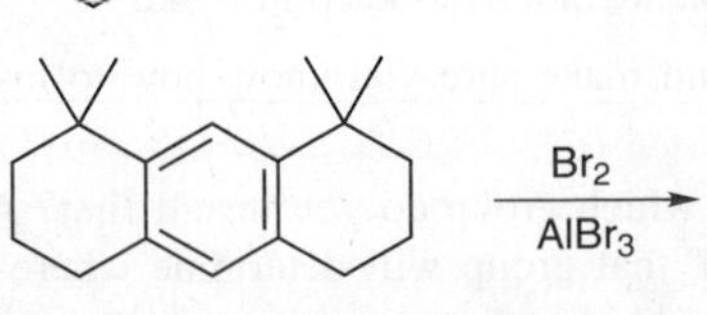

3.9 SYNTHESIS STRATEGIES

In this section, we will discuss some strategies for the toughest problems you can expect to see— synthesis problems. Let's begin with a quick review of the reactions we have seen earlier in this chapter. We have seen how to install many different groups on a benzene ring:

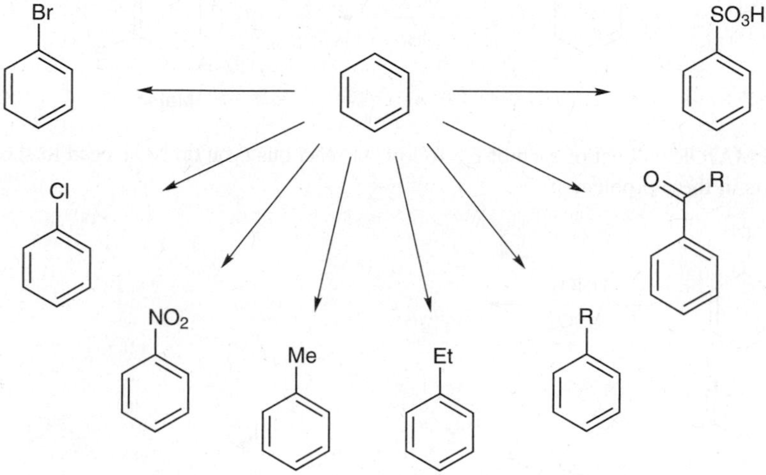

Carefully look at the chart above and make sure that you know the reagents that you would use to achieve each of these transformations. If you are not familiar with the reagents, then you will be totally unable to do synthesis problems.

It would be nice if all synthesis problems were just one-step problems, like this one:

Usually, however, synthesis problems require a few steps, where you must install two or more groups on a ring, like this:

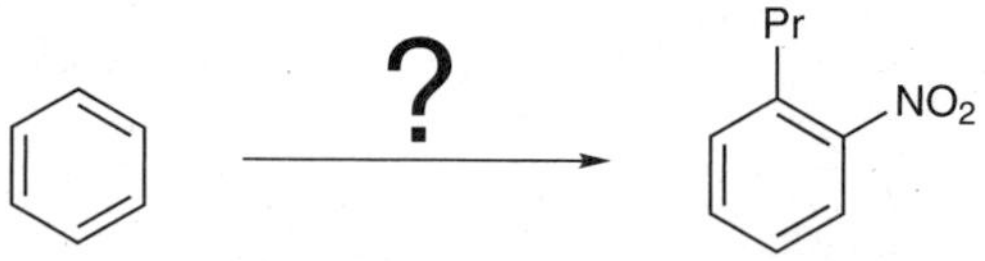

When dealing with such problems, there are many considerations to keep in mind:

- Take a close look at the groups on the ring and make sure you know how to install each group individually.

- Consider the order of events. In other words, which group do you install first? After you install the first group, the directing effects of that group will determine where the next group will be installed. This is an important consideration because it will affect the relative position of the two groups in the product. In the example above, the two groups are *ortho* to each other. So we must choose a strategy that installs the two groups *ortho* to each other.

- Take steric effects into account (and determine when you need to use sulfonation as a blocking group).

There are certainly other considerations, but these will help you begin to master synthesis problems. The first consideration above is just a simple knowledge of the reagents necessary to install any group on a ring. The last two considerations can be summarized like this: electronics and sterics (hopefully, this will make it easy for you to remember these considerations). Whenever you are solving any problem, you must always consider electronic effects and steric effects. As you move through this course, you will find the same theme in every chapter. You will find that you must always consider steric effects and electronic effects.

Let's try to use these considerations to solve the problem we just saw:

Let's begin by making sure we know how to install both of these groups individually. There are two groups that we need to install on the ring: a propyl group and a nitro group. The nitro group is easy—we just perform a nitration (using sulfuric acid and nitric acid). The propyl group is a bit trickier because we *cannot* use a Friedel-Crafts alkylation (remember carbocation rearrangements). Instead, we must use a Friedel-Crafts acylation, followed by a reduction to remove the C=O double bond. All-in-all, we have a total of three steps: one step to install the nitro group and two steps to install the propyl group.

Now let's focus on electronic considerations. In this case, we can begin to appreciate the importance of "order of events." Imagine that we install the nitro group first. The nitro group is a *meta* director, so the next group will end up being installed *meta* to the nitro group. That doesn't work for us because we want the groups to be *ortho* to each other in the final product. So, we have decided that we cannot install the nitro group first. Instead, let's try to install the propyl group first. That should work because the propyl group is an *ortho-para* director. So the propyl group will direct the incoming nitro group into the correct position (*ortho*). BUT the propyl group will also direct to the *para* position. And this is where sterics comes into the picture.

When we look at the steric effects, we encounter a difficulty. The steric effects are not in our favor here. We should expect the following results:

Notice that the compound we want to make is the *minor* product. But, we need a way to obtain the *ortho* product as our major product. And we have seen exactly how to do that. We just use sulfonation to block the *para* position. So, our overall synthesis goes like this:

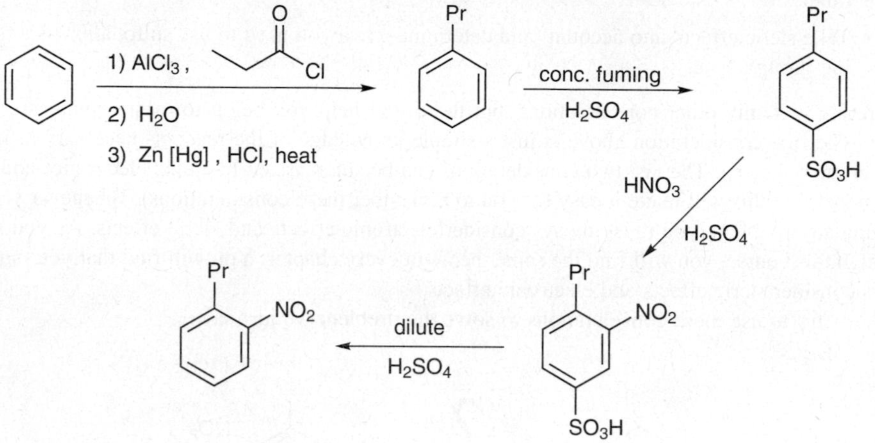

The answer that we just developed can be summarized like this:

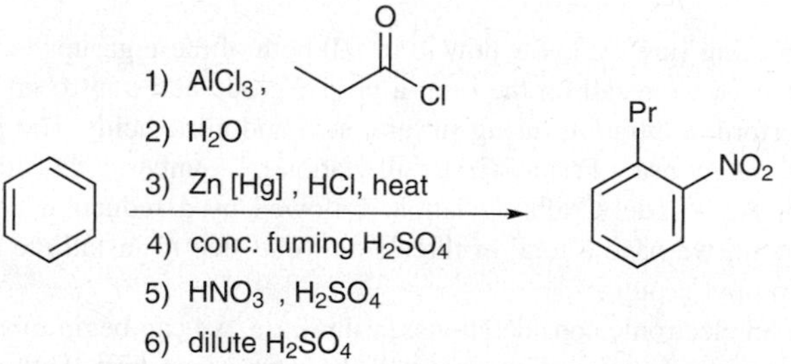

Now let's consider another example in which order of events is especially relevant. Consider how you might achieve the following transformation:

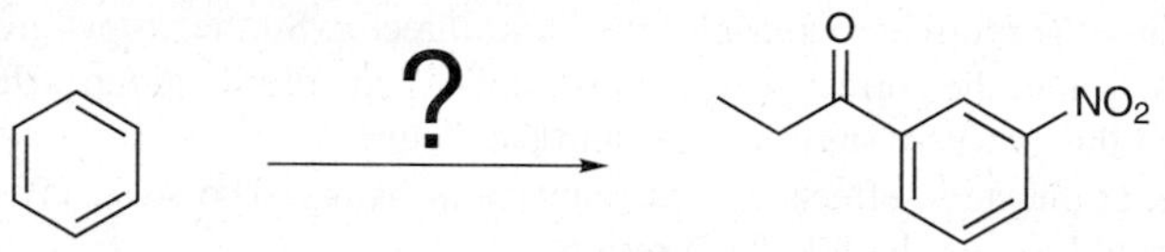

In this case, we must install an acyl group and a nitro group, and the two groups must be *meta* to each other. Both groups are *meta* directing, so it might seem like we can install them in either order (first acylation and then nitration, or vice versa). However, there is a serious limitation to Friedel-Crafts reactions that we have not mentioned until now. And that limitation will dictate the order of events that we must follow in this case. It turns out that a Friedel-Crafts reaction cannot be performed on a ring that is either moderately deactivated or strongly deactivated. You *can* perform a Friedel-Crafts on a weakly deactivated ring (and certainly on an activated ring). But not on a significantly deactivated ring—the reaction just doesn't work (you can perform other reactions with deactivated rings, such as bromination, but not Friedel-Crafts reactions). With that in mind, the nitro group cannot be installed first.

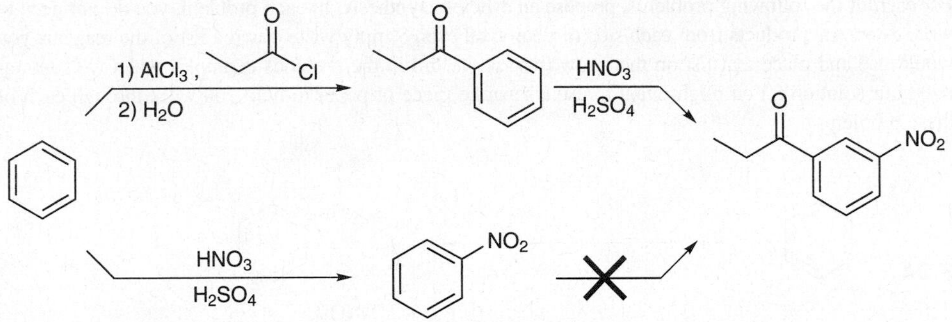

This synthesis teaches us the importance of "order of events." Whenever you are trying to solve a synthesis problem, you must always consider the order of events.

EXERCISE 3.93 Propose an efficient synthesis for the following transformation:

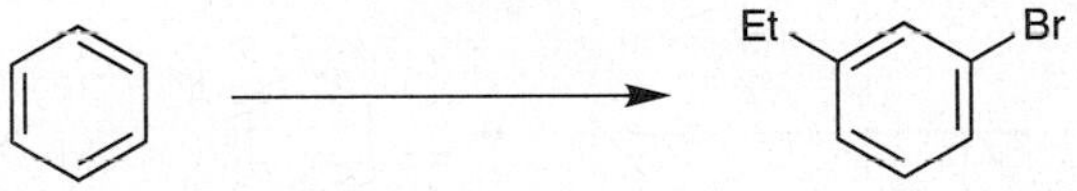

Answer We must install two groups on the ring; an ethyl group and bromine. Let's first make sure that we know what reagents we would use to install each group individually. To install bromine, we would use Br_2 and a Lewis acid. To install the ethyl group, we would use a Friedel-Crafts alkylation (or acylation, followed by a reduction). Whenever we install an ethyl group on a ring, we don't need to worry about carbocation rearrangements, so we can use a simple alkylation (rather than an acylation followed by reduction).

But we immediately see a serious issue when we consider the directing effects of each group. The bromine is *ortho-para* directing, so we can't install the bromine on the ring first (if we did, we would not get the groups to be *meta* to each other). And the ethyl group is also *ortho-para* directing. So, whichever group we put on first, there would seem to be no way to get these two groups to be *meta* to each other.

UNLESS we use an acylation (rather than an alkylation). If we do that, we will install an acyl group on the ring first. *And acyl groups are meta directing*. That would allow us to install the bromine in the correct spot. So our strategy would go like this:

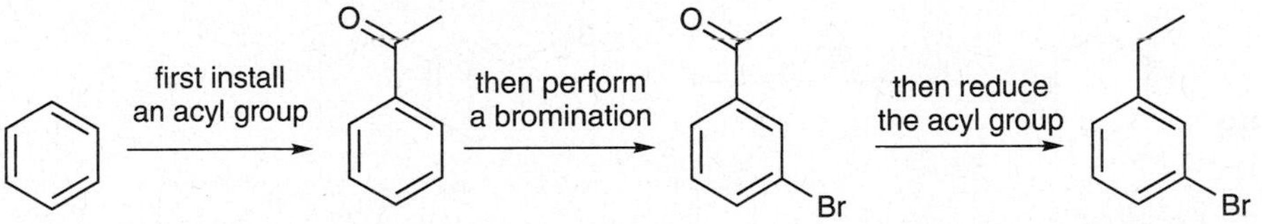

The reagents for our proposed synthesis are as follows:

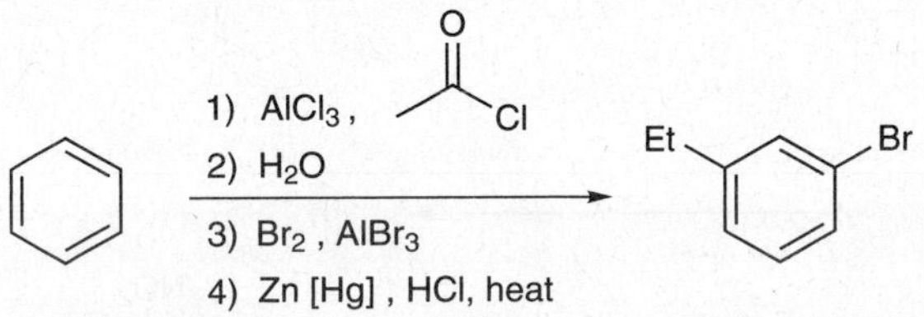

For each of the following problems, propose an efficient synthesis. In each problem, you do **not** need to write down the products from each step of your synthesis. Simply write down a list of the reagents you would use and place that list on the arrow (just as we did in the previous examples when we summarized our solution). You might want to use a separate piece of paper to help you work through each of these problems.

3.94

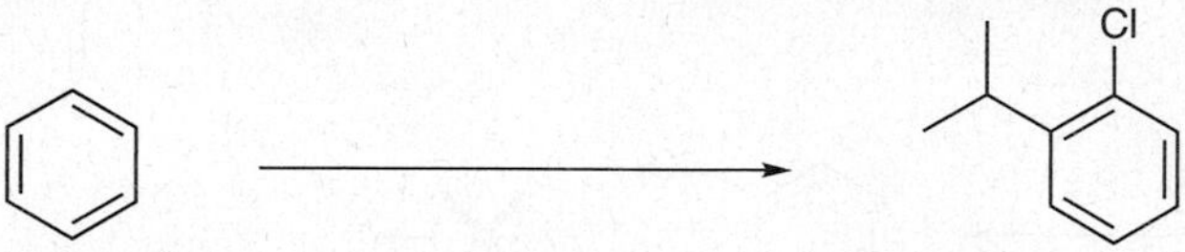

3.95

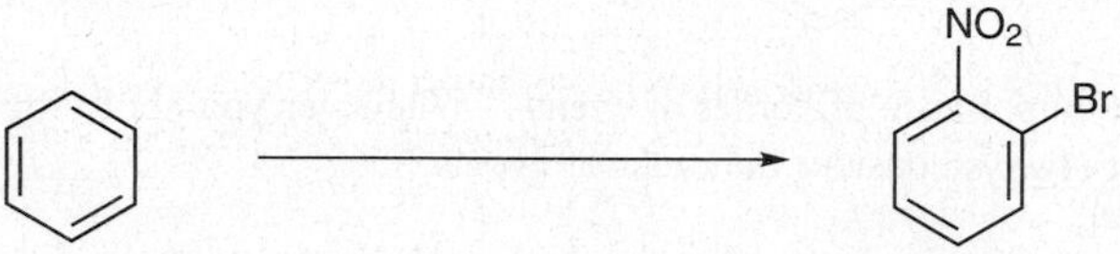

3.96

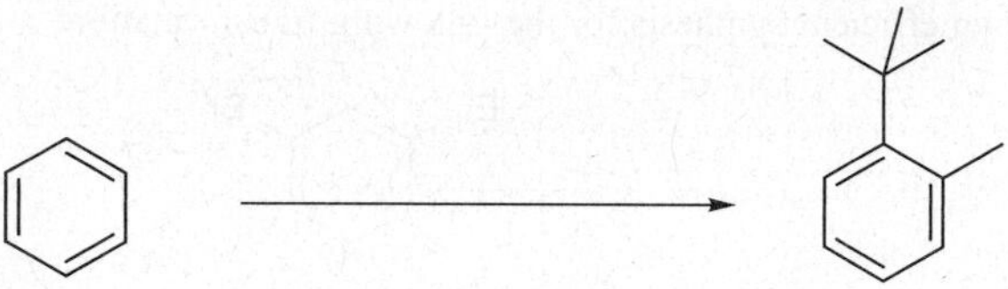

3.97

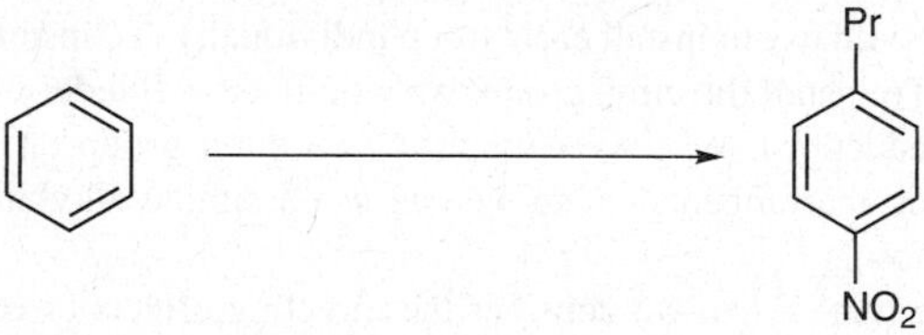

3.98

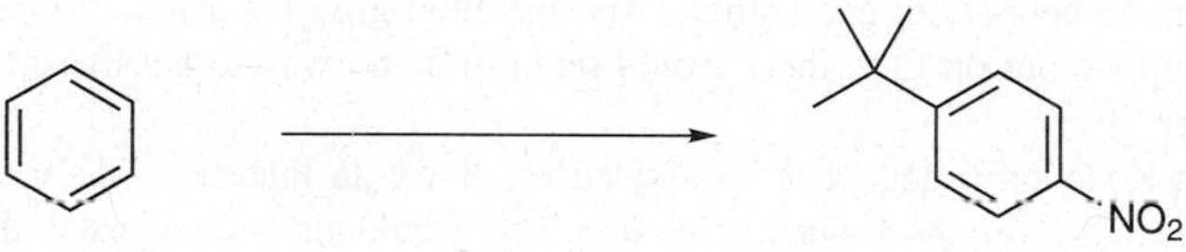

3.99

3.100

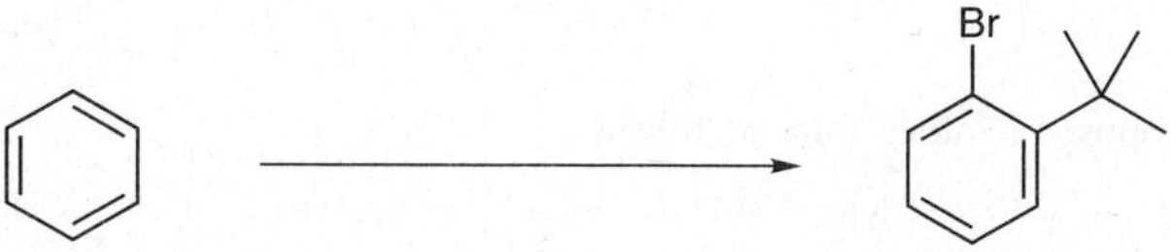

3.101

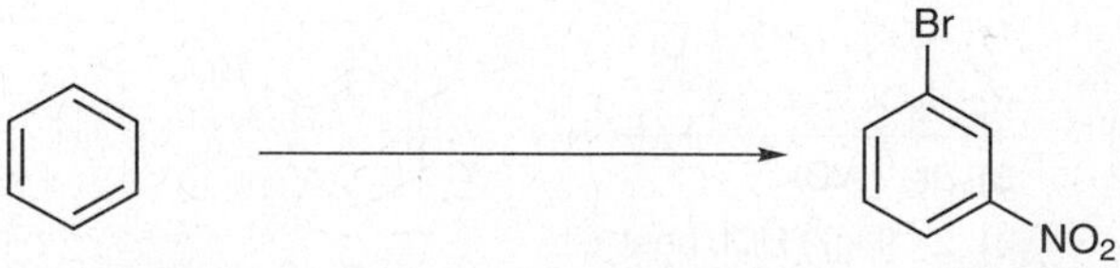

3.102

Before we end this chapter, it is important that you realize what we have covered here and, more importantly, what we have *not* covered. We did not cover everything in your textbook chapter on electrophilic aromatic substitution. As you go through your lecture notes and your textbook chapter, you will find a few reactions that we did not cover here. You will need to go through your textbook and your notes carefully to make sure that you learn those reactions. You should find that we covered 80%, or even 90%, of what you read in your textbook.

The purpose of this chapter was not to cover everything but rather to serve as a foundation for your mastery of electrophilic aromatic substitution. If you went through this chapter, then you should feel comfortable with the steps involved in proposing mechanisms, predicting products, and proposing a synthesis. You should know how directing effects work and how to use them when proposing syntheses. You should also know about steric effects and how to use them when proposing syntheses.

With all of that as a foundation, you should now be ready to go through your textbook and lecture notes, and polish off the rest of the material that you must know for your exam. Through the foundation that we have developed in this chapter, you should find (hopefully) that the content in your textbook will seem easy.

Do the problems in your textbook. Do all of them. Good luck.

CHAPTER 4

NUCLEOPHILIC AROMATIC SUBSTITUTION

4.1 CRITERIA FOR NUCLEOPHILIC AROMATIC SUBSTITUTION

In the previous chapter, we learned all about *electrophilic* aromatic substitution reactions.

In this short chapter, we will look at the flipside: is it possible for an aromatic ring to function as an electrophile and react with a nucleophile? In other words, is it possible for the aromatic ring to be so electron-poor that it is subject to attack by a nucleophile? The answer is: yes.

But in order to observe this kind of reaction, called nucleophilic aromatic substitution, we will need to meet three very specific criteria. Let's look closely at each one of these criteria:

1. The ring must have a very powerful electron-withdrawing group. The most common example is the nitro group:

We saw in the previous chapter that the nitro group is a strong deactivator toward electrophilic aromatic substitution because the nitro group very powerfully withdraws electron density from the ring (by resonance). This causes the electron density in the ring to be very poor:

At the time, we wanted the aromatic ring to function as a nucleophile. And we saw that the effect of a nitro group is to *deactivate* the ring. But now, in this chapter, we want the ring to act as an *electrophile*. So the effect of the nitro group is a very good thing. In fact, it is ***necessary*** to have a powerful electron-withdrawing group if you want the ring to function as an electrophile. The presence of the electron-withdrawing group is the first criterion that must be met in order for the ring to function as an electrophile. Now, let's look at the second criterion:

2. There must be a leaving group that can leave.

| reactive toward | NOT reactive toward |
| nucleophilic aromatic substitution | nucleophilic aromatic substitution |

To understand this, let's think back to what happened in the previous chapter, when the ring always functioned as a nucleophile. We saw that all the reactions from the previous chapter could be summarized like this: E^+ comes on the ring, and then H^+ comes off (or, in other words: attack, then deprotonate). But now, in this chapter, we want the ring to function as an electrophile. So we are trying to see if we can get a nucleophile (Nuc^-) to attack the ring. If we can make it happen, and a nucleophile (with a negative charge) actually does attack the ring, then something with a negative charge is going to have to come off of the ring. We can summarize it like this: Nuc^- comes on the ring, and X^- comes off.

There is one main difference between the mechanism here and the one we saw in Chapter 3. The difference is in the kind of charges we are dealing with. In the previous chapter, we dealt with something positively charged coming onto the ring to form a positively charged sigma complex, and then H^+ came off the ring to restore aromaticity. In those mechanisms, everything was positively charged. But now, we are dealing with negative charges. A nucleophile with a negative charge will attack the ring to form some kind of negatively charged intermediate. That intermediate must then expel something negatively charged. And that explains the second criterion for this reaction to occur: we need the ring to have some leaving group that can leave with a negative charge.

If there is no leaving group that can leave with a negative charge, then the ring will have no way of reforming aromaticity. And we cannot just kick off H^- because H^- is a terrible leaving group. NEVER kick off H^-. If you are a bit rusty on leaving groups, you might want to go back to first-semester material and quickly review which groups are good leaving groups.

3. The final criterion is: the leaving group must be *ortho* or *para* to the electron-withdrawing group:

reactive toward
nucleophilic aromatic substitution

NOT reactive toward
nucleophilic aromatic substitution

To understand why, we will need to take a closer look at the accepted mechanism. In the upcoming section, we will explore the mechanism of this reaction so that we can understand this last criterion. For now, let's just make sure that we can identify when all three criteria have been met. Once again, the three criteria are:

1. There must be an electron-withdrawing group on the ring.
2. There must be a leaving group on the ring.
3. The leaving group must be *ortho* or *para* to the electron-withdrawing group.

Now let's get some practice looking for all three criteria:

EXERCISE 4.1 Predict whether the following compound can function as a suitable electrophile in a nucleophilic aromatic substitution reaction.

Answer   In order to have a nucleophilic aromatic substitution, all three criteria must be met.

We look at the ring, and we see that it does have a nitro group. Therefore, the first criterion has been met.

We then look for a leaving group. There is NO leaving group here. Methyl is NOT a leaving group. Why not? Because a carbon with a negative charge is a *terrible* leaving group. Never kick off C⁻. So criterion 2 has not been met.

Therefore, we conclude that this compound will not function as an electrophile in a nucleophilic aromatic substitution reaction.

Determine whether each of the following compounds can function as a suitable electrophile in a nucleophilic aromatic substitution reaction. If you determine that not all three criteria are met, then simply write "no reaction."

4.2

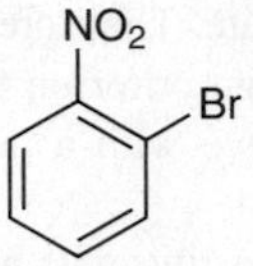

4.3

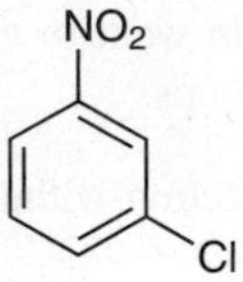

4.4

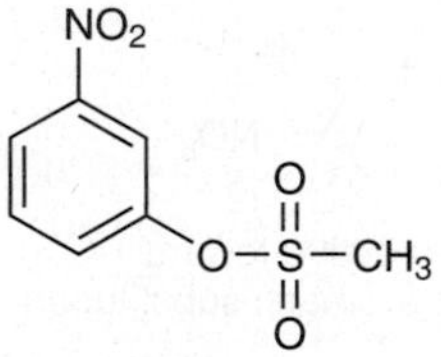

4.5

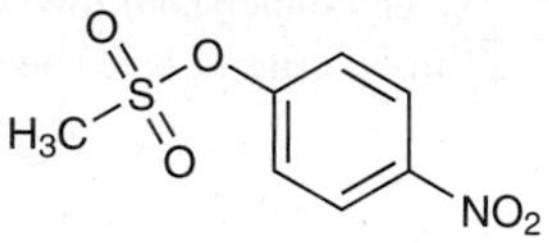

4.6

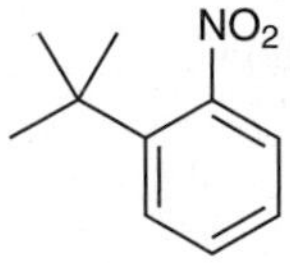

4.7

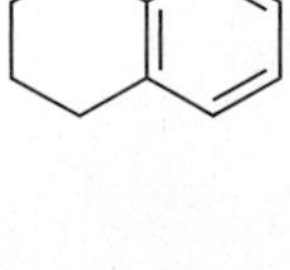

4.2 S_NAr MECHANISM

In the previous section, we saw the three criteria that are necessary in order for an aromatic ring to undergo a nucleophilic aromatic substitution reaction. The following transformation is an example:

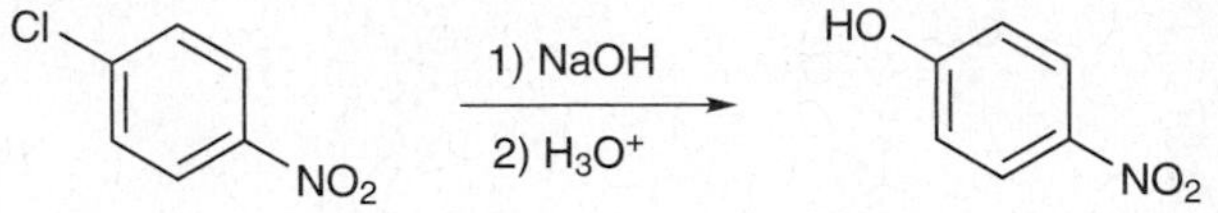

Let's explore some possible mechanisms for this process. It cannot be an S$_N$2 process because S$_N$2 processes do not readily occur at an sp^2-hybridized center:

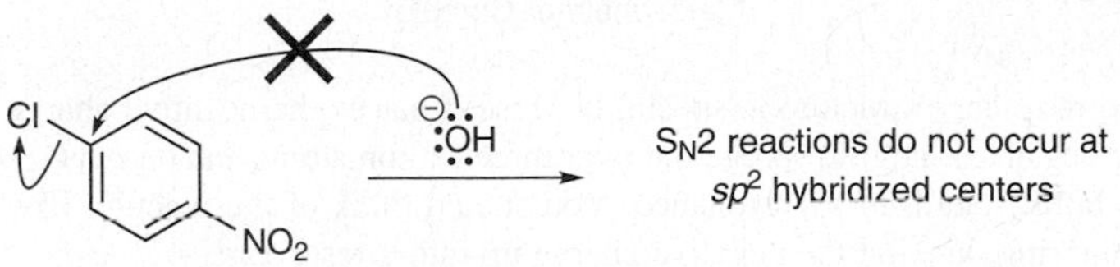

S$_N$2 reactions do not occur at sp^2 hybridized centers

S$_N$2 processes are only effective with sp^3-hybridized centers. So our reaction cannot be an S$_N$2 mechanism. What about S$_N$1? That would require the loss of the leaving group *first* to form a carbocation:

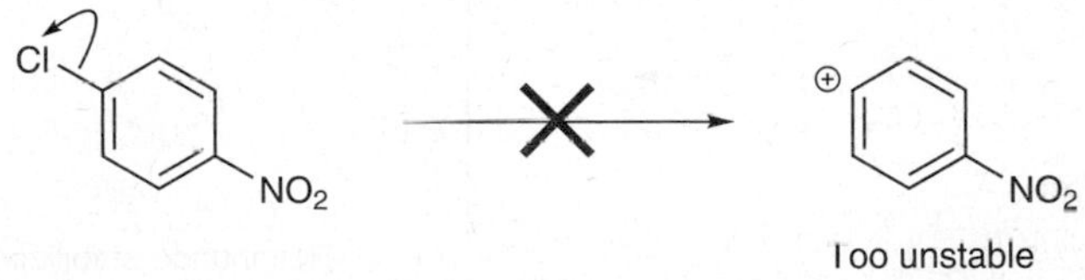

I oo unstable

This kind of carbocation is not stabilized by resonance. Since it is unstable, it is therefore a very high-energy intermediate. So, we don't expect the leaving to leave if it means creating an unstable intermediate. Therefore, we don't expect the mechanism to be an S$_N$1 mechanism either.

So, if it's not S$_N$2 and its not S$_N$1, then what is it? And the answer is: it's a new mechanism, called S$_N$Ar. In many textbooks, it is called an ***addition-elimination*** mechanism. In the first step of the mechanism, the ring is attacked by a nucleophile, generating a resonance-stabilized intermediate, called a Meisenheimer complex:

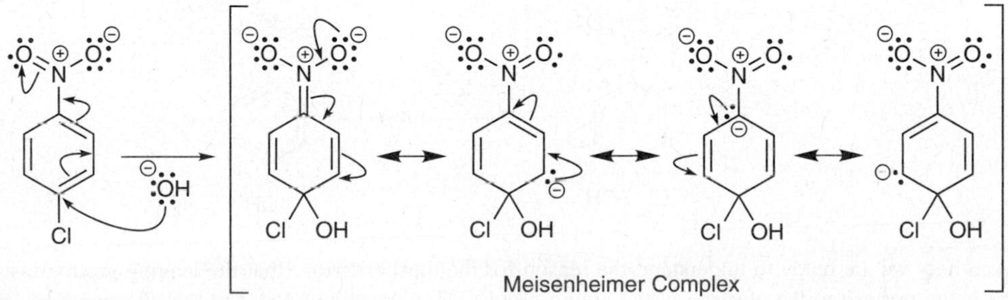

Meisenheimer Complex

This intermediate should remind us of the intermediate in an electrophilic aromatic substitution reaction (the sigma complex), but the main difference is that a Meisenheimer complex is *negatively* charged (a sigma complex is positively charged). Let's take a close look at the Meisenheimer complex, and let's focus our attention on one particular resonance structure:

Meisenheimer Complex

The highlighted resonance structure is special because it places the negative charge on an *oxygen* atom. Since the negative charge is spread out over three carbon atoms *and an oxygen atom*, the negative charge is fairly stabilized by resonance. You should think of the reaction like this: A nucleophile attacks the ring, kicking the negative charge up into a reservoir:

This negative charge goes up onto the oxygen atom of the nitro group

Reservoir

Resonance-stabilized Meisenheimer complex

Then, in the second step of the mechanism, the reservoir releases its load by pushing the electron density back down onto a leaving group, thereby restoring aromaticity to the ring:

Negative charge leaves reservoir to be expelled with the leaving group

And now we are ready to understand the reason for the third criterion (that the leaving group must be *ortho* or *para* to the electron-withdrawing group). Now we can understand that the reservoir is available only if the nucleophile attacks at the *ortho* or *para* positions. If attack occurs at the *meta* position, there is no way to place the negative charge up onto the reservoir:

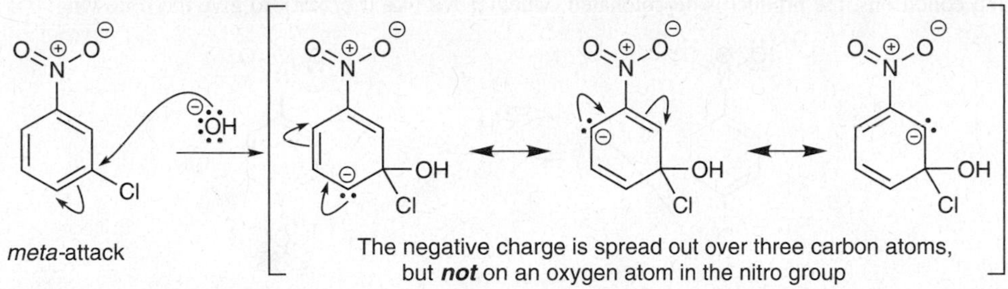

meta-attack

The negative charge is spread out over three carbon atoms, but **not** on an oxygen atom in the nitro group

Therefore, the intermediate is not stabilized. So the reaction doesn't happen. And that is why the leaving group must be *ortho* or *para* to the electron-withdrawing group.

Before you get practice drawing the complete mechanism of an S$_N$Ar process, there is one subtle point that deserves attention. Let's first summarize what we have seen so far:

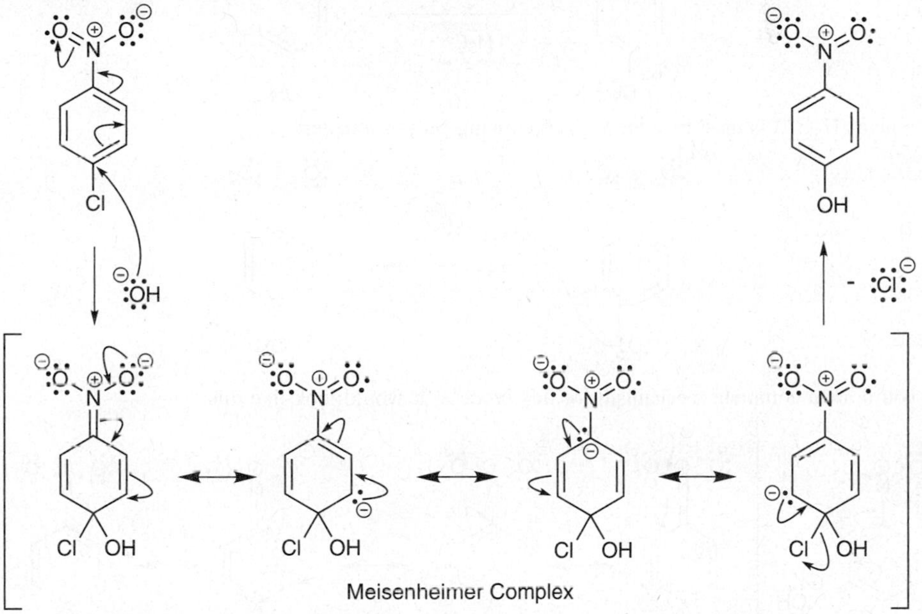

Meisenheimer Complex

There are two steps: 1) a nucleophile attacks the ring to generate a Meisenheimer complex, followed by 2) loss of a leaving group to restore aromaticity. But inspect the product very carefully. It contains a phenolic proton, highlighted below:

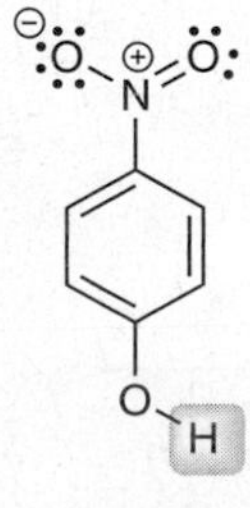

This proton is mildly acidic and cannot survive the strongly basic conditions being employed (hydroxide is a strong base, and hydroxide is present in the reaction flask). So, under these reaction conditions, the product is deprotonated (whether we like it or not), to give the following:

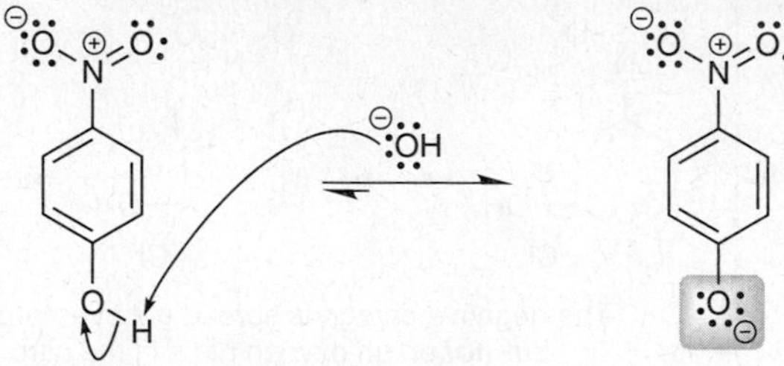

Therefore, in order to regenerate the desired product, a source of proton must be introduced into the reaction flask (after the reaction is complete), which is shown like this:

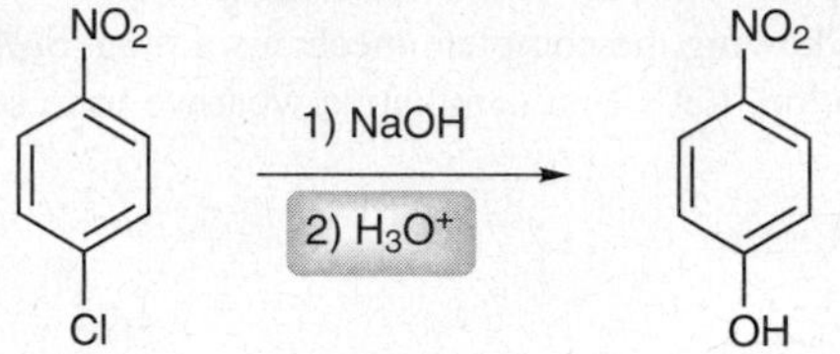

The acid (H_3O^+) is used to achieve the following proton transfer:

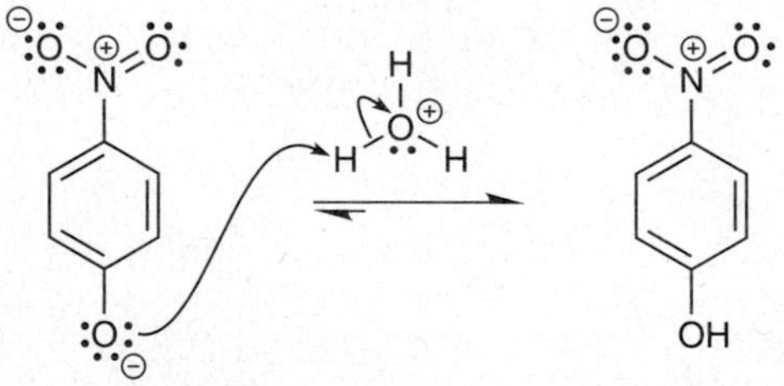

If you draw a complete mechanism for this process, it would look like this:

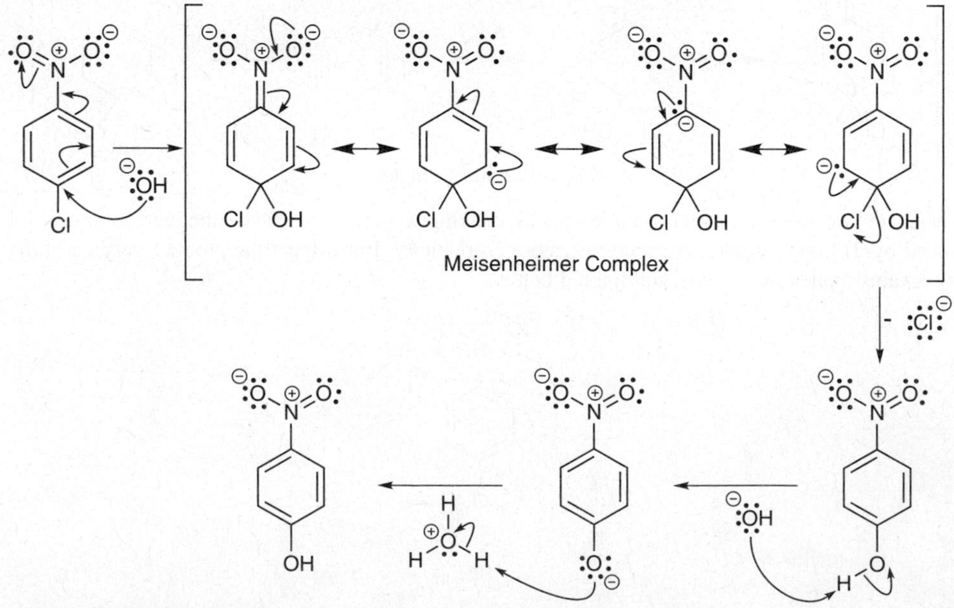

Meisenheimer Complex

When you stare at this mechanism, you might be surprised by the last two steps (deprotonation, followed by protonation). Specifically, you might wonder why the mechanism needs to show these last two steps. After all, if we delete these last two steps from the mechanism, isn't the product still correct? Yes, that is true, but these last two steps are necessary. Why? Because under the reaction conditions employed (strongly basic conditions), the phenolic proton does not survive. Deprotonation occurs, whether we like it or not. The mechanism, as drawn, indicates that we understand that subtle point. By drawing the last two steps of the mechanism, you are demonstrating that you understand why a proton source (like H$_3$O$^+$) must be added to the reaction flask after the reaction is complete.

Now we are ready to get some practice drawing an S$_N$Ar mechanism.

EXERCISE 4.8 Predict the product of the following reaction and draw a plausible mechanism for its formation:

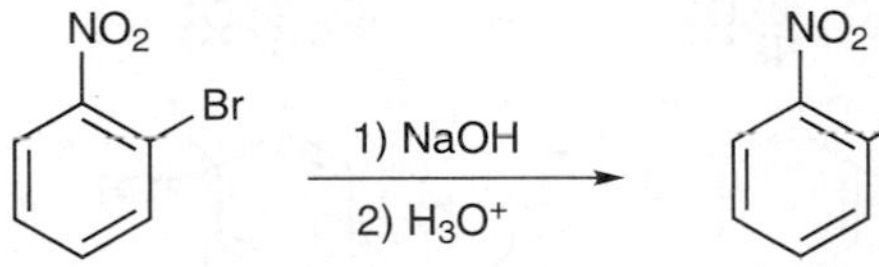

Answer The reagent is a strong nucleophile (hydroxide), and the starting compound has all three criteria for an S$_N$Ar mechanism: an electron withdrawing group (NO$_2$) and a leaving group (Cl) that are *ortho* to each other.

In a nucleophilic aromatic substitution reaction, the nucleophile (hydroxide) attacks at the position bearing the leaving group, and electrons are pushed up into the reservoir:

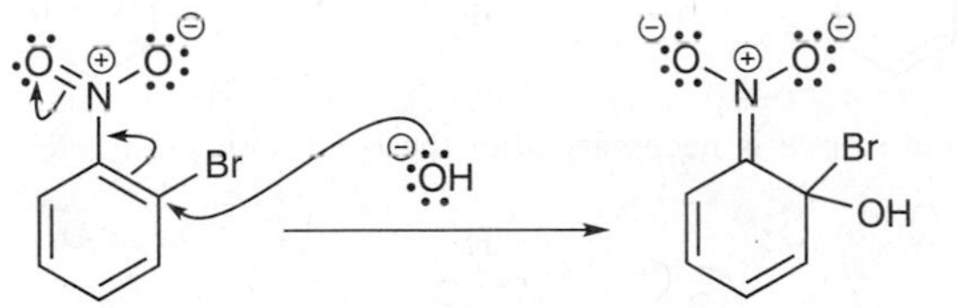

The resulting intermediate is a Meisenheimer complex, and it has resonance structures that should be drawn:

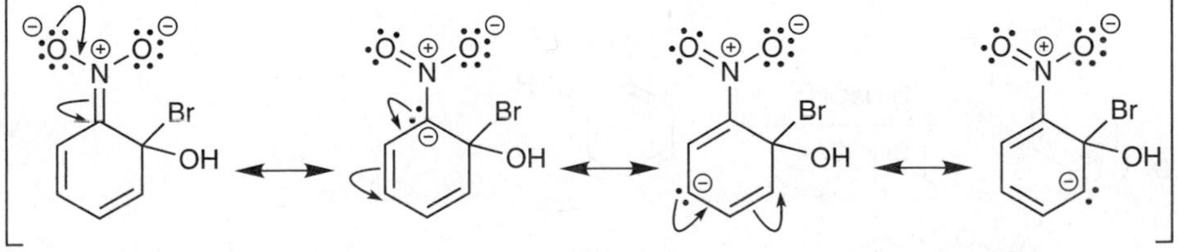

Then, in the second step of the mechanism, the leaving group is expelled to give the product:

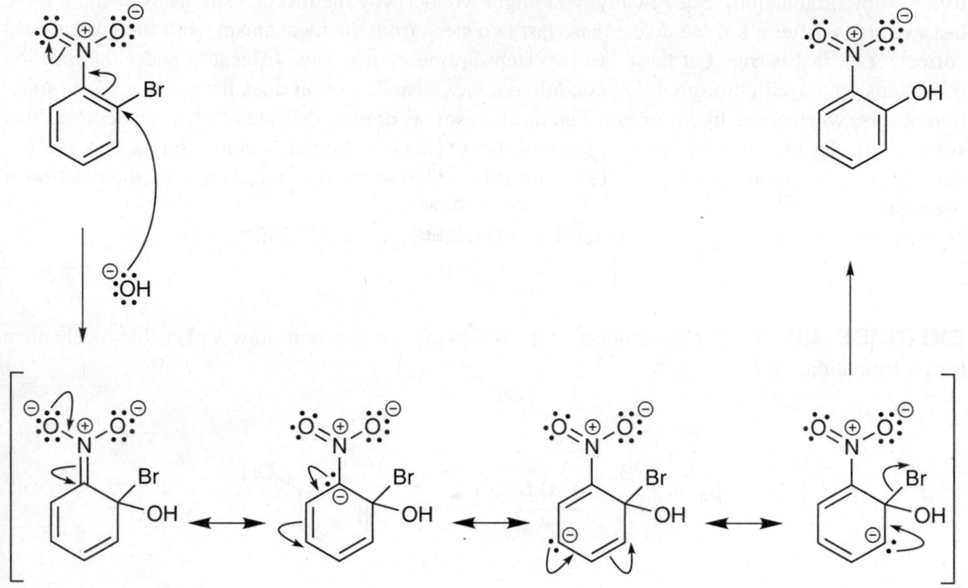

This might appear to be a complete mechanism. But remember, that under basic conditions, the product is deprotonated:

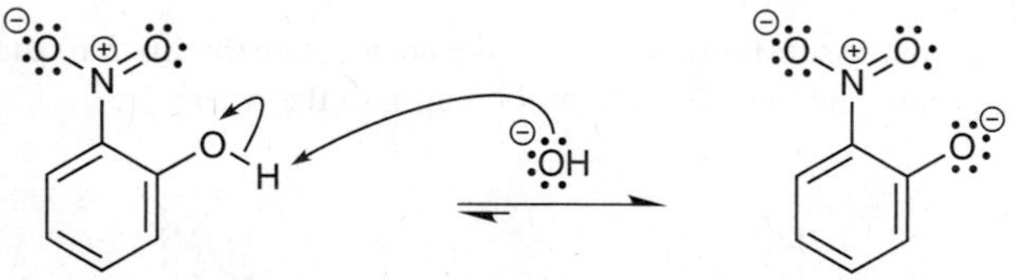

And that is why an acid source is necessary after the reaction is complete:

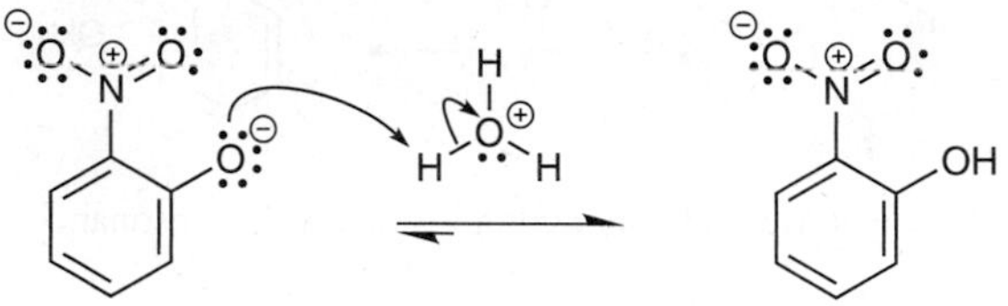

Propose a mechanism for each of the following transformations:

4.9

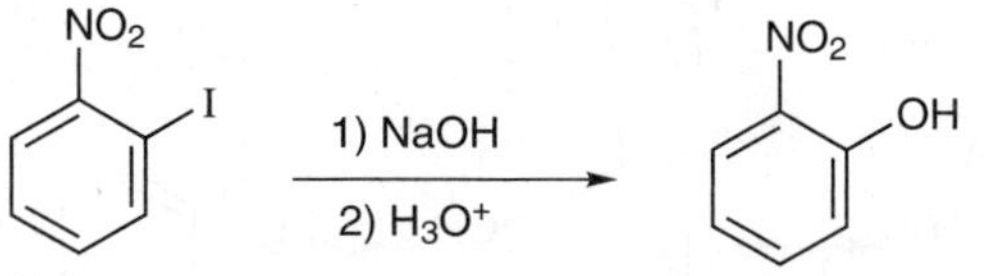

$$\text{1) NaOH} \quad \text{2) H}_3\text{O}^+$$

4.10

$$\text{1) NaOH} \quad \text{2) H}_3\text{O}^+$$

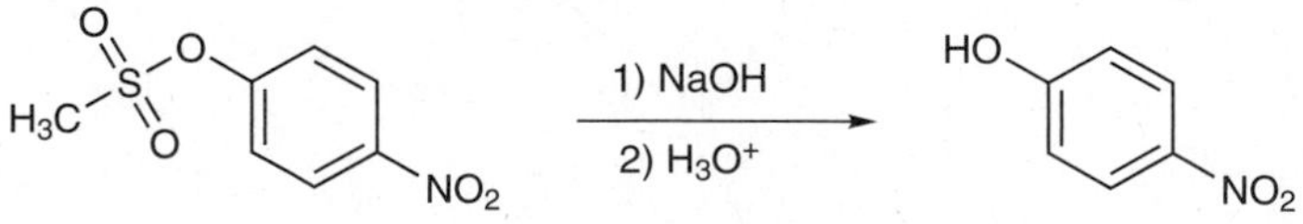

4.11

$$\underset{\text{NO}_2,\ \text{Cl}}{} \xrightarrow[\text{2) H}_3\text{O}^+]{\text{1) NaOH}} \underset{\text{NO}_2,\ \text{OH}}{}$$

4.3 ELIMINATION-ADDITION

In the previous section, we discussed the three criteria that you need in order to get an S_NAr mechanism. The obvious question is: can it occur without all three criteria? For example, what if there is no electron-withdrawing group?

If we treat chlorobenzene with hydroxide, no reaction is observed:

$$\underset{\text{Cl}}{} \xrightarrow{\text{NaOH}} \text{no reaction}$$

The hydroxide ion does not attack to kick off the leaving group because there is no "reservoir" to hold the electron density for a moment. In fact, if we try to apply heat, there is still no reaction.

However, at much higher temperatures, such as 350 °C, a reaction is in fact observed:

$$\underset{\text{Cl}}{} \xrightarrow[\text{2) H}_3\text{O}^+]{\text{1) NaOH, 350 °C}} \underset{\text{OH}}{}$$

This reaction, often called the Dow process, is commercially important because it is an excellent way of making phenol.

We can use this same process to make *aniline* (that is the common name for aminobenzene):

$$\underset{\text{Cl}}{} \xrightarrow[\text{2) H}_3\text{O}^+]{\text{1) NaNH}_2,\ \text{NH}_3\ (\textit{liq})} \underset{\text{NH}_2}{}$$

aniline

We don't even need high temperatures to make aniline. We just use H_2N^- in liquid ammonia. So we have a serious question: if there is no "reservoir" to temporarily hold the electron density, then how does this reaction work? What is the mechanism?

To understand the mechanism, chemists have used an important technique called isotopic labeling. All elements have isotopes (for example, deuterium is an isotope of hydrogen because deuterium has an extra neutron in the nucleus). Carbon also has some important isotopes. ^{13}C is an important isotope because we can easily determine the position of a ^{13}C atom in a compound using NMR spectroscopy. So if we enrich a specific spot with ^{13}C, then we can follow where that carbon atom goes during the reaction. For example, let's say we take chlorobenzene, and we enrich one particular site with ^{13}C:

$$\underset{\text{Cl}}{}$$

The site with the asterisk is the location where we placed the ^{13}C. When we say that we enriched that spot with ^{13}C, we mean that most of the molecules in the flask have a ^{13}C atom in that position.

Now let's see what happens to that isotopic label as the reaction proceeds. After running the reaction, here are the results that are observed:

50 %   50 %

This seems quite strange: how does the isotopic label "move" its position? We will not be able to explain this with a simple nucleophilic aromatic substitution. Even if we could somehow ignore the issue of not having a reservoir for the electron density during the reaction, we would still not be able to explain the isotopic labeling results.

So here is a proposal that explains the isotopic labeling experiments. Imagine that in the first step, the hydroxide ion acts as a base (rather than a nucleophile), giving an elimination reaction:

Benzyne

This generates a very strange-looking (and very reactive) intermediate, which we will call benzyne. Then, another hydroxide ion is involved, this time acting as a nucleophile to attack benzyne. But there are two places it can attack:

Hydroxide can attack like this

or like this

And there is no reason to prefer one site over the other, so we must assume that these two pathways occur with equal probability. That would give a 50–50 mixture of the two anions above, which would undergo proton transfers to generate phenol, with the isotopic labels in the appropriate locations:

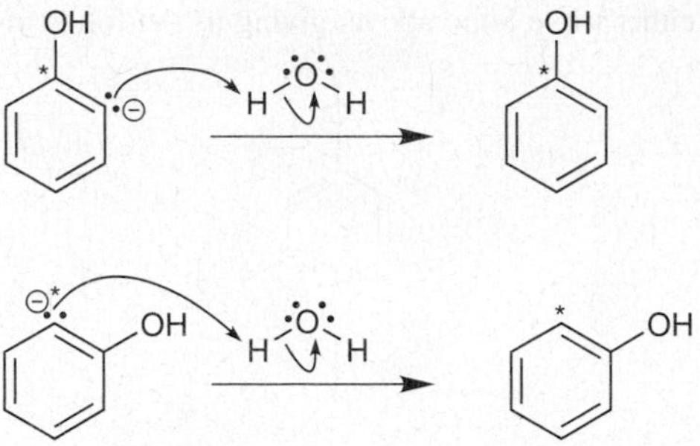

Just as we saw in the previous section, phenol has an acidic proton and will therefore undergo deprotonation as a result of the basic reaction condition (hydroxide). So, just as we saw in the previous section, a proton source (such as H_3O^+) must be introduced into the reaction flask after the reaction is complete.

This proposed mechanism is essentially an elimination followed by an addition. So, it makes sense that we call this process an *elimination-addition* reaction (as compared with the S_NAr mechanism, which was called *addition-elimination*). This mechanism certainly seems a bit off the wall when you think about. Benzyne? It looks like a terrible intermediate. But chemists have been able to show (with other experiments) that benzyne is in fact the intermediate of this reaction. Your textbook or instructor will most likely provide the additional evidence for the short-lived existence of benzyne (we use a trapping technique involving a Diels-Alder reaction). If you are curious about the evidence, you can look in your textbook. For now, let's make sure that we can predict the products of elimination-addition reactions.

EXERCISE 4.12 Predict the products for the following reaction:

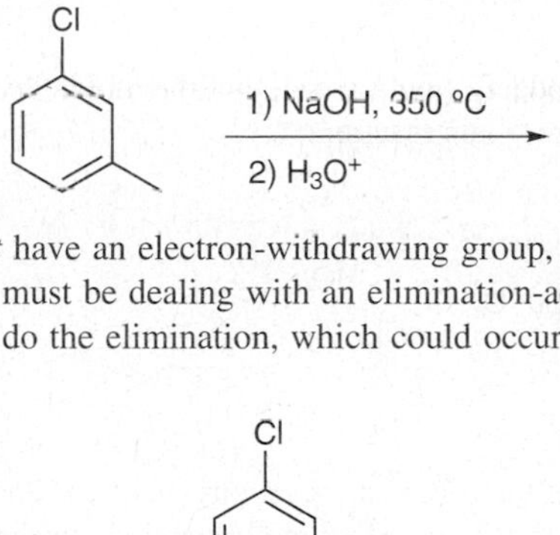

Answer The ring does *not* have an electron-withdrawing group, so we are not dealing with an S_NAr mechanism. Rather, we must be dealing with an elimination-addition mechanism.

The first step would be to do the elimination, which could occur on either side of the chlorine atom:

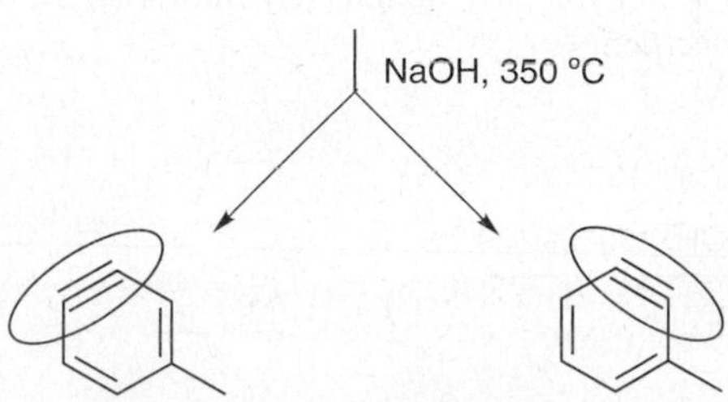

Then, we can add across either triple bond above, giving us the following products:

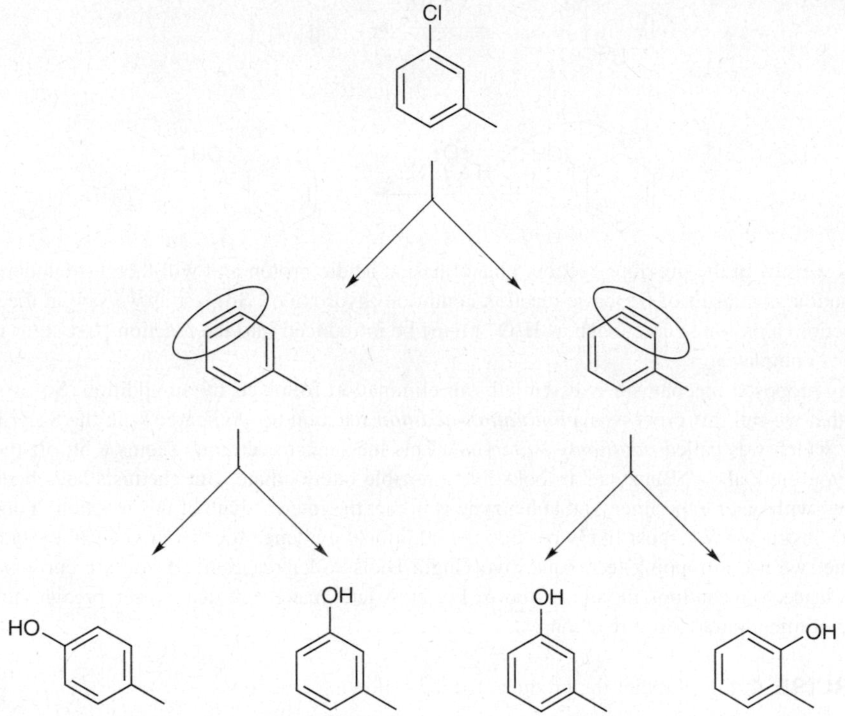

If you look closely at these products, you will see that the middle two are the same. So, we expect the following three products from this reaction:

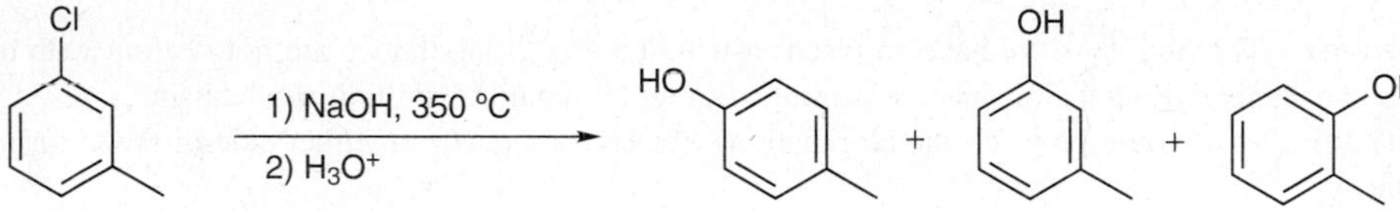

Predict the products for each of the following reactions. Just to keep you on your toes, I will throw in some problems that go through an addition-elimination mechanism (S$_N$Ar), rather than an elimination-addition. In each case, you will have to decide which mechanism is responsible for the reaction (based on whether or not you have all three criteria for an S$_N$Ar mechanism). Your products will be based on that decision.

4.13

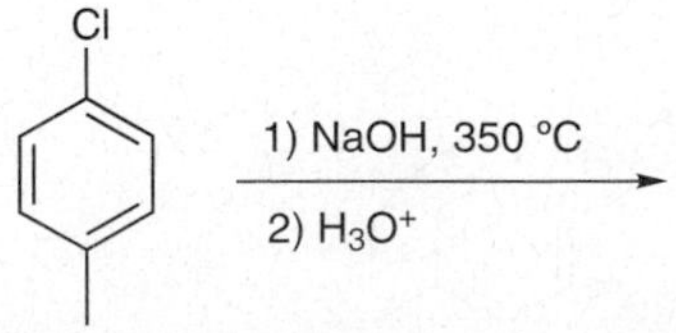

4.14 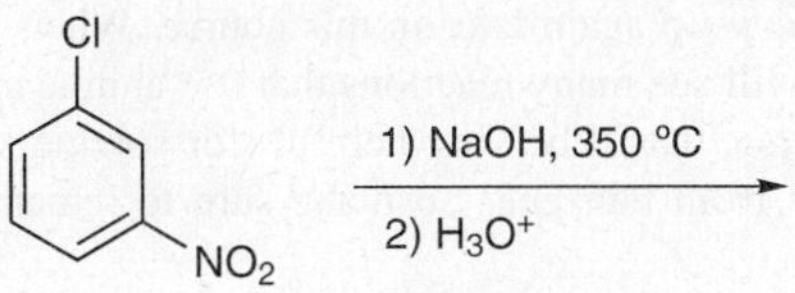
Cl
NO2
1) NaOH, 350 °C
2) H3O+

4.15 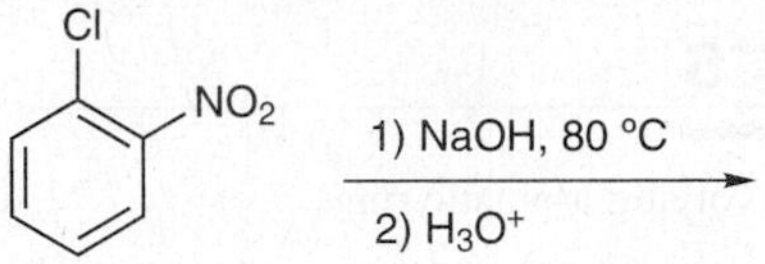
Cl
NO2
1) NaOH, 80 °C
2) H3O+

4.16 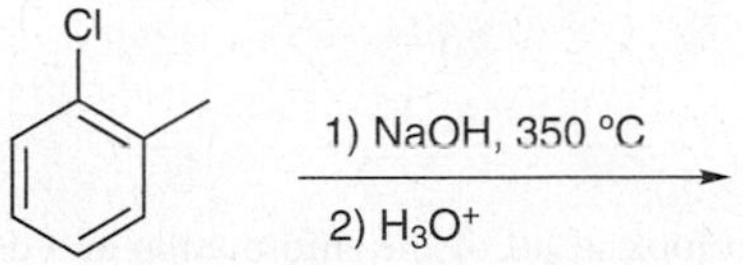
Cl
1) NaOH, 350 °C
2) H3O+

4.17 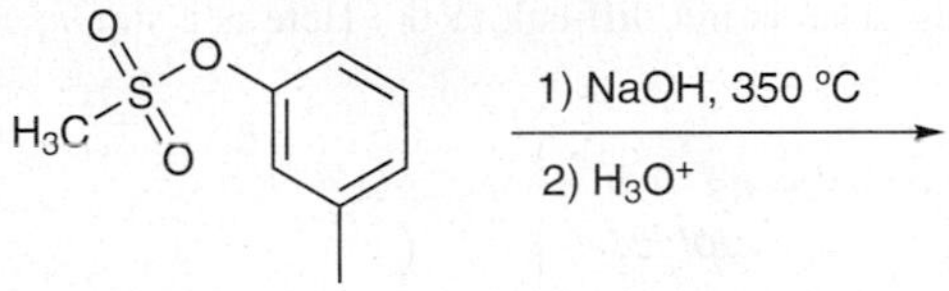
O
O
H3C
S
O
1) NaOH, 350 °C
2) H3O+

4.18 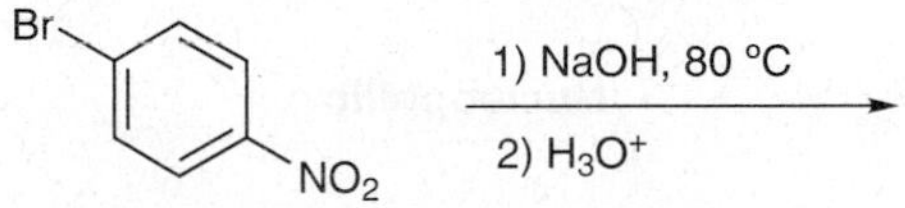
Br
NO2
1) NaOH, 80 °C
2) H3O+

In this section, we have seen how to install an OH group or an NH_2 group on an aromatic ring. This is important because we did not see how to achieve either of these two transformations in the previous chapter. Here is a summary of how to install an OH group or NH_2 group on an aromatic ring. In each case, it is a two-step process that starts with installing a Cl on the ring:

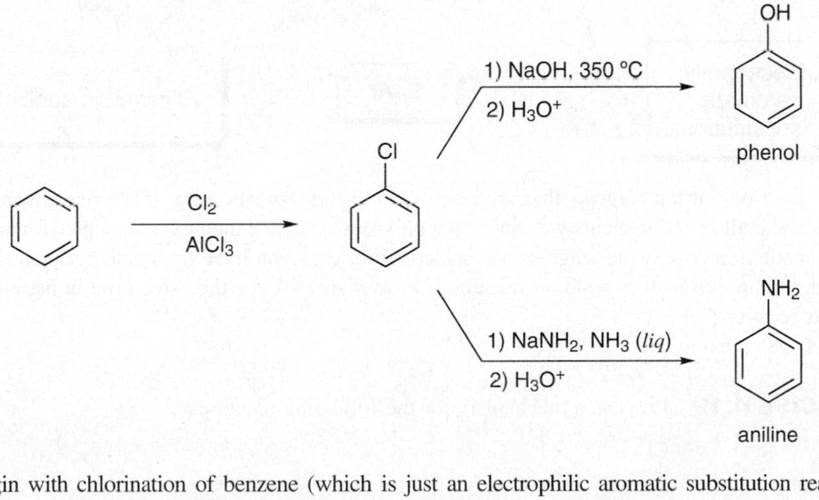
Cl2
AlCl3
Cl
1) NaOH, 350 °C
2) H3O+
OH
phenol
1) NaNH2, NH3 (liq)
2) H3O+
NH2
aniline

We begin with chlorination of benzene (which is just an electrophilic aromatic substitution reaction), followed by an elimination-addition reaction. When we perform the elimination-addition process, we must carefully choose the reagents. If we use NaOH followed by H_3O^+, the product will be phenol. If we use $NaNH_2$ followed by H_3O^+, the product will be aniline (shown in the scheme above).

The synthesis of aniline, shown above, will show up again later in this course. When we learn about the chemistry of amines (in Chapter 8), we will see many reactions that use aniline as a starting material. When we get to that part of the course, it will be very helpful (for solving synthesis problems) if you remember how to make aniline from benzene. So make sure to remember this reaction. You will definitely see it again later.

4.4 MECHANISM STRATEGIES

So far, we have seen three different mechanisms involving aromatic rings:

1. Electrophilic aromatic substitution
2. S_NAr (sometimes called addition-elimination)
3. Elimination-addition

When you are given a problem, you must be able to look at all of the information and determine which of the three mechanisms is operating. This is not difficult to do. Here is a simple chart that shows the thought processes involved:

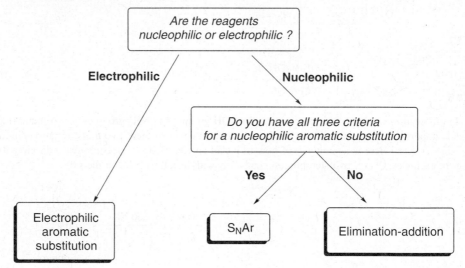

You first look at the reagents that are reacting with the aromatic ring. If the reagents are electrophilic (like all of the reagents we saw in the previous chapter), then expect an electrophilic aromatic substitution. But if the reagents are nucleophilic, then you have to decide between an S_NAr reaction and an elimination-addition reaction. To do that, look for the three criteria necessary for an S_NAr reaction.

Let's see an example:

EXERCISE 4.19 Propose a mechanism for the following reaction:

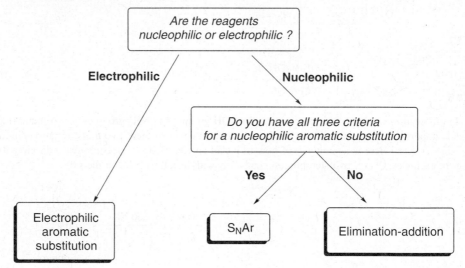

Answer   We begin by looking at the reagents. Hydroxide is a nucleophile, so we do NOT get an electrophilic aromatic substitution. We must decide between an S_NAr mechanism and an elimination-addition mechanism. We look for the three criteria that we need for an S_NAr reaction. (1) We do have an electron-withdrawing group, and (2) we do have a leaving group, BUT (3) the electron-withdrawing group and the leaving group are NOT *ortho* or *para* to each other. That means that an S_NAr mechanism cannot occur. (When the electron-withdrawing group and the leaving group are *meta* to each other, we don't have the "reservoir" to use.) Therefore, the mechanism must be an elimination-addition:

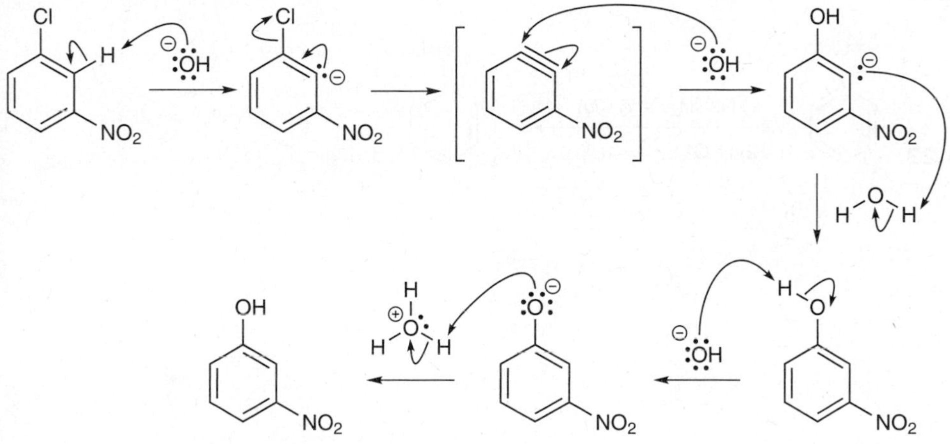

In this particular example, it is true that we would expect a total of three products from an elimination-addition mechanism:

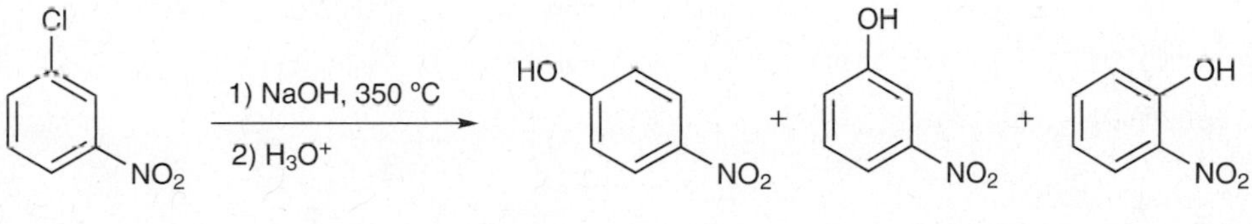

But keep in mind that mechanism problems do not always show you all of the products. The problem will typically show you just one product, and you will need to show the mechanism for forming that product (and only that product). In some cases, it might even be a minor product. But the problem is not making any claims that the product is major or minor. A mechanism problem is simply asking you to justify "how" the product was formed, regardless of how much of it was actually obtained from the reaction.

Propose a mechanism for each of the following reactions:

4.20

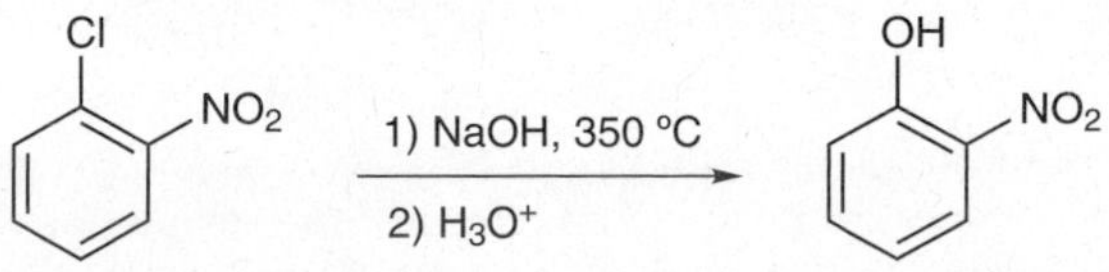

4.21

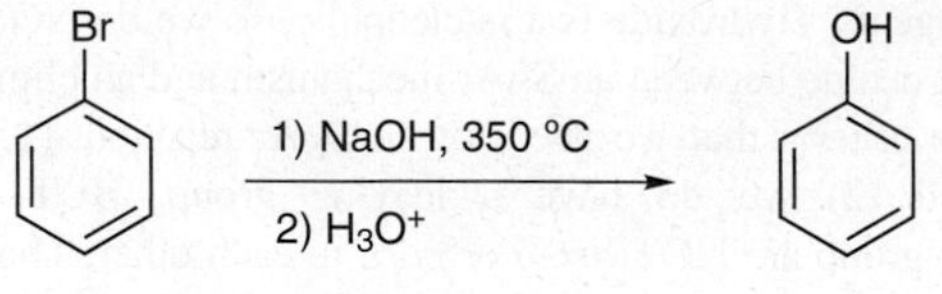
Br
1) NaOH, 350 °C
2) H3O+
OH

4.22

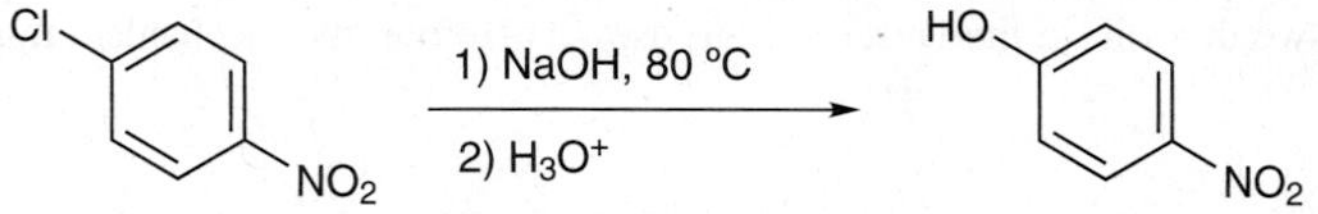
Cl
NO2
1) NaOH, 80 °C
2) H3O+
HO
NO2

4.23

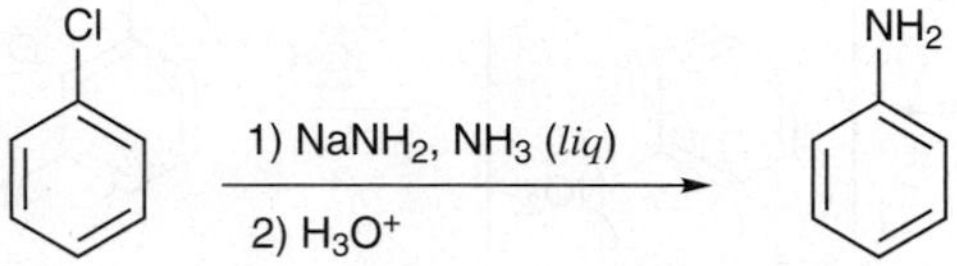
Cl
1) NaNH2, NH3 (liq)
2) H3O+
NH2

KETONES AND ALDEHYDES

5.1 PREPARATION OF KETONES AND ALDEHYDES

Before we can explore the reactions of ketones and aldehydes, we must first make sure that we know how *to make* ketones and aldehydes. That information will be vital for solving synthesis problems.

Ketones and aldehydes can be made in many ways, as you will see in your textbook. In this book, we will only see a few of these methods. These few reactions should be sufficient to help you solve many synthesis problems in which a ketone or aldehyde must be prepared.

The most useful type of transformation is forming a C=O bond from an alcohol. *Primary* alcohols can be oxidized to form aldehydes:

And *secondary* alcohols can be oxidized to form ketones:

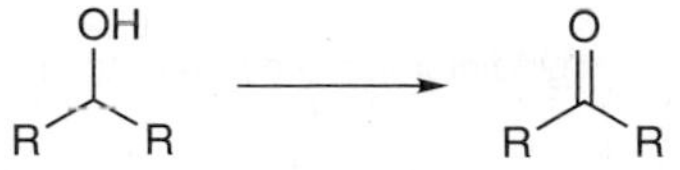

Tertiary alcohols cannot be oxidized, because carbon cannot form five bonds:

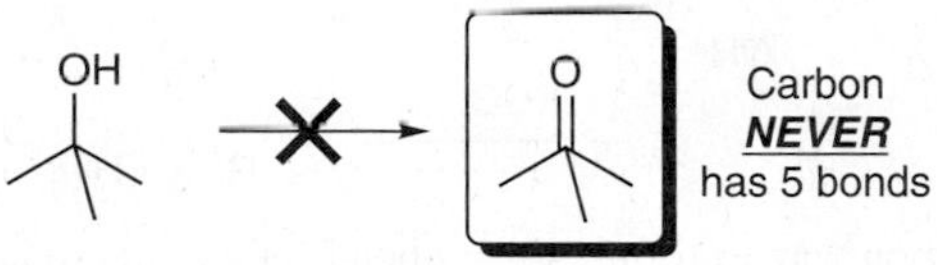

So, we need to be familiar with the reagents that will oxidize primary and secondary alcohols (to form aldehydes or ketones, respectively). Let's start with secondary alcohols.

A secondary alcohol can be converted into a ketone upon treatment with sodium dichromate and sulfuric acid:

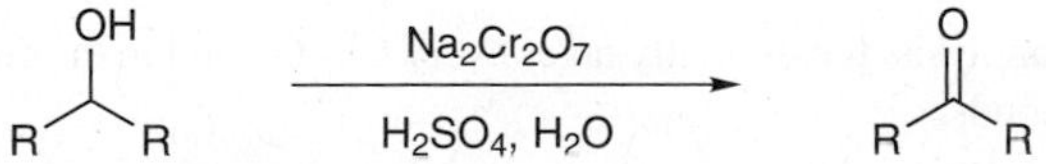

Alternatively, the Jones reagent can be used, which is formed from CrO_3 in aqueous acetone:

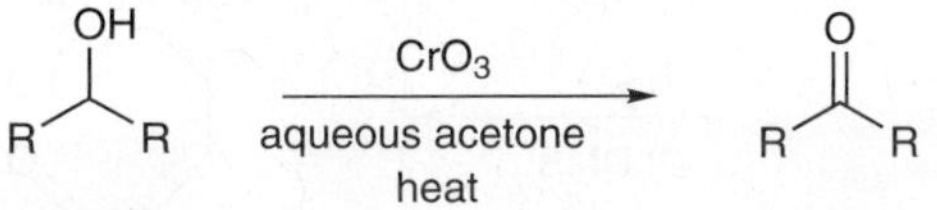

Whether you use sodium dichromate or the Jones reagent, you are essentially performing an oxidation that involves a chromium oxidizing agent (the alcohol is being oxidized and the chromium reagent is being reduced). You should look through your lecture notes and textbook to see if you are responsible for the mechanisms of these oxidation reactions. Whatever the case, you should definitely have these reagents at your fingertips, because you will encounter many synthesis problems that require the conversion of an alcohol into a ketone or aldehyde.

Chromium oxidations work well for secondary alcohols, but we run into a problem when we try to perform a chromium oxidation on a primary alcohol. The initial product is indeed an aldehyde:

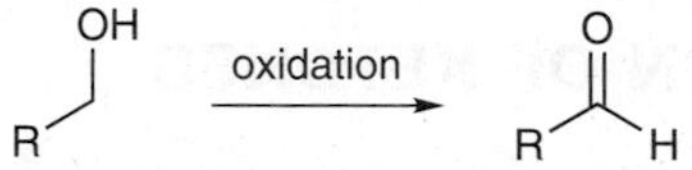

But under these strong oxidizing conditions, the aldehyde does not survive. The aldehyde is further oxidized to give a carboxylic acid:

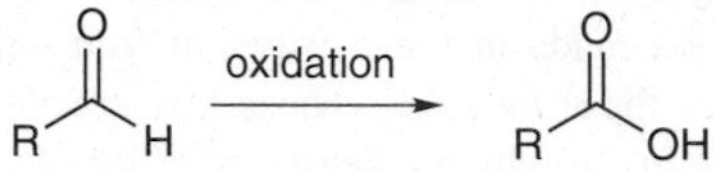

So clearly, we need a way to oxidize a primary alcohol into an aldehyde, under conditions that will *not* further oxidize the aldehyde. This can be accomplished with a reagent called pyridinium chlorochromate (or PCC):

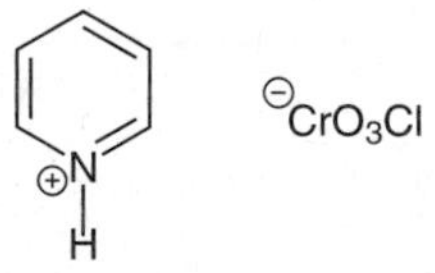

pyridinium chlorochromate
(PCC)

This reagent provides milder oxidizing conditions, and therefore, the reaction stops at the aldehyde. That is, PCC will oxidize a primary alcohol to give an aldehyde:

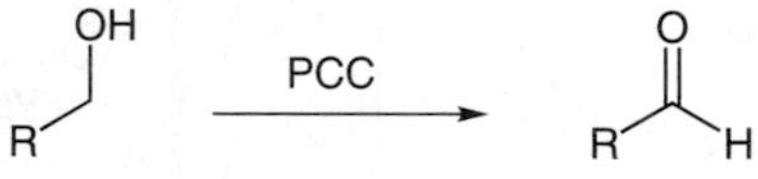

There is another common way to form a C=O bond (other than oxidation of an alcohol). You might remember the following reaction from last semester:

This reaction is called ozonolysis. It essentially takes every C=C bond in the compound, and breaks it apart into two C=O bonds:

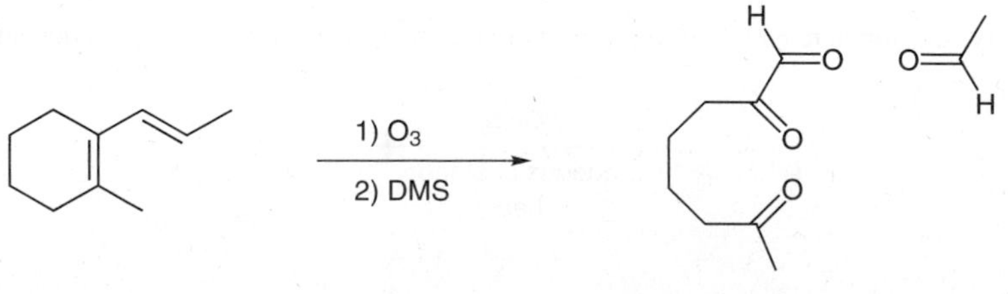

There are many reagents that can be used for the second step of this process (other than DMS). You should look in your lecture notes to see what reagents your instructor (or textbook) used for step 2 of an ozonolysis.

So far, this section has covered only a few ways to make a C=O bond. We saw that ketones can be made by treating a secondary alcohol with sodium dichromate (or the Jones reagent), and aldehydes can be made by treating a primary alcohol with PCC. We also saw that ketones and aldehydes can be made via ozonolysis. Let's get some practice with these reactions.

EXERCISE 5.1 Predict the major product of the following reaction:

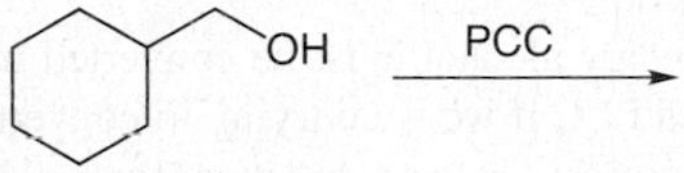
OH
PCC

Answer The oxidizing agent in this case is PCC, and we have seen that PCC will convert a primary alcohol into an aldehyde:

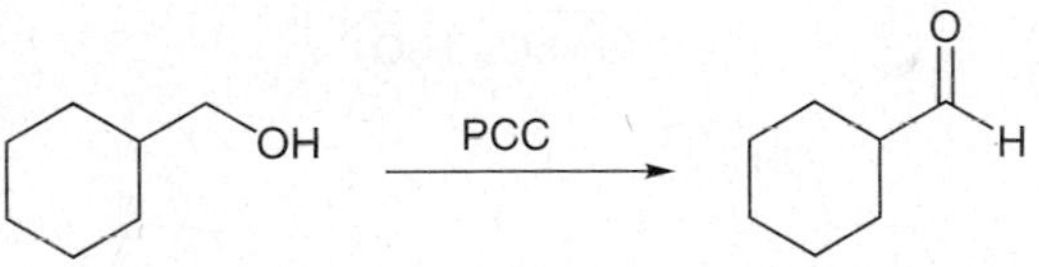
OH
PCC
O
H

PROBLEMS Predict the major product for each of the following reactions:

5.2

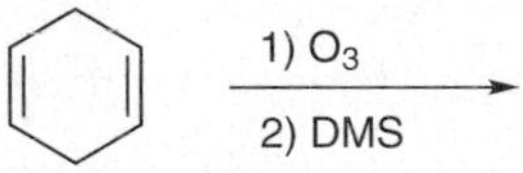
1) O_3
2) DMS

5.3

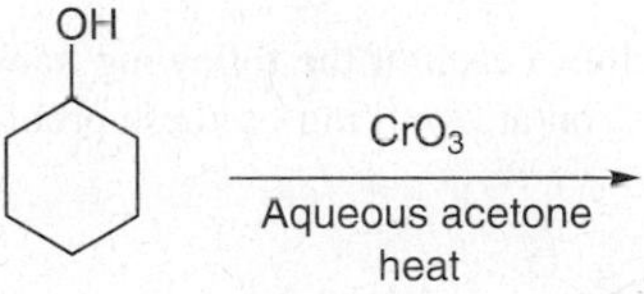
OH
CrO_3
Aqueous acetone
heat

5.4

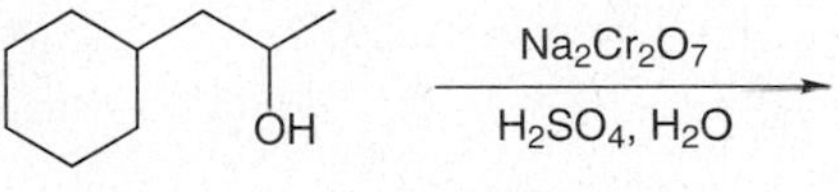
OH
$Na_2Cr_2O_7$
H_2SO_4, H_2O

5.5

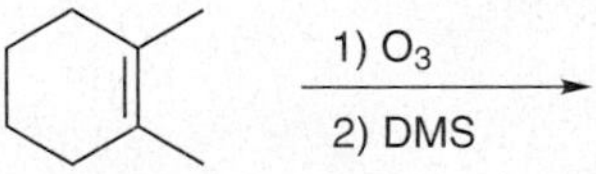
1) O_3
2) DMS

5.6

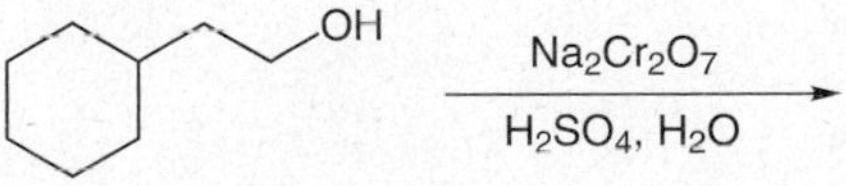
OH
$Na_2Cr_2O_7$
H_2SO_4, H_2O

5.7

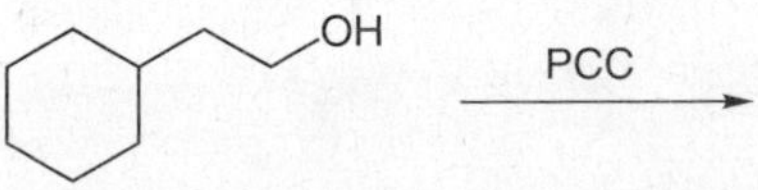
OH
PCC

It is not enough to simply "recognize" the reagents when you see them (like we did in the previous problems). But you actually need to know the reagents well enough to write them down when they are not in front of you. Let's get some practice:

EXERCISE 5.8 Identify the reagents you would use to achieve the following transformation:

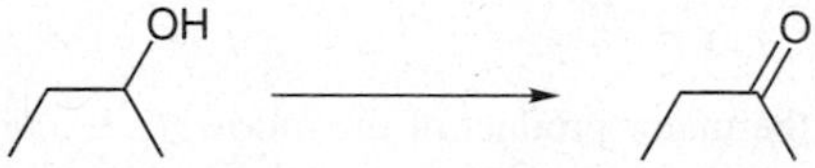
OH
O

Answer In this case, a secondary alcohol must be converted into a ketone. So, we don't need to use PCC. We would only need PCC if we were trying to convert a primary alcohol into an aldehyde. In this case, PCC is unnecessary. Instead, we would use either sodium dichromate and sulfuric acid or the Jones reagent:

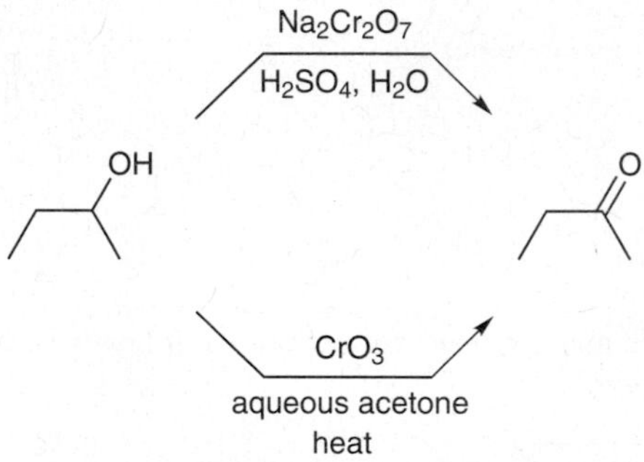
Na2Cr2O7
H2SO4, H2O
OH
O
CrO3
aqueous acetone
heat

PROBLEMS Identify the reagents you would use to achieve each of the following transformations. Try not to look back at the previous problems while you are working on these problems.

5.9

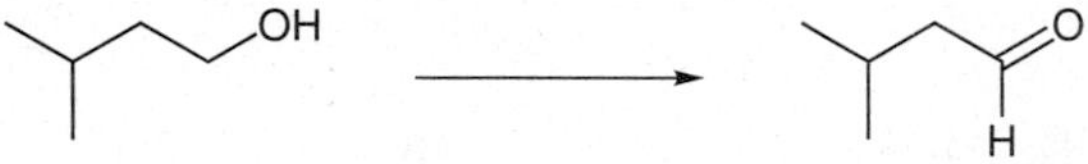
OH
O
H

5.10

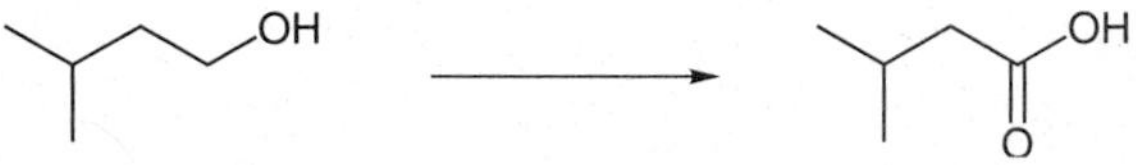
OH
OH
O

5.11

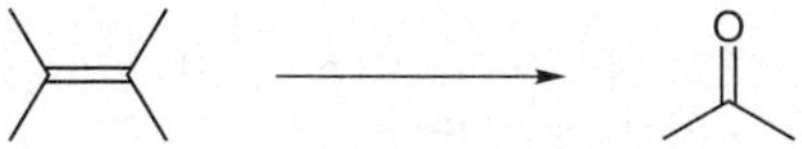
O

5.12

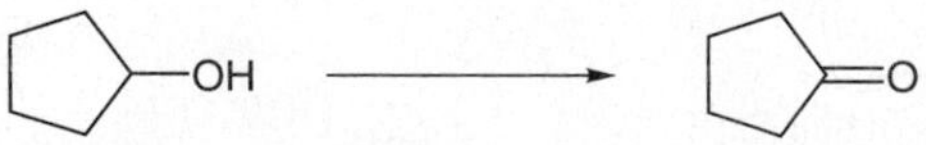
OH
O

5.2 STABILITY AND REACTIVITY OF C=O BONDS

Ketones and aldehydes are very similar to each other in structure:

ketone   aldehyde

Therefore, they are also very similar to each other in terms of reactivity. Most of the reactions that we see in this chapter will work for both ketones and aldehydes. So, it makes sense to learn about ketones and aldehydes in the same breath.

But before we can get started, we need to know some basics about C=O bonds. Let's start with a bit of terminology that we will use throughout the entire chapter. Instead of constantly using the expression "C=O double bond," we will call it a *carbonyl group*. This term is NOT used for nomenclature. You will never see the term "carbonyl" appearing in the IUPAC name of a compound. Rather, it is just a term that we use when we are talking about mechanisms, so that we can quickly refer to the C=O bond without having to say "C=O double bond" all of the time.

Don't confuse the term "carbonyl" with the term "acyl." The term "acyl" is used to refer to a carbonyl group *together with* one alkyl group:

carbonyl   acyl

We will use the term "acyl" in the next chapter. But in this chapter, we will focus on the carbonyl group.

If we want to know how a carbonyl group will react, we must first consider electronic effects (the locations of $\delta+$ and $\delta-$). There are always two factors to explore: induction and resonance. If we start with induction, we notice that oxygen is more electronegative than carbon, and therefore, the oxygen atom will withdraw electron density:

As a result, the carbon atom of the carbonyl group is $\delta+$ and the oxygen atom is $\delta-$.

Next, we look at resonance:

And we see, once again, that the carbon atom is $\delta+$ and the oxygen atom is $\delta-$, this time because of resonance. So, both induction and resonance paint the same picture:

This means that the carbon atom is very electrophilic, and the oxygen atom is very nucleophilic. While there are many reactions involving the oxygen atom functioning as a nucleophile, you probably won't see any of those reactions in your organic chemistry course. Accordingly, we will focus all of our attention in this chapter on the carbon atom of a carbonyl group. We will see *how* the carbon atom functions as an electrophile, *when* it functions as an electrophile, and *what happens after* it functions as an electrophile.

The geometry of a carbonyl group facilitates the carbon atom functioning as an electrophile. We saw in the first semester of organic chemistry that sp^2-hybridized carbon atoms have trigonal planar geometry:

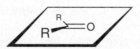

This makes it easy for a nucleophile to attack the carbonyl group, because there is no steric hindrance that would block the incoming nucleophile:

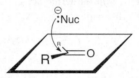

In this chapter, we will see many different kinds of nucleophiles that can attack a carbonyl group. In fact, this entire chapter will be organized based on the kinds of nucleophiles that can attack. We will start with hydrogen nucleophiles and continue with oxygen nucleophiles, sulfur nucleophiles, nitrogen nucleophiles, and, finally, carbon nucleophiles. This approach (dividing the chapter based on the kinds of nucleophiles) might be somewhat different than your textbook. But hopefully, the order that we use here will help you appreciate the similarity between the reactions.

There is one more feature of carbonyl groups that must be mentioned before we can get started. Carbonyl groups are thermodynamically very stable. In other words, forming a carbonyl group is generally a process that is downhill in energy. On the flipside, converting a C=O bond into a C—O bond is generally a process that is uphill in energy. As a result, the formation of a carbonyl group is often the driving force for a reaction. We will use that argument many times in this chapter, so make sure you are prepared for it. The mechanisms in this chapter will be explained in terms of the stability of carbonyl groups.

Now let's just quickly summarize the important characteristics that we have seen so far. The carbon atom (of a carbonyl group) is electrophilic, and it is readily attacked by a nucleophile (and there are MANY different kinds of nucleophiles that can attack it). We have also seen that a carbonyl group is very stable. So, the formation of a carbonyl group can serve as a driving force.

These principles will guide us throughout the rest of the chapter, and they can be summarized like this:

- A carbonyl group can be attacked by a nucleophile, and
- After a carbonyl group is attacked, it will try to re-form, if possible.

5.3 H-NUCLEOPHILES

We will now explore the various nucleophiles that can attack ketones and aldehydes. We will divide all nucleophiles into categories, and in this section, we will focus on hydrogen nucleophiles. I call them "hydrogen" nucleophiles, because they are a source of a negatively charged hydrogen atom (which we call a "hydride" ion) that can attack a ketone or aldehyde. The simplest way to get a hydride ion is from sodium hydride (NaH). This compound is ionic, so it is composed of Na^+ and H^- ions (very much the way NaCl is composed of Na^+ and Cl^- ions). So, NaH is certainly a good source of hydride ions.

However, you will not see any reactions where we use NaH as a source of hydride *nucleophiles*. As it turns out, NaH is a very strong base, but it is not a strong nucleophile. This is an excellent example of how basicity and nucleophilicity do NOT completely parallel each other. The reason for this goes back to something from the first semester of organic chemistry. Try to remember back to the difference between basicity and nucleophilicity. Let's review it real quickly.

The strength of a base is determined by the *stability* of the negative charge. An unstable negative charge corresponds with a strong base, while a stabilized negative charge corresponds with a weak base. But nucleophilicity is NOT based on stability. Nucleophilicity is based on *polarizability*. Polarizability describes the ability of an atom or molecule to distribute its electron density unevenly in response to external influences. Larger atoms are more polarizable, and are therefore strong nucleophiles; while smaller atoms are less polarizable, and are therefore weak nucleophiles.

With that in mind, we can understand why H^- is a strong base, but not such a strong nucleophile. It is a strong base, because hydrogen does not stabilize the charge well. But when we consider the nucleophilicity of H^-, we must look at the polarizability of the hydrogen atom. Hydrogen is the smallest atom, and therefore, it is the least polarizable. Therefore, H^- is generally not observed to function as a nucleophile.

Now we can understand why we don't use NaH as a source for a hydrogen nucleophile. It is true that it is an excellent base, and you will see NaH used several times this semester. But it will always be used as a strong *base*; never as a *nucleophile*. So, how do we form a hydrogen nucleophile?

Although H^- itself cannot be used as a nucleophile, there are many reagents that can serve as a "delivery agent" of H^-. For example, consider the structure of sodium borohydride (NaBH$_4$):

$$\overset{\oplus}{Na} \qquad H-\overset{\overset{H}{|}}{\underset{\underset{H}{|}}{B}}{}^{\ominus}-H$$

If we look at the periodic table, we see that boron is in Column 3A, and therefore, it has three valence electrons. Accordingly, it can form three bonds. But in sodium borohydride (above), the central boron atom has *four* bonds. So it must be using one extra electron, and therefore, it has a negative formal charge (you can ignore the sodium ion, Na^+, because it is just the counter ion). This reagent can serve as a delivery agent of H^-, as seen in the following example:

Notice that H⁻ never really exists by itself in this reaction. Rather, H⁻ is "delivered" from one place to another. That is a good thing, because H⁻ by itself would not serve as a nucleophile (as we saw earlier). But sodium borohydride can serve as a source of a hydrogen nucleophile, because the central boron atom is somewhat polarizable. The polarizability of the boron atom allows the entire compound to serve as a nucleophile, and *deliver* a hydride ion to attack the ketone. Now, it is true that boron is not so large, and therefore, it is not very polarizable. As a result, $NaBH_4$ is a somewhat tame nucleophile. In fact, we will soon see that $NaBH_4$ is selective in what it reacts with. It will not react with all carbonyl groups (for example, it will not react with an ester). But it will react with ketones *and* with aldehydes (and that is our focus in this chapter).

There is another common reagent that is very similar to sodium borohydride, but it is much more reactive. This reagent is called lithium aluminum hydride ($LiAlH_4$, or even just LAH):

$$Li^{\oplus} \quad \overset{\overset{\displaystyle H}{|}}{\underset{\underset{\displaystyle H}{|}}{H-\overset{\ominus}{Al}-H}}$$

This reagent is very similar to $NaBH_4$ because aluminum is also in Column 3A of the periodic table (directly beneath boron). So, it also has three valence electrons. In the structure above, the aluminum atom has four bonds, which is why it has a negative charge. Just as we saw with $NaBH_4$, LAH is also a source of nucleophilic H⁻. But compare these two reagents to each other—aluminum is larger than boron. That means that it is more polarizable, and therefore, LAH is a much better nucleophile than $NaBH_4$. LAH will react with almost any carbonyl group (not just ketones and aldehydes).

It will soon become very important that LAH is more reactive than $NaBH_4$. But for now, we are talking about nucleophilic attack of ketones and aldehydes; and both $NaBH_4$ and LAH will react with ketones and aldehydes.

In addition to $NaBH_4$ and LAH, there are other sources of hydrogen nucleophiles as well, but these two are the most common reagents. You should look through your textbook and lecture notes to see if you are responsible for being familiar with any other hydrogen nucleophiles.

Now let's take a close look at what can happen after a hydrogen nucleophile attacks a carbonyl group. As we have seen, the reagent (either $NaBH_4$ or LAH) can deliver a hydride ion to the carbonyl group, like this:

In the beginning of this chapter, we covered two important rules that govern the behavior of a carbonyl group:

- it is easily attacked by nucleophiles (as we just saw in the step above), and
- after a carbonyl group is attacked, it will try to re-form, if possible. Now we need to understand what we mean when we say: "if possible."

In trying to re-form the carbonyl group, we realize that the central carbon atom cannot form a fifth bond:

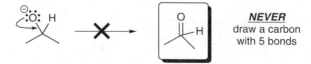

NEVER
draw a carbon
with 5 bonds

That would be impossible, because carbon only has four orbitals to use. So, in order for the carbonyl group to re-form, a leaving group must be expelled, like this:

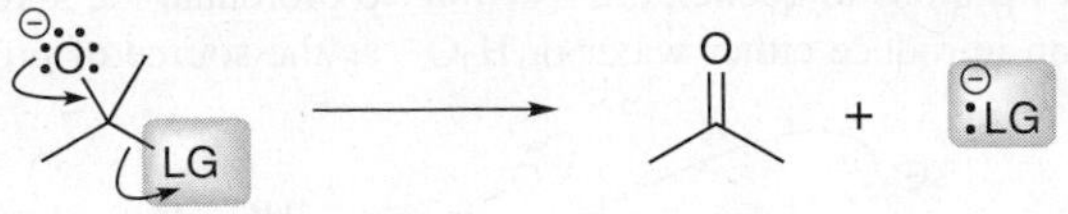

So we just need to know what groups can function as leaving groups. Fortunately, there is one simple rule that can guide you: NEVER expel H$^-$ or C$^-$ (there are a few exceptions to this rule, which we will see later, but unless you recognize that you are dealing with one of the rare exceptions, do NOT expel H$^-$ or C$^-$). For example, never do this:

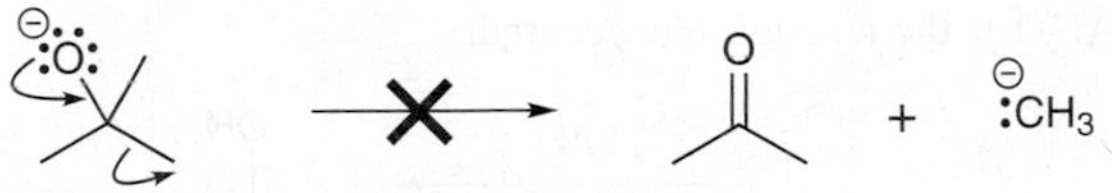

And never do this:

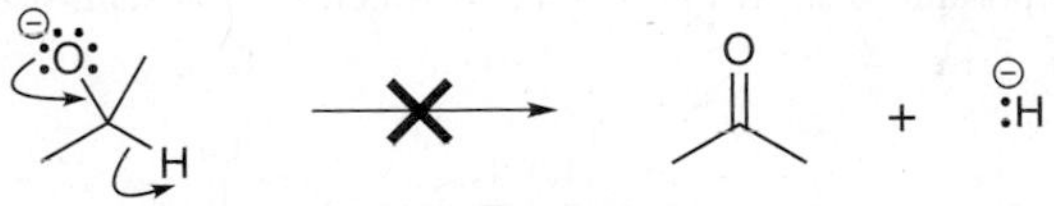

We have just learned a simple general rule. Now let's try to apply this rule to determine the outcome that is expected when a ketone or aldehyde is treated with a hydrogen nucleophile. Once again, the first step was for the hydrogen nucleophile to attack the carbonyl group:

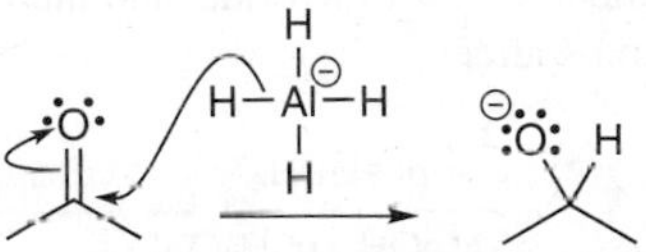

Now let's consider what can possibly happen next. In order for the carbonyl group to re-form, a leaving group must be expelled. But there are no leaving groups in this case. The carbonyl cannot re-form by expelling C$^-$:

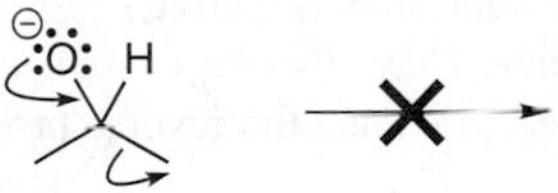

And it cannot re-form by expelling H$^-$:

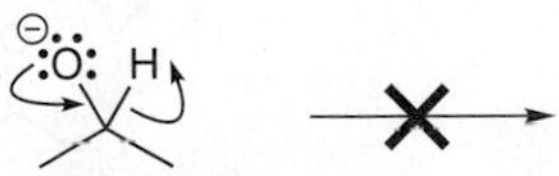

And it cannot re-form by expelling C$^-$:

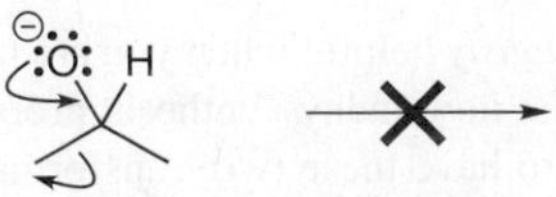

So we are stuck. Once a hydrogen nucleophile delivers H⁻ to the carbonyl group, then it will not be possible for the carbonyl group to re-form. So the reaction is complete, and it just waits for us to introduce a source of protons to quench the reaction (to protonate the alkoxide ion). To achieve this protonation, we can introduce either water or H_3O^+ as the source of protons:

Regardless of the identity of the proton source that we add to the reaction flask after the reaction is complete, the product of this reaction will be an alcohol.

Whenever you are using this transformation in a synthesis, you must clearly show that the proton source is added AFTER the reaction has occured:

In other words, it is important to show that LAH and water are **two separate steps**. Do **not** show it like this:

This would mean that LAH and H_2O are present at the same time, and that is not possible. LAH would react violently with water to form H_2 gas (because H⁺ and H⁻ would react with each other).

As it turns out, $NaBH_4$ is a milder source of hydride, and therefore, $NaBH_4$ can actually be present at the same time as the proton source:

Common proton sources include MeOH and water (sometimes you might see EtOH). Notice that we didn't show it as two separate steps. When you are dealing with LAH, you must show two steps (one step for LAH and another step for the proton source); but when you are dealing with $NaBH_4$, you should show the proton source in the same step as $NaBH_4$.

LAH and $NaBH_4$ are very useful reagents. They allow us to *reduce* a ketone or aldehyde, which is important when you realize that we already learned the reverse process:

Oxidation

Reduction

These two transformations will be *tremendously* helpful when you are trying to solve synthesis problems later on. You would be surprised just how many synthesis problems involve the conversion between alcohols and ketones. You need to have these two transformations at your fingertips.

EXERCISE 5.13 Predict the major product of the following reaction:

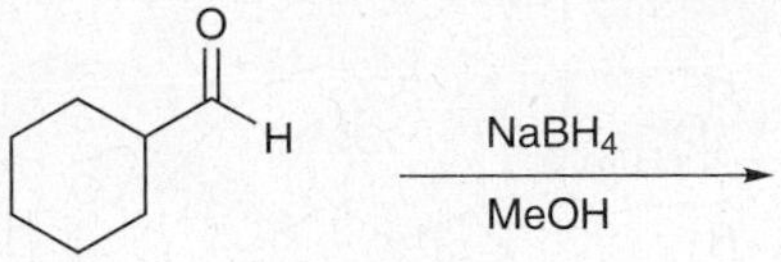
O
H
$NaBH_4$
MeOH

Answer The starting compound is an aldehyde, and it is being treated with sodium borohydride. This hydrogen nucleophile will *deliver* H^- to the aldehyde, and the carbonyl group will not be able to re-form, because there is no leaving group. In this case, methanol serves as the proton source, and an alcohol will be obtained:

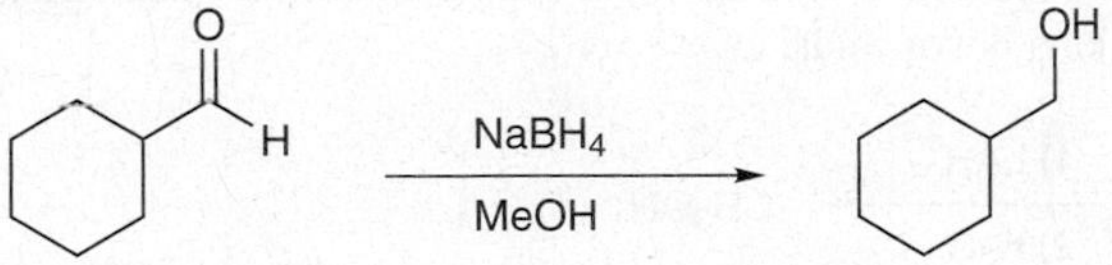
O
H
$NaBH_4$
MeOH
OH

PROBLEMS Predict the major product for each of the following reactions:

5.14

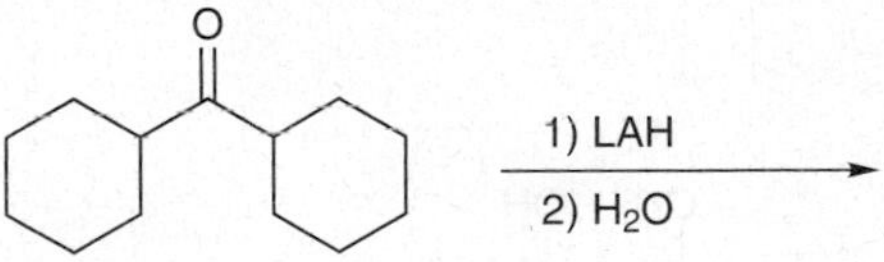
O
1) LAH
2) H_2O

5.15

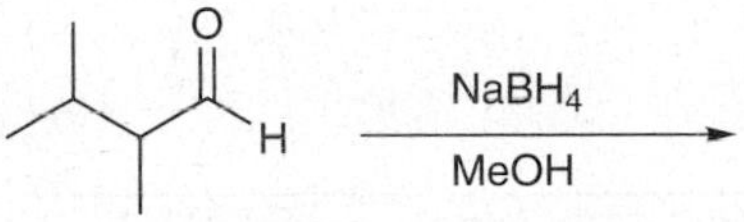
O
H
$NaBH_4$
MeOH

5.16

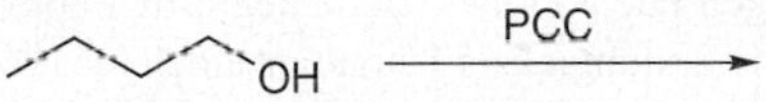
OH
PCC

5.17

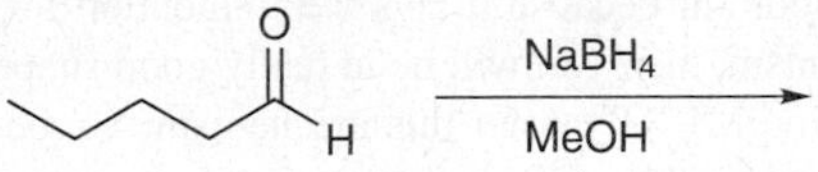
O
H
$NaBH_4$
MeOH

5.18

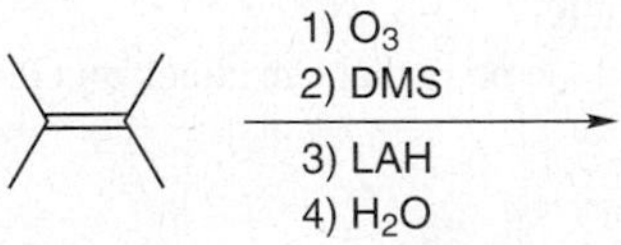
1) O_3
2) DMS
3) LAH
4) H_2O

EXERCISE 5.19 Draw a mechanism for the following transformation:

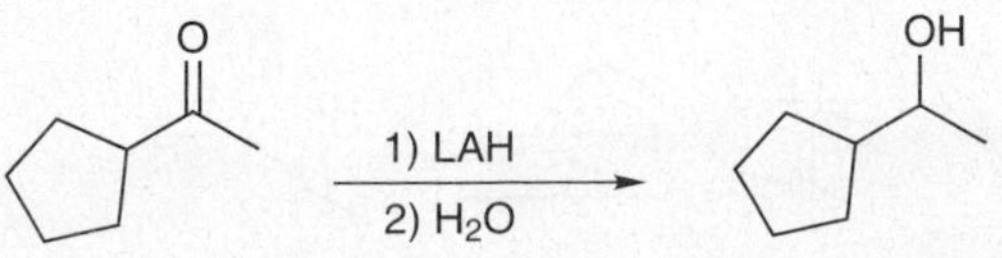
O
1) LAH
2) H_2O
OH

Answer First, LAH delivers a hydride ion to the ketone. Then, the carbonyl group is not able to re-form, so the intermediate waits for a proton from water, in the next step:

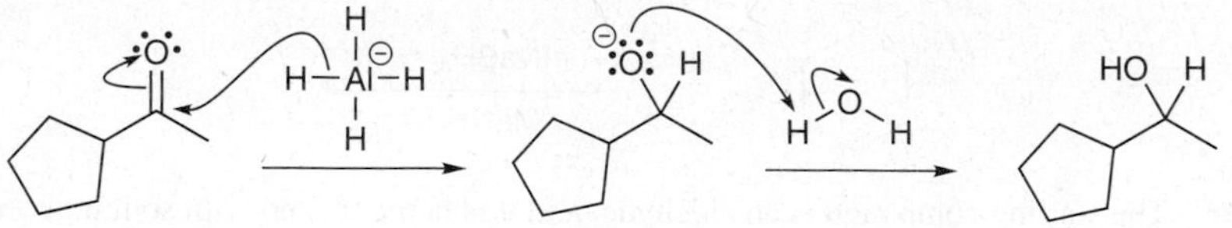

PROBLEMS Propose a mechanism for each of the following transformations. The following problems will probably seem too easy—but just do them anyway. These basic arrows need to become *routine* for you, because we will step up the complexity in the next section, and you will want to have these basic skills down cold:

5.20 H—(C=O)—H → 1) LAH 2) H$_2$O → CH$_3$OH

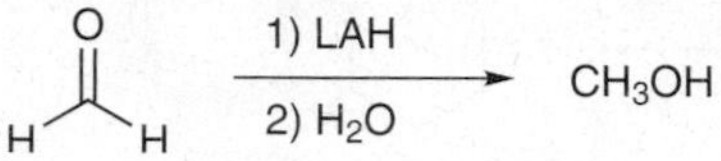

5.21 → NaBH$_4$ MeOH → OH

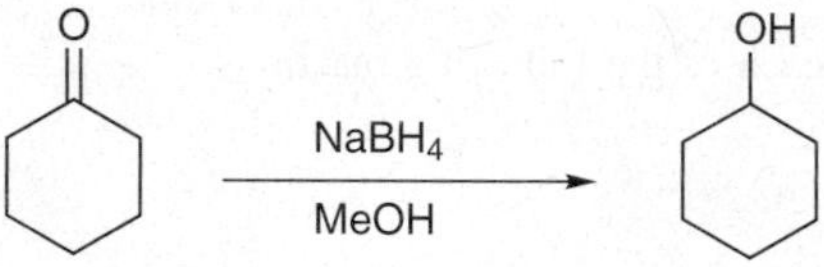

5.22 → 1) excess LAH 2) excess H$_2$O → OH OH

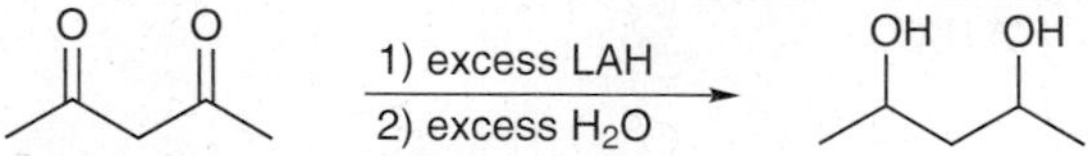

5.4 O-NUCLEOPHILES

In this section, we will focus our attention on oxygen nucleophiles. Let's begin by exploring what happens when an alcohol functions as a nucleophile and attacks a ketone or aldehyde.

Be warned: the mechanism we are about to see is one of the longer mechanisms that you will encounter in this course. But it is incredibly important because it lays the foundation for so many other mechanisms. If you can master this mechanism, then you will be in really good shape to move on. And to be honest, there is no other option; you MUST master this mechanism. So, be prepared to read through the next several pages slowly, and then be prepared to reread those pages as many times as necessary until you know this mechanism intimately.

Alcohols are nucleophilic because the oxygen atom has lone pairs that can attack an electrophile:

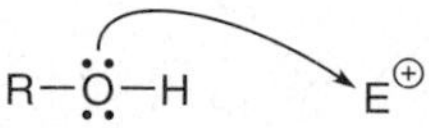

$$R-\ddot{O}-H \qquad E^{\oplus}$$

When an alcohol attacks a carbonyl group, an intermediate is generated that should remind us of the intermediate that was formed in the previous section:

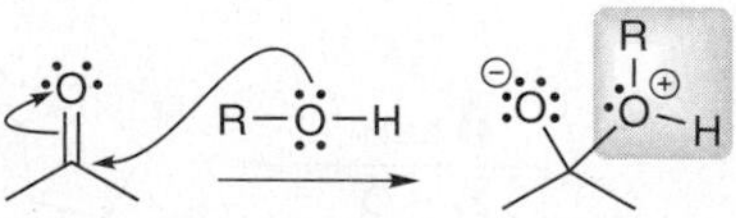

Notice how similar this is to the hydride attack we explored in the previous section:

But there is one major difference here. When we saw the attack of a hydrogen nucleophile in the previous section, we argued that the carbonyl group could not re-form after the attack because there was no leaving group. But here, in this section (with an alcohol functioning as a nucleophile), there is a leaving group. So, it *is* possible for the carbonyl group to re-form:

The attacking nucleophile (ROH) can function as the leaving group. But, of course, that gets us right back to where we started. As soon as a molecule of alcohol attacks the carbonyl group, it just gets expelled immediately, and there is no net reaction.

So, let's explore other possible avenues, to see if there is a reaction that can occur. First of all, we should realize that the attack of an alcohol is much slower than the attack of a hydrogen nucleophile, because alcohols do not have a negative charge and are not strong nucleophiles. So, if we want to speed up this reaction, we would want to make the nucleophile more nucleophilic (for example, using RO$^-$ instead of ROH):

Theoretically, this would speed up the reaction, but under these conditions, we would have the same problem that we just had a moment ago. We cannot prevent the carbonyl group from re-forming. The initial intermediate will just eject the nucleophile, and we would get right back to where we started:

So, we will take a slightly different approach. Rather than making the nucleophile more nucleophilic, we will focus on making the electrophile more electrophilic. So, let's focus on the electrophile of our reaction:

How do we make a carbonyl group even more electrophilic? By introducing a small quantity of catalytic acid into the reaction flask:

The resulting protonated ketone is significantly more electrophilic (this entity bears a full positive charge, rendering the carbonyl group more electron-poor). This is VERY IMPORTANT, because we will see this many times throughout this chapter. Many acids can be used for this purpose, including H_2SO_4. When drawing a mechanism for the protonation of a ketone in the presence of an acid catalyst, we should recognize that the identity of the acid (H—A$^+$) is most likely a protonated alcohol, which received its extra proton from H_2SO_4.

So protonation of the ketone most likely occurs in the following way:

As we continue to discuss this mechanism, we will just show H—A$^+$ as the proton source, and it is expected that you will understand that the identity of H—A$^+$ is likely a protonated alcohol.

Now that the ketone has been protonated, rendering it more electrophilic, let's consider what happens if an alcohol molecule functions as a nucleophile and attacks the protonated ketone:

This gives an intermediate that has a tetrahedral geometry (the starting ketone was sp^2 hybridized, and therefore trigonal planar; but this intermediate is now sp^3 hybridized, and therefore tetrahedral). So, we will refer to this intermediate as a "tetrahedral intermediate."

Doesn't this tetrahedral intermediate give us the same problem? Doesn't it simply expel a leaving group to re-form the protonated ketone?

Yes, this *can* happen. In fact, it *does* happen—most of the time. That is in fact why we are using equilibrium arrows, highlighted below:

So it is true that there is an equilibrium between the forward and reverse processes. But every now and then, there is something else that can happen to the tetrahedral intermediate. There is a different way in which the carbonyl group can be re-formed:

What about expelling this leaving group ?

In other words, we are exploring whether HO^- can be expelled as a leaving group, which should theoretically work because we said before that anything can be expelled except for H^- and C^-. Nonetheless, *we cannot expel HO^- in acidic conditions*. Rather, it will have to be protonated first, which converts it into a better leaving group (this is a BIG DEAL—make sure that this rule becomes part of the way you think—NEVER expel HO^- into acidic conditions—always protonate it first). So we draw the following proton transfer steps:

Notice that we first deprotonated to form an intermediate with no charge, and only then protonated. We specifically chose this order (first protonate, then deprotonate) to avoid having an intermediate with two positive charges. This is another important rule that you should make part of the way you think from now on. Avoid intermediates with two similar charges. Now, there are always some clever students who try to combine the two steps above into one step, by transferring a proton intramolecularly, like this:

While it might make sense, it actually doesn't occur that way because the oxygen atom and the proton are simply too far apart to transfer a proton intramolecularly. So, you must first remove a proton, and only then, do you protonate (and it is probably not going to be the same exact proton that was removed).

The result of our two separate proton transfer steps is the following intermediate:

And now we are ready to expel the leaving group (which is now H_2O, rather than HO^-) to re-form a carbonyl group, like this:

This new intermediate now *does* have a carbonyl group, *but* there is no easy way to remove the charge. You can't just lose R^+ the way you can lose a proton:

But there is another way for the charge to be removed. This intermediate can be attacked by *another* molecule of alcohol, just like the protonated ketone was attacked at the beginning of the mechanism:

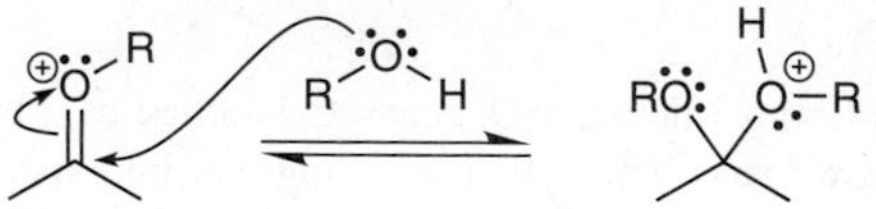

And finally, removal of a proton gives our product:

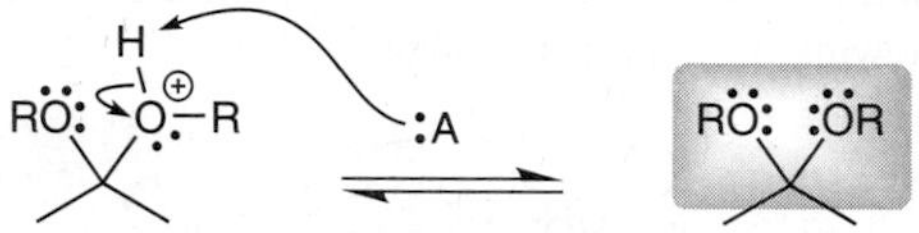

The overall process can be summarized as follows:

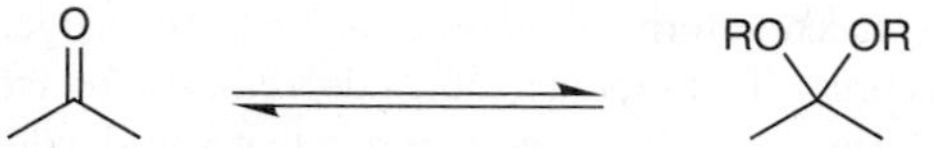

To make sure that we understand some of the key features of this mechanism, let's take a close look at the whole thing all at once. There are seven steps:

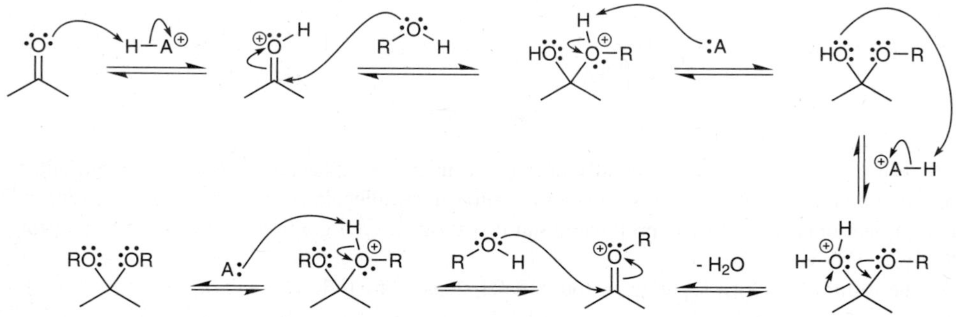

First let's focus our attention on all of the proton transfers in the entire mechanism. Four of the steps above are proton transfer steps. Two of them involve protonation and two involve deprotonation. So, in the end, the acid is not consumed by the reaction. It is a *catalyst* here. From now on, we will place brackets around the acid to indicate that its function is catalytic:

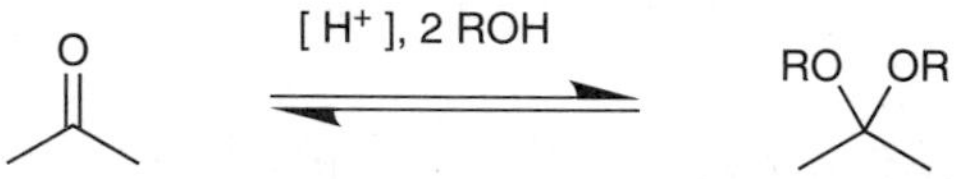

$$[H^+], 2 ROH$$

It is interesting to realize that *most* of steps in the mechanism above are just proton transfer steps. There are only three steps other than proton transfers, and they are: nucleophilic attack (with ROH as the nucleophile), loss of a leaving group (H_2O), and another nucleophilic attack (again, with ROH as the nucleophile). All of the proton transfers are simply used to facilitate these three steps (we use proton transfers to make the carbonyl group more electrophilic, to produce water as a leaving group instead of hydroxide, and to avoid multiple charges). It is important that you see the reaction in this way. It will greatly simplify the whole mechanism in your mind.

The drawing below is NOT a mechanism—the arrows in this drawing are only being used to help you review all three critical steps at once:

Oxygen of ketone
ends up leaving
as H_2O

Functions as
a nucleophile ROH
and attacks

Functions as
ROH a nucleophile
and attacks

H_2O

RO OR

The product of this reaction is called an *acetal*. When we form an acetal from a ketone, there is one intermediate that gets a special name, because it is the only intermediate that does not have a charge. It is called a *hemiacetal*, and you can think of it as "half-way" toward making an acetal:

RO OH

hemiacetal

RO OR

acetal

We give it a special name because it is theoretically possible to isolate it and store it in a bottle (although in many cases, this is very difficult to actually do), and because this type of intermediate will be important if/when you learn biochemistry.

Notice that an acetal does not have a carbonyl group. This means that the equilibrium will lean toward the starting materials, rather than the products:

$$+ \quad 2\ ROH \quad \rightleftharpoons \quad [H^+] \quad RO \quad OR$$

In other words, if we try to perform this reaction in a lab, we will obtain very little (if any) product. So, the question is: how can we force the reaction to form the acetal? There is a clever trick for doing this, and it involves removing water from the reaction as the reaction proceeds. If we remove water as it is formed, we will essentially stop the reverse path at a particular step (highlighted in the following mechanism). It is like putting up a brick wall that prevents the reverse reaction from occurring:

$- H_2O$

By removing water as it is being formed, we force the reaction to a certain point. Now let's focus on all of the steps (highlighted below) that come after the water-removal step:

In the highlighted area, we see three structures in equilibrium with each other. Two of them are positively charged, and one of them (the product) is uncharged. This equilibrium now favors formation of the uncharged product.

In summary, formation of the acetal can be favored by depriving the system of water. Re-forming the carbonyl group would require water, but there is no water present because it has been removed. This very clever trick allows us to force the equilibrium to favor the products even though they are less stable than the reactants.

In your textbook and in your lectures, you will probably explore the way that chemists remove water from the reaction as it proceeds. It is called azeotropic distillation, and there is a special piece of glassware that is used (called a Dean-Stark trap). I will not go into the details of azeotropic distillation here, but I wanted to just briefly mention it, because you should know how to indicate the removal of water. There are two ways to show it:

or like this:

By just writing the words "Dean-Stark," you are indicating that you understand that it is necessary to remove water in order to form the acetal.

Now we can also appreciate how you would reverse this reaction. Suppose you have an acetal, and you want to convert it back into a ketone. You would just add water with a catalytic amount of acid, and the acetal would be converted back into a ketone:

$$RO \underset{\text{}}{\overset{OR}{\bigwedge}} \quad \xrightarrow[\text{H}_2\text{O}]{[\text{H}^+]} \quad \overset{O}{\underset{\text{}}{\bigwedge}}$$

Under these conditions, the equilibrium will favor formation of the ketone. So, now we know how to convert a ketone into an acetal, and we know how to convert the acetal back into a ketone:

$$\overset{O}{\underset{\text{}}{\bigwedge}} \quad \underset{\begin{array}{c}\text{2 ROH}\\\text{Dean-Stark}\end{array}}{\overset{[\text{H}^+]}{\rightleftharpoons}} \quad RO \underset{\text{}}{\overset{OR}{\bigwedge}}$$

$$\underset{\text{H}_2\text{O}}{\overset{[\text{H}^+]}{}}$$

It is very important that we are able to control the conditions to push the reaction in either direction. We will soon see why this is so important. But first, let's make sure we are comfortable with the mechanism of acetal formation:

EXERCISE 5.23 Propose a mechanism for the following reaction.

$$\text{HO}\diagdown\diagup\diagdown\overset{O}{\underset{\text{}}{\bigvee}}\diagdown\diagup\diagdown\text{OH} \quad \xrightarrow[\text{Dean-Stark}]{[\text{H}^+]} \quad \text{[acetal structure]}$$

Answer Notice that we are starting with a ketone, and we are ending up with an acetal. It is a bit tricky to see, because it is all happening in an intramolecular fashion. In other words, the two alcoholic OH groups are *tethered* to the ketone:

So, the mechanism should follow the same order of steps as the mechanism we have already seen. Namely, there are three critical steps (nucleophilic attack, loss of water, and another nucleophilic attack) surrounded by many proton transfer steps. The proton transfer steps are just there to facilitate these three steps. We use a proton transfer in the very first step to render the carbonyl group more electrophilic. Then, we use proton transfers to form water (so that it can leave). And finally, we use a proton transfer to remove the charge and generate the product.

Perhaps you should try to draw the mechanism for this reaction on a separate piece of paper. Then, when you are done, you can compare your work to the following answer:

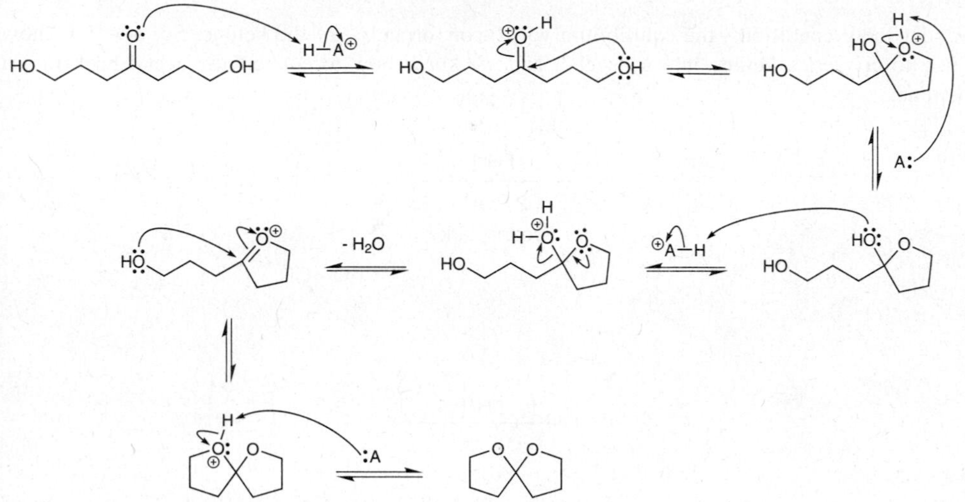

We said before that this type of mechanism is so incredibly important because there will be so many more reactions that build upon the concepts that we developed in this mechanism. To get practice, you should work through the following problems slowly and methodically.

PROBLEMS Propose a plausible mechanism for each of the following transformations. You will need a separate piece of paper for each mechanism.

5.24

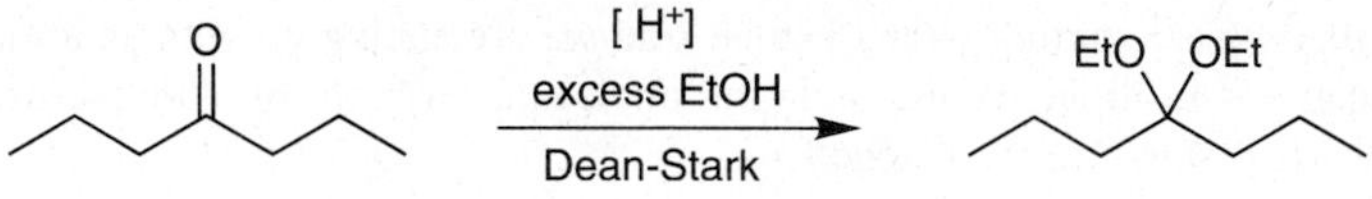

5.25

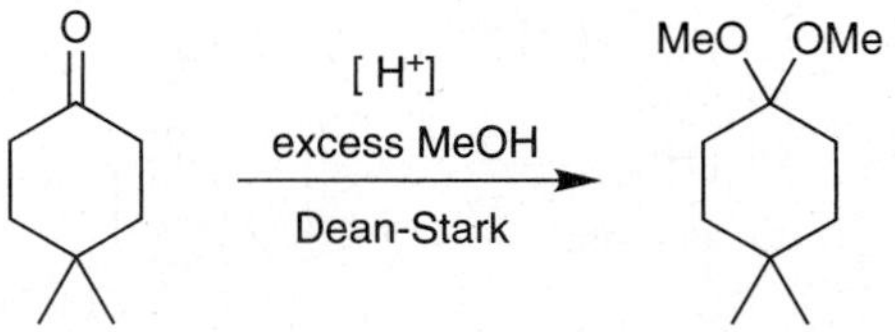

5.26

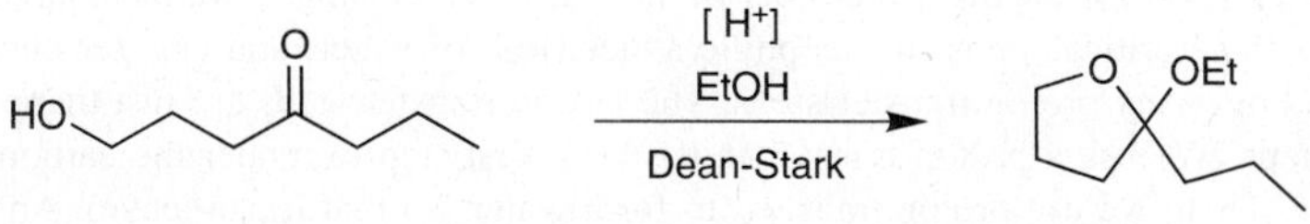

5.27 There is one sure way to know whether or not you have mastered a mechanism forwards and backwards—you should try to actually draw the mechanism backwards. That's right, backwards. For example, draw a mechanism for the following transformation. Make sure to first read the advice below before attempting to draw a mechanism.

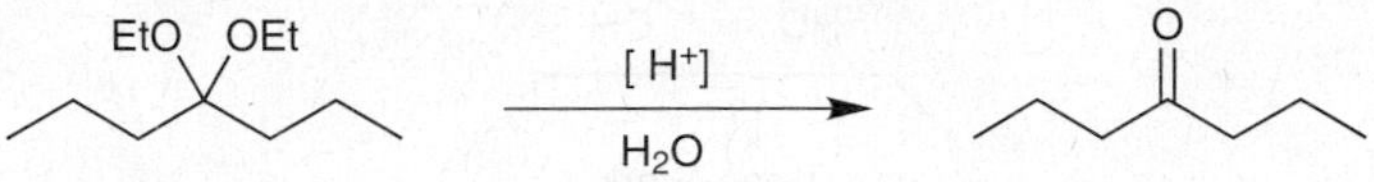

My advice for this mechanism is to start at the end of the mechanism (with the ketone), and then draw the intermediate you would get if you were converting the ketone into an acetal, like this:

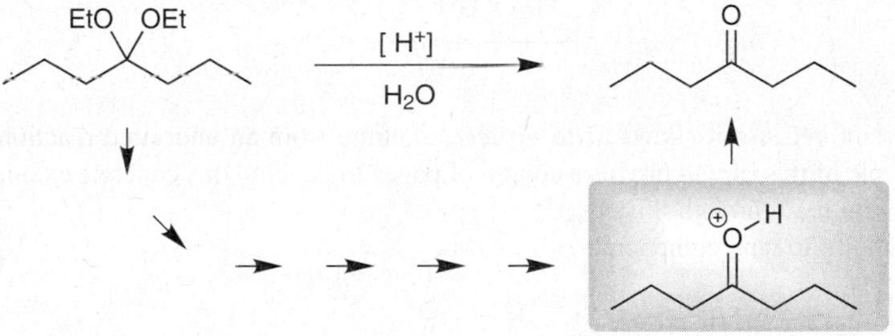

Keep drawing only the intermediates, working your way backwards, until you arrive at the acetal. But *don't* draw any curved arrows yet. Draw only the intermediates, working backwards from the ketone to the acetal. Then, once you have all of the intermediates drawn, then come back and try to fill in arrows, starting at the beginning, with the acetal. Use a separate sheet of paper to draw your mechanism. When you are finished, you can compare your answer to the answer in the back of the book.

In this section, we have seen the reaction that takes place between a ketone and *two* molecules of ROH, in the presence of an acid catalyst and under Dean-Stark conditions:

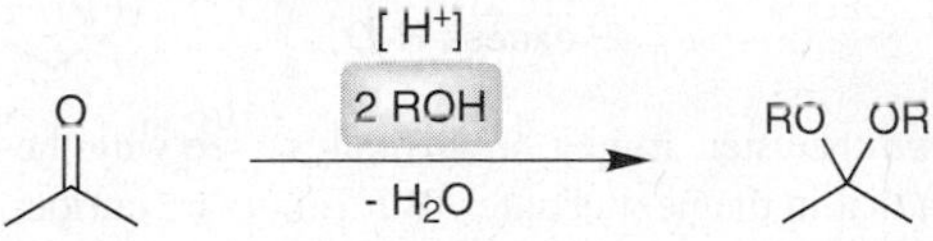

We saw a mechanism, in which the ketone is attacked twice. This same reaction can occur when both alcoholic OH groups are in the same molecule. This produces a *cyclic* acetal:

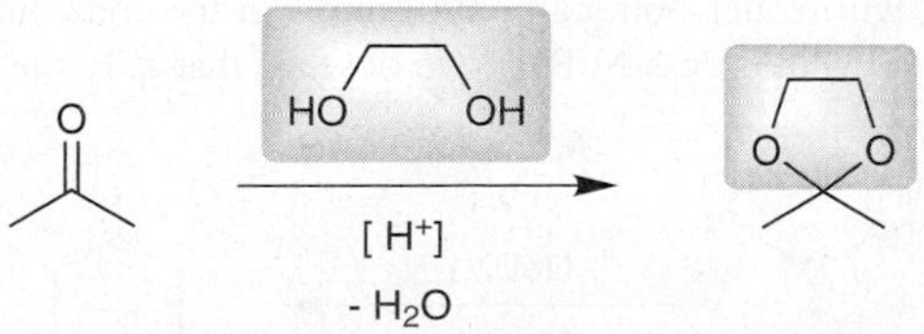

This type of reaction might appear several times throughout your lectures and textbook, so it would be wise to be familiar with this process. The diol in the reaction above is called ethylene glycol, and the transformation can be extremely useful. Let's see why.

We have seen before that we can manipulate the conditions of this reaction to control whether the ketone is favored or whether the acetal is favored. The same is true when we use ethylene glycol to form a cyclic acetal:

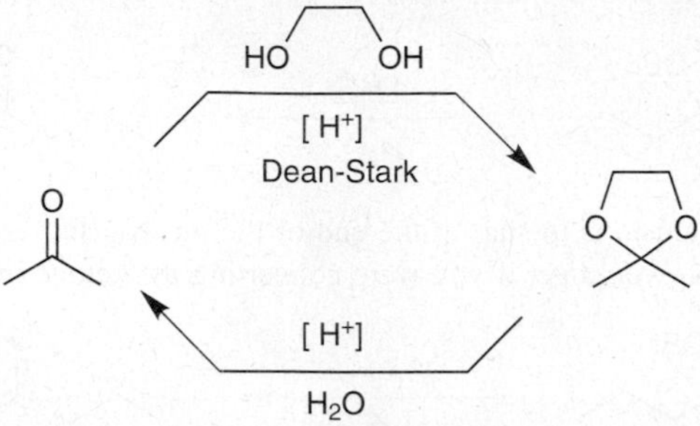

This is important because it allows us to *protect* a ketone from an undesired reaction. Let's see a specific example of this (it will take us a couple of pages to develop this concrete example, so please be patient as you read through this).

Consider the following compound:

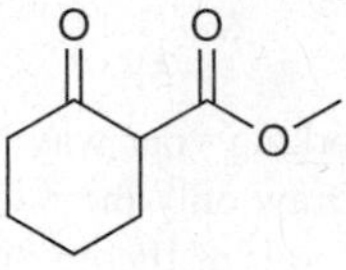

When this compound is treated with excess LAH, followed by water, both carbonyl groups are reduced:

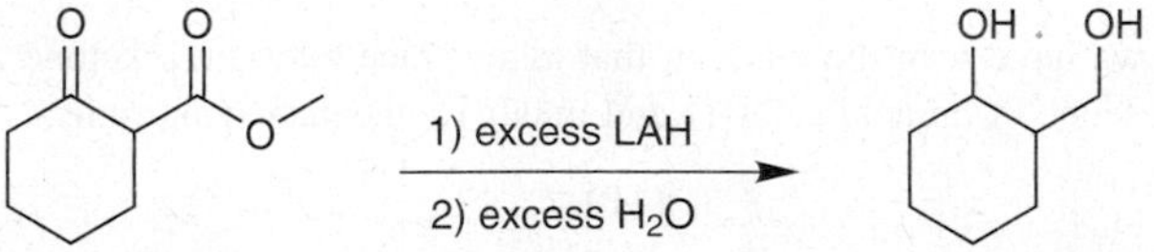

1) excess LAH

2) excess H_2O

LAH attacks the ketone *and* the ester. It may be difficult to see why the ester is converted into an alcohol—we will focus on that in the next chapter. But if you are curious to test your abilities, you have actually learned everything you need in order to figure out how an ester is converted into an alcohol in the presence of excess LAH (remember that you should always re-form a carbonyl group if you can, but never expel H^- or C^-).

So, we see that LAH will reduce both carbonyl groups in the compound above. If instead, we treat the starting compound with excess $NaBH_4$, we observe that only the ketone is reduced:

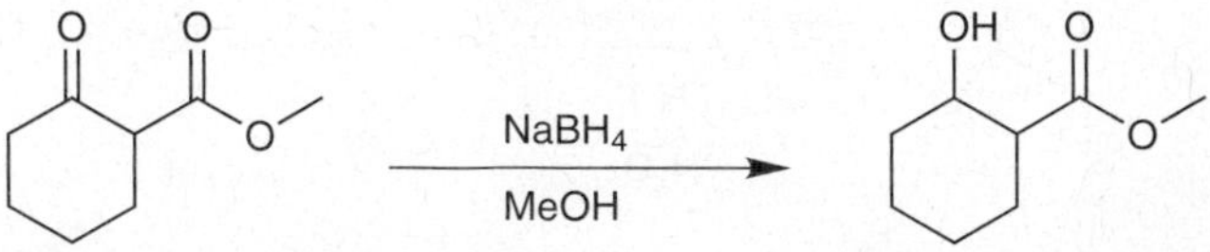

$NaBH_4$

MeOH

The ester is *not* reduced, because $NaBH_4$ is a milder source of hydride (as we have explained earlier). We will see in the next chapter that $NaBH_4$ will not react with esters (only with ketones and aldehydes) because the carbonyl group of an ester is less reactive than the carbonyl group of a ketone.

Now suppose you want to achieve the following transformation:

Essentially, you want to reduce the ester, *but not* the ketone. That would seem impossible, because esters are less reactive than ketones. Any reagent that reduces an ester should also reduce a ketone.

But there is a way to achieve the desired goal. Suppose we "protect" the ketone by converting it into an acetal:

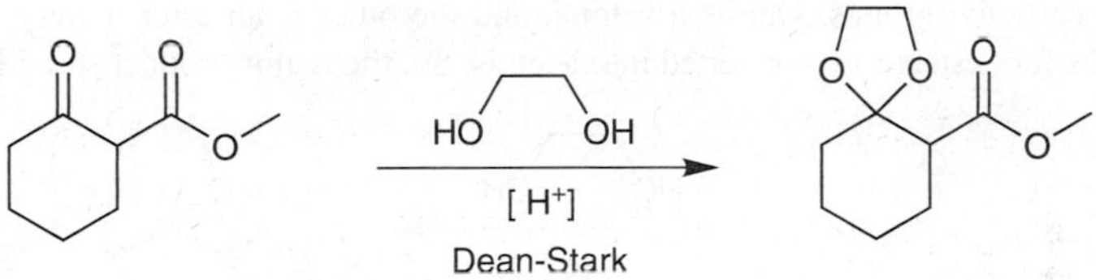

Only the ketone is converted into an acetal. The carbonyl group of the ester is *not* converted into an acetal (because esters are less reactive than ketones). So, we are using the reactivity of the ketone to our advantage, by selectively "protecting" the ketone. Now, we can treat this compound with excess LAH, followed by water, and the acetal will not be affected (acetals do not react with bases or nucleophiles under basic conditions):

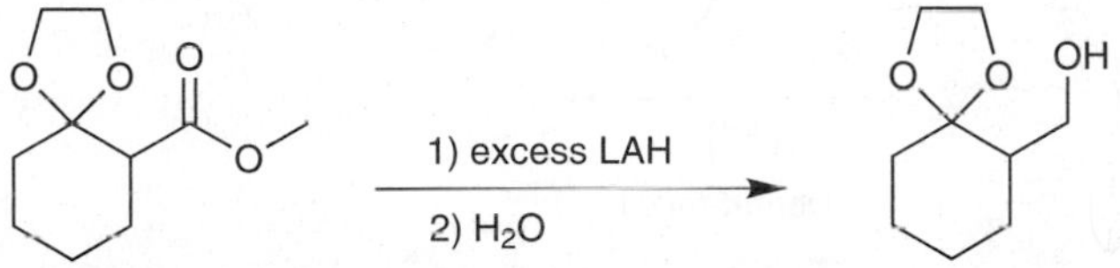

Notice that we use water above in the second step (as we have done every other time that we used LAH). In the presence of water, the acetal is removed but only if the conditions are acidic. So, to remove the acetal in this case, we would use aqueous acid, rather than H_2O, after the reduction:

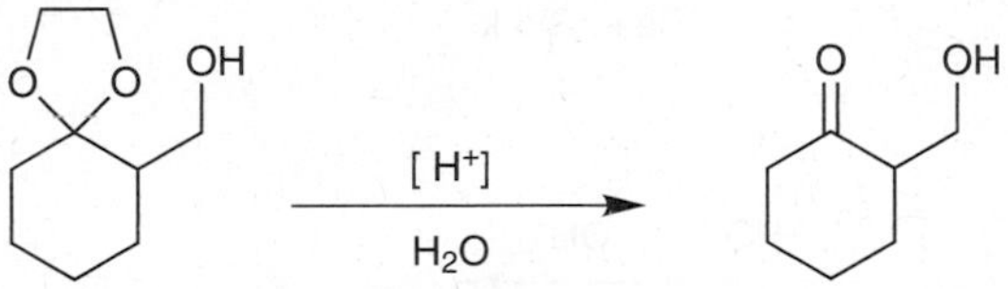

In the end, we have a 3-step process for reducing the ester moiety *without affecting* the ketone moiety:

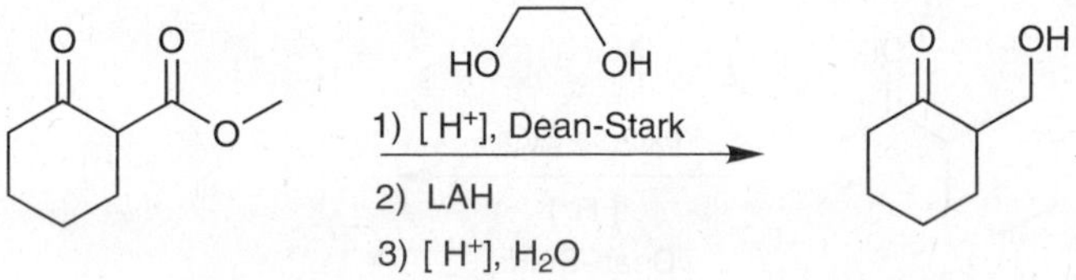

We will talk more about this strategy in the next chapter. For now, let's just focus on knowing the reactions well enough to predict products.

EXERCISE 5.28 Predict the major product of the following reaction:

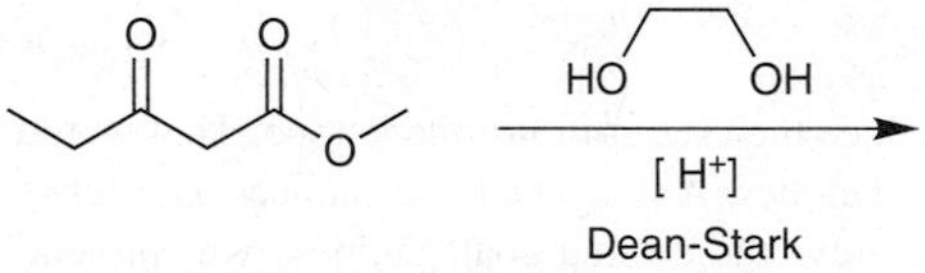

Answer This reaction utilizes ethylene glycol, so we expect a cyclic acetal. Our starting compound has two carbonyl groups. One is a ketone, and the other is an ester moiety. We have seen that only ketones (not esters) are converted into acetals. So, the major product should be as follows:

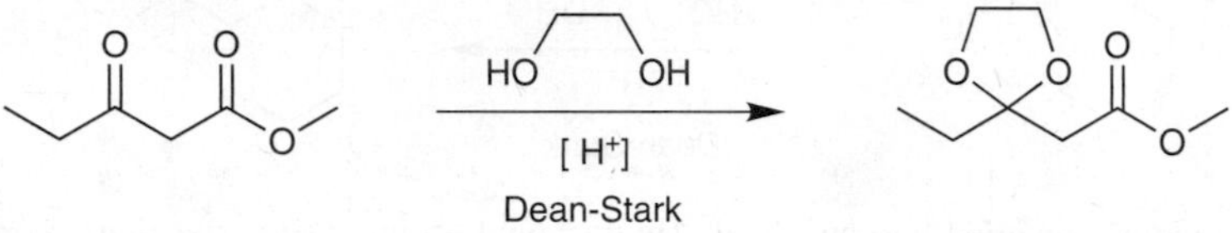

PROBLEMS Predict the major product of each of the following reactions:

5.29

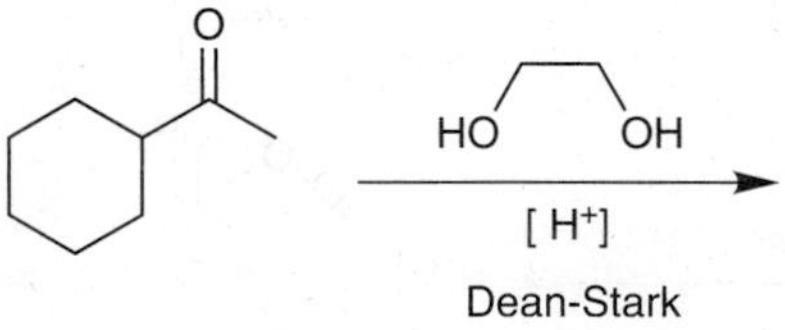

5.30

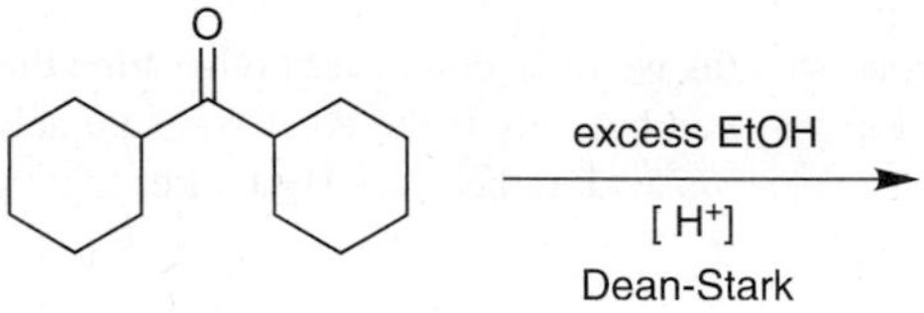

5.31

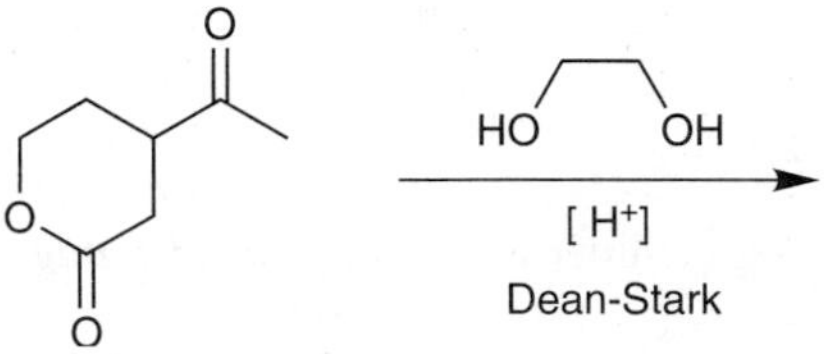

5.32

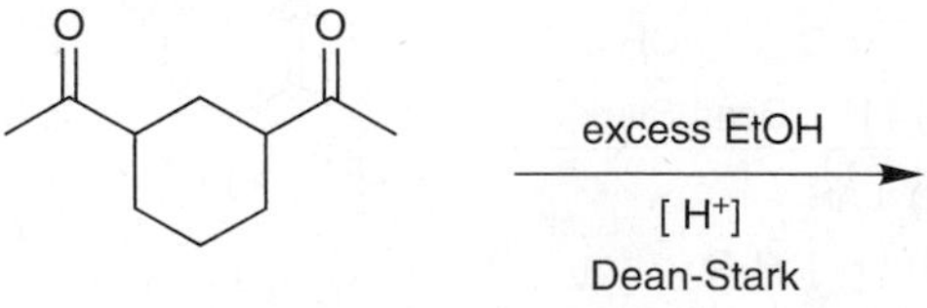

5.5 S-NUCLEOPHILES

Sulfur is directly below oxygen on the periodic table (in Column 6A). Therefore, the chemistry of sulfur-containing compounds is very similar to the chemistry of oxygen-containing compounds. In the previous section, we saw a method for converting a ketone into an acetal:

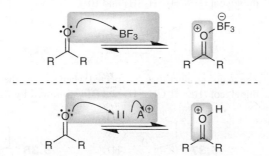

In much the same way, a ketone can also be converted into a *thio*acetal (thio means sulfur instead of oxygen):

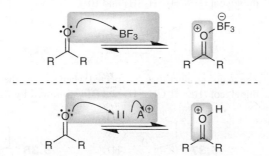

thioacetal

The main difference is that we use BF_3 instead of H^+ to make the carbonyl group more electrophilic:

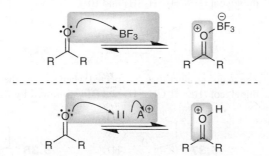

Other than this small difference, making a *thio*acetal is very similar to making an acetal. After all, they are very similar in structure:

acetal   thioacetal

But thioacetals will undergo a transformation not observed for acetals. Specifically, thioacetals are reduced when treated with Raney nickel:

Raney nickel is finely divided nickel that has hydrogen atoms adsorbed to it. The mechanism for this reduction process is beyond the scope of this course. But, this is a VERY useful synthetic transformation. So it is worth remembering, even if you don't know the mechanism. It provides a way to completely reduce a ketone down to an alkane:

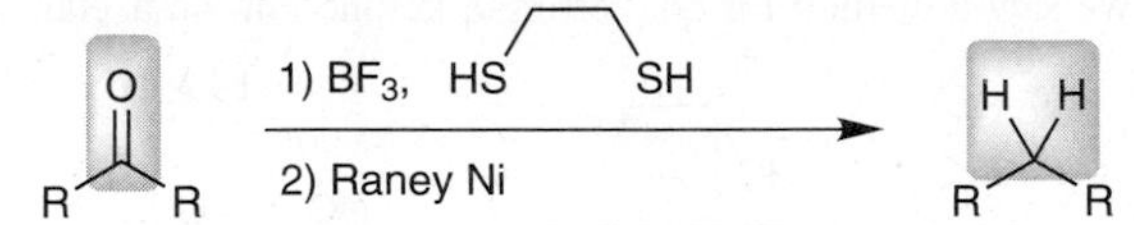

We have actually already seen one way to achieve this kind of transformation. It was called the Clemmensen reduction, which we explored in Chapter 3 (electrophilic aromatic substitution). We will also see one more way to achieve this transformation in the upcoming section.

Why do we need three different ways to do the same thing? Because each of these methods involves a different set of conditions. The Clemmensen reduction employs *acidic* conditions. The method we learned just now (desulfurization with Raney nickel) employs *neutral* conditions. And the method in the upcoming section will employ *basic* conditions. As we move through the course, we will see times when it won't be good to subject an entire compound to acidic conditions, and we will see other times when it won't be good to subject an entire compound to basic conditions. When in doubt whether it is bad to use acidic conditions or basic conditions, you can always just use a desulfurization with Raney nickel, which employs neutral conditions.

PROBLEMS Predict the major product that is expected when each of the following compounds is treated with ethylene thioglycol ($HSCH_2CH_2SH$) and BF_3.

5.33 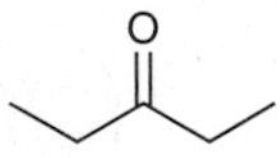 **5.34** 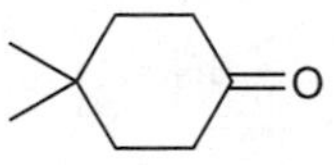 **5.35**

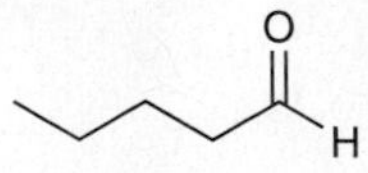

PROBLEMS Predict the major product that is expected when each of the following compounds is treated with ethylene thioglycol ($HSCH_2CH_2SH$) and BF_3, followed by Raney nickel.

5.36 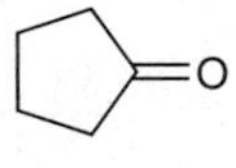 **5.37** 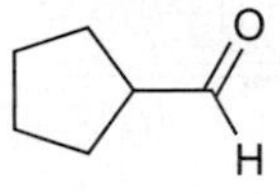 **5.38**

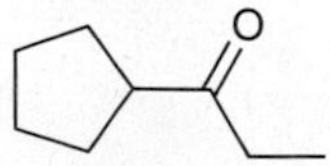

PROBLEMS Identify the reagents you would use to achieve each of the following transformations:

5.39

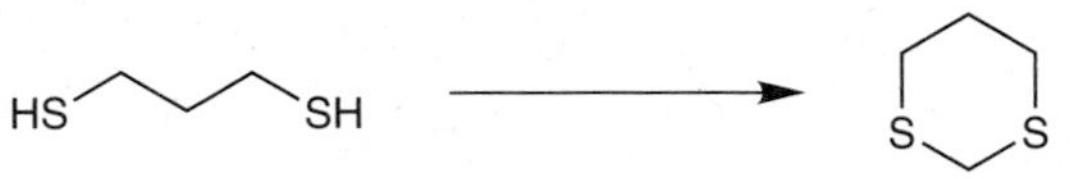

5.40

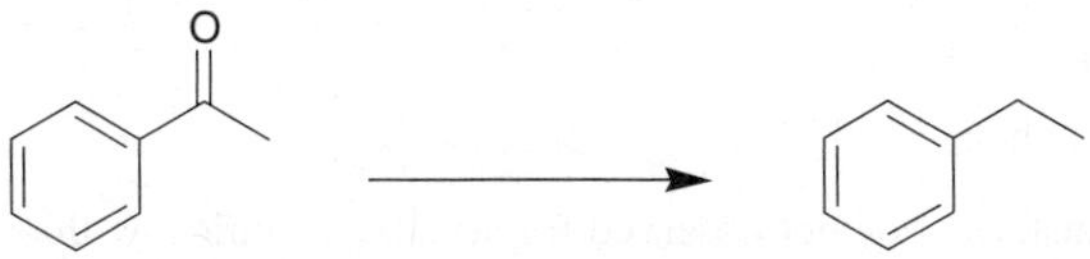

5.41

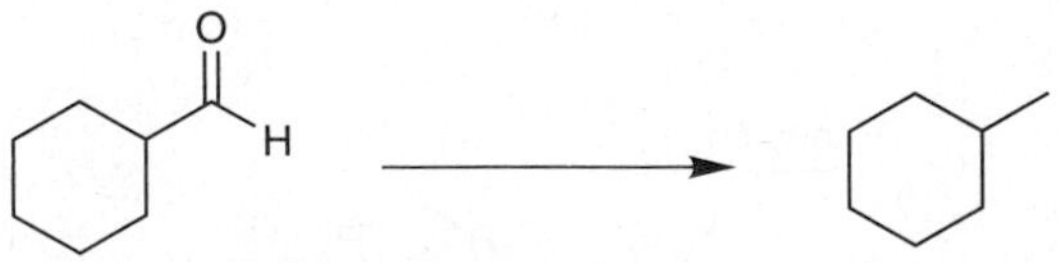

5.42

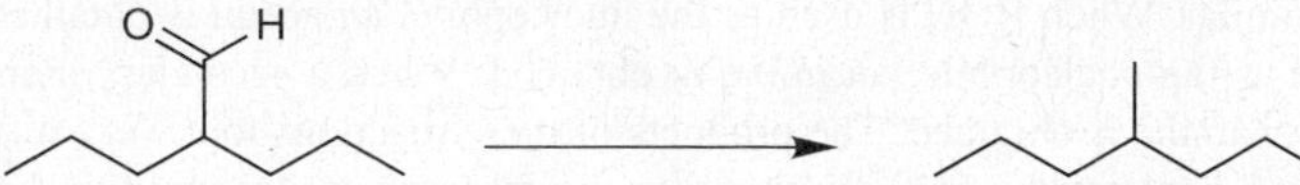

5.43

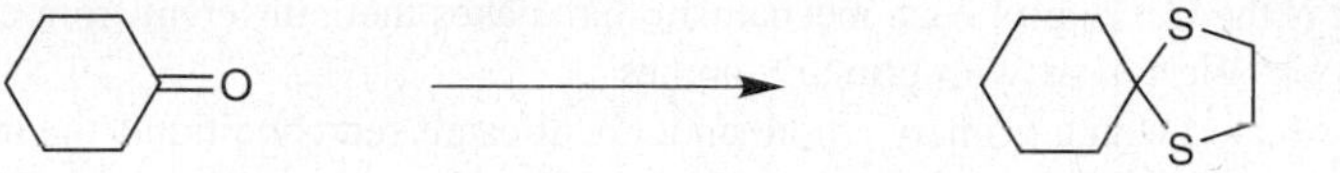

5.44

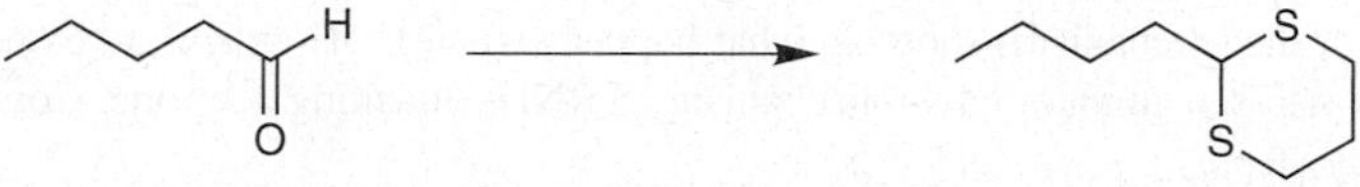

5.45

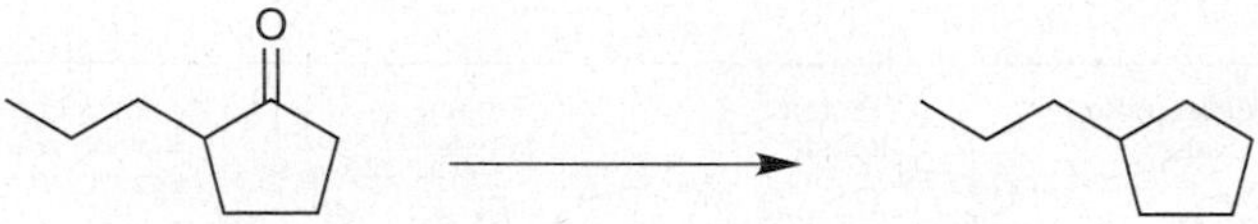

5.46

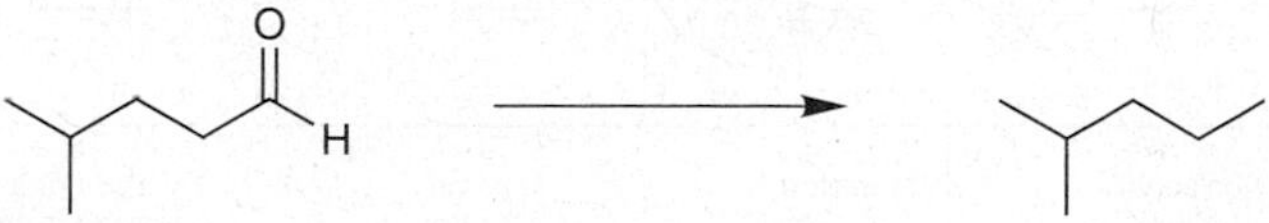

5.6 N-NUCLEOPHILES

When we first learned about acetal formation, we said that the mechanism would serve as a foundation for other reactions in this chapter. Now, we will see the power of mechanisms in helping us understand the similarities between reactions.

Compare the products of the following three reactions:

Nucleophile	Reaction		
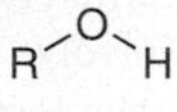 $R\overset{O}{\diagdown}H$		[H⁺], ROH → Dean-Stark	ketal
$R\overset{H}{\underset{H}{\diagup N}}$		[H⁺], RNH_2 → Dean-Stark	imine
$R\overset{R}{\underset{H}{\diagup N}}$		[H⁺], R_2NH → Dean-Stark	enamine

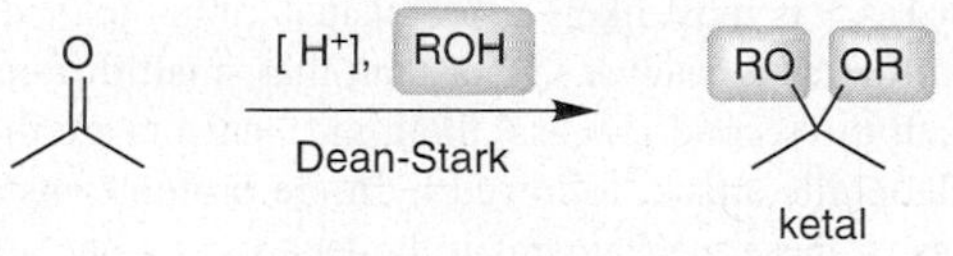

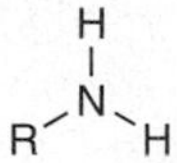

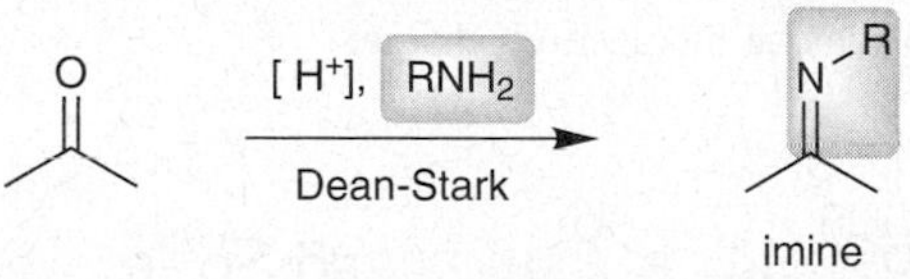

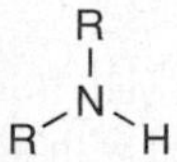

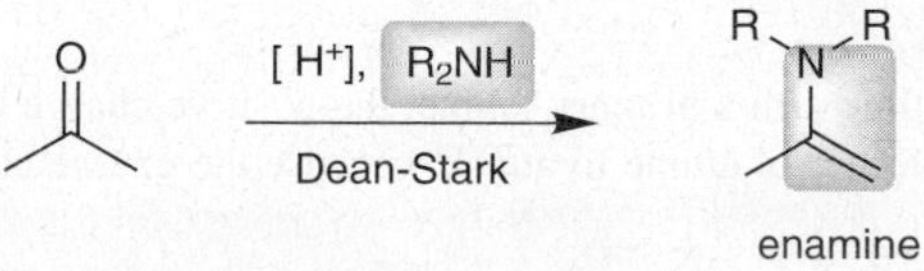

The products are not similar. When ROH is used as the nuclcophile, an acetal is obtained. When a primary amine is used as the nucleophile, an imine is obtained. When a secondary amine is used as the nucleophile, an enamine is obtained. The products of these reactions look very different, but when we analyze the mechanisms, we will see that they are all very similar up until the very end of the mechanism. It is the last step of each mechanism that makes them different from each other. Let's take a closer look. We'll start with primary amines.

When a ketone is treated with a primary amine under acid-catalyzed conditions, the mechanism begins just like the mechanism of acetal formation. Shown below is an incomplete mechanism (only the first two-thirds of the mechanism) showing what happens when ROH attacks a ketone. And directly below it, you will see an incomplete mechanism of RNH$_2$ attacking a ketone. Compare both mechanisms, step-by-step:

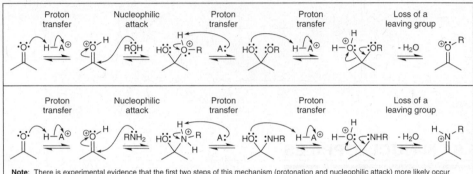

Note: There is experimental evidence that the first two steps of this mechanism (protonation and nucleophilic attack) more likely occur either simultaneously or in the reverse order of what is shown above. Most nitrogen nucleophiles are sufficiently nucleophilic to attack a carbonyl group directly, before protonation occurs. Nevertheless, the first two steps of the mechanism above have been drawn in the order shown (which only rarely occurs), because this sequence enables a more effective comparison of all acid-catalyzed mechanisms in this chapter. Be sure to look in your lecture notes to see the order of events that your instructor used for the first two steps of the mechanism.

In the first process above, the identity of HA$^+$ (the proton source) is most likely a protonated alcohol, which received its proton from the acid catalyst. Similarly, in the second process above, the identity of HA$^+$ is most likely a protonated amine (called an ammonium ion), which received its proton from the acid catalyst. Other than that small difference (and the difference described in the note beneath the second process), both mechanisms are the same. Both involve a proton transfer and a nucleophilic attack, followed by more proton transfer steps, and then loss of water. But the conclusions of these mechanisms truly depart from one another. Let's try to understand why.

In the first mechanism (acetal formation), another molecule of ROH attacked, because there was no other way to remove the positive charge:

But in the reaction with a primary amine, the positive charge is easily removed. It is not necessary for another molecule of amine to attack, because the charge can be removed with a proton transfer step:

And this is our product. It is called an *imine*, because it has a C═N bond. So the mechanism of this reaction is almost identical to the mechanism of acetal formation, except for the very end. And the difference at the end makes sense when you really think about it.

When we perform this reaction, we just need to take special notice of whether or not the starting ketone is symmetrical:

symmetrical   unsymmetrical

If the starting ketone is *un*symmetrical, then we should expect two diastereomeric imines:

So far, we have seen what happens when a *primary* amine attacks a ketone. Now, let's see what happens when a ketone is treated with a *secondary* amine under acid-catalyzed conditions. Let's compare it to the mechanisms that we have seen so far:

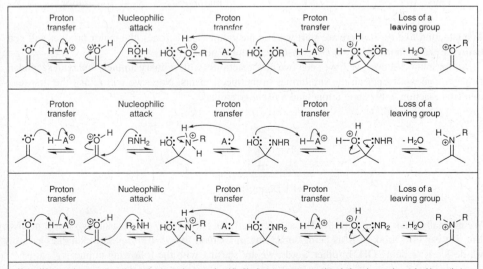

Note: When a primary or secondary amine is used as a nucleophile (the last two processes above), there is experimental evidence that the first two steps of the mechanism (protonation and nucleophilic attack) more likely occur either simultaneously or in the reverse order of what is shown above. Most nitrogen nucleophiles are sufficiently nucleophilic to attack a carbonyl group directly, before protonation occurs. Nevertheless, the first two steps of the mechanism above have been drawn in the order shown (which only rarely occurs), because this sequence enables a more effective comparison of all acid-catalyzed mechanisms in this chapter. Be sure to look in your lecture notes to see the order of events that your instructor used for the first two steps of the mechanism.

None of these mechanisms are complete. All three of them are missing the last steps. But compare the steps that are shown above. Notice that, once again, these mechanisms are identical up until the very end of each mechanism. And it is right at the end where we see differences in the final products. In the first mechanism (ROH as the nucleophile), we saw that another molecule of ROH attacks. In the second mechanism, we saw that deprotonation led to formation of an imine. But in the third reaction, we cannot just lose a proton the way we did in the mechanism of imine formation. So, we might be tempted to do what we did in the first mechanism (acetal formation). We might be tempted to say that another molecule of amine should attack. But there is something else that happens instead:

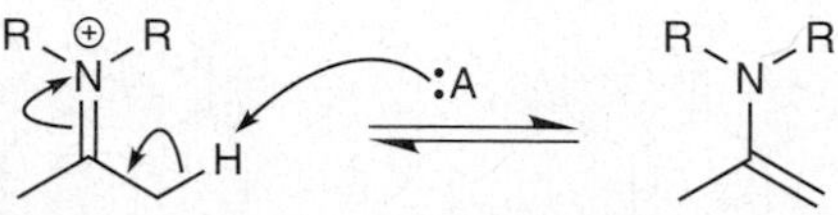

Another molecule of the secondary amine functions as a base, rather than a nucleophile, so a proton is in fact removed. This gives a product called an *enamine* ("en" because there is a double bond, and "amine" because there is an NH_2 group).

Once again, we need to be careful to check if the ketone is unsymmetrical. If it is, then there will be two ways to form the double bond in the last step of the mechanism. This will give two different enamine products. Here is an example:

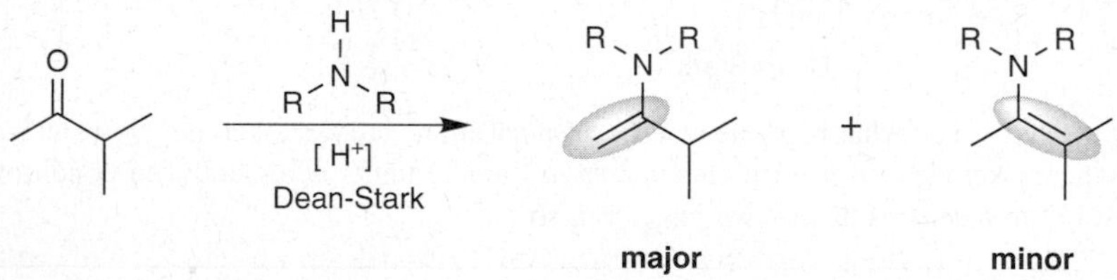

In a situation like this (where we start with an unsymmetrical ketone), the major product will generally be the enamine with the *less*-substituted double bond.

So far in this section, we have seen two new reactions (with primary amines, and with secondary amines), and we have seen the similarities in the mechanisms. Now, we will revisit the first reaction (the reaction between a ketone and a *primary* amine):

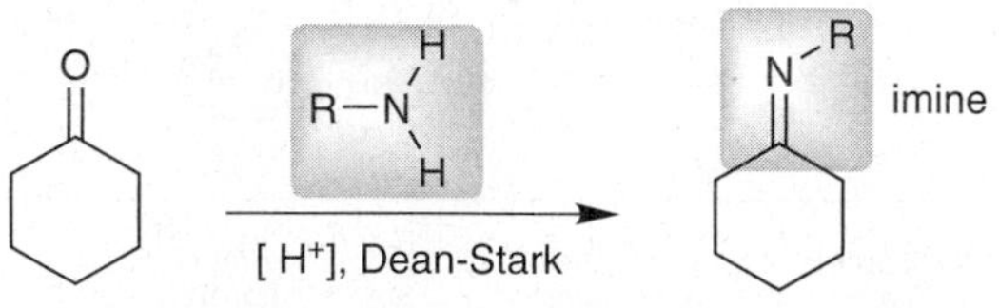

We normally think of the R (in RNH_2) as referring to an alkyl group (that is usually what R means). But, we can also think of R as being something *other than an alkyl group*. For example, let's say we define R as being OH. In other words, we are starting with the following amine:

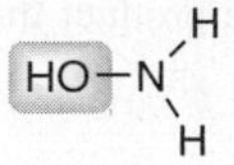

$$HO-N\overset{H}{\underset{H}{}}$$

This compound is called hydroxylamine, and the product that it forms (when it reacts with a ketone) is not surprising at all:

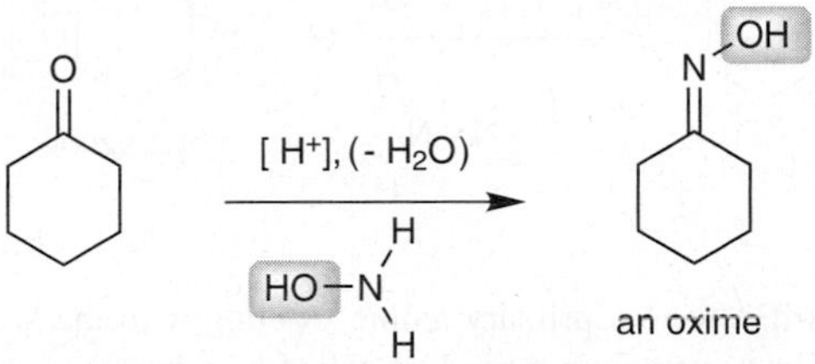

[H⁺], (- H₂O)

HO—N

an oxime

It is the same reaction as if it were a primary amine reacting with the ketone. But, instead of getting an imine, we get something that we call an oxime. Remember to always look if the starting ketone is unsymmetrical. If it is, we should expect to form two diastereomeric oximes:

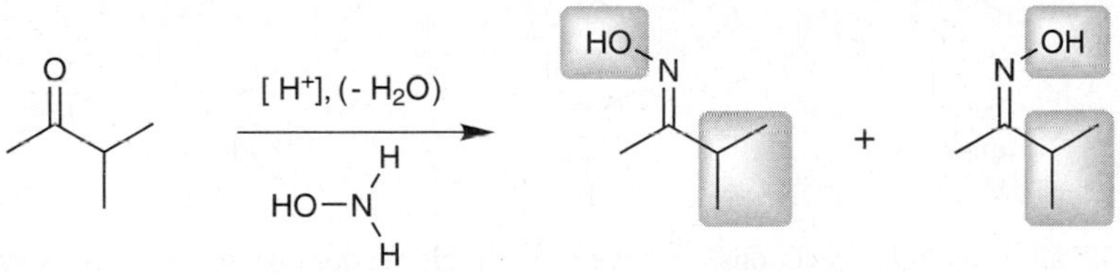

[H⁺], (- H₂O)

HO—N

+

When you see this type of reaction, there are several ways that the presence of hydroxylamine can be indicated:

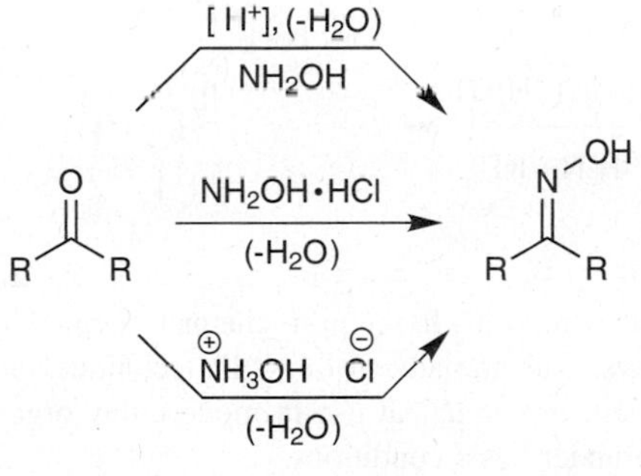

[H⁺], (-H₂O)
NH₂OH

NH₂OH•HCl
(-H₂O)

$\overset{\oplus}{NH_3OH}$ $\overset{\ominus}{Cl}$
(-H₂O)

All of these representations are just different ways of showing the same reagent.

Now that we have seen a special N-nucleophile (RNH₂ where R is OH), let's take a close look at one more special N-nucleophile. Let's look at a case where R is NH₂. In other words, we are using the following nucleophile:

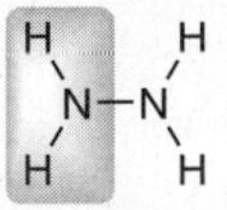

$$\overset{H}{\underset{H}{}}N-N\overset{H}{\underset{H}{}}$$

This compound is called hydrazine, and the product that it forms (when it reacts with a ketone) is not surprising at all:

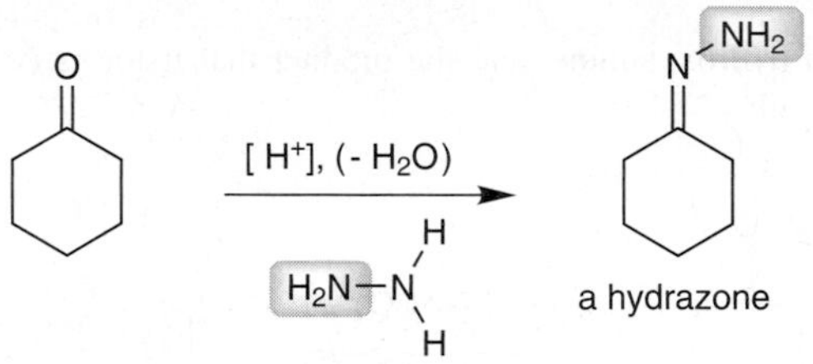

It is the same reaction as if it were a primary amine reacting with the ketone. But, instead of obtaining an imine or an oxime, we obtain a product called a hydrazone.

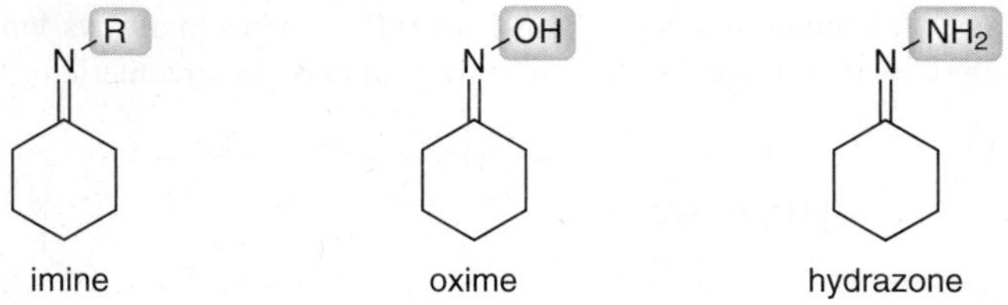

imine   oxime   hydrazone

Just like with all of the other reactions we have seen in this section, we need to take special notice of whether or not the starting ketone is symmetrical. If the starting ketone is unsymmetrical, then we should expect to form two diastereomeric hydrazones:

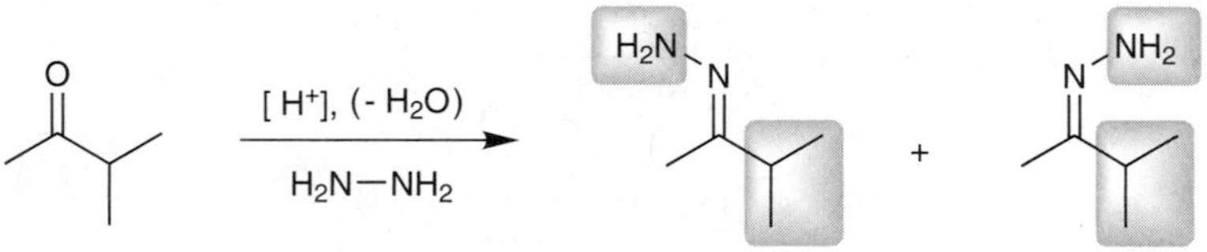

Hydrazones are useful for many reasons. In the past, chemists formed hydrazones as a way of identifying ketones, but nowadays, with the advent of NMR techniques, no one uses hydrazones that way anymore. But there is still one practical use in modern-day organic chemistry. A hydrazone can be reduced to an alkane under basic conditions:

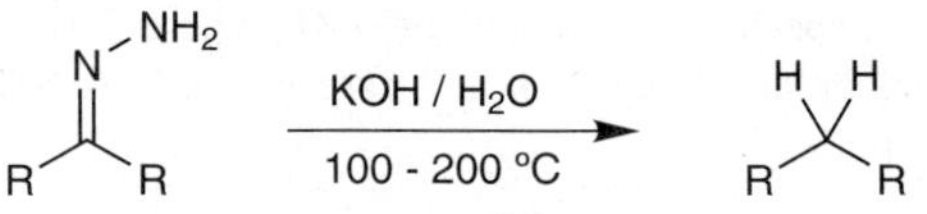

The following is a mechanism for this process:

The formation of a carbanion (highlighted in the second-to-last step) certainly creates an uphill battle (in terms of energy), so we might expect very little product to form. However, notice that the formation of the carbanion is accompanied by loss of N_2 gas (also highlighted in the mechanism above). This explains why the reaction goes to completion. The small amount of nitrogen gas (produced by the equilibrium) will bubble out of the solution and escape into the atmosphere. That forces the equilibrium to produce a little bit more nitrogen gas, which also then escapes into the atmosphere. And the process continues until the reaction reaches completion. Essentially, a reagent is being removed as it is being formed, and that is what pushes the equilibrium over the high energy barrier created by the instability of the carbanion. If you think about it, this concept is not so different from the previous sections where we removed water from a reaction as it was being formed (as a way of pushing the equilibrium toward formation of the acetal).

This now provides a two-step method for reducing a ketone to an alkane:

1) [H^+], $H_2N\text{-}NH_2$, ($-H_2O$)
2) KOH / H_2O
   100 - 200 °C

We have already seen two other ways to do this kind of transformation (the Clemmensen reduction and desulfurization with Raney Nickel). This is now our third way to reduce a ketone to an alkane, and it is called a Wolff-Kishner reduction.

In this section, we have only seen a few reactions involving nitrogen nucleophiles. Here is a short summary. We first saw how a ketone can react with a *primary amine* to form an *imine* (and we saw that the mechanism was very similar to acetal formation, except for the very end). Then, we saw how a ketone can react with a *secondary amine* to form an *enamine* (once again, the mechanism was very similar up until the very end). We also saw two special N-nucleophiles (NH_2OH and NH_2NH_2) both of which gave us products that we would have expected. The reaction with NH_2NH_2 was of special interest, because it provided a new method for reducing ketones to alkanes.

Now let's do some problems to make sure that you are familiar with the reagents and the mechanisms for the reactions that we have seen in this section. Let's start with mechanisms:

EXERCISE 5.47 Propose a plausible mechanism for the following transformation:

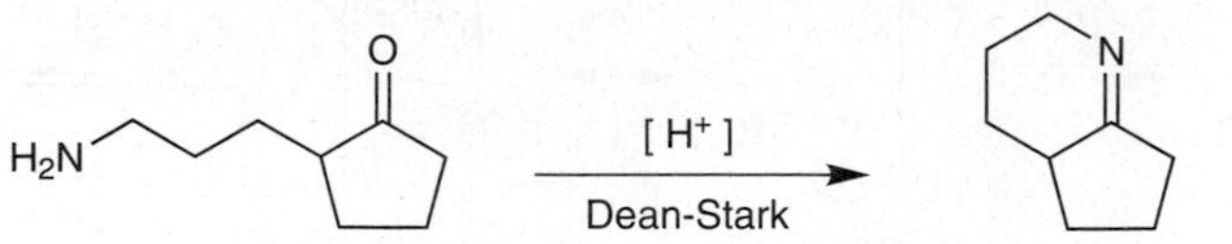

Answer We begin by looking at the starting material. It has two functional groups. This compound is a primary amine, *and* it is a ketone. That means that it could theoretically attack itself, in an *intramolecular* reaction. Then we look at the reagents (acid catalysis and Dean-Stark conditions), and we notice that we are missing a nucleophile. This further supports the idea that an intramolecular reaction will occur. The starting material can function as both the nucleophile and the electrophile. Finally, we look at the product, and we see that it is an imine, which is the type of product that is produced from the reaction between a primary amine and a ketone. With all of this information, we conclude that it is in fact an intramolecular process.

The mechanism will have the same order of steps as any other mechanism involving a primary amine attacking a ketone (protonate, attack, deprotonate, protonate, lose water, and then deprotonate):

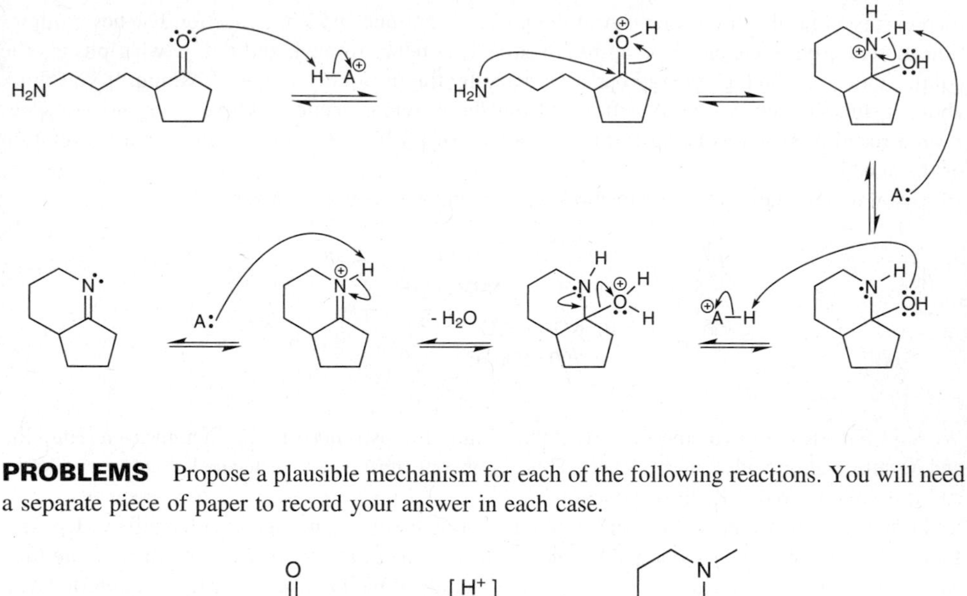

PROBLEMS Propose a plausible mechanism for each of the following reactions. You will need a separate piece of paper to record your answer in each case.

5.48

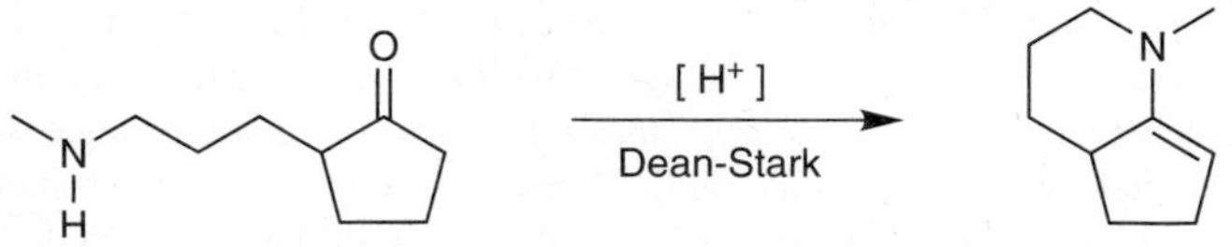

5.49

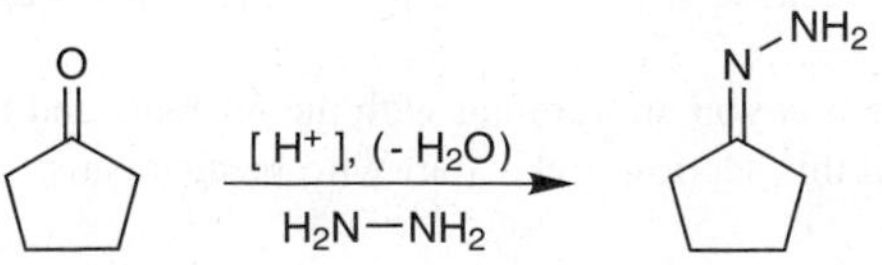

5.50

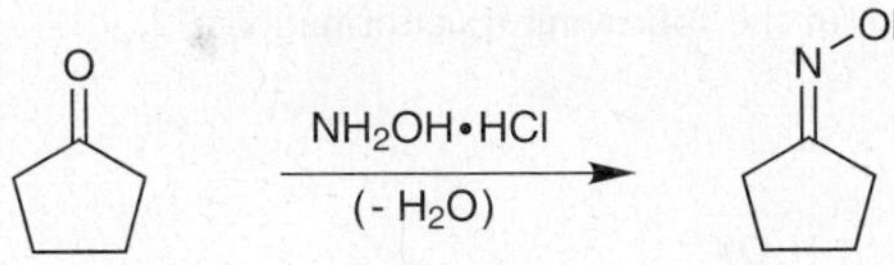

5.51

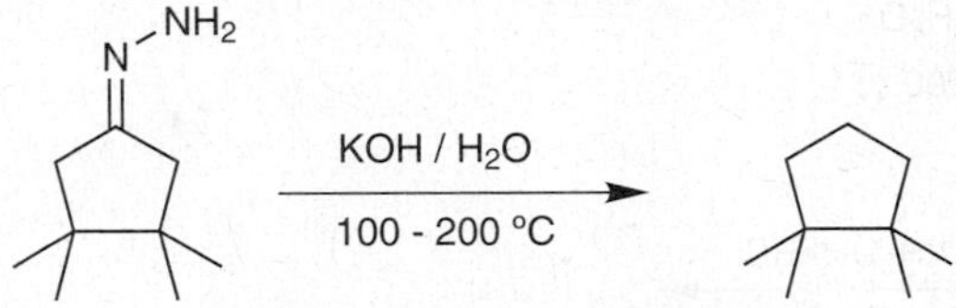

5.52

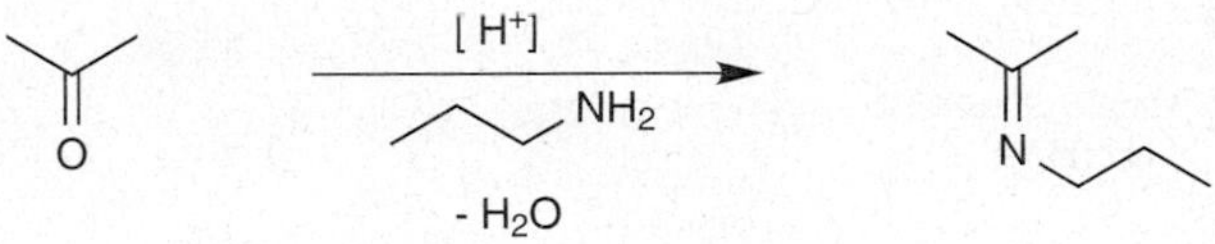

5.53

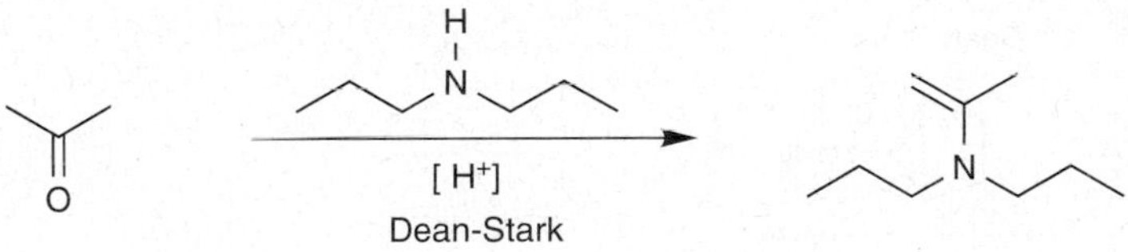

Now let's get some practice predicting products.

EXERCISE 5.54   Predict the products of the following reaction:

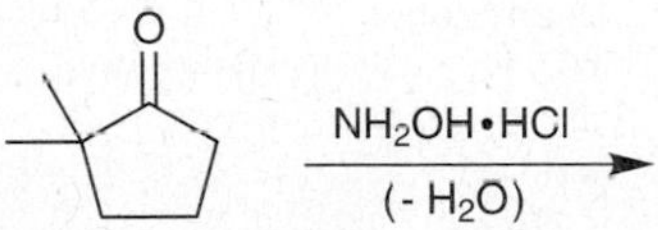

Answer   The starting material is a ketone, and the reagent is hydroxyl amine. So, as we have seen in this section, the product of this reaction should be an oxime. Since the starting ketone is unsymmetrical, we would expect two diastereomeric oximes:

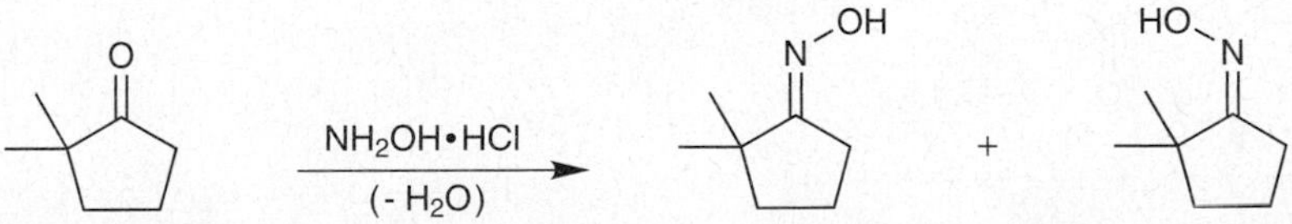

We saw several reactions in this chapter. You must be able to recognize the reagents for these reactions, so that you will be able to predict products. If you were not able to recognize that the reagent here is hydroxyl amine, then you would have not been able to predict the products in this case.

PROBLEMS Predict the products for each of the following transformations:

5.55

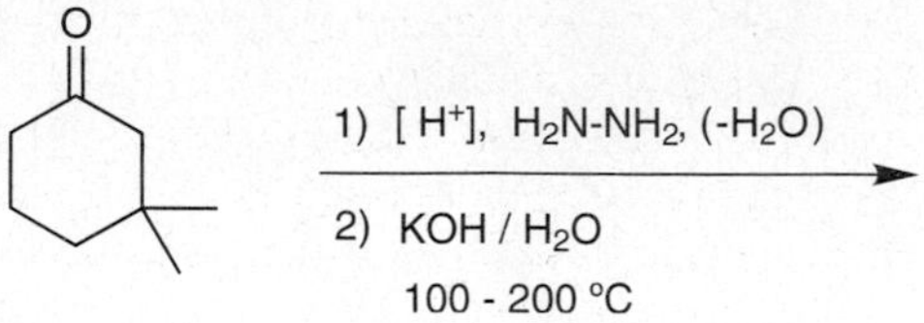

1) [H$^+$], H$_2$N-NH$_2$, (-H$_2$O)

2) KOH / H$_2$O
100 - 200 °C

5.56

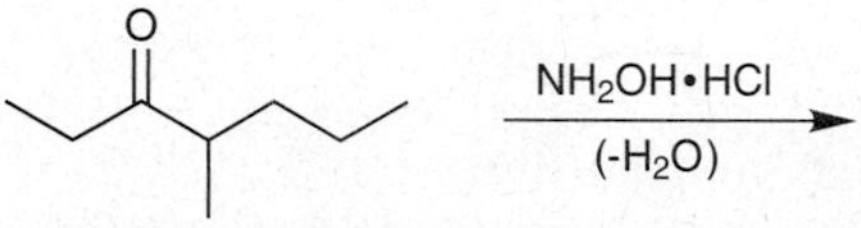

NH$_2$OH•HCl
(-H$_2$O)

5.57

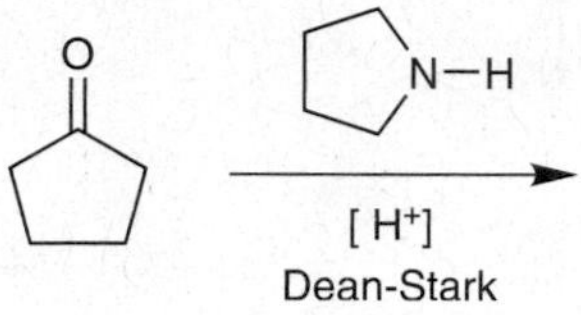

N–H

[H$^+$]
Dean-Stark

5.58

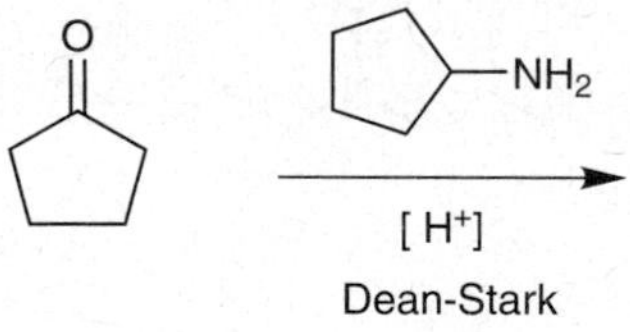

–NH$_2$

[H$^+$]
Dean-Stark

5.59

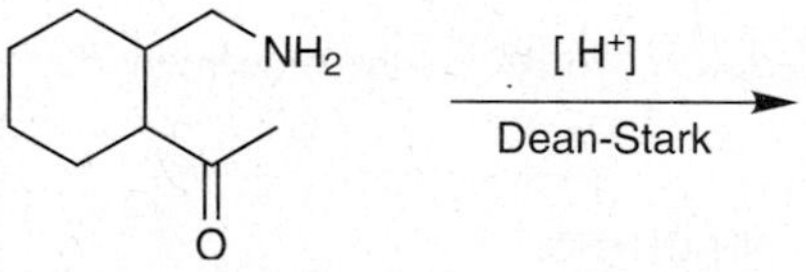

NH$_2$

[H$^+$]
Dean-Stark

5.60

N
H

[H$^+$]
Dean-Stark

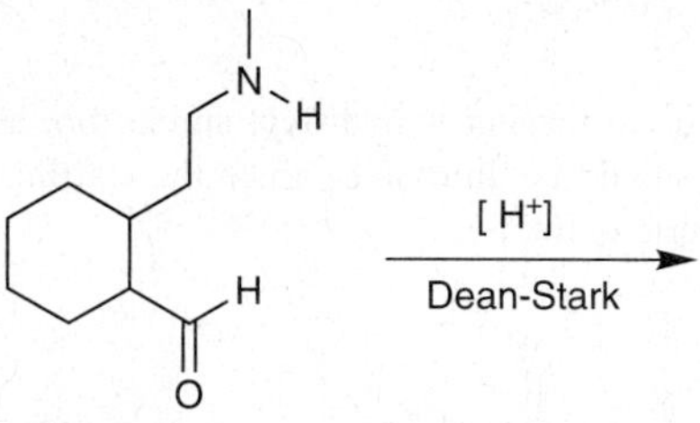

5.7 C-NUCLEOPHILES

In this chapter, we have seen many different kinds of nucleophiles that can attack ketones and aldehydes. We started with hydrogen nucleophiles. Then we moved on to oxygen nucleophiles and sulfur nucleophiles. In the previous section, we covered nitrogen nucleophiles. In this section we will discuss carbon nucleophiles. We will see three types of carbon nucleophiles.

Our first carbon nucleophile is the Grignard reagent. You may have been exposed to this reagent in the first semester. If you weren't, here is a quick overview:

Alkyl halides will react with magnesium in the following way:

Essentially, an atom of magnesium inserts itself in between the C—Cl bond (this reaction works with other halides as well, such as Br or I). This magnesium atom has a significant electronic effect on the carbon atom to which it is attached. To see the effect, consider the alkyl halide (before Mg entered the picture):

The carbon atom (connected to the halogen) is poor in electron density, or $\delta+$, because of the inductive effects of the halogen. But after magnesium is inserted between C and Cl, the story changes very drastically:

Carbon is much more electronegative than magnesium. Therefore, the inductive effect is now reversed, placing a lot of electron density on the carbon atom, making it very $\delta-$. The C—Mg bond has significant ionic character, so for purposes of simplicity, we will just treat it like an ionic bond:

Carbon is not very good at stabilizing a negative charge, so this reagent (called a Grignard reagent) is highly reactive. It is a very strong nucleophile and a very strong base. Now, let's see what happens when a Grignard reagent attacks a ketone or aldehyde.

In the previous section, we always started each mechanism by protonating the ketone (turning it into a better electrophile). That is not necessary here, because the Grignard reagent is such a strong nucleophile that it has no problem attacking a carbonyl group directly. In fact, we could *not* use acid catalysis here (even if we wanted to), because protons destroy Grignard reagents. For example, consider what happens when a Grignard reagent is exposed even to a very mild acid, such as water:

The Grignard reagent acts as a base and removes a proton from water, to form a more stable hydroxide ion. The negative charge is MUCH more stable on an electronegative atom (oxygen), and as a result, the reaction essentially goes to completion. This means that you can never use a Grignard reagent to attack a compound that has acidic protons. For example, the following reaction would not work:

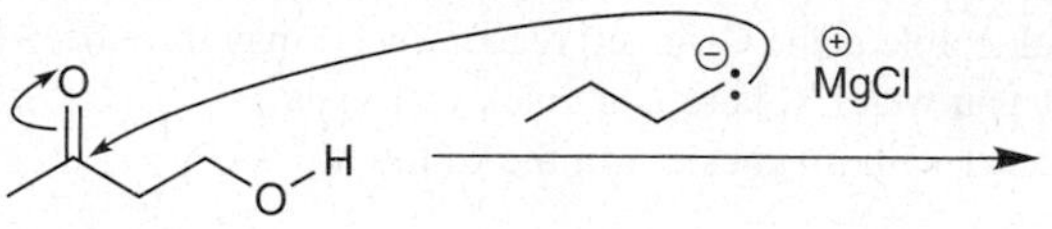

Because this would happen instead:

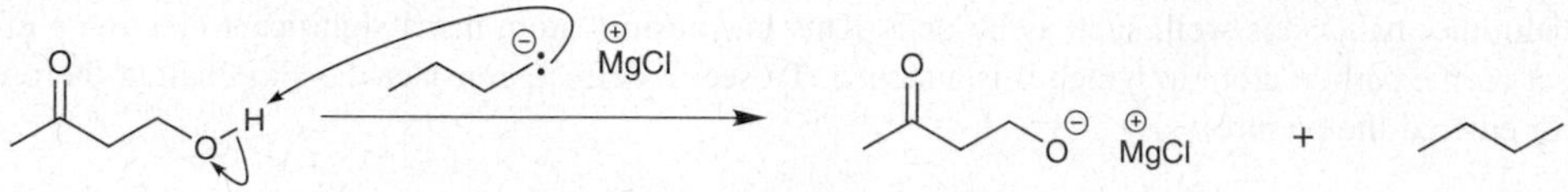

In general, proton transfers are faster than nucleophilic attack. And when the Grignard reagent removes a proton, it irreversibly destroys the Grignard reagent. Similarly, you could never prepare the following kinds of Grignard reagents:

These reagents could not be formed, because each of these reagents could react with itself to remove the negative charge on the carbon atom, for example:

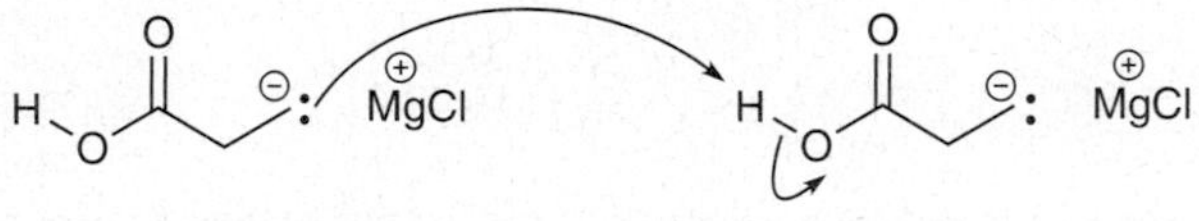

All of that was a quick review of Grignard reagents. Now let's see how Grignard reagents can attack a ketone or aldehyde. In the first step, the Grignard reagent attacks the carbon atom of the carbonyl group:

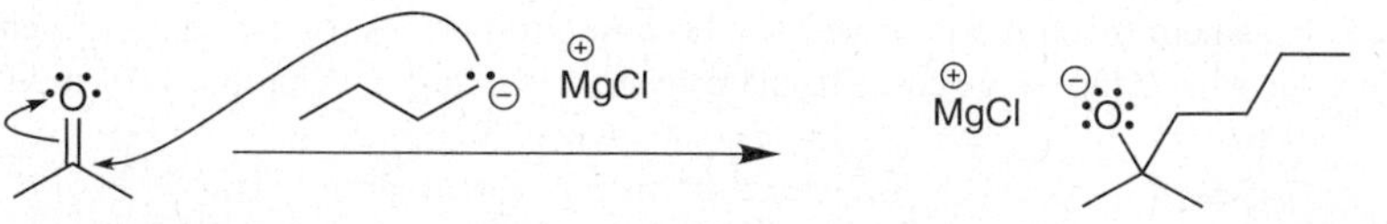

This intermediate then will attempt to re-form the carbonyl group, if it can. But let's see if it can. Remember our rules from the beginning of this chapter: re-form the carbonyl if you can, but never expel H⁻ or C⁻. This intermediate is NOT able to re-form the carbonyl group, because there are no leaving groups to expel. This is true whether the Grignard reagent attacks a ketone or an aldehyde:

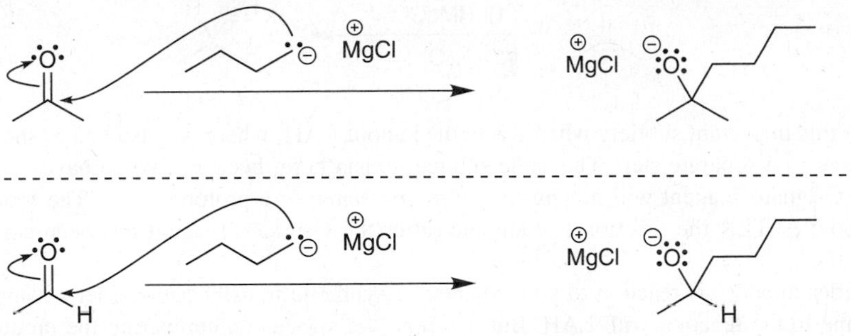

So, in either case, the reaction is complete, and we must now give the intermediate a proton to obtain the final product, which is an alcohol:

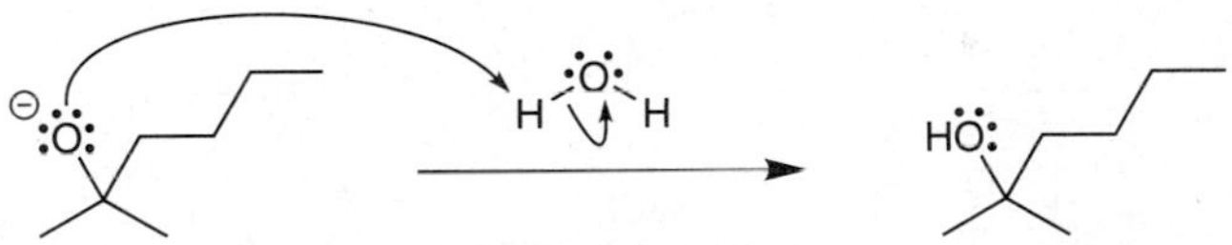

This reaction is not so different from the reactions we saw earlier in this chapter when we explored hydrogen nucleophiles (NaBH₄ and LAH). We saw a similar scenario there: the nucleophile attacked, and then the carbonyl group was NOT able to re-form because there was no leaving group. Compare one of those reactions to this reaction:

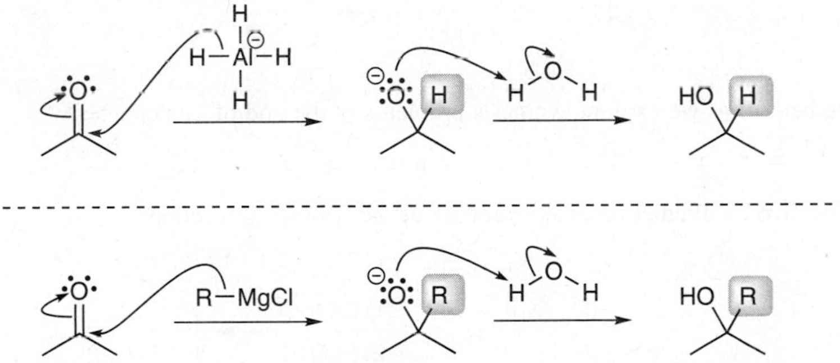

Notice that the mechanisms are identical. And it is worth a minute of time to think about why these reactions are so similar (while the other reactions in this chapter were different from these two reactions). What is special about these two reactions that makes them so similar? Remember our golden rule: never expel H⁻ or C⁻. So, if we attack a ketone (or aldehyde) with either H⁻ or C⁻, then the carbonyl group will be unable to re-form. And that is what these two reactions have in common.

When you write down the reagents of a Grignard reaction (in a synthesis problem), make sure you show the proton source *as a separate step*:

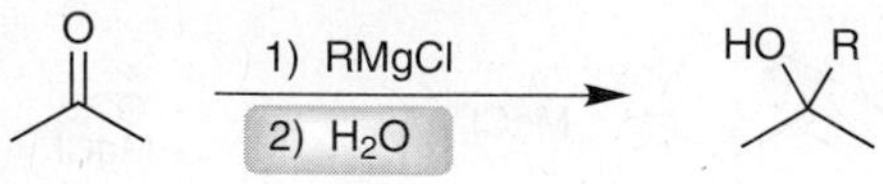

We saw this important subtlety when we learned about LAH, where we also had to show the proton source as a separate step. The same subtlety exists here, because (as we have very recently seen) a Grignard reagent will not survive *in the presence* of a proton source. The proton source must come AFTER the reaction is complete (after the Grignard reagent has been consumed by the reaction).

In order to add this reaction to your toolbox of synthetic transformations, let's compare it one more time to the reaction with LAH. But this time, let's focus on comparing the products, rather than comparing the mechanisms:

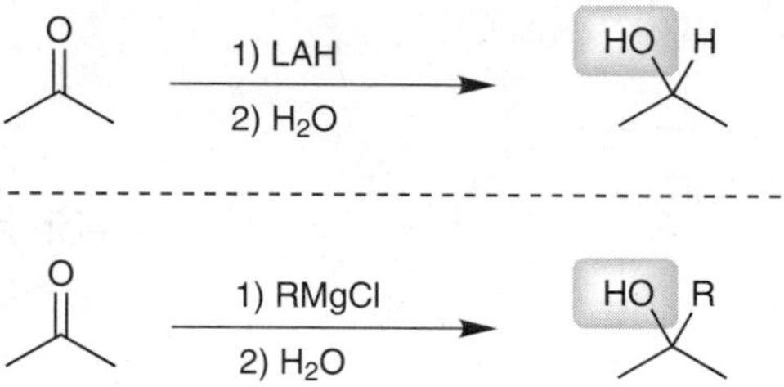

Notice that in both reactions, we are reducing the ketone to an alcohol. But in the case of a Grignard reaction, the reduction is accompanied by the installation of an alkyl group:

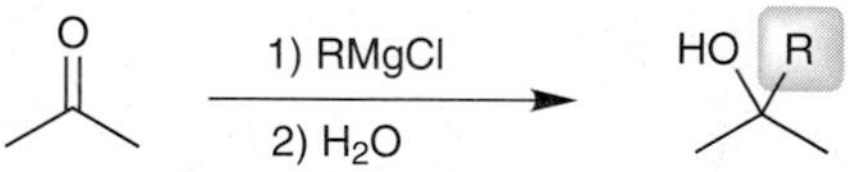

This will be helpful as we explore synthesis problems at the end of this chapter.

EXERCISE 5.61 Predict the major product of the following reaction:

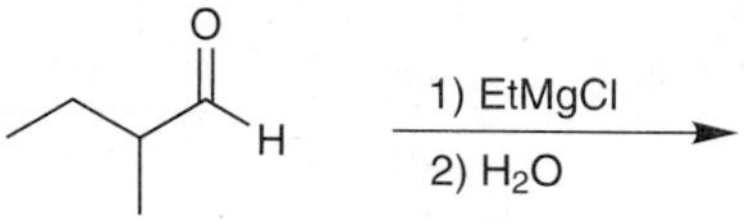

Answer The starting material is an aldehyde, and it is going to react with a Grignard reagent. First, the Grignard attacks:

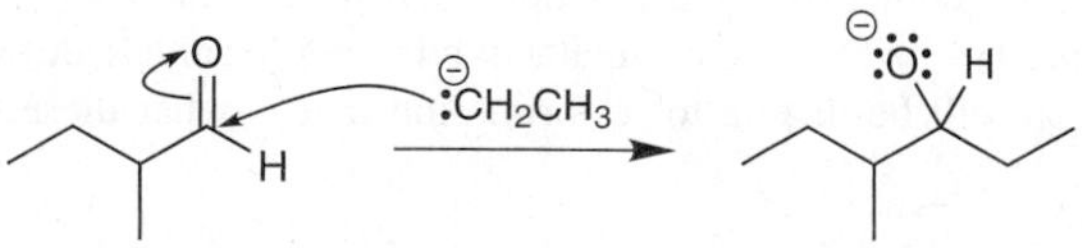

We cannot re-form the carbonyl group, because we cannot expel H^- or C^-, and there is nothing else to expel in this case. Our product (an alcohol) is obtained when a proton source is introduced, such as H_2O:

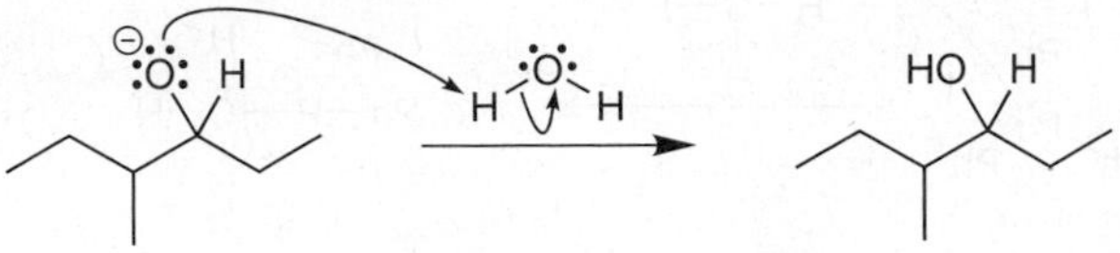

Remember that bond-line drawings don't have to show hydrogen atoms, so we can redraw the product:

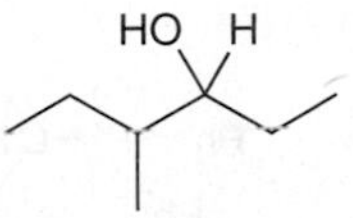

is the same as

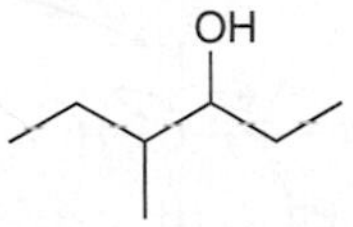

PROBLEMS   Predict the products for each of the following reactions:

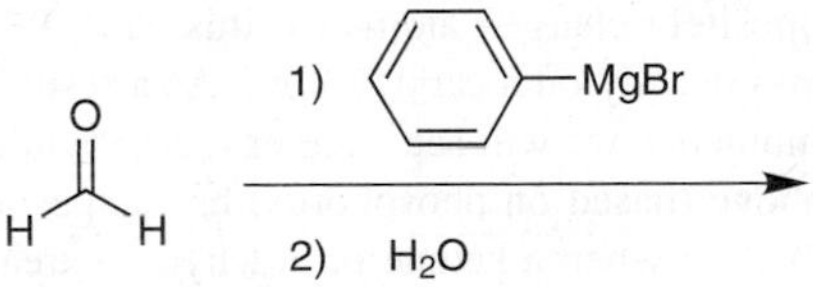

5.62

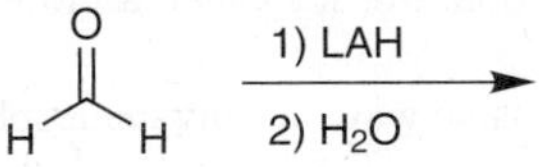

1) [phenyl]—MgBr

2) H_2O

5.63

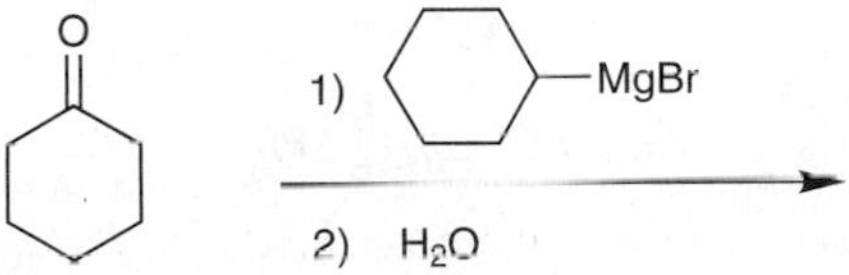

1) LAH

2) H_2O

5.64

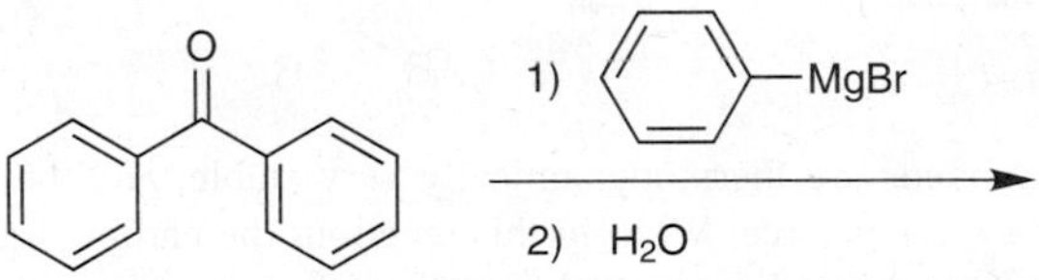

1) [cyclohexyl]—MgBr

2) H_2O

5.65

1) [phenyl]—MgBr

2) H_2O

There are two more carbon nucleophiles that we must explore. Both of them are different from the Grignard reagent. And you will certainly have to add these two new reactions to your toolbox. Both reactions involve "ylides" (pronounced "il–ids"). Let's take a close look at the first ylide, by exploring how it is prepared:

We start with a compound called triphenylphosphine:

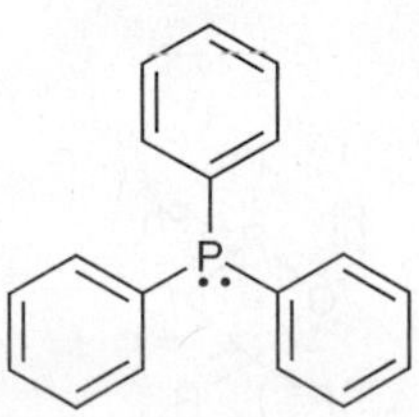

which can be drawn more quickly, like this

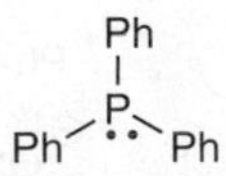

and we treat it with an alkyl halide (resulting in an S_N2 reaction):

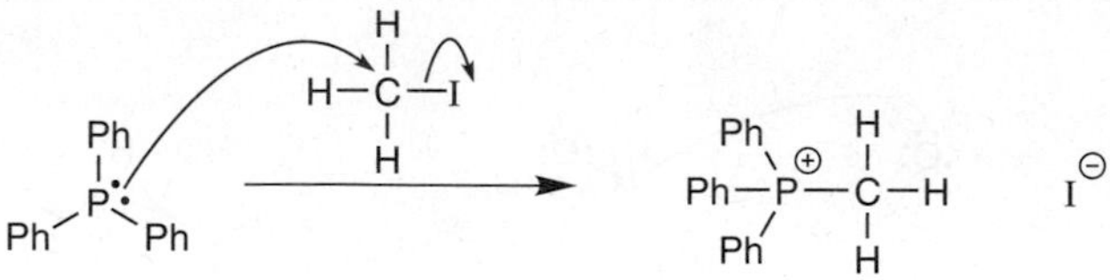

Then, we use a very strong base, such as butyllithium (BuLi) to remove a proton and form the ylide:

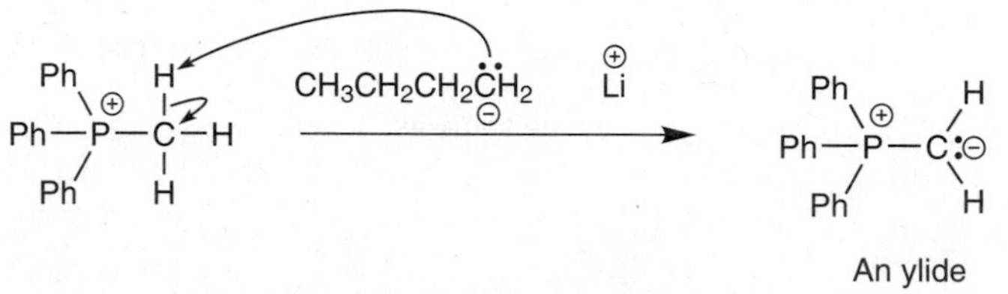

An ylide

An ylide is a compound with two adjacent, oppositely charged atoms (in this case, P^+ and C^-). Notice that this ylide has a region of high electron density on a carbon atom. As a result, this ylide can function as a carbon nucleophile. In a few moments, we will see another type of ylide (one that uses sulfur instead of phosphorus). The ylide above (based on phosphorus) has a special name. It is called a Wittig reagent (pronounced "Vittig"). And when a ketone or aldehyde is treated with a Wittig reagent, the observed reaction is called a Wittig reaction. So, let's take a close look at a mechanism for the Wittig reaction.

The Wittig reagent attacks the carbonyl group in much the same way that any nucleophile would attack a carbonyl group:

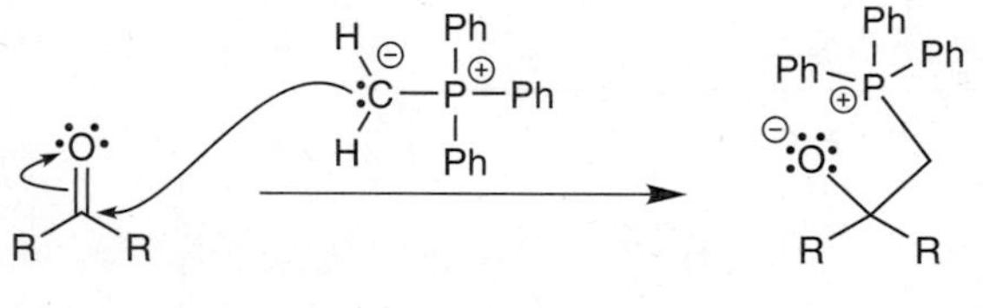

We have talked about how C=O bonds are thermodynamically very stable, and therefore, the formation of a carbonyl group is a driving force. Well, in this reaction, the carbonyl group cannot be re-formed, because H^- or C^- cannot be expelled. But there is something else that can happen.

There is another type of bond that is also very stable, and its formation can also serve as a driving force. Specifically, P—O bonds and P=O bonds are very, very strong. Chemists usually say it like this: "phosphorus is oxophilic," meaning that phosphorus demonstrates a tendency to form bonds with oxygen, if it can. And our intermediate is now perfectly set up for that to happen:

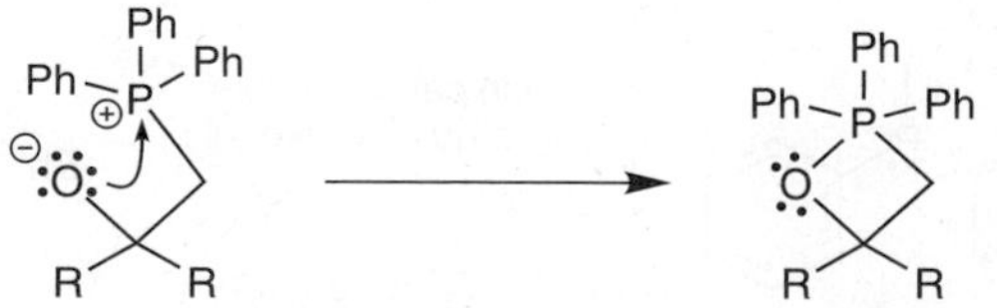

In fact, a P=O bond can now form, like this:

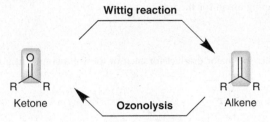

an alkene

And the product obtained is an alkene.

This reaction is incredibly useful for synthesis. We already saw how to convert an *alkene* into a *ketone*, using an ozonolysis reaction. Now, with a Wittig reaction, we can go either way:

Wittig reaction

Ketone   **Ozonolysis**   Alkene

You should always take special notice whenever you learn how to interconvert two functional groups (going in either direction), like above. We have seen several cases like this so far.

EXERCISE 5.66 Predict the product of the following reaction.

Answer We recognize the reagent to be a Wittig reagent. But this reagent is slightly different than the one we saw before. Compare this reagent to the one we saw on the previous page. This reagent has an extra methyl group:

The way to form a reagent like this is to use Et-I instead of Me-I when you are making the Wittig reagent, like this.

An ylide

The extra methyl group (highlighted above) comes along for the ride, and the final product looks like this:

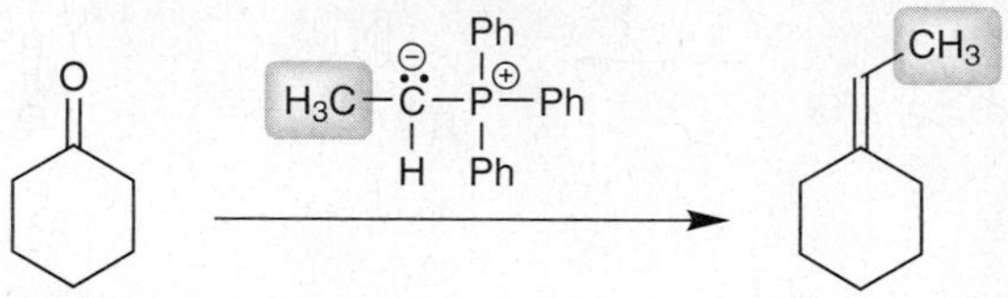

You should try to draw out a mechanism for this reaction to make sure that you can "watch" the extra carbon atom coming along for the ride.

PROBLEMS   Predict the major product for each of the following Wittig reactions:

5.67

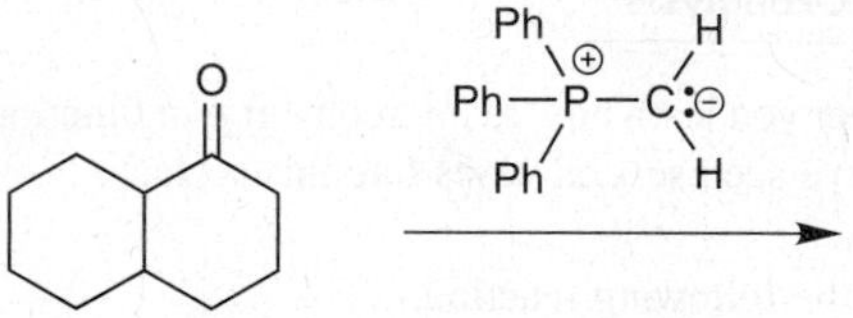

5.68

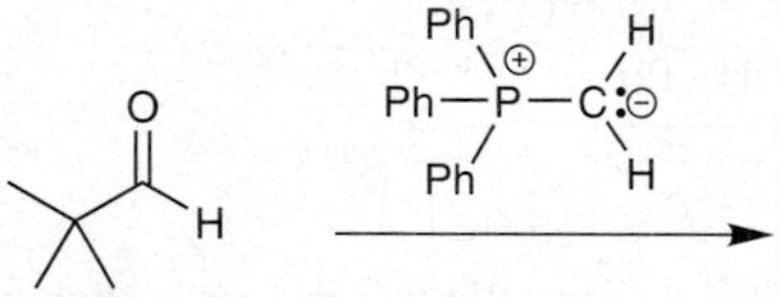

5.69

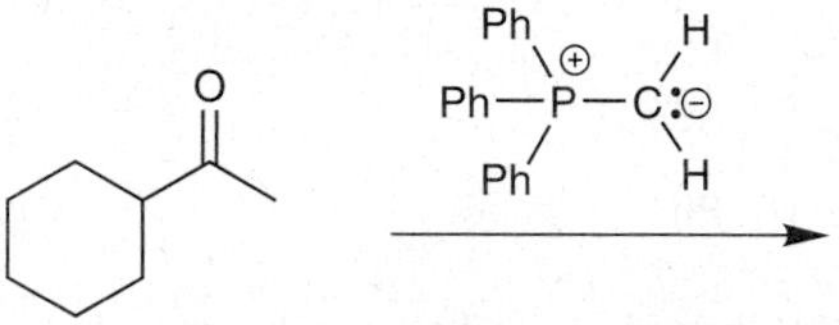

Now we will explore another kind of ylide. This time, it is a *sulfur* ylide, rather than a *phosphorus* ylide. Many instructors (and textbooks) do not cover sulfur ylides, so you might want to look through your lecture notes and textbook to find out if you are responsible for the following reaction.

The mechanism for forming a *sulfur* ylide is very similar to the mechanism for forming a *phosphorus* ylide. Let's compare them:

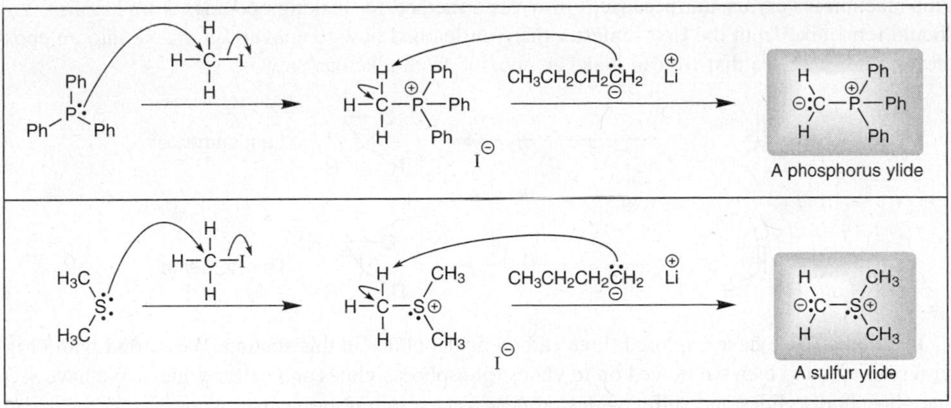

To form a phosphorus ylide, we start with dimethyl sulfide (DMS). From that point on, everything is the same: we attack an alkyl halide, and then we deprotonate with a very strong base to form the ylide.

But when a sulfur ylide attacks a ketone (or aldehyde), we observe a very different product. We **don't** obtain an alkene as our product (like we did in the Wittig reaction). Instead, we obtain an epoxide:

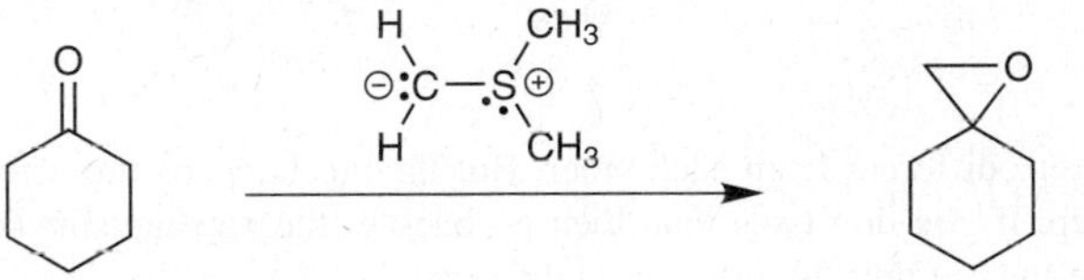

Let's see how this product is formed. The sulfur ylide attacks the carbonyl group, just like any other nucleophile attacks a carbonyl group:

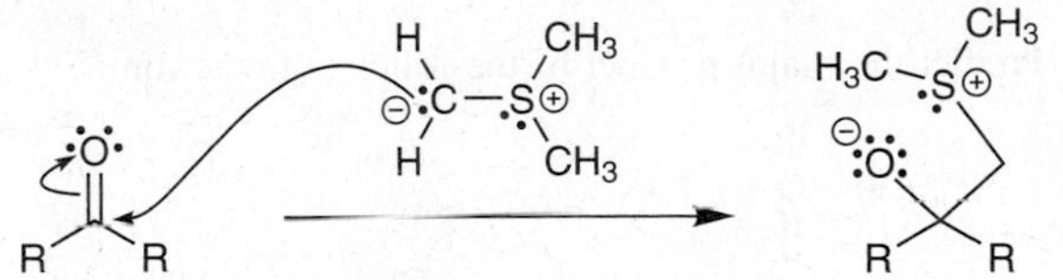

But, here is where the reaction is different than a Wittig reaction, because sulfur and oxygen do not form a bond the way phosphorus and oxygen did. Instead, the oxygen atom attacks in an intramolecular S_N2 reaction, expelling DMS as a leaving group:

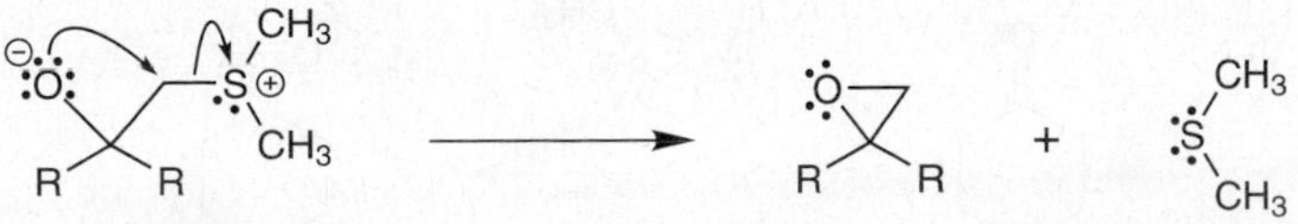

This reaction is very useful, because it provides a method for making epoxides from ketones. You should remember from the first semester that you learned how to convert an *alkene* into an epoxide. But now, we see that we can make an epoxide from a ketone as well:

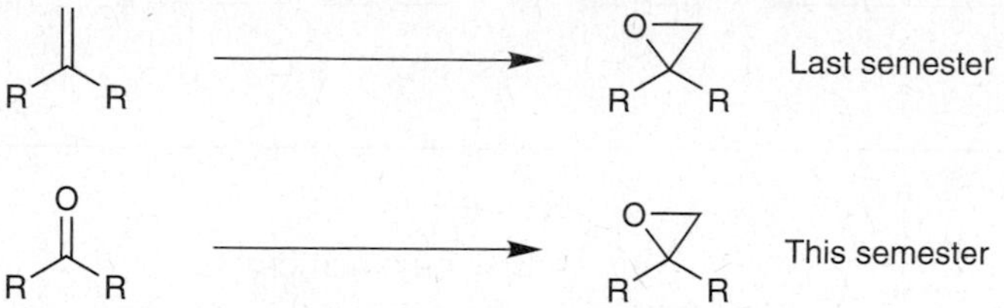

Last semester

This semester

In summary, we have explored three carbon nucleophiles in this section. We started with Grignard reagents, and then we moved on to ylides (phosphorus ylides and sulfur ylides). We have seen that phosphorus ylides and sulfur ylides produce very different products:

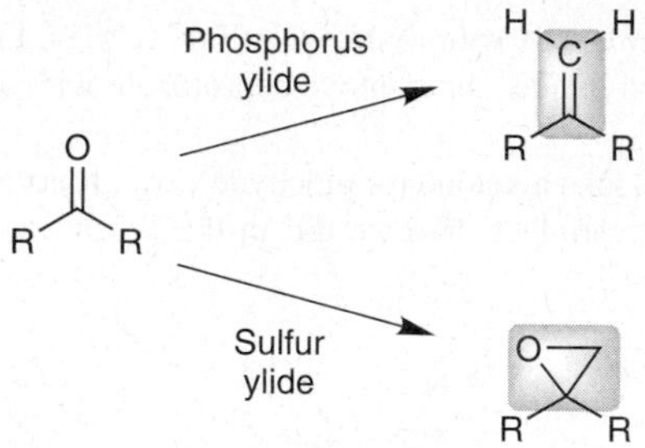

The products are very different from each other. But the mechanisms are actually identical up until the very last step. If you don't see that, then go back to the mechanisms of each of these reactions, and compare them. Only the last step is different.

We have now seen the same common idea so many times in this chapter. We have seen that you can often have two reactions that produce very different products, but when you look carefully at the mechanisms, you see that they are actually very similar.

EXERCISE 5.70 Predict the major product of the following reaction:

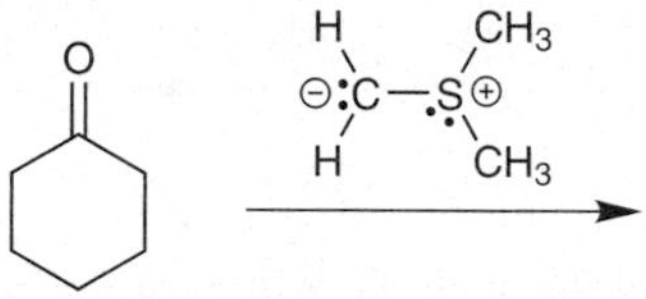

Answer This reagent is just a sulfur ylide, which is used to convert ketones into epoxides. So, our product is:

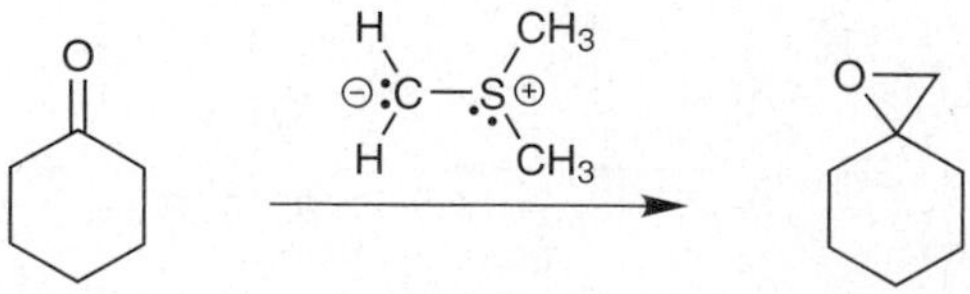

PROBLEMS   Predict the major product for each of the following reactions:

5.71

5.72

5.73

5.8   SOME IMPORTANT EXCEPTIONS TO THE RULE

In the beginning of this chapter, we saw a golden rule that helped us understand most of the chemistry that we explored. That rule was: always re-form the carbonyl if you can, but never expel H^- or C^-.

The truth is that there are a few, rare exceptions to this rule. In this section, we will look at two of these exceptions.

In the Cannizzaro reaction, it "seems" as though we are expelling H^- to re-form a carbonyl. I am not going to go into great detail on the Cannizzaro reaction, because this reaction has very little synthetic utility. It is unlikely that you will use this reaction more than once, if at all. So, I will just mention it in passing. Look up that reaction in your textbook and in your lecture notes. If you don't need to know that reaction, then you can ignore it. But if you are responsible for knowing that reaction, then you should look carefully at the mechanism in your textbook. If you focus on the step where H^- gets expelled, you will see that H^- is not really expelled by itself. Rather, it is transferred from one place to another. And in that sense, we can understand it a bit better. It is true that H^- is too unstable to ever leave as a leaving group. That is why we never expel H^- into solution. But in the Cannizzaro reaction, it never actually leaves as a leaving group.

We will now explore one other exception to the golden rule. There is a reaction where it seems like we are re-forming a carbonyl group to expel C^-. This reaction, called the Baeyer-Villiger reaction, is extremely useful. If you know how to use it properly, you will find that you might use it many times to solve synthesis problems in this course. So, we will spend some time covering that reaction now.

The Baeyer-Villiger reaction uses a peroxy acid as the reagent:

A peroxy acid has one
more oxygen atom than
a carboxylic acid

R can be anything. It can be a methyl group, or a much larger group. There are many common peroxy acids. The most common peroxy acid is MCPBA (*meta*-chloro **p**er**b**enzoic **a**cid), so when you see the letters MCPBA, you should recognize that we are talking about a peroxy acid:

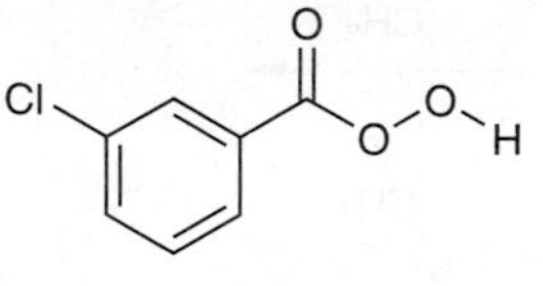

MCPBA

This reagent (or any other peroxy acid) can be used to insert an oxygen atom next to the carbonyl group of a ketone, producing an ester:

You can also use the same reaction to convert an aldehyde into a carboxylic acid:

Once again, the outcome of this reaction is to "insert" an oxygen atom next to the carbonyl group. This is a very useful synthetic trick, so let's see a mechanism for how it works. First, the peroxy acid functions as a nucleophile and attacks the carbonyl group, just like any other nucleophile:

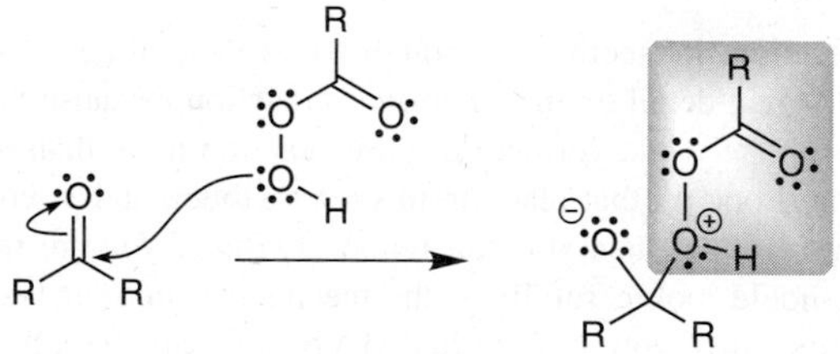

The resulting intermediate can then undergo an intramolecular proton transfer, like this:

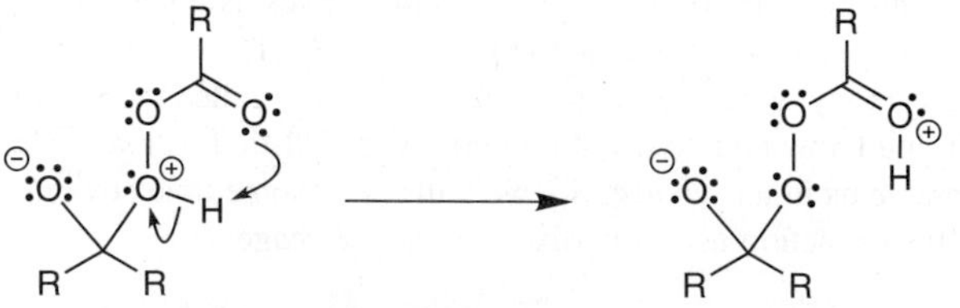

This step can occur in an intramolecular fashion (rather than requiring two separate proton transfer steps) because it involves a five-membered transition state (rather than a four-membered transition state, which would be too strained). And the driving force for this intramolecular proton transfer step can be explained in terms of the stability of the positive charge. Specifically, a localized positive charge has been replaced with a resonance-stabilized positive charge.

If we focus on the intermediate generated by the intramolecular proton transfer step shown above, we would conclude that the only way to re-form the carbonyl group would be to expel the group that was recently installed:

It is true that this probably happens, but it doesn't lead to the formation of a new product. It simply regenerates the starting ketone. So, we apply our golden rule to see if anything else can happen. In other words, we look to see if we can expel any other leaving group. Our golden rule tells us never to expel H$^-$ or C$^-$. And we don't see any other groups to expel. BUT, this is the exception to the golden rule. When a carbonyl group is attacked by a peroxy acid, there actually is something else that can happen. It is unique to this situation, and you will not see this in any other mechanism (so don't worry about trying to apply this next step to any other mechanism). A rearrangement occurs, in which an alkyl group is said to *migrate*:

Look carefully at the fate of the migrating alkyl group (highlighted above). Notice that the carbonyl group is re-forming to expel this R group, which migrates to the nearby oxygen atom. In other words, it looks like we are expelling C$^-$. But the truth is that we are not *really* expelling C$^-$ into solution by itself. After all, C$^-$ is too unstable. Rather, it is just **migrating** from one place to another (it is migrating over to attack the neighboring oxygen atom). It never really becomes C$^-$ for any period of time. And that explains how we can have an exception here.

This mechanism is truly bizarre, and you should not worry if you feel that you would not be able to predict when this could happen in other situations. This is probably the first time you are ever seeing a rearrangement that does not involve a carbocation. So, this is truly different. You will not need to apply this mechanism to any other situations. So for now, don't focus too much on this mechanism. Instead, let's focus on how to use this reaction when you are solving synthesis problems, because it will be a very useful reaction for you to have in your back pocket.

In order to use this reaction properly, you will need to know how to predict where the oxygen atom is installed. For example, consider the following ketone:

This ketone is unsymmetrical, so we must decide where the oxygen atom will be installed during a Baeyer-Villiger reaction. Which of the following two products are expected?

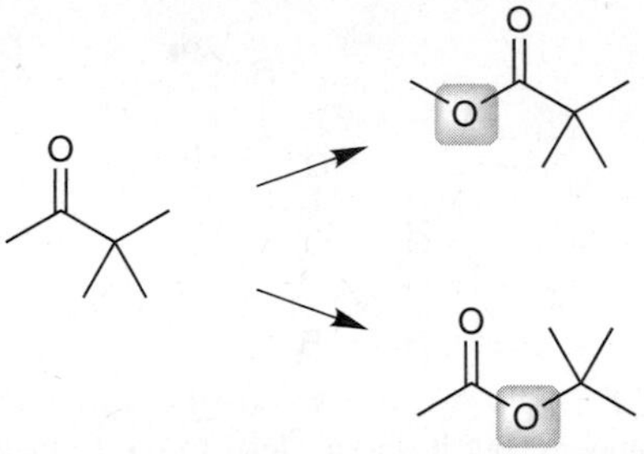

To answer this question, we need to know which alkyl group is more likely to migrate. If you look back at the mechanism above, you will see that the migrating R group is the one that ends up connected to the oxygen atom in the product. So, we just have to decide which alkyl group can migrate faster.

There is an order to how fast alkyl groups can migrate in this reaction, and we call it "migratory aptitude":

H > 3° > 2° or Ph > 1° > methyl

Notice that phenyl groups have similar migratory aptitude as secondary alkyl groups. Also notice that H migrates the fastest. This explains how we can use this reaction to convert aldehydes into carboxylic acids:

H migrates faster than any other group, so it doesn't even matter what the R group is.

If there is no H in the compound (in other words, if you are starting with a ketone instead of an aldehyde), then you should look for the most substituted alkyl group. For example:

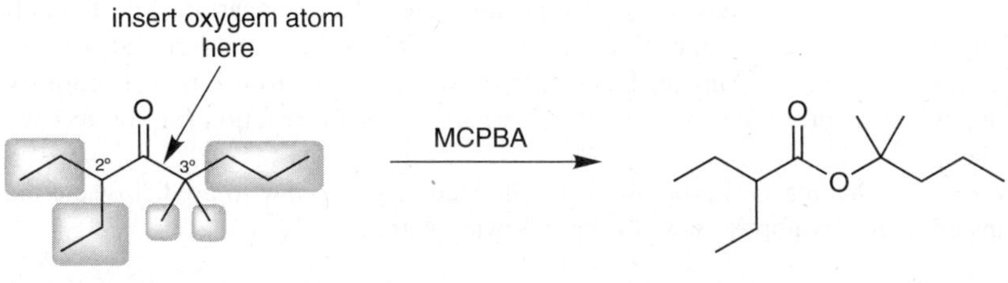

Notice that the oxygen atom is inserted on the side that is more substituted.

In order to use this reaction in synthesis, you must make sure that you can predict where the oxygen atom will insert. Let's do some problems to make sure you got it:

EXERCISE 5.74 Predict the major product of the following reaction:

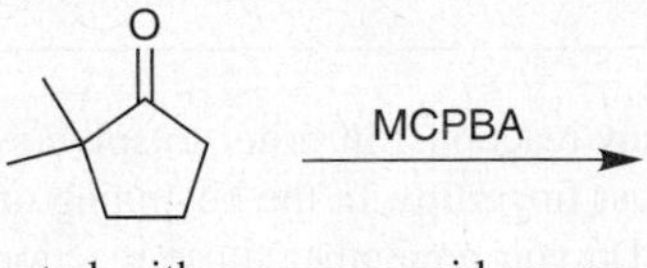

Answer A ketone is being treated with a peroxy acid, so we expect a Baeyer-Villiger reaction to occur. We look closely at our starting ketone, and we see that it is unsymmetrical. So, we must predict where the oxygen atom will insert itself. We look at both sides, and we see that the left side is more substituted. The more substituted R group will migrate faster, and that is where the oxygen atom will be inserted.

This specific case is an interesting example, because the insertion of an oxygen atom causes a ring expansion:

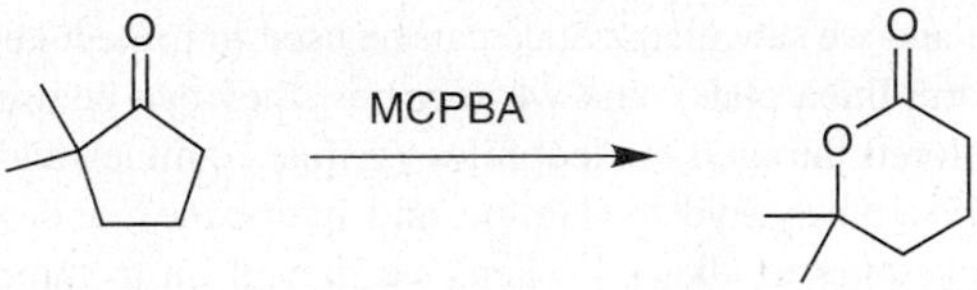

The product is not just an ester, but it is a *cyclic* ester. Cyclic esters are also called *lactones*.

PROBLEMS Predict the major product for each of the following reactions:

5.75

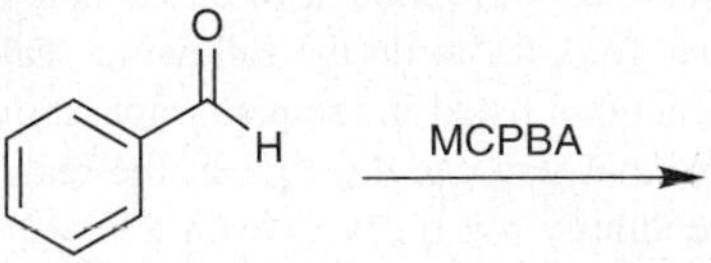

5.76

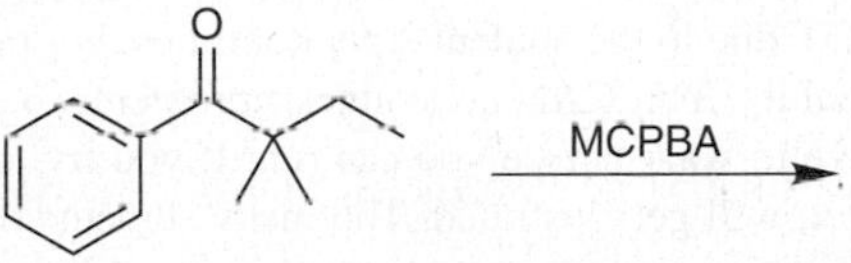

5.77

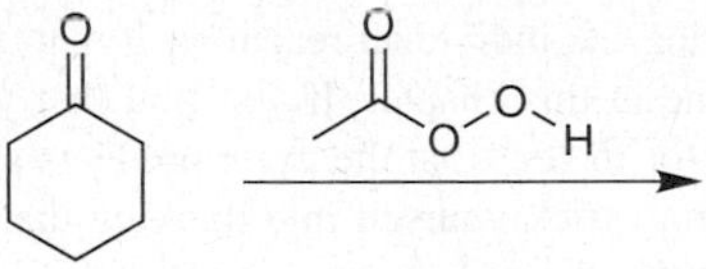

5.78

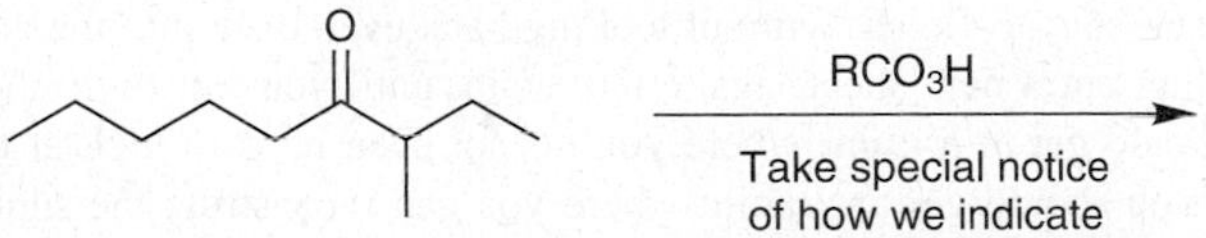

Take special notice of how we indicate the peroxy acid here. You must recognize it when you see it this way.

5.79

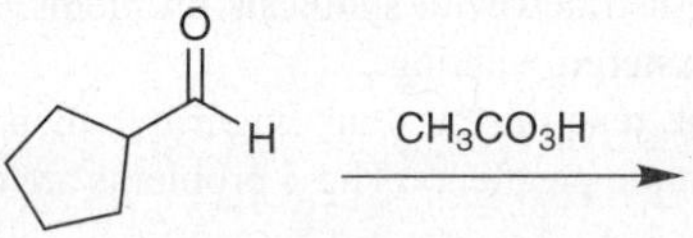

5.9 HOW TO APPROACH SYNTHESIS PROBLEMS

In this chapter, we have seen many reactions. In order to solve synthesis problems, you will need to have all of these reactions at your fingertips. In the beginning of this chapter, we saw a few ways to make aldehydes and ketones. Do you remember those reactions? If you don't, then you are in trouble. This is why organic chemistry can get tough at times. It is not sufficient to be a master of mechanisms. That is an excellent start, and it builds an excellent foundation for understanding the material. But at the end of the day, you have to be able to solve synthesis problems. And in order to do that, you must have all of the reactions organized in your mind.

Let's start with a short review of everything we saw:

We started with a few ways of making ketones and aldehydes (two ways to oxidize, and then an ozonolysis). Then, we explored hydrogen nucleophiles (NaBH$_4$ and LAH) and oxygen nucleophiles (making acetals), and we saw that acetals can be used to protect ketones. Then we explored sulfur nucleophiles (to form thioacetals), and we saw how they can be used to *reduce* ketones and aldehydes. Next, we explored nitrogen nucleophiles (primary amines and secondary amines), we examined special primary amines (hydroxylamine and hydrazine), and we saw how hydrazones could be used to reduce ketones to alkanes. Then, we moved on to three kinds of carbon nucleophiles—Grignard reagents, phosphorus ylides, and sulfur ylides. Finally, we examined the Baeyer-Villiger oxidation, focusing on its utility for synthesis. That is everything we saw in this chapter.

In this last section of the chapter, we will bring everything together to solve synthesis problems. The first step is to make sure that you know these reactions well enough to claim that you have them at your fingertips. To ensure that you achieve that goal, try to do the following. Take a separate piece of paper and try to write down all of the reactions listed in the paragraph above (without looking back in the chapter, if possible). Make sure that you can draw all of the reagents, and all of the products. If you cannot do this, then you are simply not ready to even START thinking about synthesis problems. Students often complain that they just don't know how to approach synthesis problems. But the difficulty is usually NOT due to the student's poor abilities, but rather, the difficulty arises from the student's poor study habits. You CAN do synthesis problems. You might even enjoy them, believe it or not. But, you have to walk before you can run. If you try to run before you learn how to walk, you will trip and you will get frustrated. Too many students make this mistake with synthesis problems.

So, take my advice, and focus right now on mastering the individual reactions. Try to fill out a blank sheet of paper with everything that we have done in this chapter. If you find that you have to look back into the chapter to get the exact reagents (or to see what the exact products are), then that is fine. It is part of the studying process—BUT don't trick yourself into thinking that you are ready for synthesis problems once you have filled out the sheet. You are not ready until you can fill out the entire sheet of paper, start-to-finish, without looking back even once into the chapter to get the fine details. Keep filling out a new sheet, again and again, until you can do it all without looking back. Ideally, you should get to a point where you do not even need to look at the short summary that we just gave. You should get to a point where you can reconstruct the summary in your head, and then based on that summary, you should be able to write a list of all the reactions.

It sounds like a lot of work. And it is. It will take you a while. But when you are done, you will be in an excellent position to start tackling synthesis problems. If you get lazy, and you decide to skip this advice, then don't complain later if you are frustrated with synthesis problems. It would be your own fault for trying to run before you have mastered walking.

Once you get to the point where you have all of the reactions at your fingertips, then you can come back to here, and try to prove it, by doing some simple problems. These problems are designed to test you on your ability to list the reagents that you would need in order to do simple one-step transformations. Once you have all of the reactions down cold, then we will be able to move on and conquer some multistep synthesis problems.

EXERCISE 5.80   What reagents would you use to achieve the following transformation:

Answer   By inspecting the product, we can immediately determine that we will need a nitrogen nucleophile. We just need to decide what kind of nitrogen nucleophile. Since our product is an enamine, we will need a secondary amine. When we look at the product, we can determine that we would need to use the following secondary amine:

Finally, we just need to decide if there are any special conditions that should be mentioned. And we did learn that there are special conditions for the reaction between a ketone and a secondary amine. Specifically, we need to have acid-catalysis, and Dean-Stark conditions. So, our answer is:

PROBLEMS   The following problems are designed to be simple, so that you can prove to yourself that you know these reactions cold. I highly recommend that you photocopy the following problems *before* filling them out. You might find that you get stuck on a few problems, and it might be helpful for you to come back to these problems in the near future to fill them out again. The following problems are not listed in the order in which they appear in this chapter.

5.81

5.82

5.83

5.84

5.85

5.86

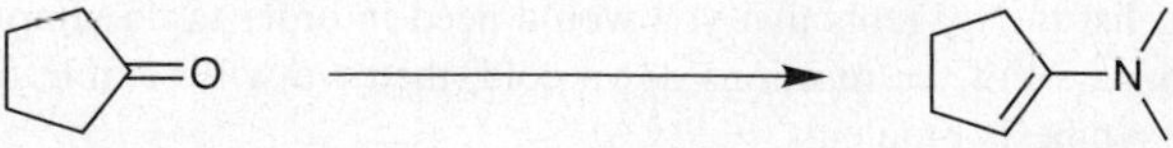

5.87

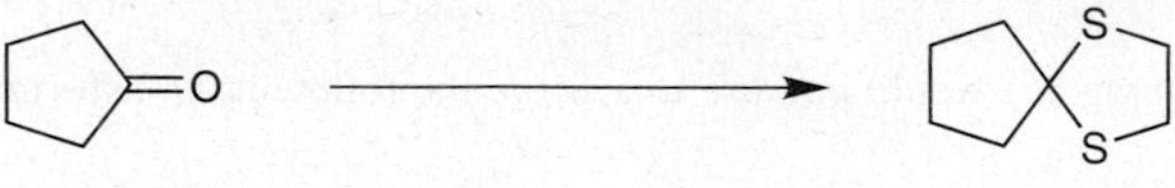

5.88

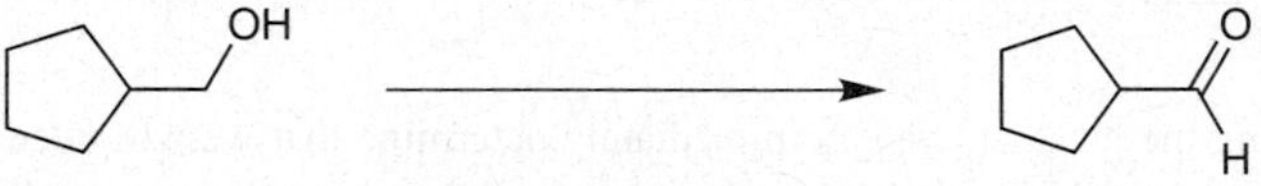

5.89

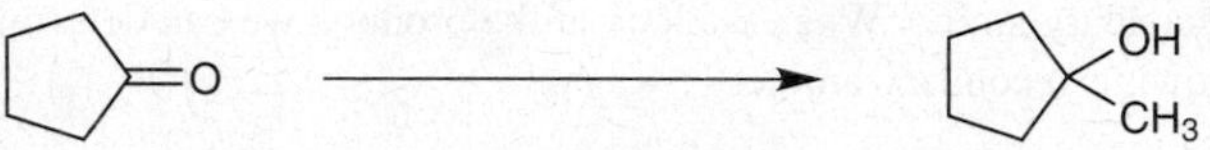

5.90

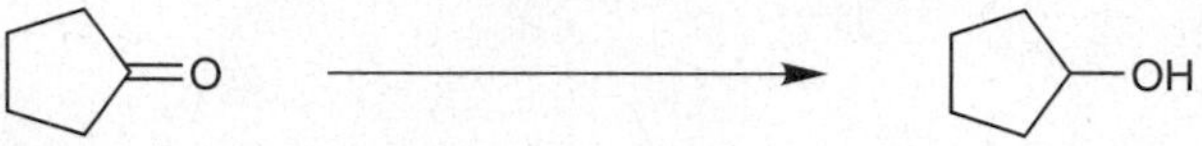

5.91

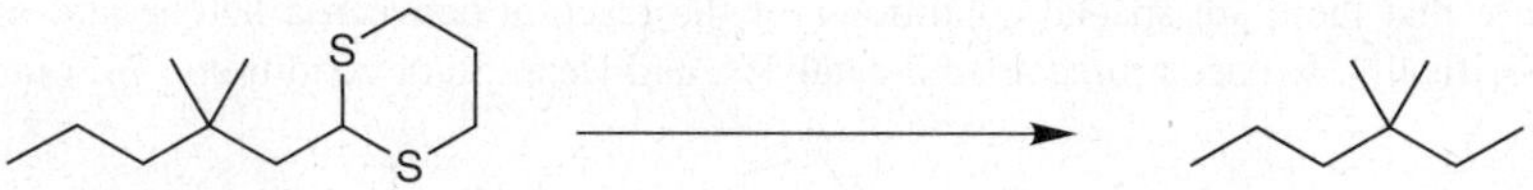

5.92

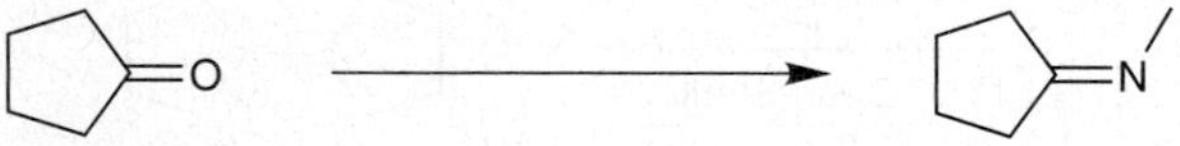

5.93

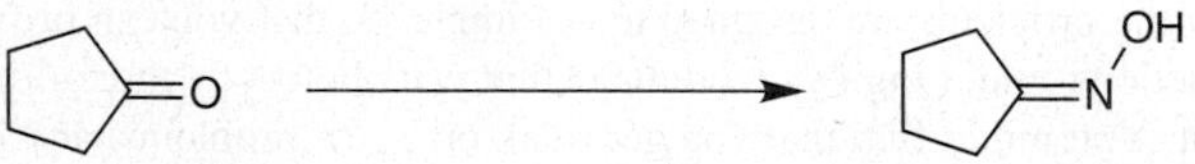

5.94

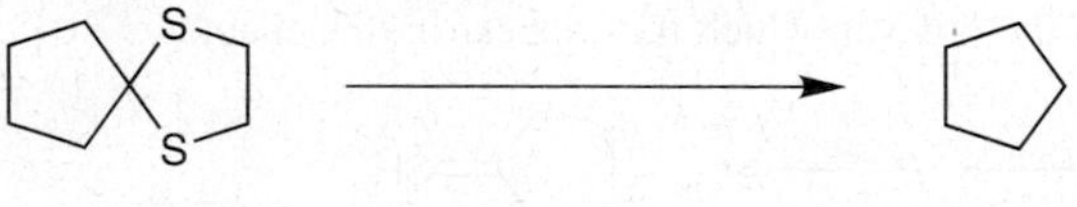

5.95

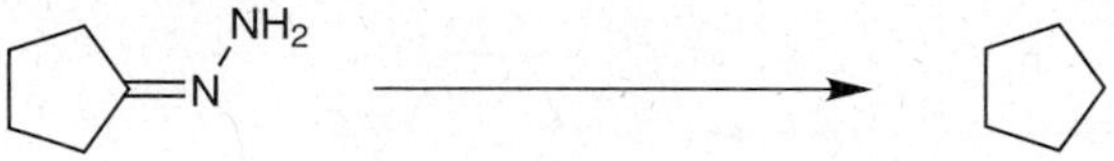

5.96

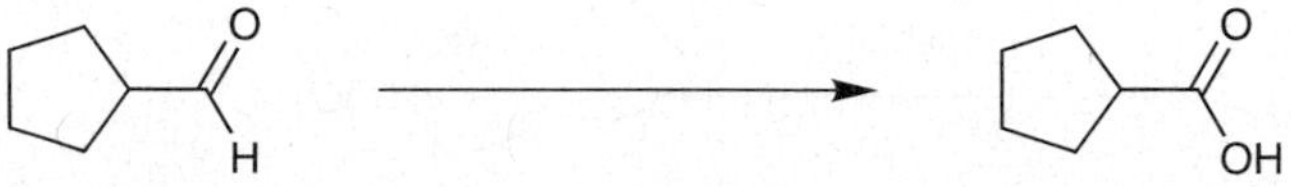

5.97

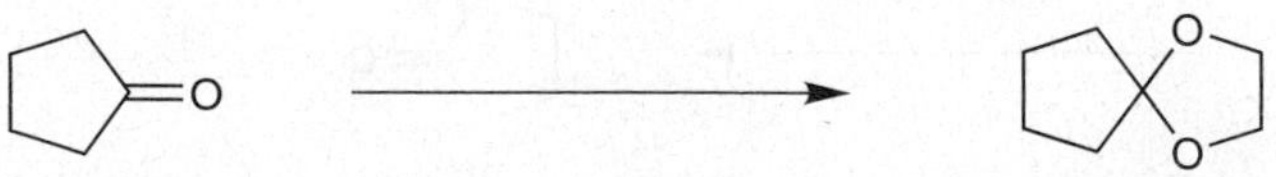

5.98

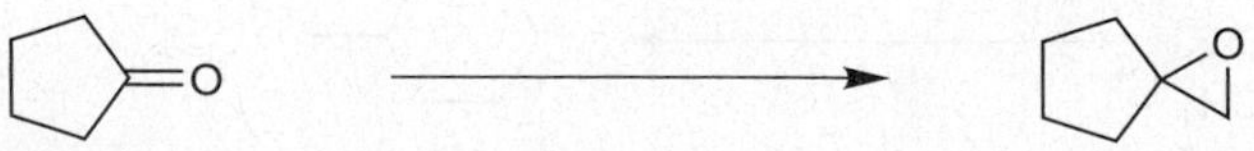

5.99

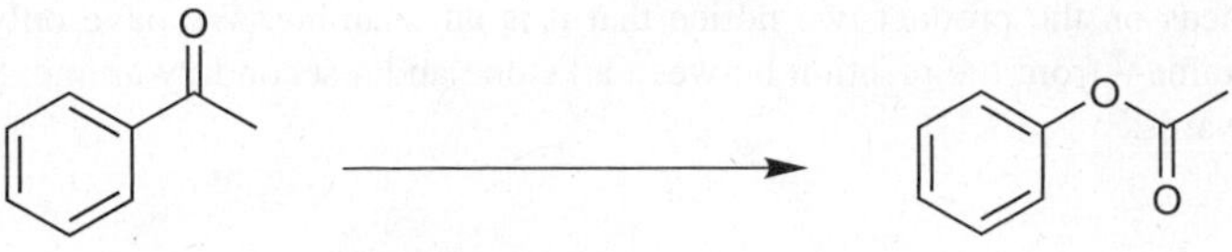

5.100

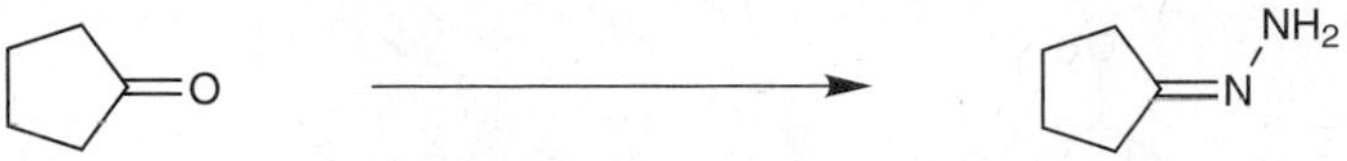

If you felt comfortable with those problems, then you should be ready to move on to solving some multistep synthesis problems. Let's see an example:

EXERCISE 5.101   Propose an efficient synthesis for the following transformation:

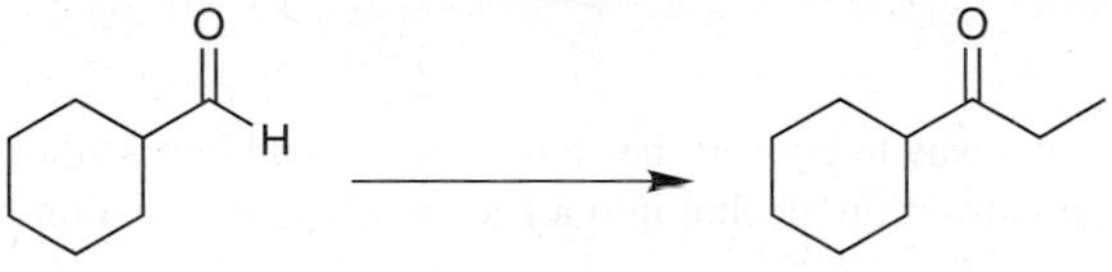

Answer   This transformation is a bit trickier than the previous problems, because it cannot be achieved in one step. We need to install an ethyl group, while maintaining the presence of the carbonyl group. If we use a Grignard reagent, we can install the ethyl group, but we will reduce the ketone to an alcohol in the process:

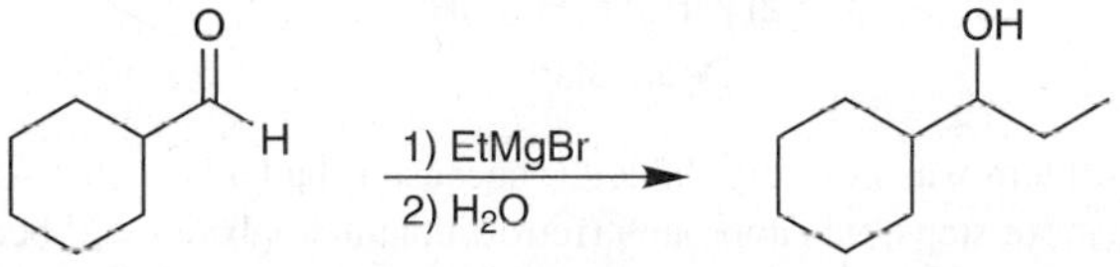

1) EtMgBr
2) H₂O

But, this issue can be easily overcome, because we can oxidize the alcohol back up to a ketone:

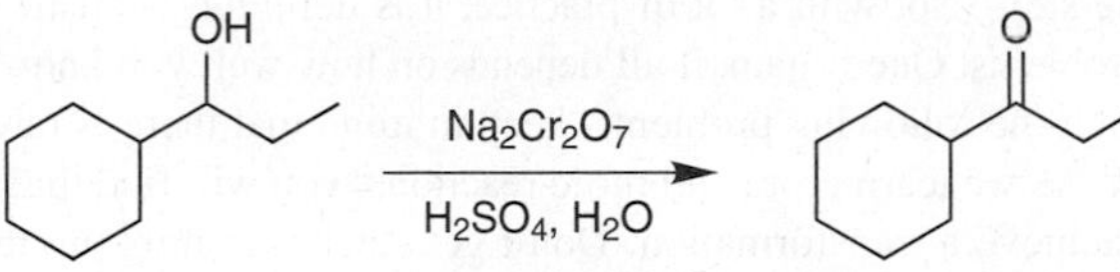

Na₂Cr₂O₇
H₂SO₄, H₂O

So, we have a two-step synthesis to accomplish this transformation.

Before you try to solve some problems yourself, there is one more important point to make about synthesis problems. Very often, it is helpful to work "backwards." We call this *retrosynthetic analysis*. Let's see an example:

EXERCISE 5.102   Propose an efficient synthesis for the following transformation:

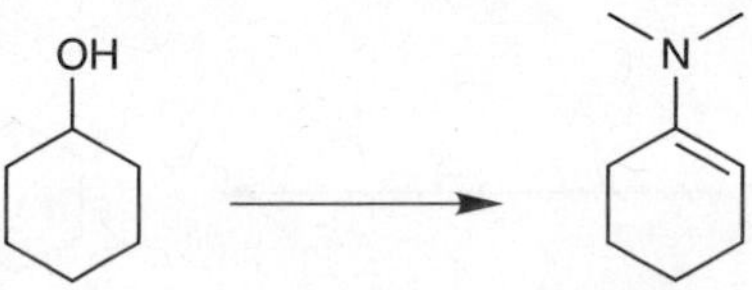

Answer If we focus on the product, we notice that it is an enamine. We have only seen one way to make an enamine—from the reaction between a ketone and a secondary amine. So, we can work our way backwards:

All we need to do is find a way to convert the starting compound into a ketone. And we have seen how to do that. We can convert an alcohol into a ketone using an oxidation reaction. So, our synthesis is:

1) Na$_2$Cr$_2$O$_7$, H$_2$SO$_4$, H$_2$O

2) [H$^+$], (CH$_3$)$_2$NH
   Dean-Stark

Of course, this last problem was not very difficult, because it had a two-step solution. As you solve problems that require more steps, this approach (retrosynthetic analysis) will become more and more important. But don't worry. You won't have to solve any problems that require ten steps. That is way beyond the scope of this course. You will generally not have to deal with syntheses that require more than three to five steps. So, with a lot of practice, it is definitely realistic to become a master of solving synthesis problems. Once again, it all depends on how well you know all of the reactions.

As you work through the following problems, keep in mind that there is rarely only one answer to a synthesis problem. As we learn more and more reactions, you will find that there are often multiple correct ways to achieve a transformation. Don't get stuck into thinking that you have to find THE answer. You may even come across a perfectly acceptable answer that no one else in the class thought of. Those are the most exciting moments. There actually is room for you to express some creativity when you solve synthesis problems. Now let's get some practice:

PROBLEMS For each of the following problems, suggest an efficient synthesis. Remember that there might be more than one correct answer for each of these problems. If you propose an answer, and it does not match the answer in the back of the book, do not be discouraged. Carefully analyze your answer, because it might also be correct.

5.103

5.104

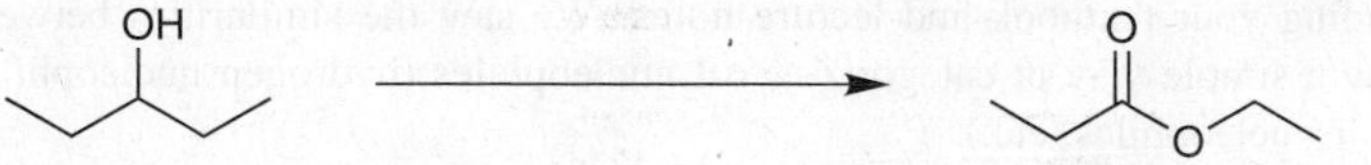

5.105

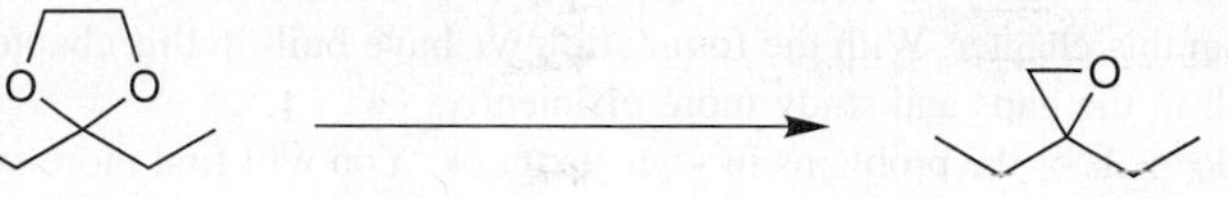

5.106

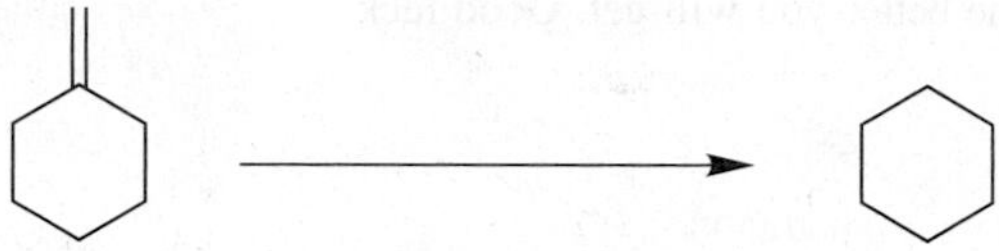

5.107

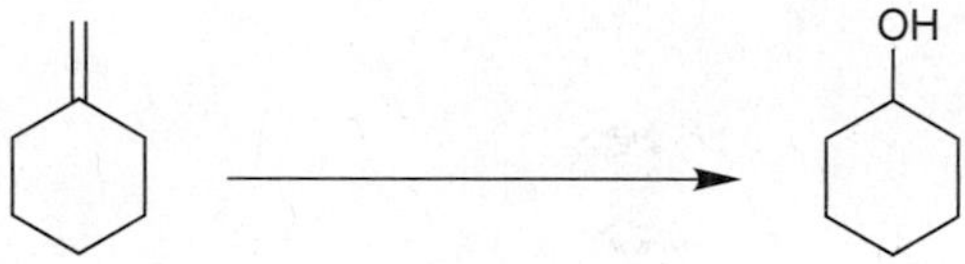

5.108

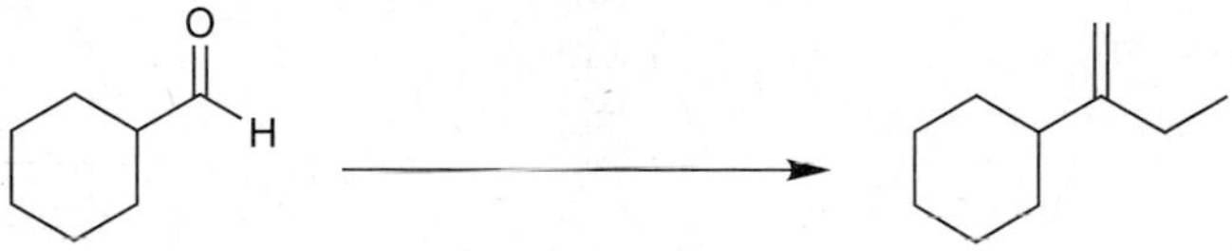

5.109

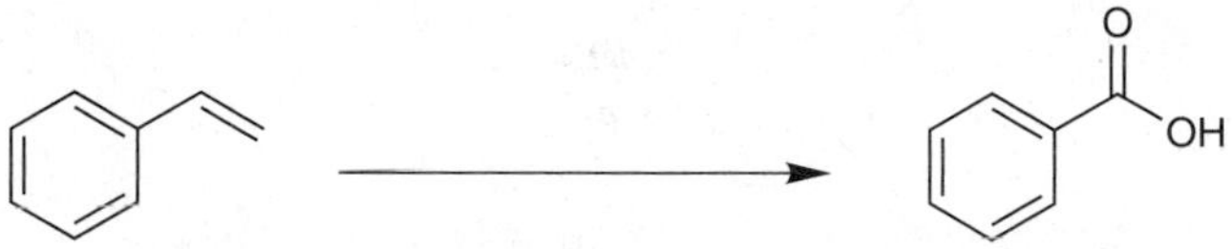

5.110

5.111

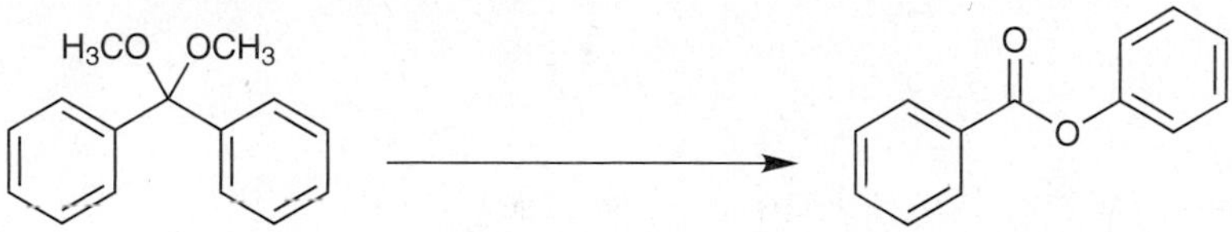

5.112

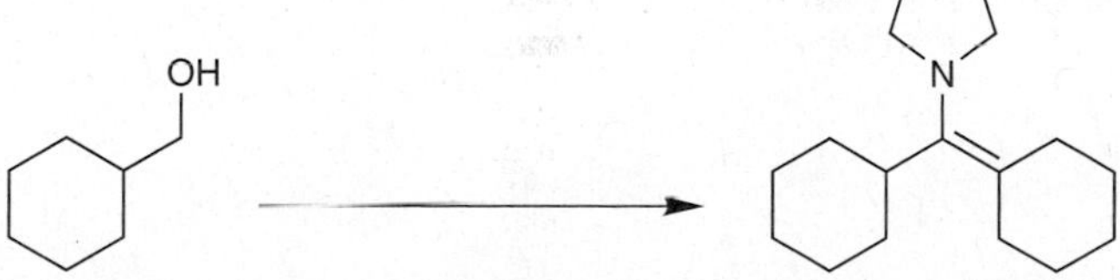

In this chapter, we did **not** see EVERY reaction that is in your textbook or lecture notes. We covered the core reactions (probably 90% or 95% of the reactions you need to know). The goal of this chapter was NOT to cover every reaction. Rather, our goal was to lay a foundation for you when you are reading your textbook and lecture notes. We saw the similarities between mechanisms, and we saw a simple way of categorizing all nucleophiles (hydrogen nucleophiles, oxygen nucleophiles, sulfur nucleophiles, etc.).

Now you can go back through your textbook and lecture notes, and look for the reactions that we did not cover here in this chapter. With the foundation we have built in this chapter, you should be in good shape to fill in the gaps and study more efficiently.

And make sure to do ALL of the problems in your textbook. You will find more synthesis problems there. The more you practice, the better you will get. Good luck.

CHAPTER 6

CARBOXYLIC ACID DERIVATIVES

6.1 REACTIVITY OF CARBOXYLIC ACID DERIVATIVES

Carboxylic acid derivatives are similar to carboxylic acids, but the OH group has been replaced with a different group (Z),

Carboxylic acid   Carboxylic acid derivative

where Z is a heteroatom (an atom other than C or H, such as Cl, O, N, etc.). The four most common types of carboxylic acid derivatives are shown below:

acid halide   acid anhydride   ester   amide

The chemistry of carboxylic acid derivatives is different from the chemistry of ketones and aldehydes, because carboxylic acid derivatives possess a built-in leaving group, which allows the carbonyl group to re-form after being attacked:

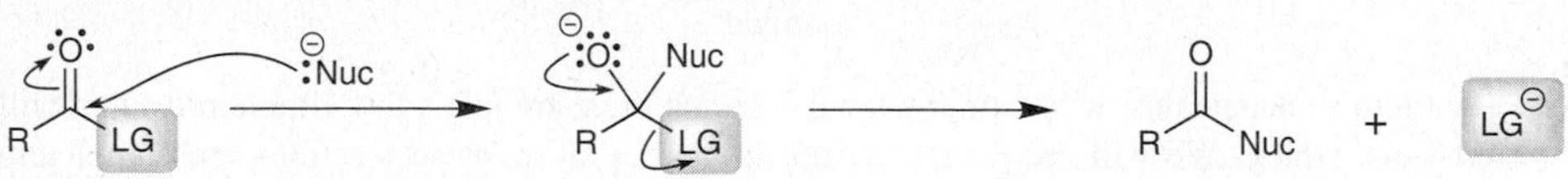

The identity of the leaving group will determine how reactive the compound is. For example, acid chlorides are extremely reactive because the built-in leaving group (chloride) is very stable:

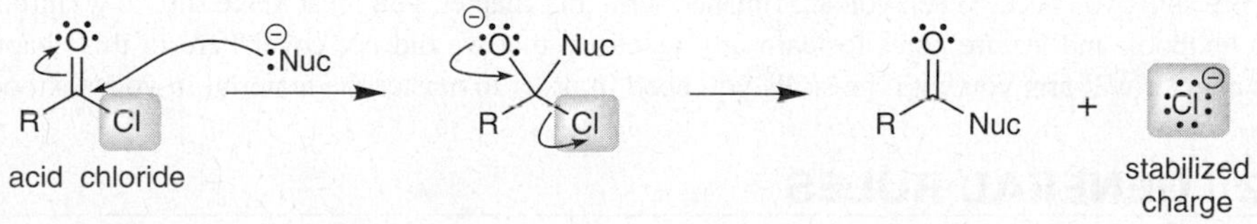

acid chloride   stabilized charge

Chloride is a great leaving group because it is a weak base. For this reason, acid halides (also called acyl halides) are the most reactive of the carboxylic acid derivatives. This is very useful, because it means that we can use acid halides to form any of the other carboxylic acid derivatives.

187

Carboxylic acid derivatives can be viewed as having a "wild card" next to the carbonyl group. I am calling it a "wild card," because it can be easily exchanged for a different group:

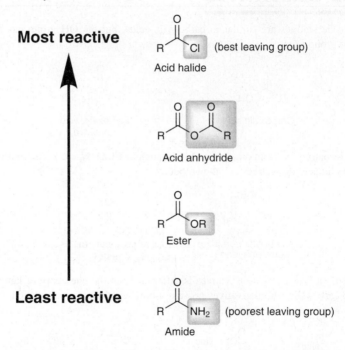

In this chapter we will learn how to exchange the groups so that we can convert one carboxylic acid derivative into another. In order to do this, we will have to know something about the order of reactivity of carboxylic acid derivatives:

As we learn to exchange the "wild card," we will see that there are just a few simple rules that will determine everything. We will see dozens of reactions, but all of these reactions are completely predictable and understandable if you know how to apply just a few simple rules, covered in the following section.

We will NOT cover every reaction in your textbook or lecture notes. Rather, we will focus on the core skills you need. When you are finished with this chapter, you must make sure to go through your textbook and lecture notes to learn any reactions that we did not cover here in this chapter. This chapter will arm you with the skills you need in order to master the material in your textbook.

6.2 GENERAL RULES

The most important rule was already covered in the previous chapter. We called it our "golden rule" and it went like this: after attacking a carbonyl group, always try to re-form the carbonyl group if you can, but never expel H^- or C^-.

With this rule, we can now appreciate that a carboxylic acid derivative will react differently with H⁻ or C⁻ than it will with any other type of nucleophile. When we use a hydrogen nucleophile or a carbon nucleophile, we find that the carboxylic acid derivative gets attacked *twice*. Let's see why:

Consider, for example, the reaction that occurs when an acid halide is treated with an excess of Grignard reagent. First, the Grignard reagent attacks the carbonyl group, just as we would expect:

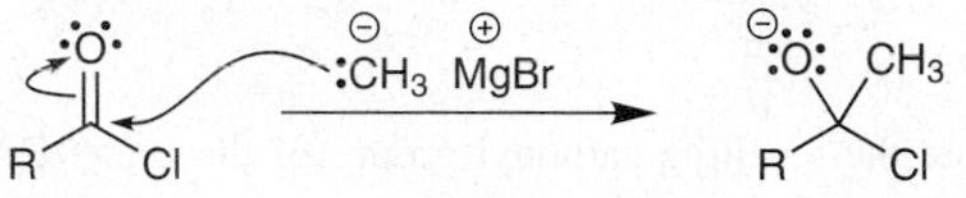

Then, we apply our golden rule: re-form the carbonyl group, if you can, but don't expel H⁻ or C⁻. In this case, Cl⁻ can be expelled, so we expect the carbonyl group to re-form:

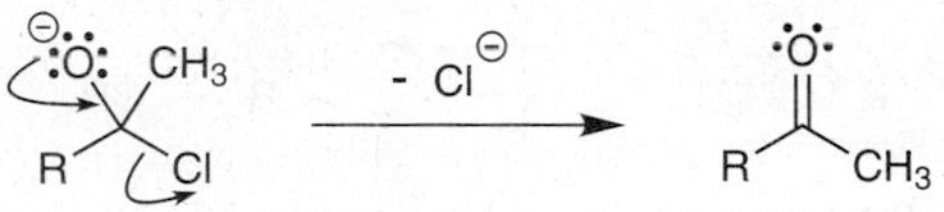

This step generates a ketone, and under these conditions, the ketone can be attacked a *second time*, generating an alkoxide ion:

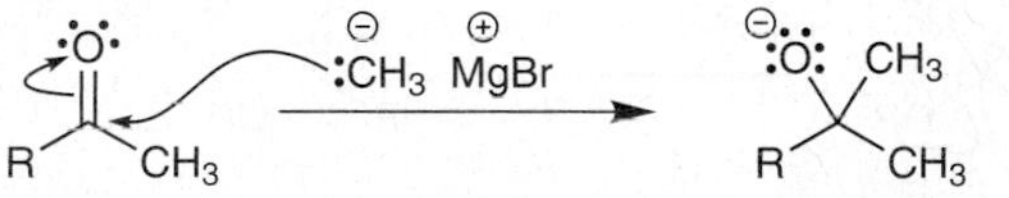

an alkoxide ion

The resulting alkoxide ion is now unable to re-form the carbonyl group, because there is no leaving group. We have already said numerous times that H⁻ or C⁻ cannot be expelled to re-form the carbonyl group. So, the reaction is complete. In order to protonate the alkoxide ion, a proton source must be introduced into the reaction flask (note that the proton source must be introduced into the reaction flask AFTER the reaction is complete):

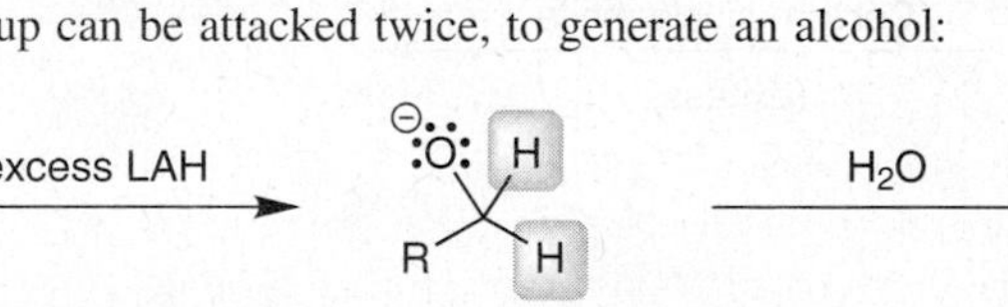

In the end, the product is an alcohol, because the nucleophile attacked *twice*.

A similar result is expected when an acid halide is attacked by a hydrogen nucleophile. Once again, the carbonyl group can be attacked twice, to generate an alcohol:

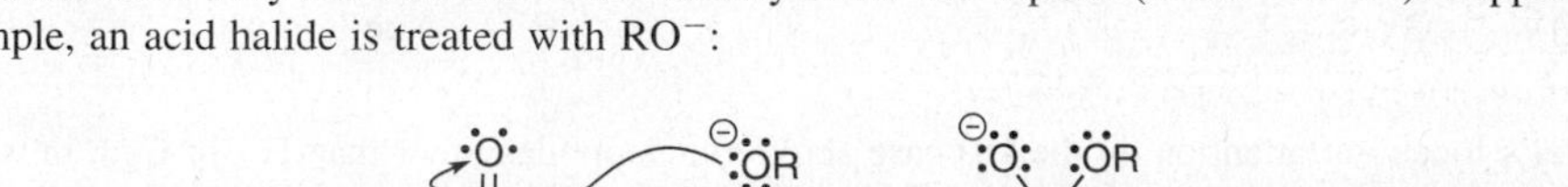

The situation is very different when we use any other nucleophile (not H⁻ or C⁻). Suppose, for example, an acid halide is treated with RO⁻:

As we might expect, the resulting tetrahedral intermediate is capable of re-forming the carbonyl group by expelling a chloride ion, generating an ester:

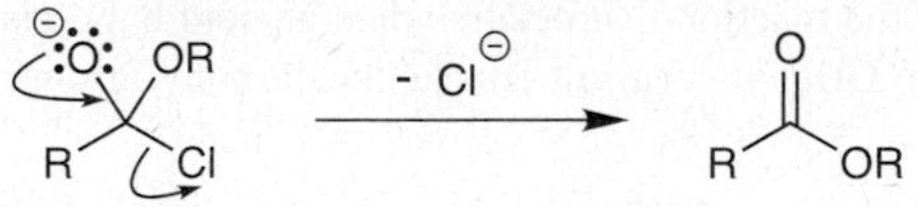

Now, it might be true that the resulting carbonyl group (of the ester) CAN be attacked a second time:

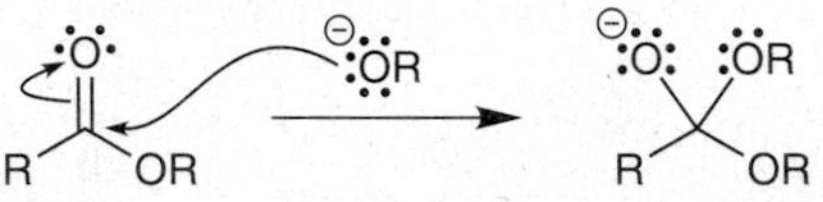

But the tetrahedral intermediate that is generated will simply re-form the carbonyl group, by expelling the second RO⁻ that just attacked. This brings us right back to the ester:

The second attack is only permanent when the nucleophile is H⁻ or C⁻. And this should make sense based on our golden rule; if the second attack is by H⁻ or C⁻, then the carbonyl group will not be able to re-form after the second attack occurs.

So, we have seen that there is a difference in the types of products we obtain when we use H⁻ or C⁻, vs. when we use any other nucleophile. Keep this in mind: H⁻ or C⁻ will attack twice, but all other nucleophiles will only attack once:

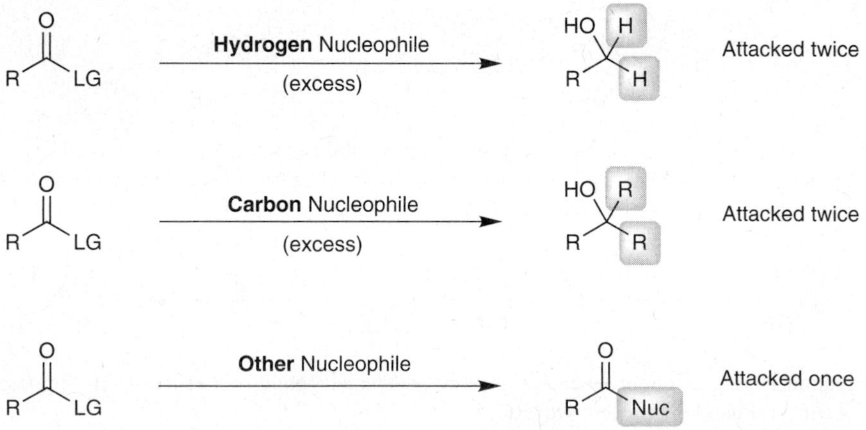

	Hydrogen Nucleophile (excess)		Attacked twice
	Carbon Nucleophile (excess)		Attacked twice
	Other Nucleophile		Attacked once

Now let's focus our attention on the last case above (nucleophiles other than H⁻ or C⁻), in which the nucleophile attacks only once. In those reactions, the outcome will be to exchange one type of carboxylic acid derivative for another. For example:

The mechanism for this process (and all others like it) involves two core steps: attack the carbonyl group, and then re-form the carbonyl group. That's it. Just two core steps. But very often, proton transfer steps are necessary when drawing a mechanism. Proton transfers can only occur at three different moments: the beginning, the middle, or the end:

1		**2**		**3**
BEGINNING	*attack* carbonyl →	MIDDLE	*re-form* carbonyl →	END

1. A proton transfer step may or may not be required at the *beginning* of the mechanism, before the nucleophile has attacked the carbonyl group.

2. Proton transfer steps may or may not be required in the *middle* of the mechanism, after the carbonyl group has been attacked, but before the carbonyl group has re-formed.

3. A proton transfer step may or may not be required at the *end* of the mechanism, after the carbonyl group has re-formed.

Sometimes, a mechanism won't involve any proton transfers at all. For example, when an acid halide is treated with an alkoxide ion, no proton transfers occur:

ATTACK CARBONYL   RE-FORM CARBONYL

And sometimes, a mechanism will require only one proton transfer. For example, when an acid halide is treated with water, one proton transfer is necessary at the end of the mechanism:

ATTACK CARBONYL   RE-FORM CARBONYL   PROTON TRANSFER

But sometimes, proton transfers are required at all three moments in the mechanism (beginning, middle, and end). For example, consider the following mechanism for the reaction in which an ester is treated with aqueous acid, generating a carboxylic acid:

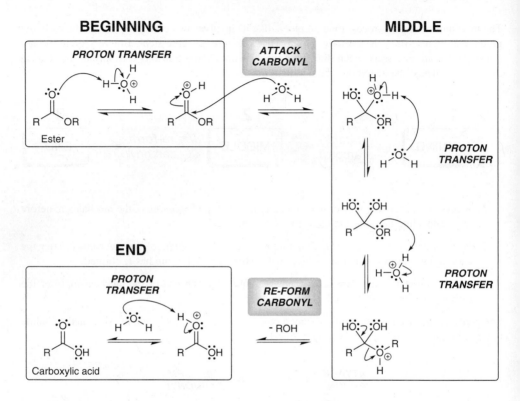

This is a long mechanism, with six steps. The two core steps (attack the carbonyl, and then re-form the carbonyl) are highlighted in gray. Notice that proton transfer steps are required at all three possible moments throughout the mechanism (beginning, middle, and end). Later in this chapter, we will study this reaction in greater detail, and we will explain the rationale behind each and every proton transfer step in this mechanism.

6.3 ACID HALIDES

As we have mentioned, acid halides are the most reactive of the carboxylic acid derivatives, because they produce the most stable leaving groups. Therefore, we can prepare any of the other carboxylic acid derivatives from acid halides. So, it is critical that you know how to make an acid halide. It is very common to encounter a synthesis problem where you will need to make an acid halide at some point in the synthesis.

To make an acid halide, it would be nice if Cl⁻ would attack a carboxylic acid directly, expelling HO⁻ as a leaving group:

But this doesn't work, because HO⁻ is less stable than Cl⁻, so this would be an uphill battle. Cl⁻ is not going to expel HO⁻. So, we must first convert the OH group into a different group that CAN be expelled by Cl⁻. So, here is our strategy:

Thionyl chloride, SOCl₂, is a common reagent used to execute both steps of this strategy:

Thionyl chloride
(SOCl₂)

An acid halide can generally be obtained in good yields by treating a carboxylic acid with thionyl chloride. This reagent achieves two objectives in one reaction flask: (1) it converts the OH group into a better leaving group, and (2) it serves as a source of chloride ions that will attack the carbonyl group and expel the newly formed leaving group. The first objective (converting OH into a better leaving group) is achieved via the following mechanism:

EXCELLENT
LEAVING GROUP

+ HCl

The carboxylic acid functions as a nucleophile and attacks the S=O bond. The S=O bond is then re-formed (by expelling a chloride ion), followed by a proton transfer step. The net result of these three steps is the conversion of an OH group into a better leaving group, thereby achieving the first objective.

The second objective is then achieved when the chloride ion (generated during the steps above), attacks the carbonyl group, expelling the leaving group:

+ SO₂ + Cl⁻

Notice that SO_2 is formed as a by-product. This is important because SO_2 is a gas, which bubbles out of solution, thereby forcing the reaction to completion.

Now that we have covered the preparation of acid halides, we are ready to explore reactions of acid halides. We will see many reactions. BUT don't try to memorize them. Instead, try to appreciate that they all follow the same general rules. We saw in the previous section that there are two core steps (attack, and re-form the carbonyl). Then, we just need to be careful about proton transfers in the three possible moments when they might be necessary:

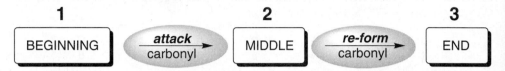

With acid halides, it gets much easier, because proton transfers generally only occur at the end of the mechanism (you can skip "1" and "2" on the diagram above). Consider the following example, in which an acid halide is treated with water to generate a carboxylic acid:

ATTACK CARBONYL   **RE-FORM CARBONYL**   **DEPROTONATE**

Carefully consider the three steps of this mechanism: attack, re-form, deprotonate. Say that out loud 10 times real fast (attack, re-form, deprotonate). You will find that this order of events keeps repeating itself in all of the reactions we are about to see.

Notice that the by-products of the reaction are Cl^- and H_3O^+ which, together, represent an aqueous solution of HCl. The build-up of acid can often produce undesired reactions (depending on what other functional groups might be present in the compound), so pyridine is used to remove the acid as it is produced:

pyridine   pyridinium chloride

Pyridine is a base that reacts with HCl to form pyridinium chloride. This process effectively traps the HCl so that it is unavailable for any other side reactions. The use of pyridine is often represented with the letters "py," like this:

$$R \overset{O}{\underset{}{\text{-}}} Cl \xrightarrow[\text{py}]{H_2O} R \overset{O}{\underset{}{\text{-}}} OH$$

When an acid halide is treated with an alcohol, in the presence of pyridine, the product is an ester:

$$R \overset{O}{\underset{}{\text{-}}} Cl \xrightarrow[\text{py}]{ROH} R \overset{O}{\underset{}{\text{-}}} OR$$

The mechanism for this process has the familiar three steps: attack, re-form, deprotonate:

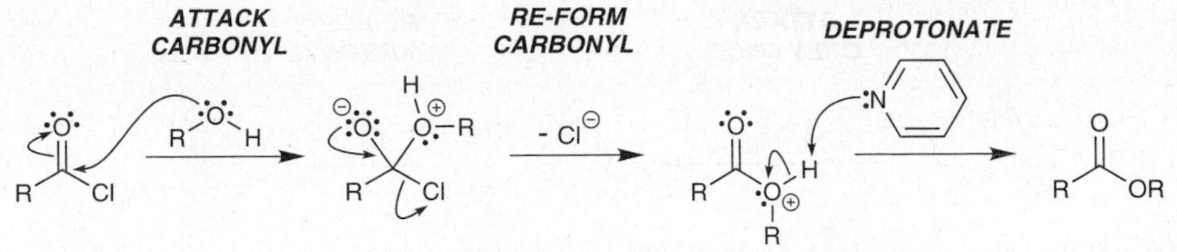

When an acid halide is treated with an amine, the product is an amide:

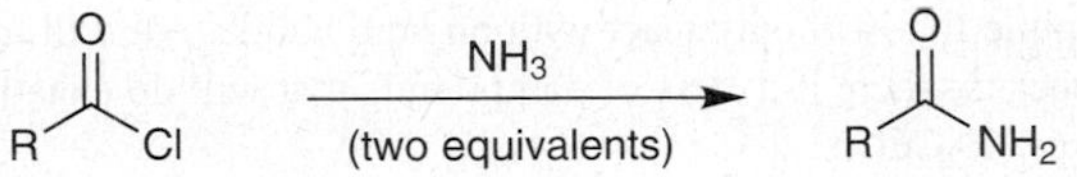

Notice that pyridine is not used in this case. Instead, we use two equivalents of ammonia (twice as much ammonia as acid chloride). One equivalent serves as a nucleophile in the first step of the mechanism, and the other equivalent serves as a base in the last step of the mechanism:

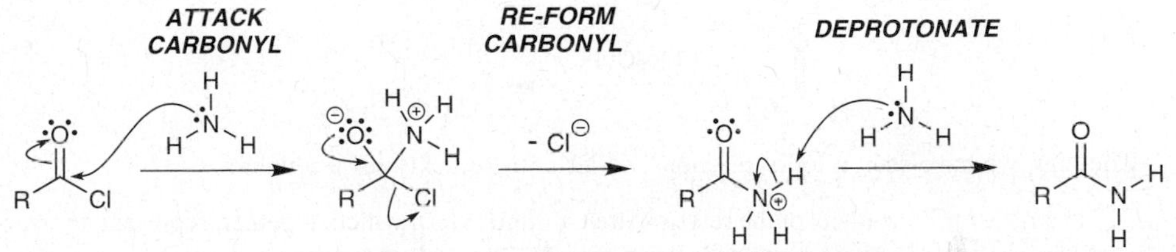

Once again, notice that this process has the familiar three steps: attack, re-form, deprotonate.

Now let's consider what happens when an acid halide is treated with H⁻ or C⁻. We have already seen that H⁻ and C⁻ are special, because they will attack twice:

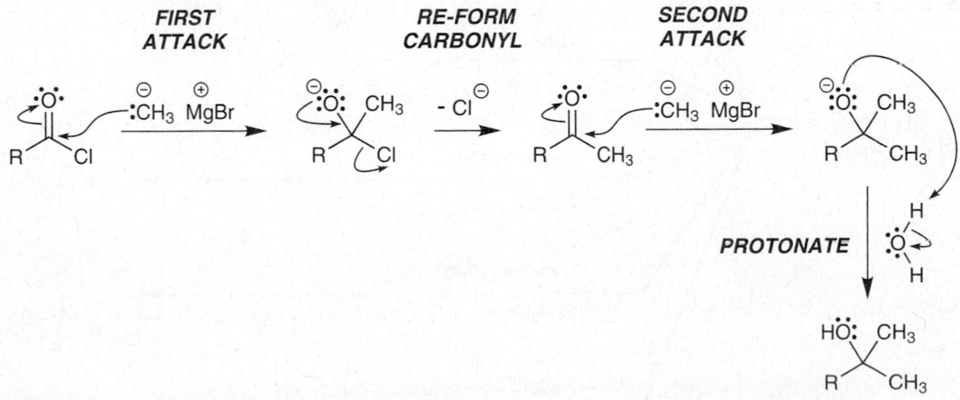

This leaves us with an obvious question. What if we wanted to attack with C⁻ just once? In other words, suppose we wanted the following outcome:

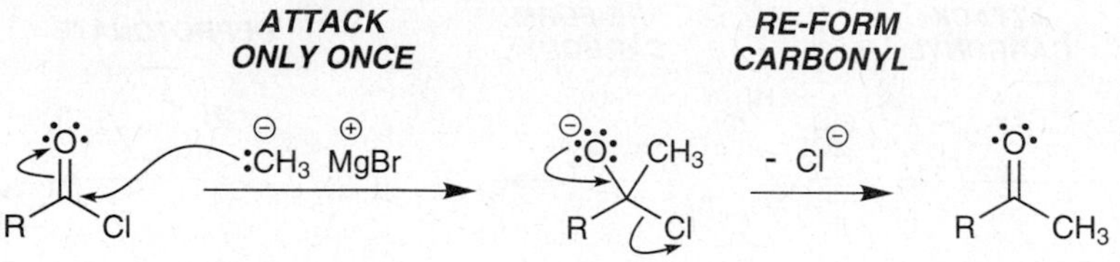

ATTACK ONLY ONCE   **RE-FORM CARBONYL**

What if the ketone is the desired product? We have a problem here, because the Grignard reagent will attack twice. We can't stop it from attacking twice. If we just try to use exactly one equivalent of the Grignard reagent, we will observe a mess of products (some molecules of the acid halide will get attacked twice, and others will not get attacked at all). In order to prepare the ketone, we need a carbon nucleophile that will only react with an acid halide, but will **not** react with a ketone. And we are in luck, because there is a class of compounds that will do exactly that. They are called lithium dialkyl cuprates (R_2CuLi).

You might have been introduced to these compounds in the first semester of organic chemistry. Lithium dialkyl cuprates are carbon nucleophiles, but they are less reactive than Grignard reagents. Lithium dialkyl cuprates will react with acid halides, but not with ketones. And therefore, we can use these reagents to convert acid halides into ketones:

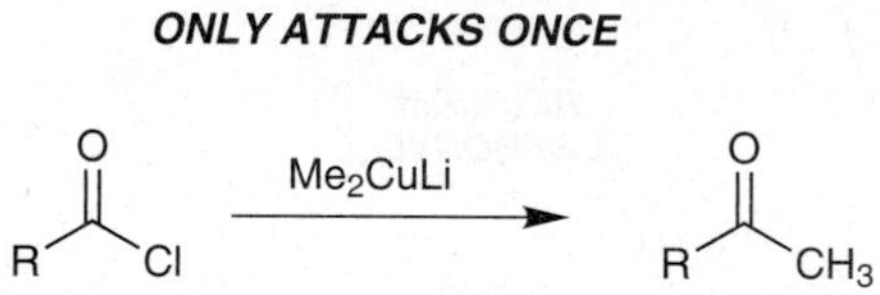

ONLY ATTACKS ONCE

Thus far, we have seen a lot of reactions, so let's just quickly review them:

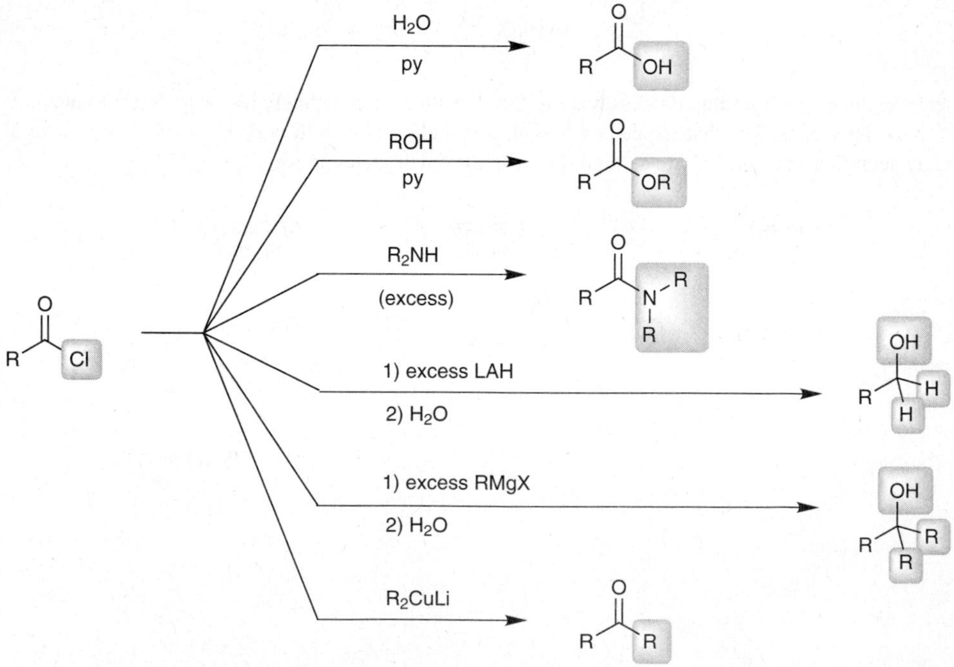

EXERCISE 6.1 Propose a mechanism for the following reaction:

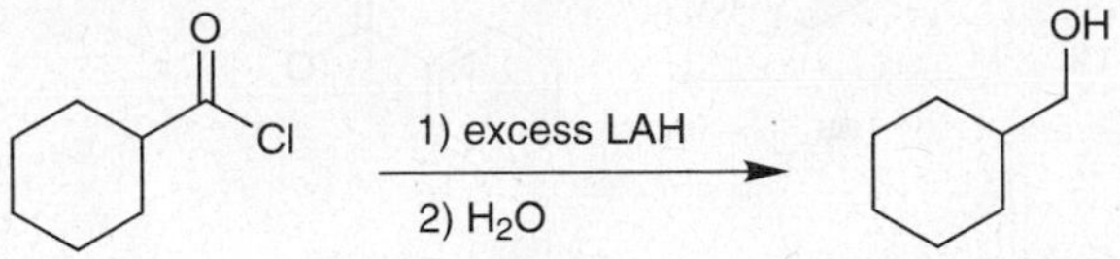

Answer The starting material is an acid chloride, so let's carefully consider the reagents. LAH is a hydrogen nucleophile (source of H⁻), and it is expected to attack the carbonyl group twice:

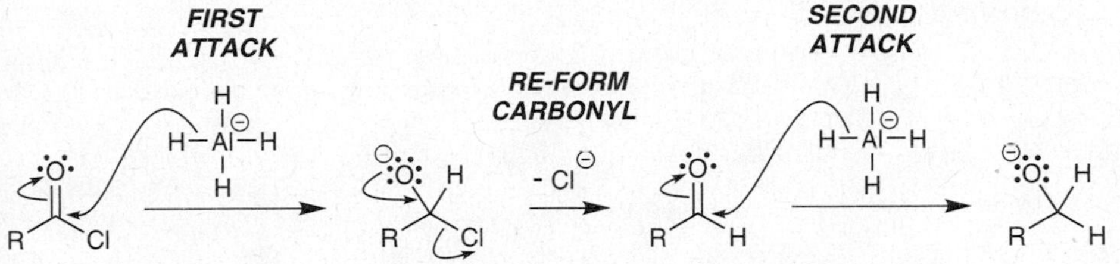

Finally, a proton transfer will be required at the end of the mechanism:

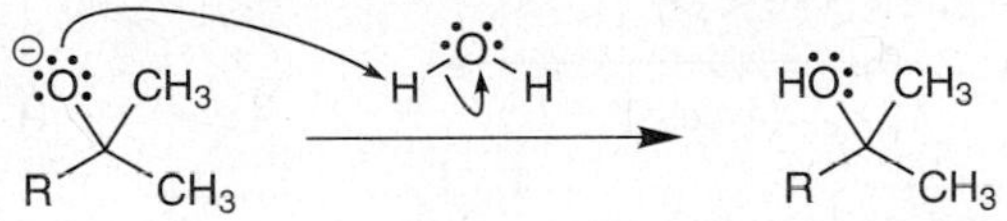

That is, in fact, the reason why H_2O is listed as a reagent (in a separate step):

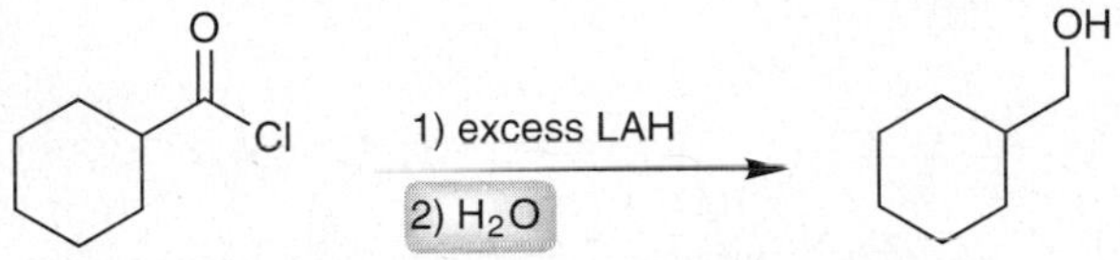

PROBLEMS Propose a plausible mechanism for each of the following reactions. Space has been provided for you to record your answer directly below each reaction:

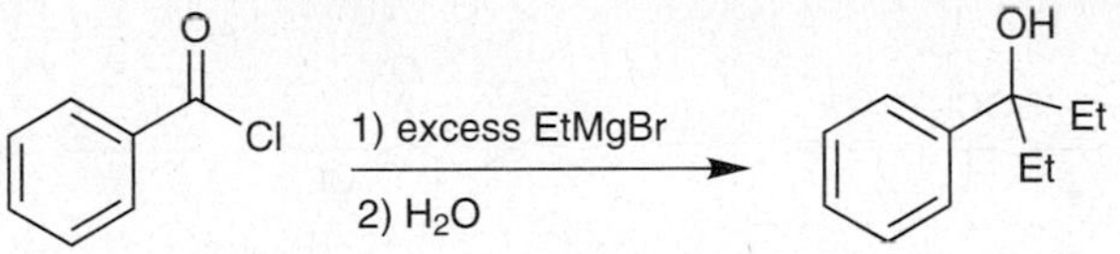

6.2

6.3

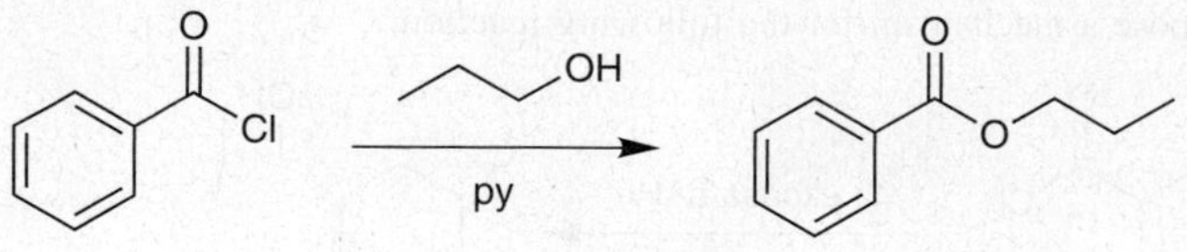

6.4

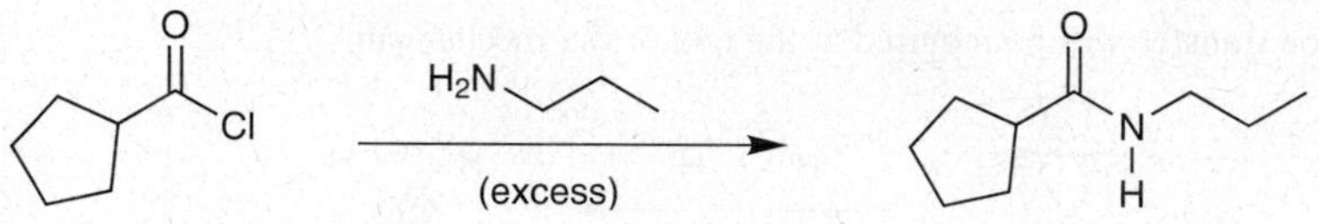

6.5

6.6 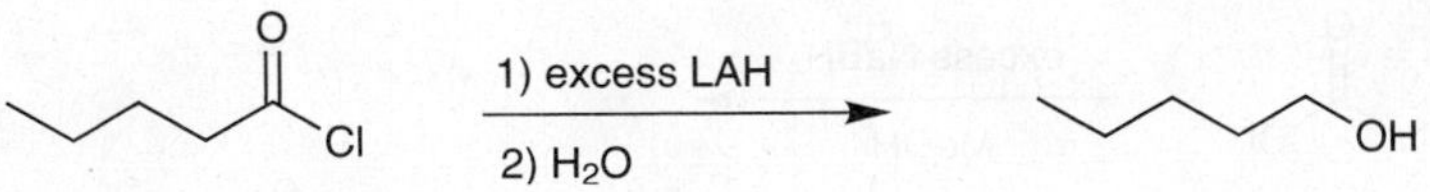
O
Cl
1) excess LAH
2) H_2O
OH

EXERCISE 6.7 Predict the major product of the following reaction:

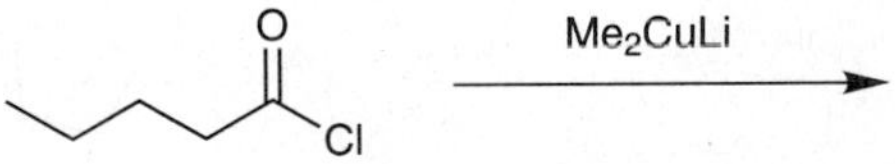
O
Cl
Me_2CuLi

Answer We are starting with an acid halide. So, in order to predict the major product of this reaction, we will have to determine if the nucleophile attacks once or twice. The reagent is a lithium dialkyl cuprate. This reagent is a carbon nucleophile, but it is not like a Grignard reagent—it does not attack twice. Instead, it will only attack once, because it is a very *tame* carbon nucleophile, as we saw earlier in this section. And the product is expected to be a ketone:

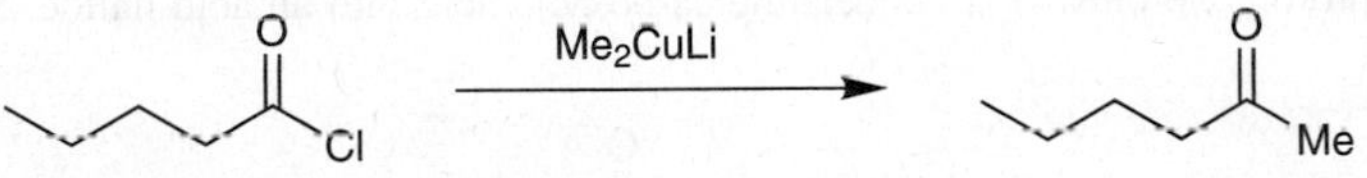
O
Cl
Me_2CuLi
O
Me

PROBLEMS Predict the major product in each of the following cases:

6.8 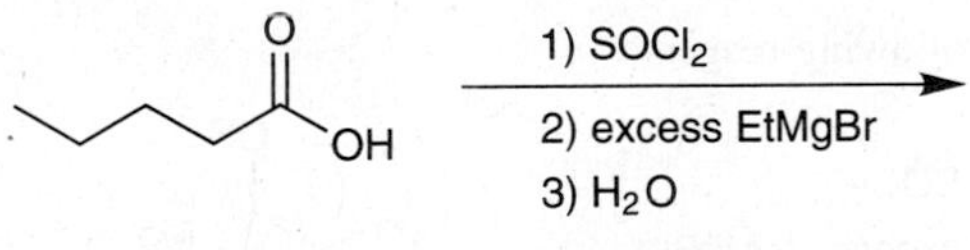
O
OH
1) $SOCl_2$
2) excess EtMgBr
3) H_2O

6.9 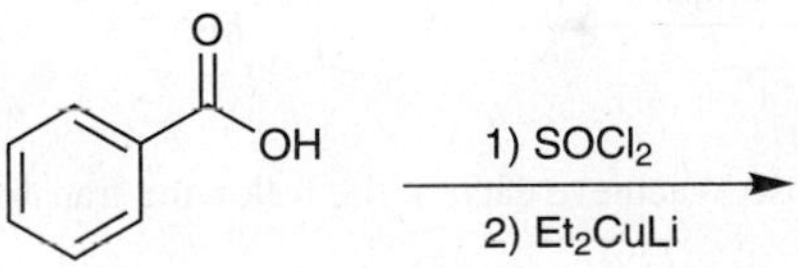
O
OH
1) $SOCl_2$
2) Et_2CuLi

6.10 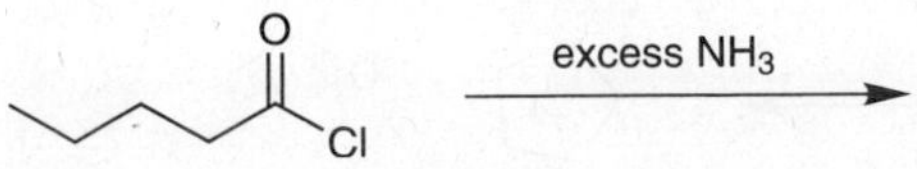
O
Cl
excess NH_3

6.11

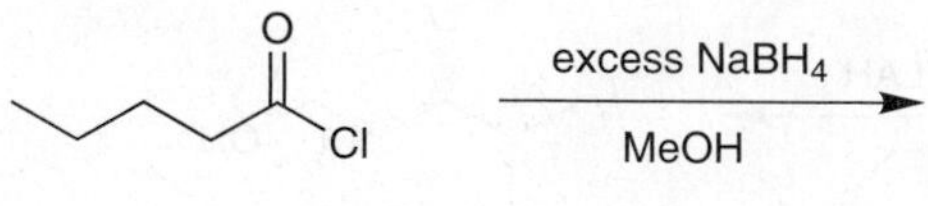

$$\text{(structure: butanoyl chloride)} \xrightarrow[\text{MeOH}]{\text{excess NaBH}_4}$$

6.12

$$\text{(structure: cyclohexanecarbonyl chloride)} \xrightarrow[\text{py}]{\text{EtOH}}$$

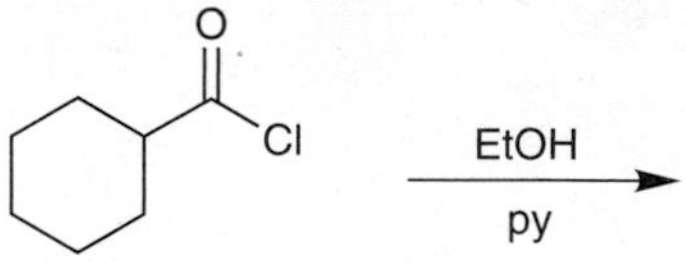

EXERCISE 6.13 Identify the reagents you would use to achieve the following transformation:

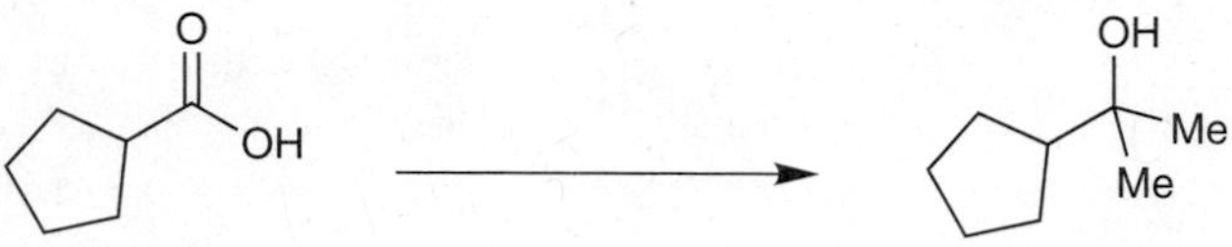

Answer We are starting with a carboxylic acid, and the final product is an alcohol. We also notice that there are two methyl groups in the product:

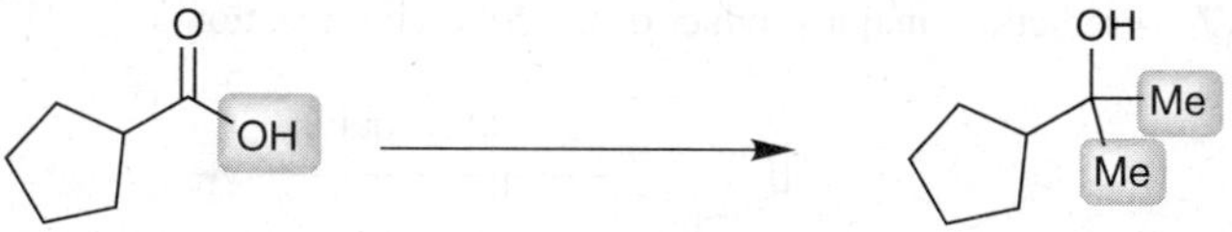

We therefore need to install *two* methyl groups, which means that we need to attack the carbonyl group *twice*. And in the process, the carbonyl group must be reduced to an alcohol. This sounds like a Grignard reaction.

But we have to be careful. We cannot perform a Grignard reaction with a carboxylic acid. Remember that a Grignard reagent is sensitive to its conditions—if there are any acidic protons available, the Grignard reagent will be destroyed. Since our starting compound (a carboxylic acid) has an acidic proton, we must first convert the carboxylic acid into an acid halide. So our strategy goes like this:

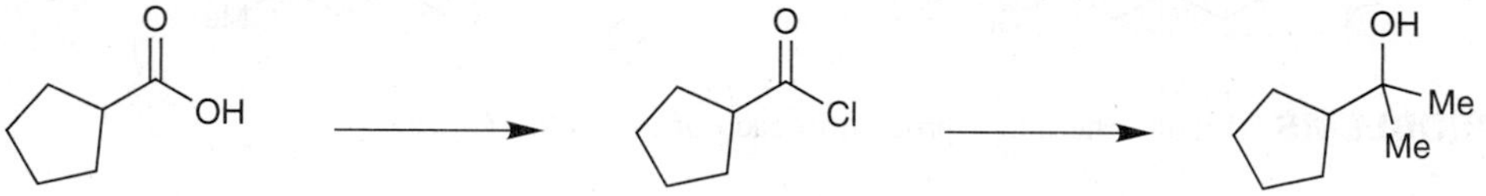

And to achieve this, we would use the following reagents:

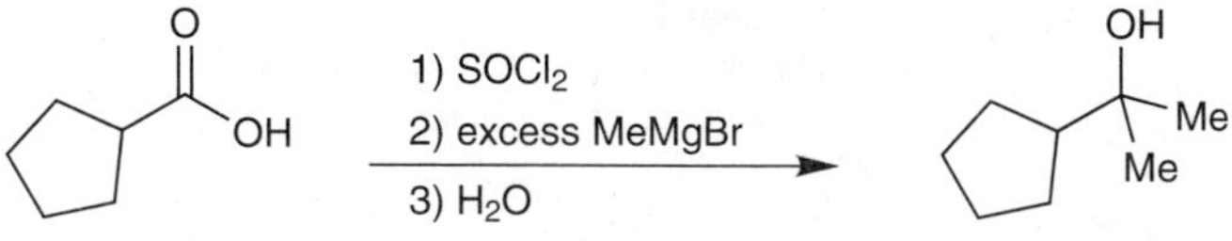

1) SOCl$_2$
2) excess MeMgBr
3) H$_2$O

PROBLEMS Identify the reagents you would use to achieve each of the following transformations:

6.14

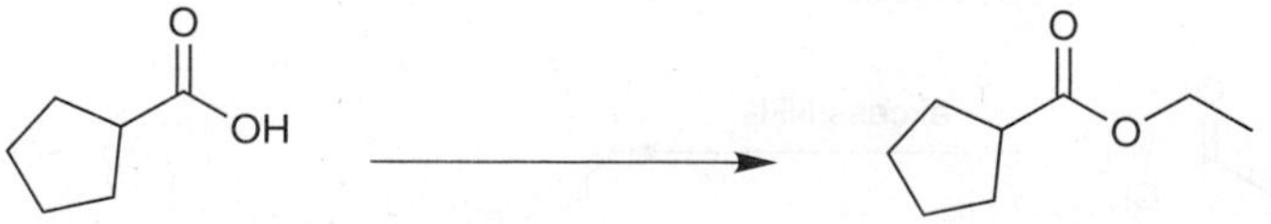

6.15

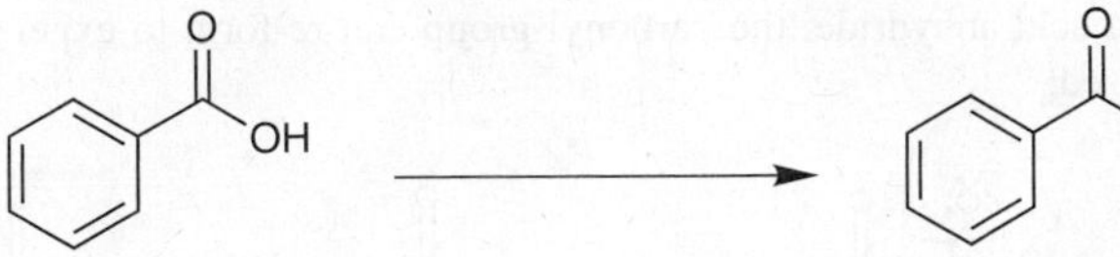

6.16

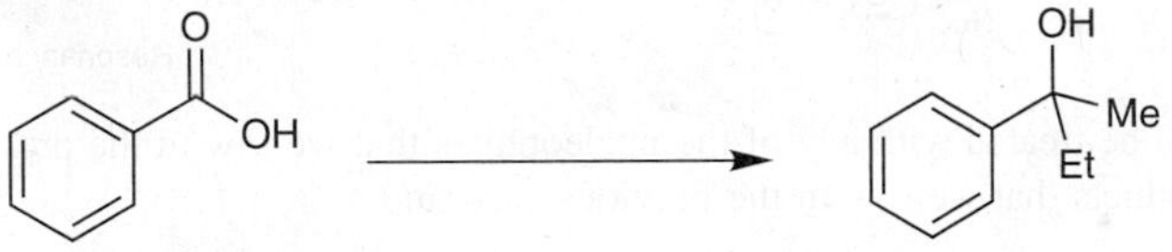

6.17

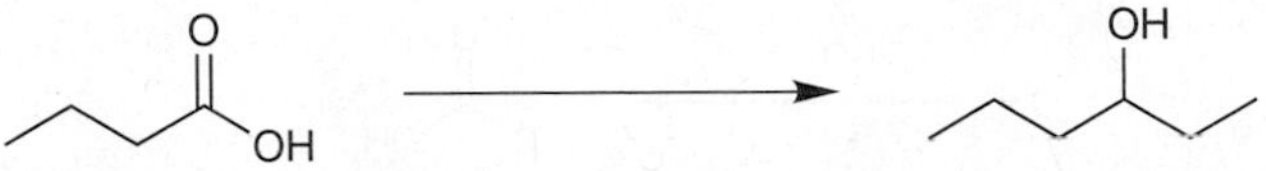

6.18

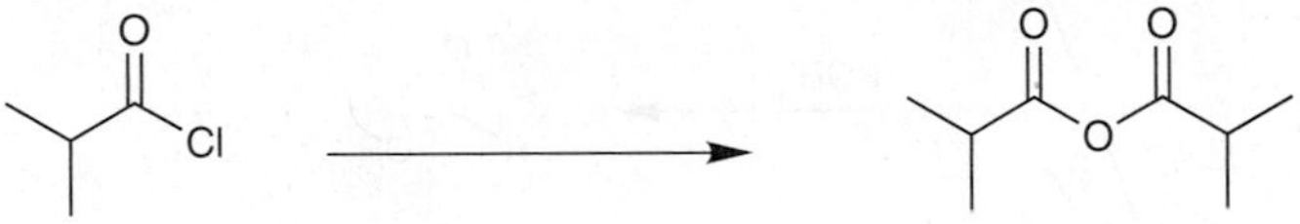

6.4 ACID ANHYDRIDES

Anhydrides can be prepared by the reaction between a carboxylic acid and an acid halide:

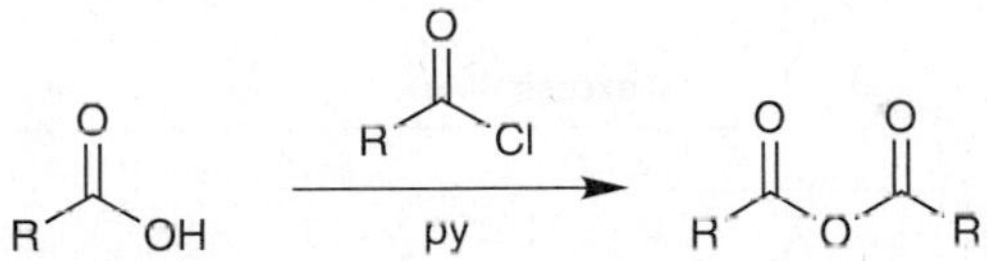

Notice that pyridine is used, once again, to remove the HCl that is formed as a by-product. We can avoid the need for pyridine by using a carboxylate ion (a deprotonated carboxylic acid) instead of a carboxylic acid:

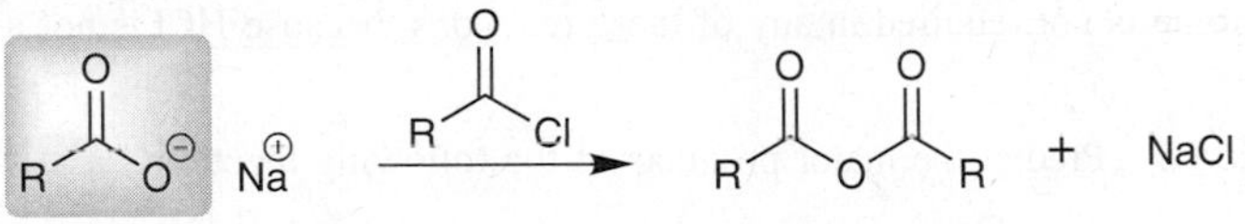

Notice that the by-product is NaCl, rather than HCl, so pyridine is no longer needed.

Acid anhydrides are almost as reactive as acid halides. So, the reactions of acid anhydrides are very similar to the reactions of acid halides. You just have to train your eyes to see the leaving group:

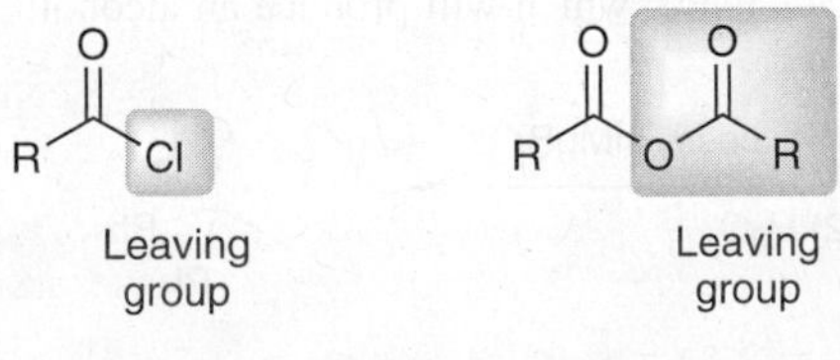

Leaving
group

Leaving
group

When a nucleophile attacks an acid anhydride, the carbonyl group can re-form to expel a leaving group that is resonance-stabilized:

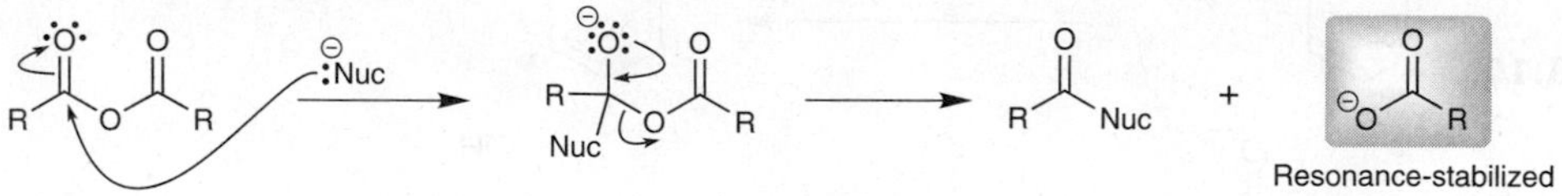

So, an acid anhydride can be treated with any of the nucleophiles that we saw in the previous section to give the same products that we saw in the previous section:

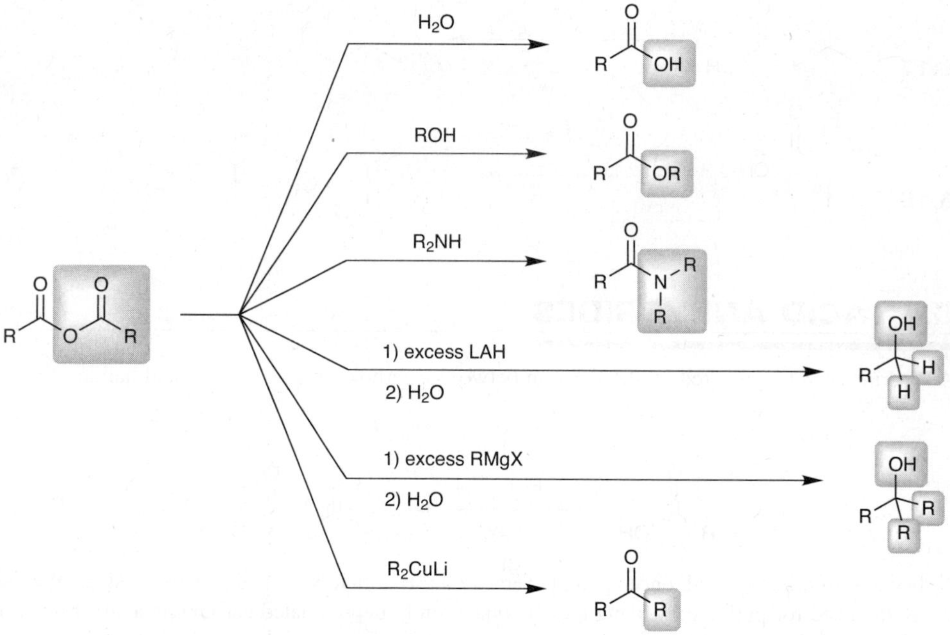

Notice that pyridine is not required in any of these reactions, because HCl is not a by-product.

EXERCISE 6.19 Predict the major products of the following reaction:

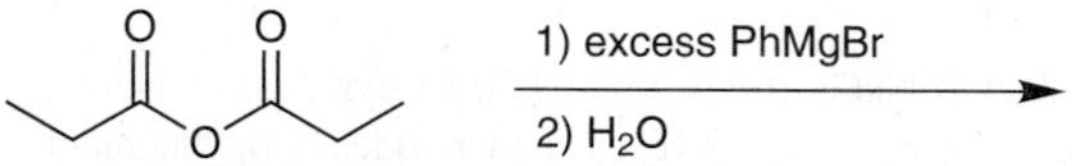

Answer In order to predict the major products of this reaction, we will have to determine if the nucleophile attacks once or twice. The reagent is a Grignard, which is a strong carbon nucleophile. Se we expect that it will attack twice, which will produce an alcohol:

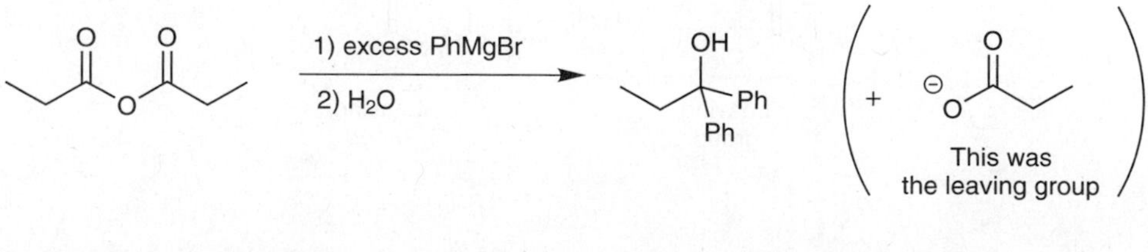

PROBLEMS Predict the major products for each of the following reactions:

6.20 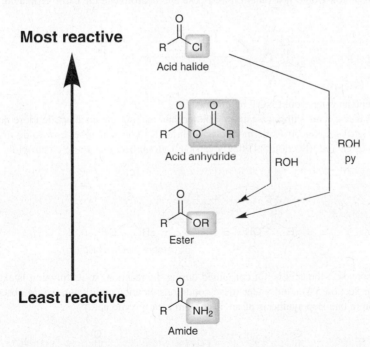 $\xrightarrow{\text{excess NH}_3}$

6.21 $\xrightarrow[\text{2) H}_2\text{O}]{\text{1) LAH}}$

6.22 $\xrightarrow{\text{H}_3\text{O}^+}$

6.5   ESTERS

Esters can be made from carboxylic acid derivatives that are more reactive than esters. In other words, we can make an ester from an acid halide, or from an anhydride:

Most reactive

R Cl
Acid halide

R O R
Acid anhydride

R OR
Ester

ROH

ROH
py

Least reactive

R NH₂
Amide

And we have already seen how to make acid halides—we can make them from carboxylic acids (using thionyl chloride). So, this provides a two-step method for making an ester from a carboxylic acid:

R OH $\xrightarrow{\text{Cl-S-Cl}}$ R Cl $\xrightarrow[\text{py}]{\text{ROH}}$ R OR

The carboxylic acid is first converted into an acid halide, which is then converted into an ester. But this begs the question: can we achieve the desired transformation *in one step*? That is, can we convert a carboxylic acid into an ester in just one step, avoiding the need to prepare an acid halide? If we simply try to mix an alcohol and a carboxylic acid, we do *not* observe a reaction:

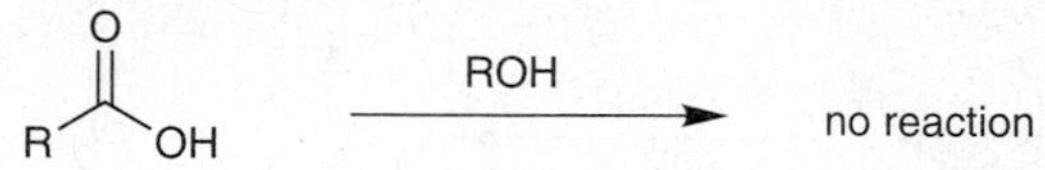

So, let's see what we can do to force this reaction along. Let's consider what happens if we try to make the nucleophile more nucleophilic. In other words, suppose we try to use RO⁻ instead of ROH:

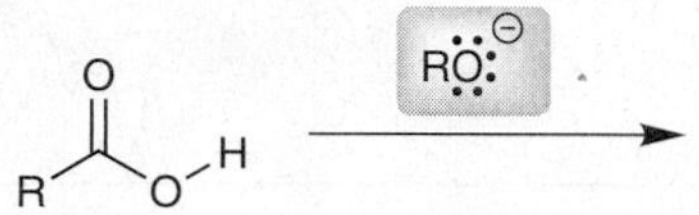

Recall that carboxylic acids have a mildly acidic proton, and alkoxide ions (RO⁻) are strong bases. So, the alkoxide ion would just function as a base and deprotonate the carboxylic acid:

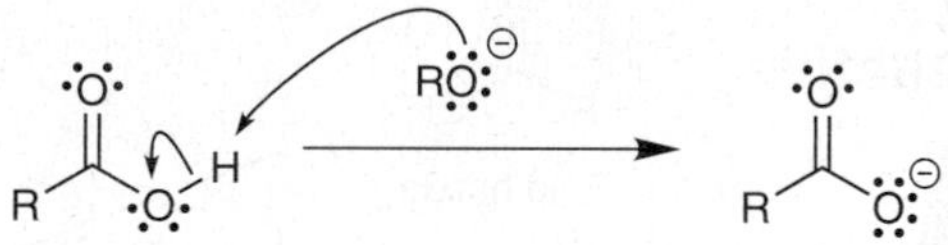

So, once again, an ester would NOT be produced.

But there is one more thing we can try. Rather than making the nucleophile more nucleophilic, we can try to make the electrophile more electrophilic. Do you remember how to do that? We saw in the previous chapter (Section 5.4) that a carbonyl group becomes more electrophilic when it is protonated:

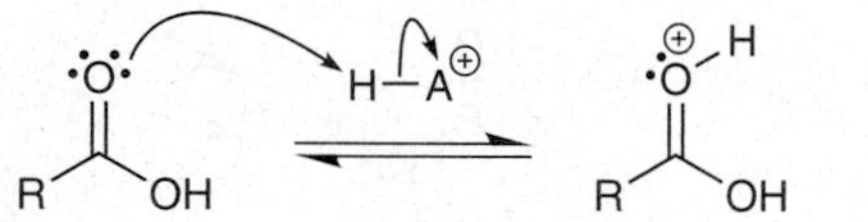

more electrophilic

As we will soon see, the acid is not consumed during the reaction, so its function is catalytic (just as we saw in Section 5.4). And under these conditions (acidic conditions), we *do* observe the desired reaction (a one-step synthesis of an ester from a carboxylic acid):

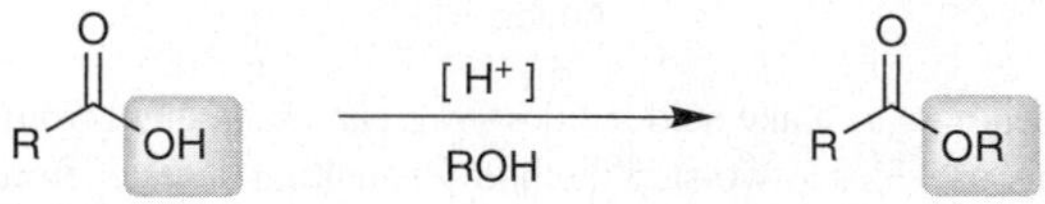

This reaction is incredibly useful and important (and it is considered to be a staple of any organic chemistry course), so let's explore the accepted mechanism, step-by-step.

The mechanism has six steps, but you should notice that there are only two core steps here: attack, and then re-form. All other steps are just proton transfers:

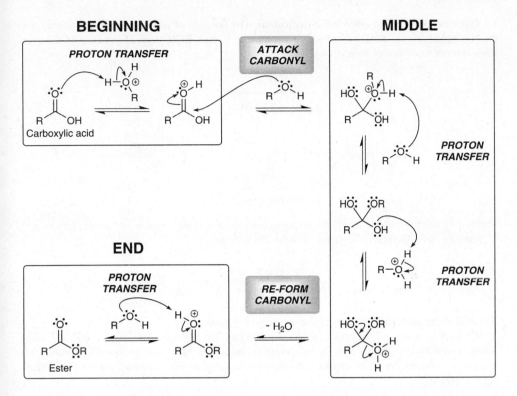

Carefully consider the mechanism above, and you will see there are three moments at which proton transfer steps are believed to occur. This exactly follows the pattern that we described in the beginning of this chapter:

1. *In the beginning of the mechanism*, a proton transfer step is necessary in order to protonate the carbonyl group, rendering it more electrophilic.

2. *In the middle of the mechanism* (after attack of the carbonyl, but before re-forming the carbonyl), two proton transfers are required to protonate the leaving group (HO^- cannot be expelled in acidic conditions). Notice that two protons are required. It would not be OK to simply protonate the OH group, because that would give an intermediate with two positive charges. So, first we deprotonate to remove the positive charge, and then, we protonate the OH group.

3. *In the end of the mechanism*, a proton transfer is required to remove the positive charge and generate the product.

This reaction is called a Fischer esterification. The position of equilibrium is very sensitive to the concentrations of starting materials and products. Excess ROH favors formation of the ester, while excess water favors the carboxylic acid:

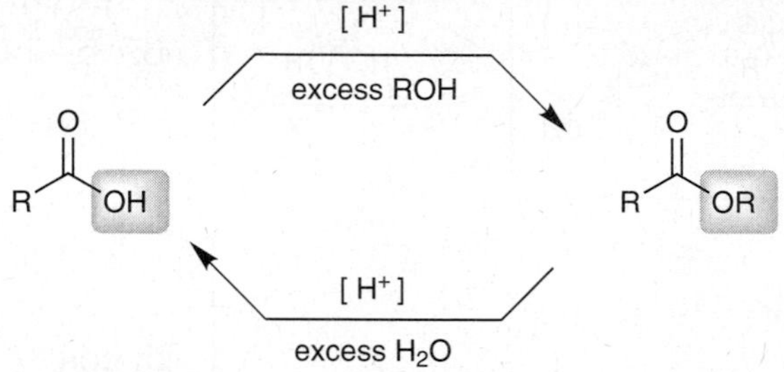

This is very helpful, because it provides a way to convert an acid into an ester, *or* an ester into an acid. For now, let's focus on converting an acid into an ester:

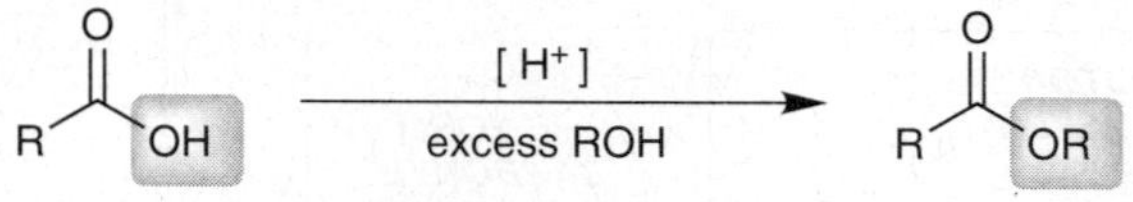

We will spend some time on the reverse process very soon.

We have seen that a Fischer esterification is the reaction between a carboxylic acid and an alcohol (with acid catalysis). If a single compound contains both functional groups (COOH and OH), it is possible to observe an intramolecular Fischer esterification. For example, consider the following compound:

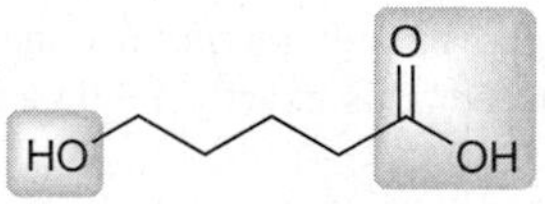

This compound has both a COOH group and an OH group. And, in this case, an intramolecular reaction is observed:

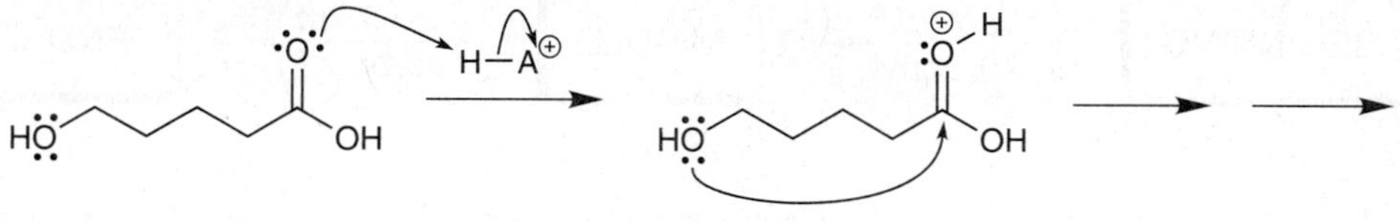

Intramolecular attack

The rest of the mechanism is directly analogous to what we have already seen (see Problem 6.23). The mechanism has two core steps (attack and re-form), with proton transfer steps in the beginning, middle, and end.

PROBLEM 6.23   In the space provided on the next page, draw a mechanism for the following transformation:

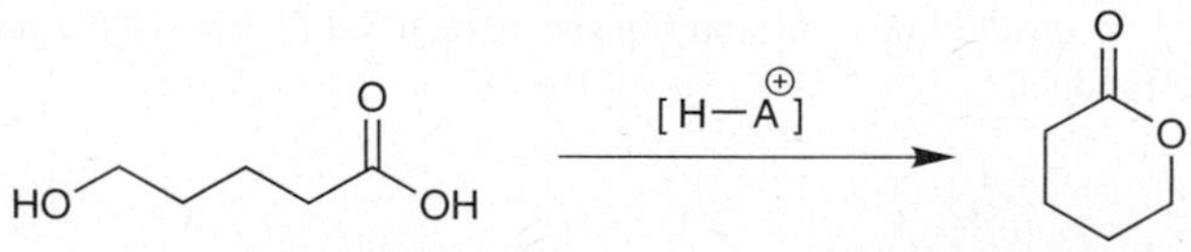

Remember that the general steps are:

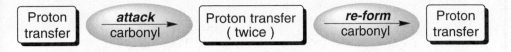

EXERCISE 6.24 Identify the reagents you would use to make the following compound via a Fischer esterification:

Answer To make an ester using a Fischer esterification, we need to start with a carboxylic acid and an alcohol. The question is: how do we decide which carboxylic acid and which alcohol to use. To do this, we must identify the bond that will be formed during the reaction:

Therefore, we will need the following reagents:

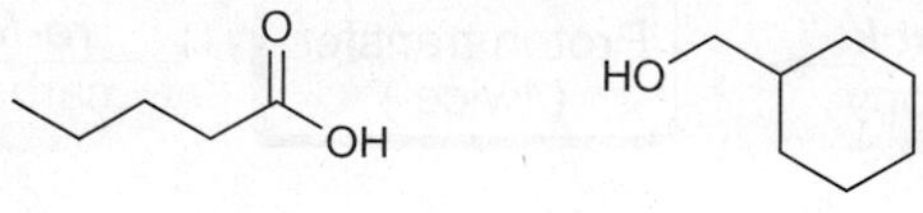

And, don't forget that we need acid catalysis. So our synthesis would look like this:

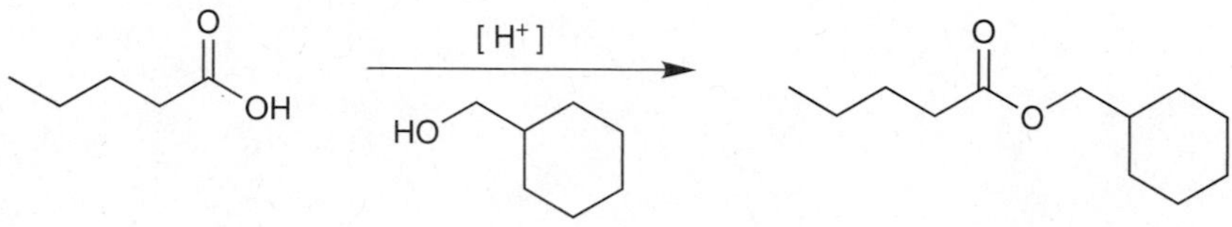

PROBLEMS Identify the reagents you would use to make each of the following esters:

6.25

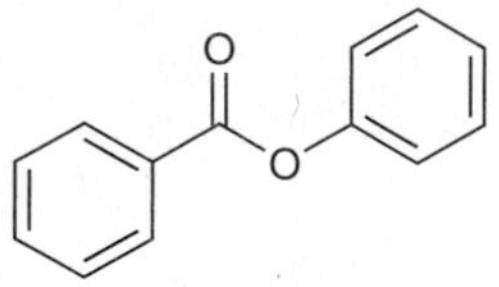

6.26

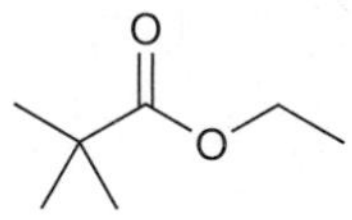

6.27

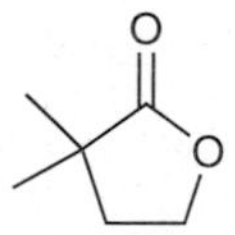

6.28

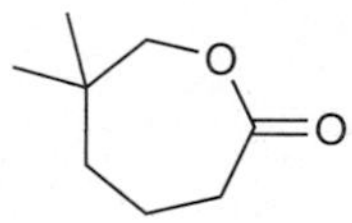

Now that we have seen how to make esters, let's focus our attention on the reactions of esters. We will center on two reactions in particular. Esters can be hydrolyzed to give carboxylic acids, under two different sets of conditions:

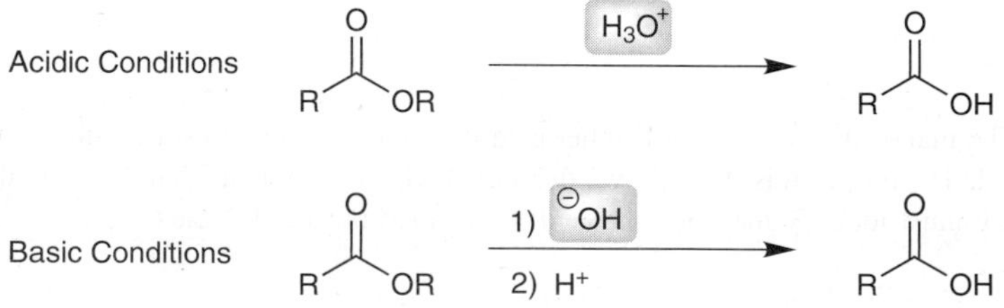

Acidic Conditions

Basic Conditions

The first set of conditions above (acidic conditions) should seem very familiar to you. This reaction is simply the reverse of a Fischer esterification. Let's explore the accepted mechanism for this process:

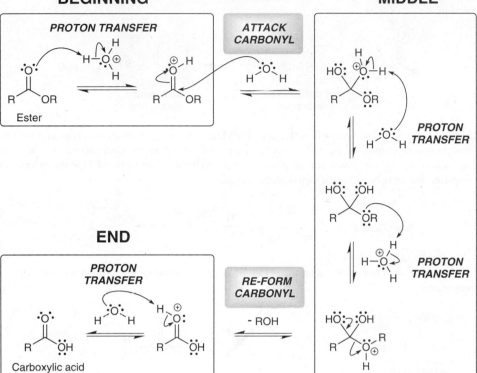

Once again, we see the same pattern again and again. Look closely at this mechanism. Ther are two core steps: attack, and re-form. All other steps in the mechanism are just proton transfers that facilitate the reaction. One proton transfer is required in the beginning (to protonate the carbonyl group), two proton transfers are required in the middle (so that the leaving group can leave as a neutral species), and one proton transfer is required at the end (to deprotonate).

The process above occurs under acidic conditions. But it is also possible to hydrolyze an ester under basic conditions as well. The following is a mechanism for that process:

ATTACK CARBONYL   RE-FORM CARBONYL   PROTON TRANSFER

Once again, we have two core steps (attack and re-form), followed by a deprotonation. This proton transfer at the end is unavoidable under these conditions. In basic conditions, a carboxylic acid will be deprotonated. In fact, formation of a more stable anion is the driving force for this reaction:

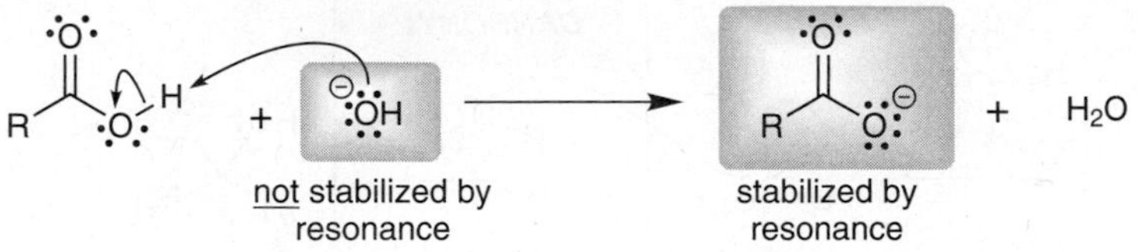

not stabilized by resonance   →   stabilized by resonance

This resonance-stabilized anion is called a carboxylate ion, and its formation serves as a driving force. In order to protonate the carboxylate ion, a proton source must be introduced into the reaction flask (note that the proton source is introduced into the reaction flask AFTER the reaction is complete, and must be indicated as a separate reagent):

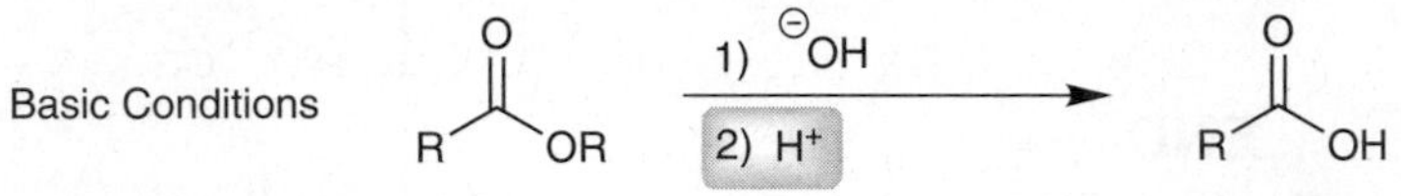

Basic Conditions

Notice that we need to use H^+ rather than H_2O, because a carboxylate ion will not remove a proton from water:

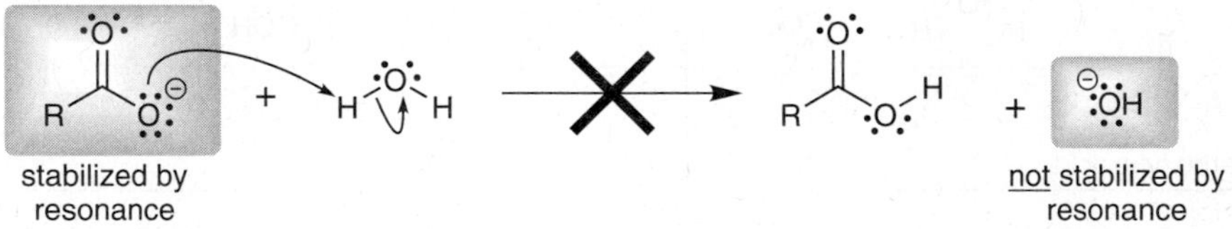

stabilized by resonance   →   not stabilized by resonance

This process (hydrolysis of an ester under basic conditions) has a special name: *saponification*.

EXERCISE 6.29 Propose a mechanism for the following transformation:

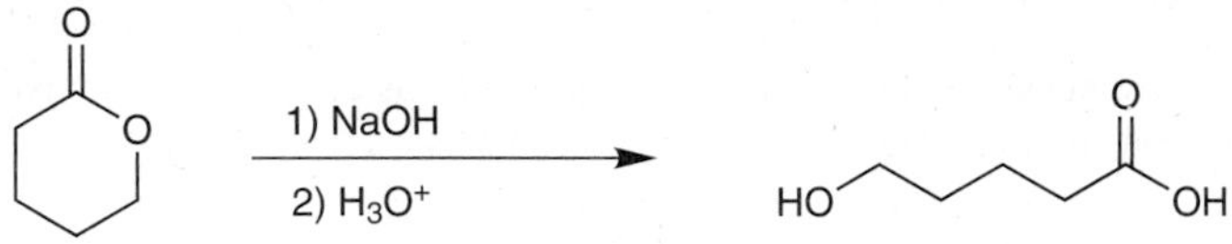

1) NaOH
2) H_3O^+

Answer This reaction utilizes basic conditions to hydrolyze an ester (a process called saponification). The only twist here is that the reaction is intramolecular, but that does not change the sequence of steps required to draw the mechanism. We begin by attacking the carbonyl group, and then re-forming it:

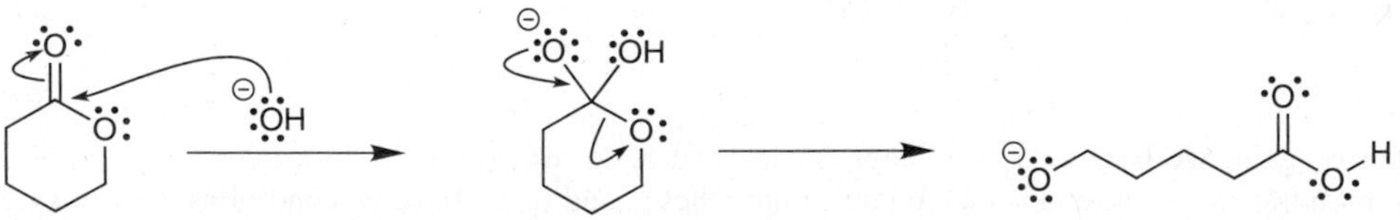

Notice that this compound is both an alkoxide (strong base) and a carboxylic acid (mild acid). Therefore, deprotonation of the carboxylic acid occurs in an intramolecular fashion:

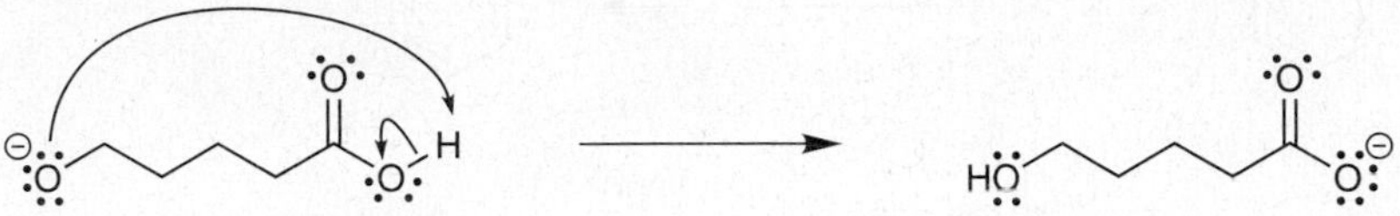

The resulting carboxylate ion is only protonated when an acid is introduced into the reaction flask:

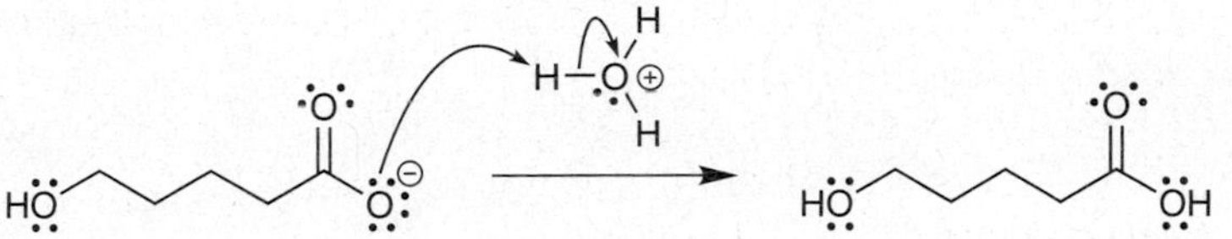

PROBLEMS   In the space provided, draw a mechanism for each of the following transformations:

6.30

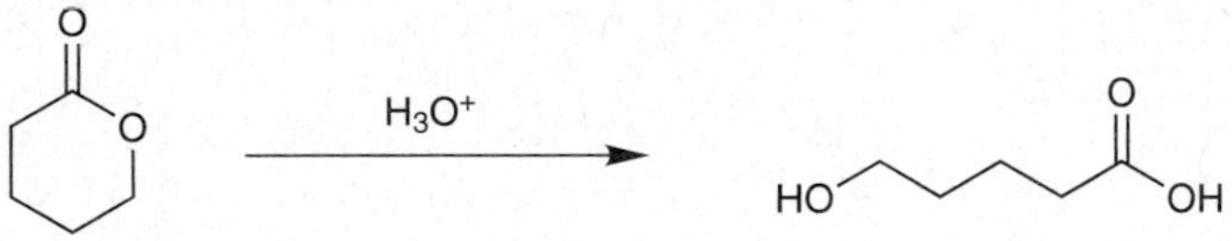

6.31

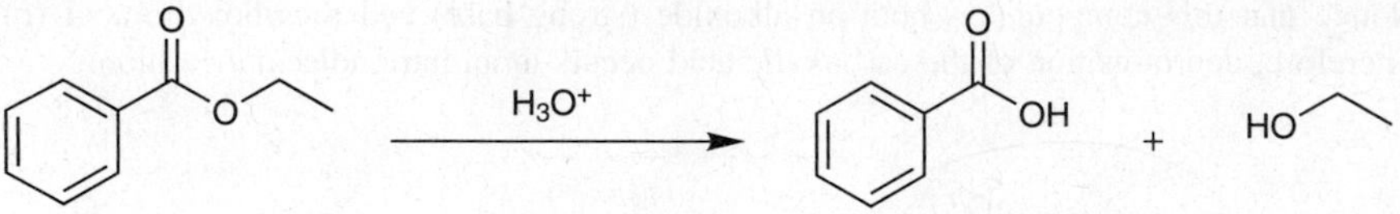

EXERCISE 6.32 Predict the products of the following reaction:

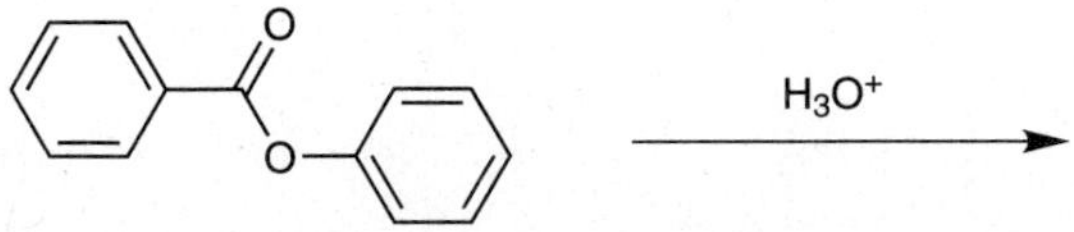

Answer We are starting with an ester, and we are subjecting it to aqueous acidic conditions. This will convert the ester into a carboxylic acid and an alcohol (the reverse of a Fischer esterfication). We expect the following products:

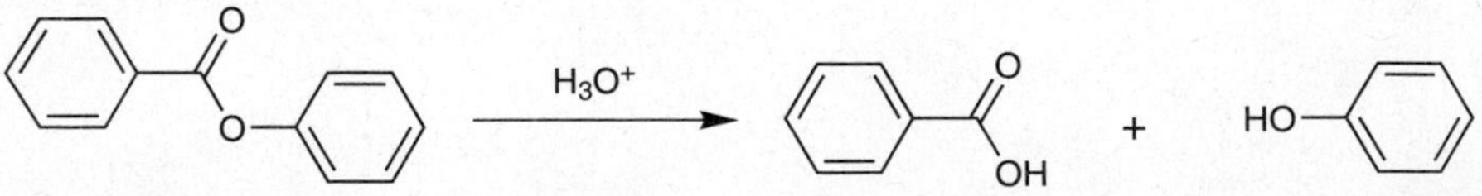

PROBLEMS   Predict the products of each of the following reactions:

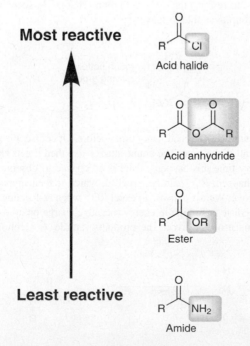

6.33

6.34

6.35

6.36

6.6   AMIDES AND NITRILES

We have said before that a carboxylic acid derivative can be prepared from any other carboxylic acid derivative that is more reactive. Let's go back to our reactivity chart to see what this means practically:

Most reactive

$$R\overset{\displaystyle O}{\underset{\displaystyle }{\|}}Cl$$

Acid halide

$$R\overset{\displaystyle O}{\|}O\overset{\displaystyle O}{\|}R$$

Acid anhydride

$$R\overset{\displaystyle O}{\|}OR$$

Ester

Least reactive

$$R\overset{\displaystyle O}{\|}NH_2$$

Amide

Since amides are the least reactive of the carboxylic acid derivatives (shown on the chart above), we can therefore make amides from any carboxylic acid derivatives that are higher on the chart. In other words, we can make amides from acid halides, from anhydrides, or from esters.

Earlier in this chapter, we saw how to make amides from acid halides or anhydrides:

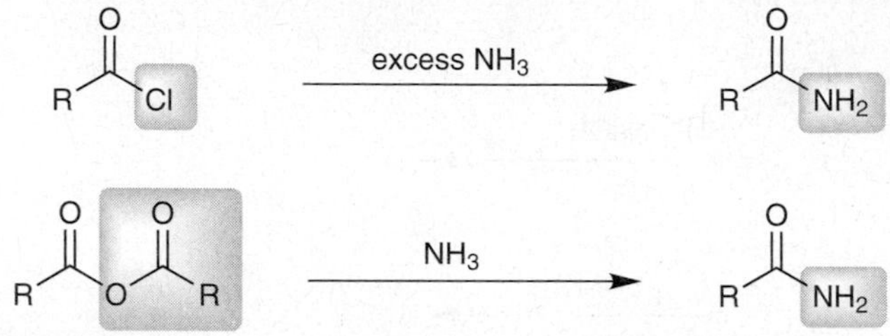

But now the question is: how do we make amides from esters? Esters are less reactive than acid halides or anhydrides. So, we have to use some kind of trick to coax the reaction along. We cannot use acid or base (an acid would just protonate the attacking amine, rendering it useless; and a base would cause other side reactions that we will learn in the next chapter). Instead, we use brute force and patience. We just heat the reaction for a long time, and a reaction is observed, which can occur via the following mechanism:

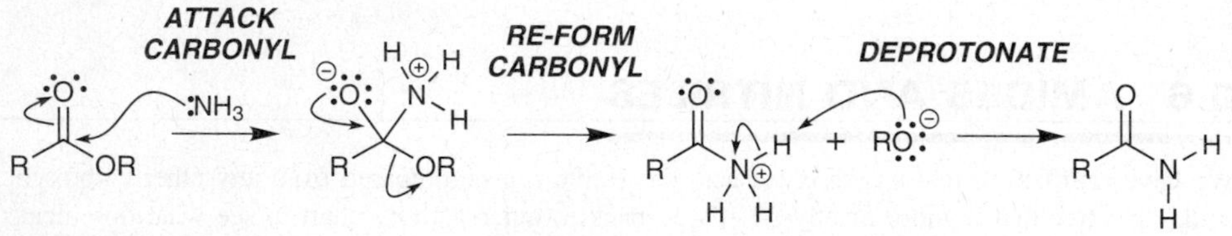

Notice that RO⁻ is expelled to re-form the carbonyl group. That might seem strange, because there is a much better leaving group available to leave:

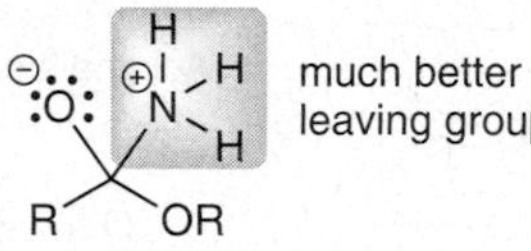

much better leaving group

But this gets back to something we have said many times before. Of course it is possible for the amine to leave. In fact, it happens all of the time. The amine attacks, and then it gets expelled. It attacks, and is then expelled again. Every time this happens, there is no change to observe. But every once in a while, something else can happen. RO⁻ can be expelled, which is then immediately protonated, as shown in the mechanism above. We are allowed to expel RO⁻ when re-forming a carbonyl group, because the tetrahedral intermediate is so high in energy (negative charge on an oxygen atom).

The equilibrium for this process favors the products (amide + alcohol) over the reactants (ester + amine):

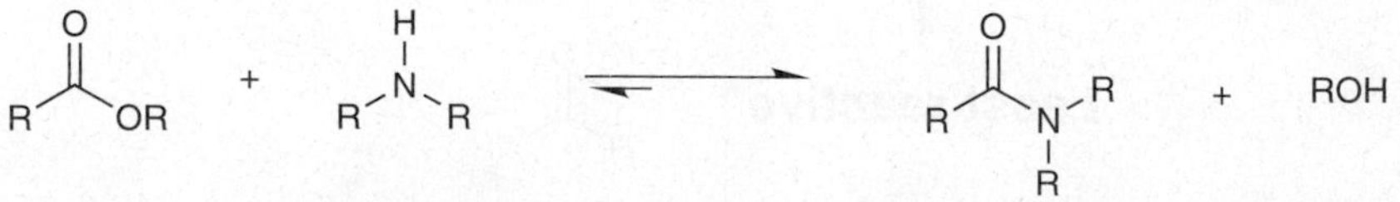

So, this is another method for making amides (and when we use this method, an alcohol is formed as a by-product).

So far, in this section, we have seen that we can make amides from acid halides, from acid anhydrides, or from esters. Now that we know how to make amides, let's explore some important reactions of amides. Specifically, we will explore hydrolysis of amides (under acidic or basic condition). It is worth mentioning that much of biochemistry is dependent on how, when, and why amides will undergo hydrolysis. So, if you plan on taking biochemistry, you should certainly be familiar with the hydrolysis of amides, which can occur under either basic conditions or acidic conditions:

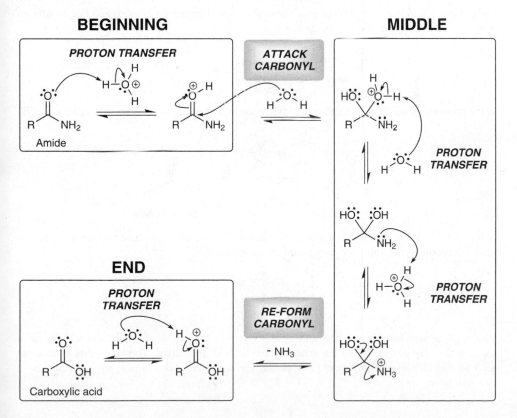

Let's begin with acid-catalyzed conditions. This reaction is really no different than the other acid-catalyzed reactions we have seen. Take a close look at the following mechanism:

Notice that it follows the exact same pattern that we have seen again and again:

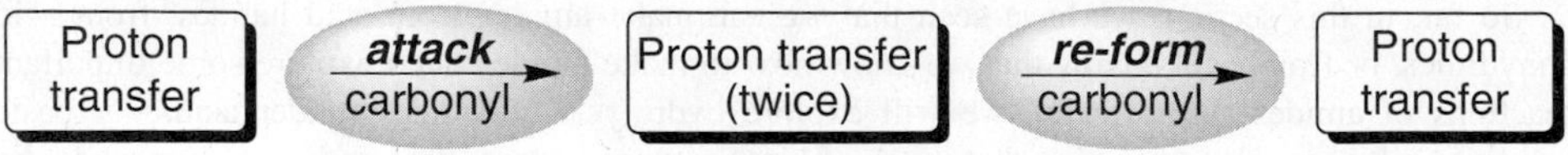

| Proton transfer | *attack* carbonyl | Proton transfer (twice) | *re-form* carbonyl | Proton transfer |

This pattern is common among the acid-catalyzed reactions that we have seen so far in this chapter.

Now let's explore base-catalyzed conditions for the hydrolysis of an amide:

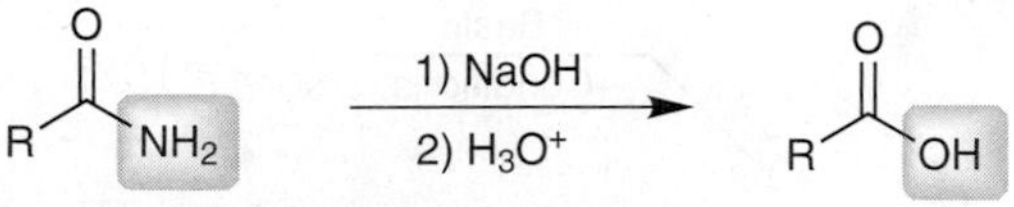

This transformation can occur via the following mechanism:

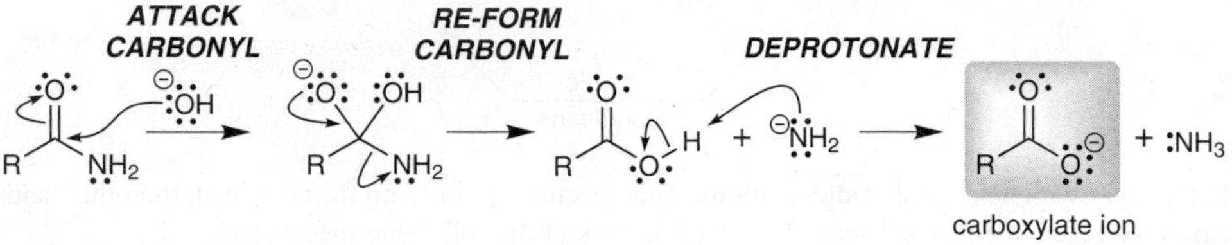

carboxylate ion

Under basic conditions, the product is a carboxylate ion (highlighted above), and that is why an acid is listed as a reagent (in a separate step):

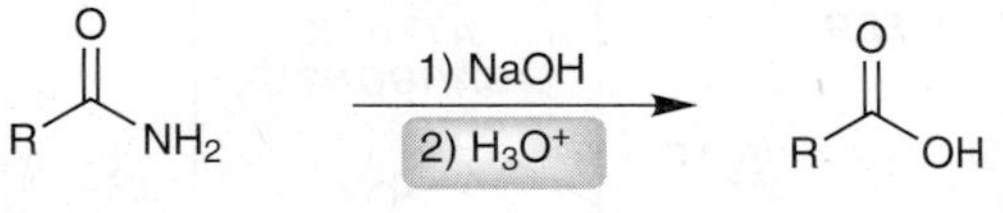

In the presence of acidic protons, the carboxylate ion is protonated to generate the carboxylic acid:

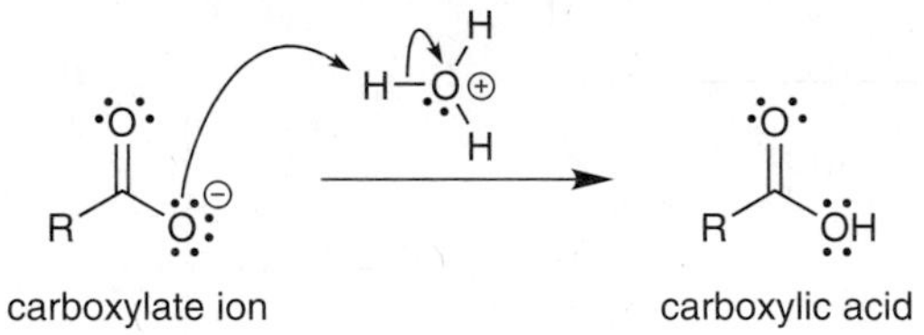

carboxylate ion   carboxylic acid

Before we do some problems, let's look at one last carboxylic acid derivative that we have not yet seen. Compounds containing a cyano group are called nitriles:

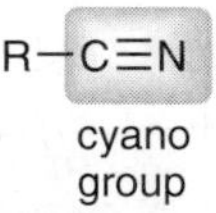

$$R-C\equiv N$$

cyano group

You might be wondering why nitriles are considered to be carboxylic acid derivatives. After all, a nitrile looks very different from the other carboxylic acid derivatives. To make sense of this, we need to consider oxidation states. Each of the carboxylic acid derivatives has three bonds to electronegative atoms:

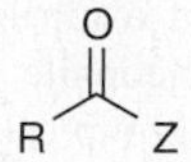

The carbon atom of the carbonyl group has two bonds with oxygen, and it also has one more bond with some heteroatom, Z (O, N, Cl, etc.). That gives a total of three bonds to heteroatoms. The carbon atom of a cyano group also has three bonds to a heteroatom. So, nitriles are at the same oxidation level as the other carboxylic acid derivatives.

Nitriles can be prepared using cyanide as a nucleophile to attack an alkyl halide:

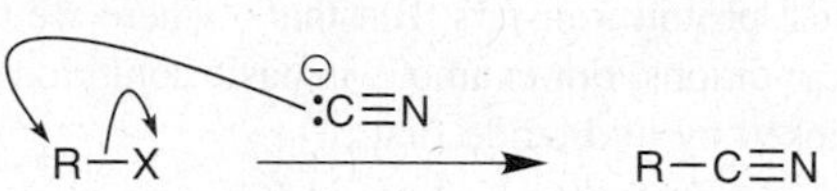

This is an S_N2 process, so you can only use this method with primary or secondary alkyl halides (primary halides are much better). Don't use this method with tertiary alkyl halides. There are other ways to make nitriles. Some textbooks cover other methods to prepare nitriles. You should look through your textbook (and your lecture notes) to see if you are responsible for knowing any other ways to make nitriles.

We can easily see that nitriles really are at the same oxidation level as other carboxylic acid derivatives, because hydration (which is *not* an oxidation-reduction reaction) produces an amide:

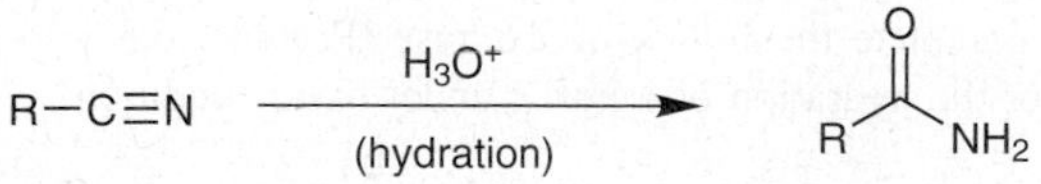

Hydration can occur either under acidic conditions or under basic conditions.

Whether we perform an acid-catalyzed hydration or a base-catalyzed hydration, the core steps are slightly different than the core steps of the mechanisms we have seen in this chapter. So far, all of our reaction mechanism have had at least *two* core steps (attack the carbonyl, and then expel a leaving group to re-form the carbonyl), with all other steps being proton transfers. But now, we will see a mechanism that has just *one* core step (attack the carbonyl). When it comes to the hydration of nitriles, we don't need to expel a leaving group. We can actually re-form the carbonyl group simply through proton transfers:

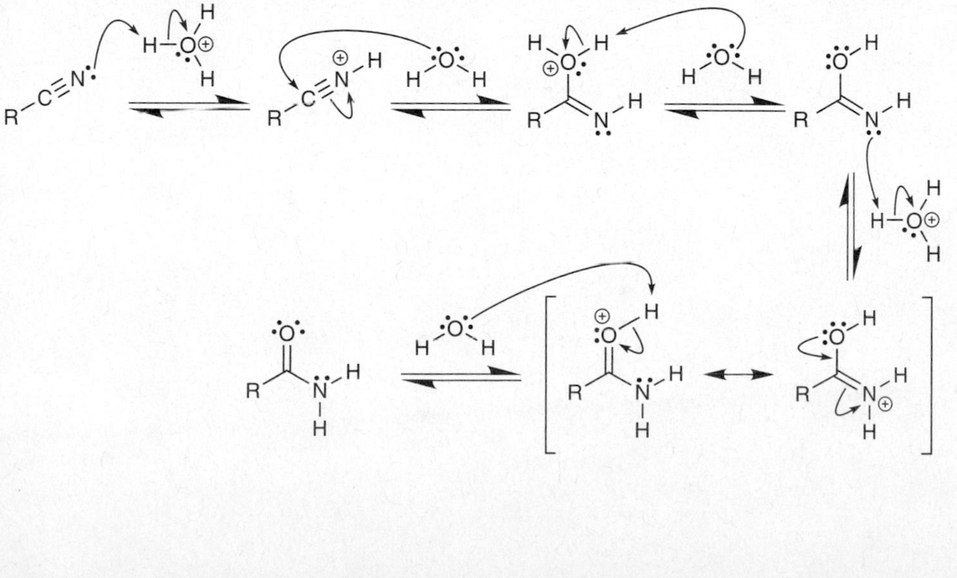

This is the mechanism for the hydration of a nitrile under *acidic* conditions. Notice that the second step in the mechanism shows a nucleophile (H_2O) attacking a protonated cyano group (very much the way a protonated carbonyl group can be easily attacked). All other steps in the mechanism are just proton transfers. When you think of it this way, it greatly simplifies the mechanism. The many proton transfers are necessary to avoid forming intermediates with negative charges. Notice that in acidic conditions, all of the intermediates are either positively charged or neutral.

Now let's consider the hydration of nitriles under *basic* conditions. The mechanism is actually VERY similar to the mechanism above. There is also only *one* core step (attacking the cyano group), and all the other steps are just proton transfers. But that's where we have our difference between acidic conditions and basic conditions. For example, in basic conditions, the cyano group is *not* first protonated. Rather, it is attacked by hydroxide first:

The rest of the mechanism is all just proton transfer steps. In order to draw the proton transfers properly, you must keep one thing in mind: stay consistent with the conditions. In acidic conditions, all intermediates should be either positively charged or neutral. In basic conditions, all intermediates should be either negatively charged or neutral.

With this in mind, complete the following exercise (Problem 6.37) to see if you can draw a plausible mechanism for the hydration of a nitrile under basic conditions.

PROBLEM 6.37 Based on everything we have just seen, propose a mechanism for the hydration of a nitrile under basic conditions:

Remember, there is just one core step (attacking the cyano group with hydroxide). After that, all other steps are just proton transfers. Use the space below to record your answer. When you have finished, you can look in the back of the book (or in your textbook) to see if you got it right.

EXERCISE 6.38 Draw a plausible mechanism for the following transformation:

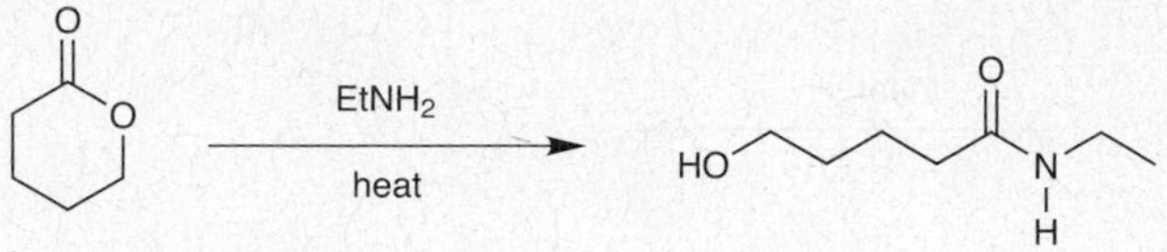

Answer In this reaction, an ester is being treated with an amine under conditions of heating. We have seen these conditions before. A source of protons (an acid) has not been indicated among the reagents, so the carbonyl group is not protonated. The first step of the mechanism will involve the amine directly attacking the carbonyl group of the ester:

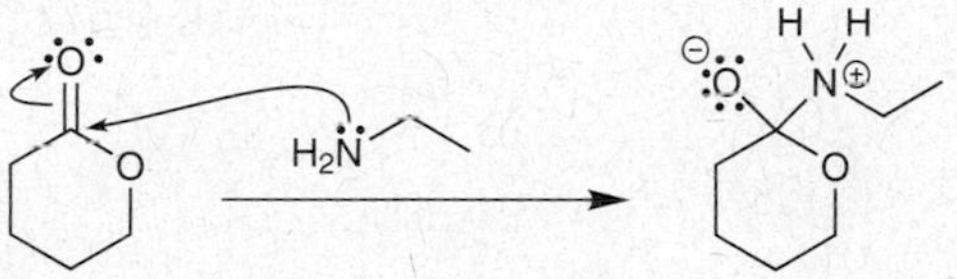

Then the carbonyl group is re-formed via expulsion of an alkoxide ion (RO⁻) as a leaving group:

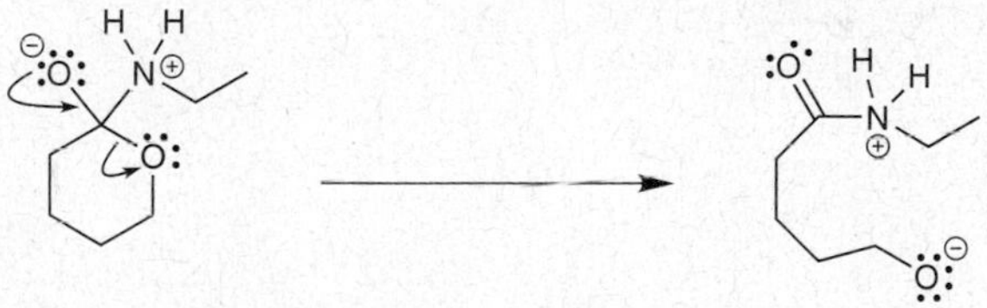

A proton transfer then generates the product:

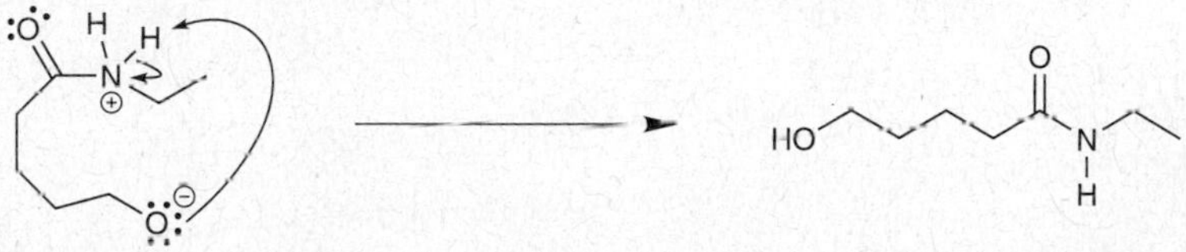

PROBLEMS Propose a plausible mechanism for each of the following transformations.

6.39

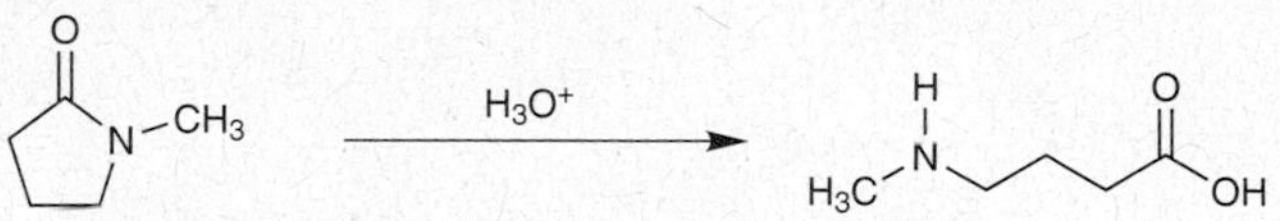

6.40

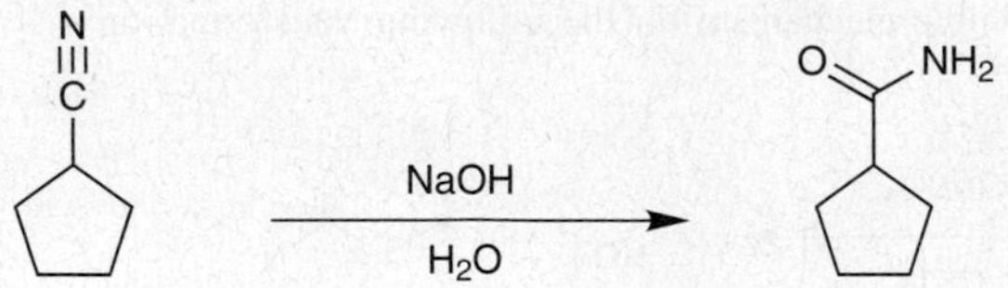

$$\xrightarrow[\text{H}_2\text{O}]{\text{NaOH}}$$

6.41

$$\xrightarrow[\text{H}_2\text{O}]{\text{NaOH}}$$

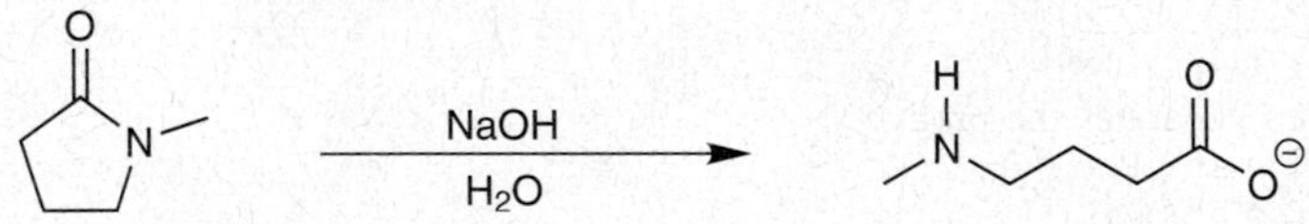

And now let's just do one more challenging mechanism. I say "challenging" not because it is difficult, but because you have not seen this exact mechanism before. Rather, you should be able to work your way through the mechanism, using all of the skills we have developed in this chapter:

PROBLEM 6.42 Propose a mechanism for the following reaction:

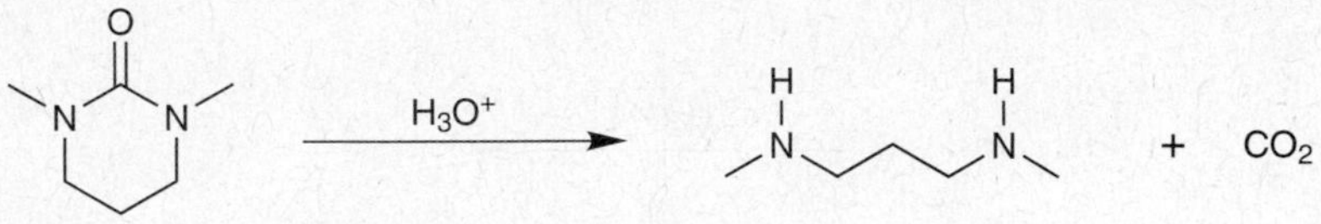

EXERCISE 6.43 Predict the products of the following reaction:

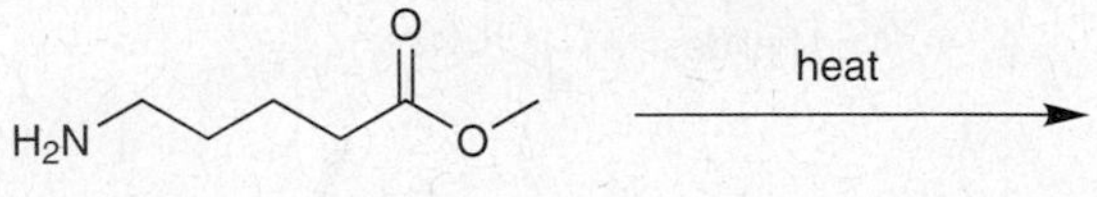

Answer We were not given any reagents here (just conditions of heat), so we look carefully at the starting material to see if we can have an intramolecular reaction. We notice that there are two functional groups in our starting compound (an ester and an amine). And we have seen that an ester can react with an amine under conditions of heating. The products should be an amide and an alcohol:

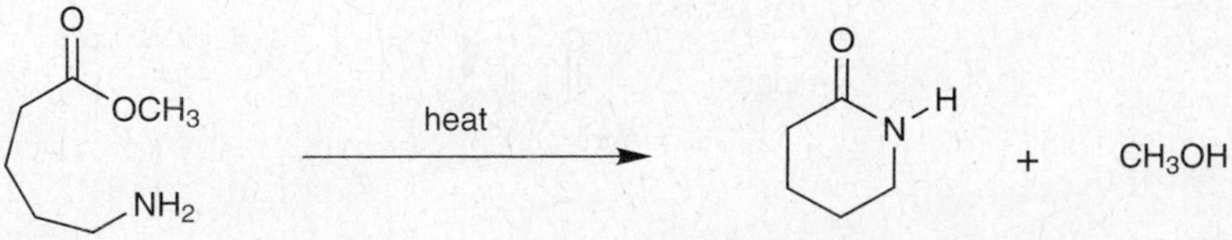

PROBLEMS Predict the products for each of the following reactions:

6.44

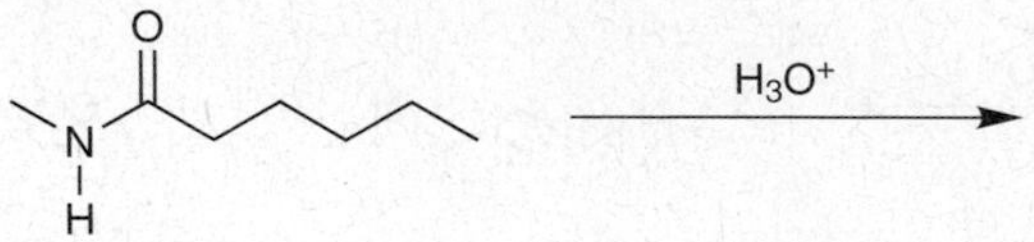

$$\text{(amide structure)} \xrightarrow{\text{H}_3\text{O}^+}$$

6.45

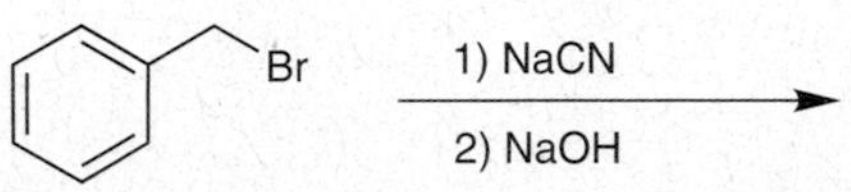

$$\text{(benzyl bromide)} \quad \begin{array}{l} \text{1) NaCN} \\ \text{2) NaOH} \end{array} \longrightarrow$$

6.46

$$\text{(lactam structure)} \xrightarrow{\text{H}_3\text{O}^+}$$

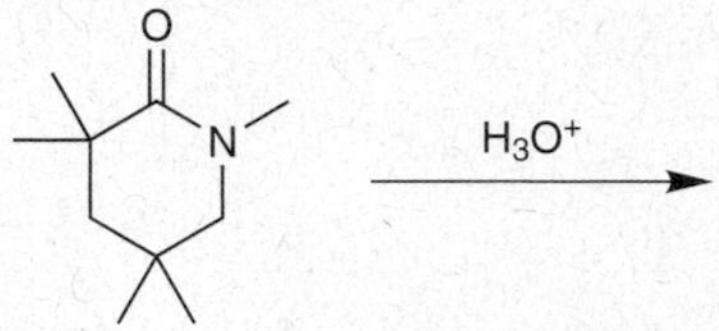

6.7 SYNTHESIS PROBLEMS

We have seen a lot of reactions in this chapter. Almost all of them involved the conversion of one carboxylic acid derivative into another. We saw that you can make a carboxylic acid derivative from any other derivative that is more reactive. In other words, you can always step your way *down* the following chart:

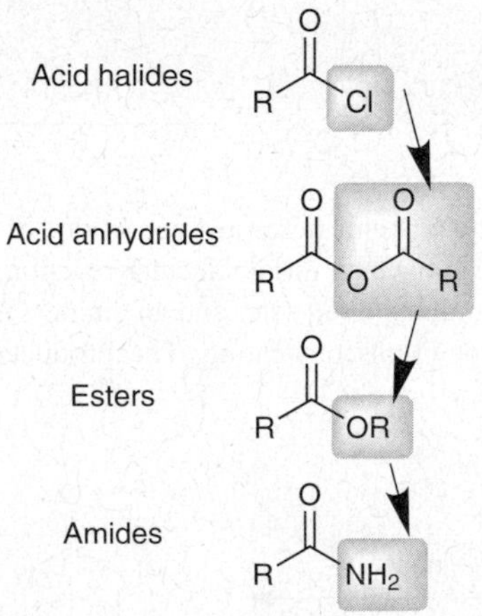

Acid halides

Acid anhydrides

Esters

Amides

You can even jump down the chart if you want:

Acid halides

$$R \overset{O}{\underset{}{\overset{}{\bigwedge}}} Cl$$

Acid anhydrides

$$R \overset{O}{\underset{}{\bigwedge}} O \overset{O}{\underset{}{\bigwedge}} R$$

Esters

$$R \overset{O}{\underset{}{\bigwedge}} OR$$

Amides

$$R \overset{O}{\underset{}{\bigwedge}} NH_2$$

Acid halides

$$R \overset{O}{\underset{}{\bigwedge}} Cl$$

Acid anhydrides

$$R \overset{O}{\underset{}{\bigwedge}} O \overset{O}{\underset{}{\bigwedge}} R$$

Esters

$$R \overset{O}{\underset{}{\bigwedge}} OR$$

Amides

$$R \overset{O}{\underset{}{\bigwedge}} NH_2$$

But travelling *up* the chart is more difficult:

Acid halides

$$R \overset{O}{\underset{}{\bigwedge}} Cl$$

Acid anhydrides

$$R \overset{O}{\underset{}{\bigwedge}} O \overset{O}{\underset{}{\bigwedge}} R$$

Cannot go up

Esters

$$R \overset{O}{\underset{}{\bigwedge}} OR$$

Amides

$$R' \overset{O}{\underset{}{\bigwedge}} NH_2$$

So, how do you travel *up* the chart, if you need to? Here is the way to do it: You can exit this chart by converting into a carboxylic acid, and then come back into the chart, like this:

$$R \overset{O}{\underset{}{\bigwedge}} OH$$

$$R \overset{O}{\underset{}{\bigwedge}} Cl$$

$$R \overset{O}{\underset{}{\bigwedge}} O \overset{O}{\underset{}{\bigwedge}} R$$

$$R \overset{O}{\underset{}{\bigwedge}} OR$$

$$R \overset{O}{\underset{}{\bigwedge}} NH_2$$

Let's get some practice with this:

EXERCISE 6.47 Propose an efficient synthesis for the following transformation:

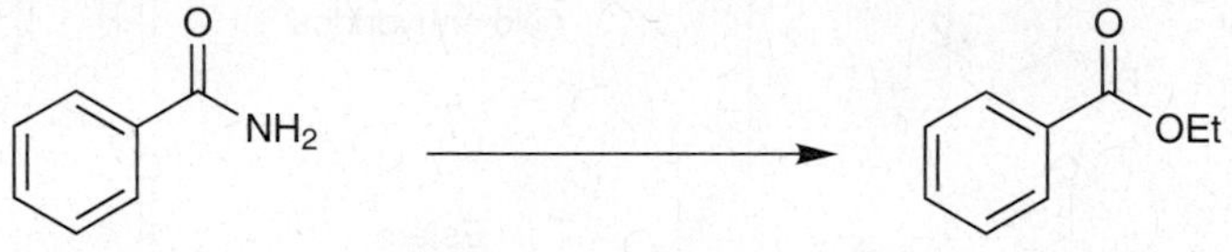

Answer We must convert an amide into an ester, but we have not learned a way to do this directly in one step (because that would involve going *up* the chart). Amides are less reactive than esters, so we cannot go directly from an amide to an ester. Instead, we can first convert the amide into a carboxylic acid, and then we can convert the carboxylic acid into an ester:

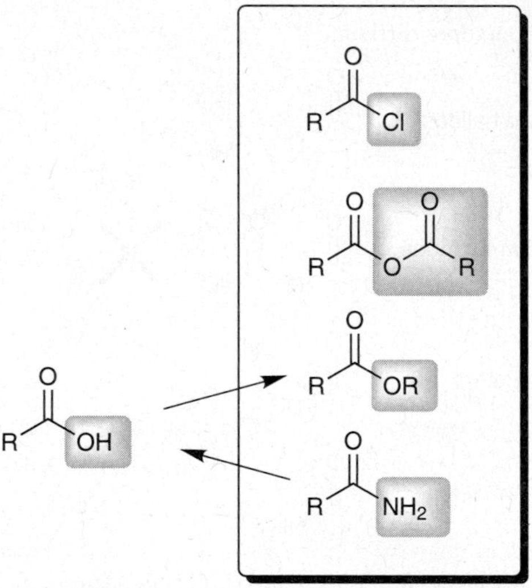

To achieve this transformation, we would use the following reagents:

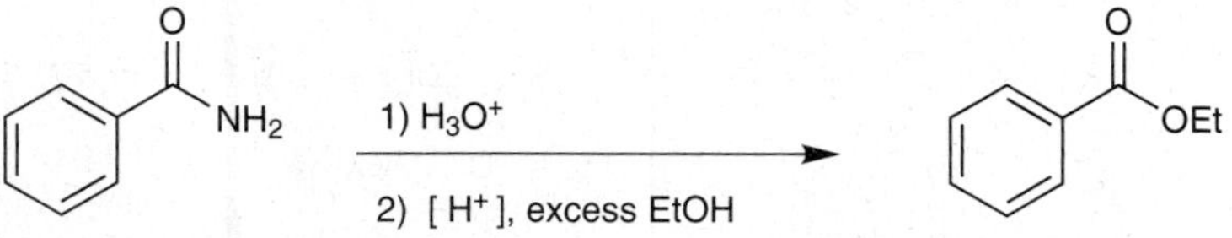

1) H₃O⁺

2) [H⁺], excess EtOH

PROBLEMS Propose an efficient synthesis for each of the following transformations:

6.48

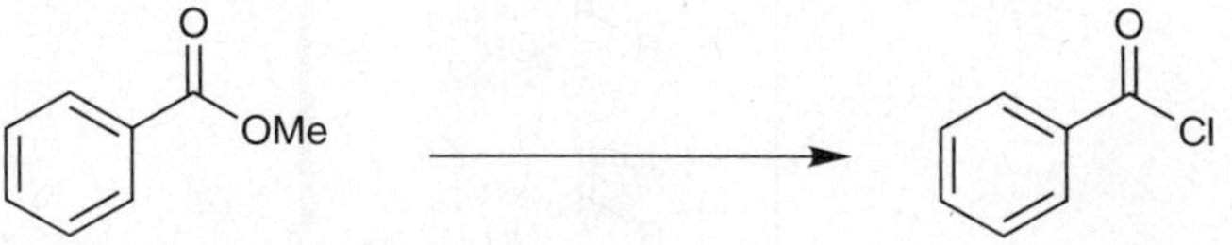

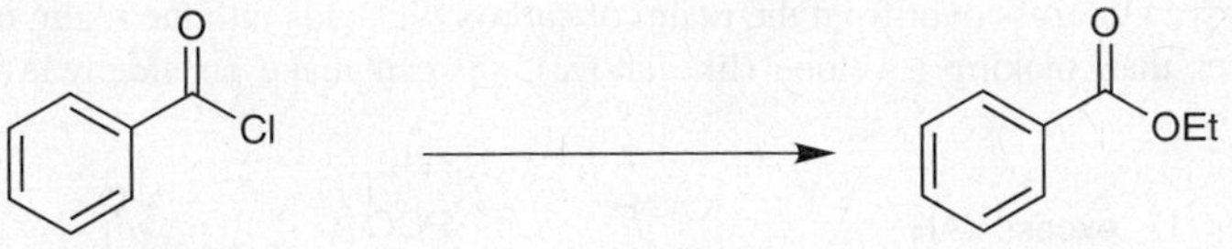

6.49

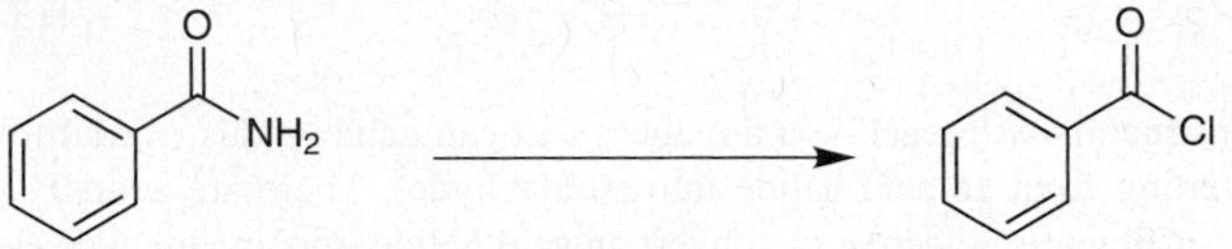

6.50

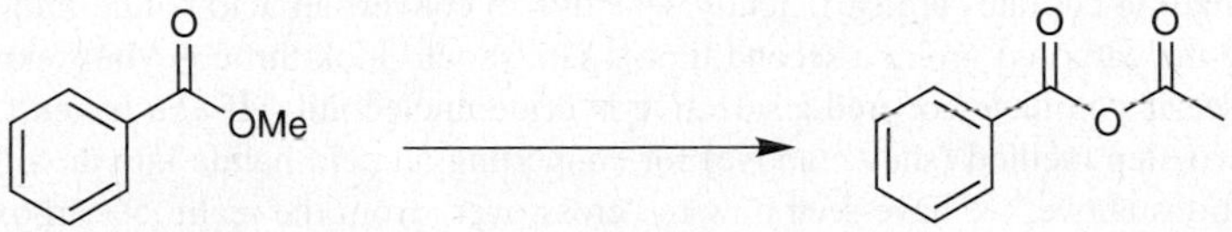

6.51

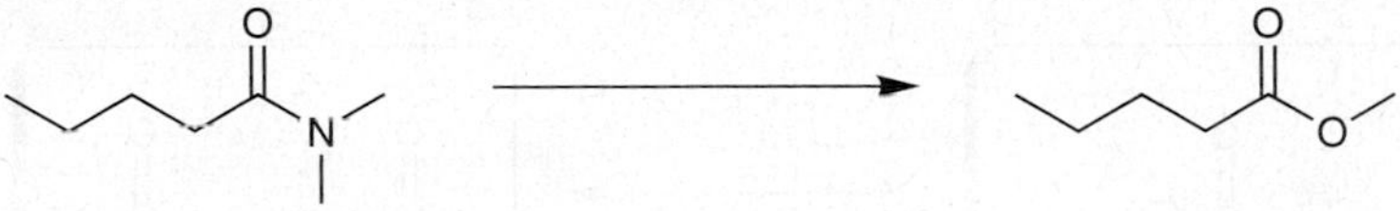

6.52

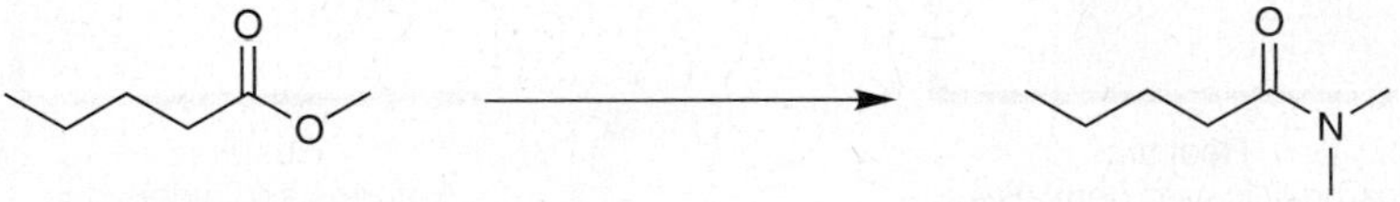

6.53

There is one other important strategy to keep in mind when you are proposing a synthesis. In this chapter, we explored the chemistry of carboxylic acid derivatives; and in the previous chapter, we explored the chemistry of ketones/aldehydes. These two chapters represent two different realms:

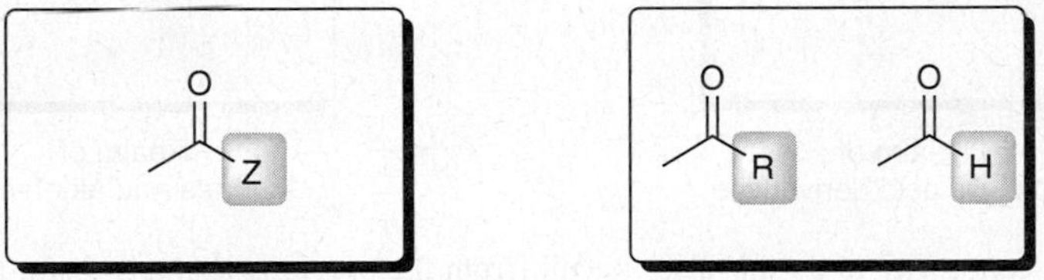

Realm of
carboxylic acid derivatives

Realm of
ketones and aldehydes

But these realms are not completely isolated from one another, because we have seen ways to convert from one realm into another. In this chapter, we saw how to convert an acid halide into a ketone:

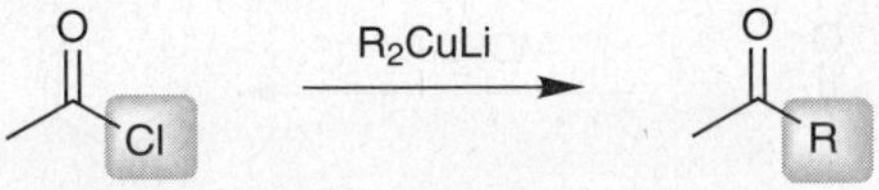

There is also another way to cross over from the realm of carboxylic acids into the realm of ketones and aldehydes. Rather than making a ketone (like above), we can make an aldehyde using the following two steps:

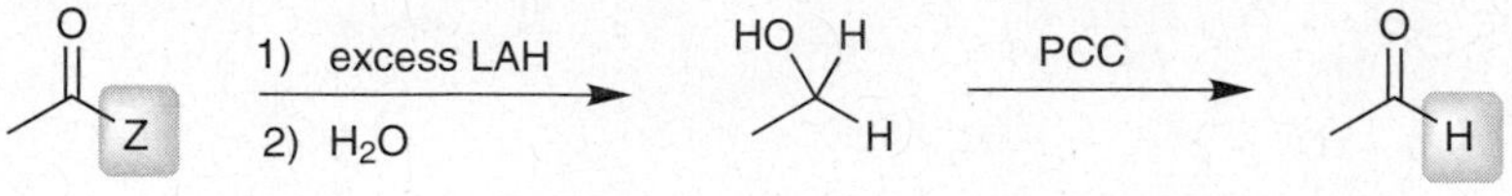

Some textbooks and instructors will teach you a reagent that can achieve this overall transformation in one step (converting from an acid halide into an aldehyde). There are actually many hydride reagents that are sufficiently selective to convert an acid halide into an aldehyde (very much the way lithium dialkyl cuprates are sufficiently selective to convert an acid halide into a ketone, without attacking the carbonyl group a second time). You should look through your textbook and lecture notes to see if you have covered a selective hydride nucleophile. If you haven't, you can always use the two-step method (shown above) for converting an acid halide into an aldehyde.

With the reactions above, we have seen how to "cross over" from the realm of carboxylic acid derivatives into the realm of ketones and aldehydes:

Realm of
carboxylic acid derivatives

Realm of
ketones and aldehydes

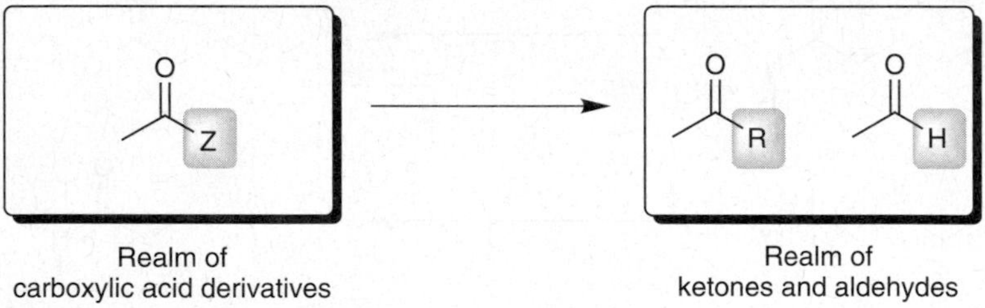

But what about the reverse direction? Do we have a way to "cross over" from the realm of ketones and aldehydes into the realm of carboxylic acid derivatives?

Realm of
carboxylic acid derivatives

Realm of
ketones and aldehydes

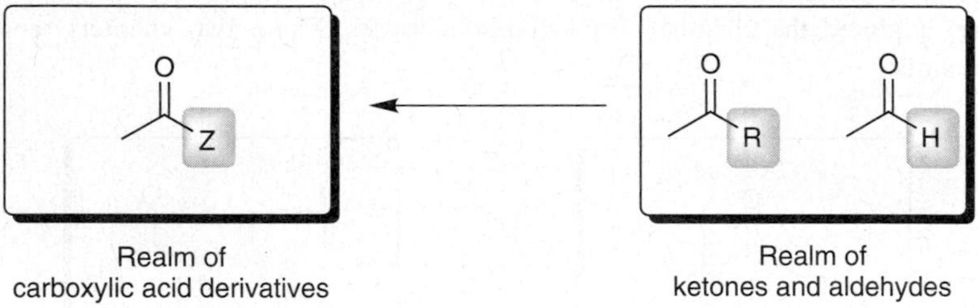

Yes, we have seen a way to do this also. Recall (from the previous chapter) that a Baeyer-Villiger oxidation will convert a ketone into an ester:

We can also use a Baeyer-Villiger oxidation to convert an aldehyde into a carboxylic acid (remember migratory aptitude?):

So, we now have reactions that allow us to "cross over" from one realm into the other (in either direction). Let's see a concrete example of how to use this:

EXERCISE 6.54 Propose an efficient synthesis for the following transformation:

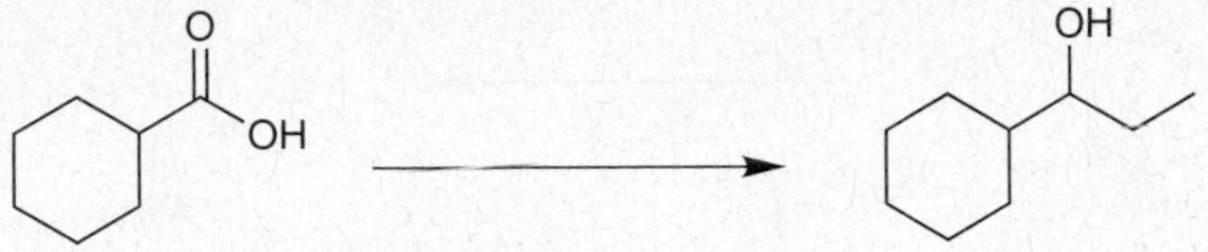

Answer The final product is an alcohol, and in the process of converting the carboxylic acid into an alcohol, an ethyl group must be installed. It might be hard to see, at first glance, how this transformation can be achieved. But don't get discouraged. You are not expected to know how to solve problems like this instantly. Synthesis problems require thought and strategizing. Remember that you always want to try to go backwards as much as possible (retrosynthetic analysis). So, let's work our way backwards.

Did we learn any simple ways to make alcohols? In the previous chapter, we learned how to make an alcohol from a ketone, using LAH:

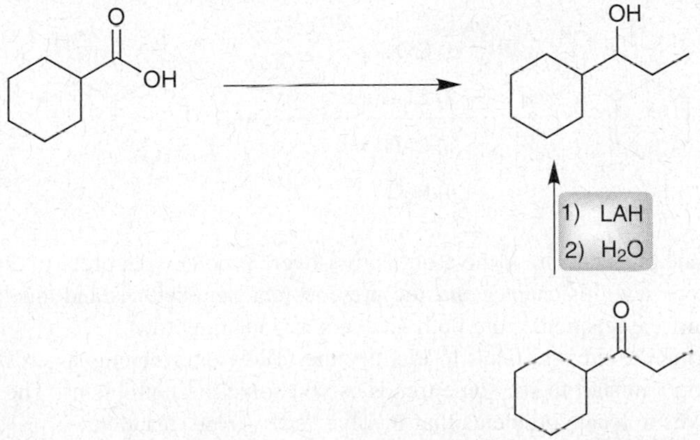

With this one important step, we are now in a position to realize that this problem can be thought of as a "cross-over" problem. The starting material is a carboxylic acid, and we need to turn it into a ketone:

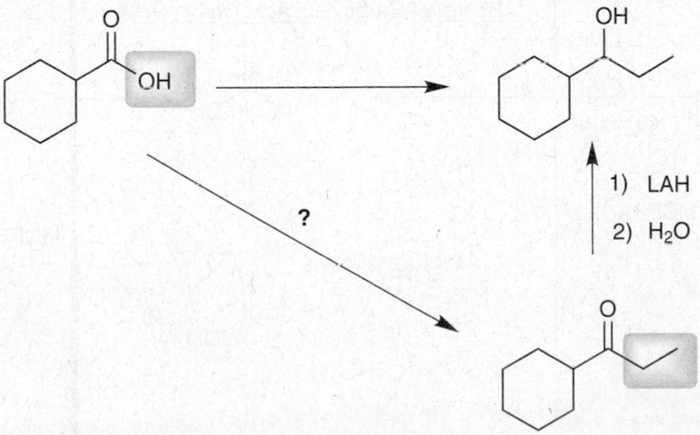

In other words, we need to cross over from the realm of carboxylic acids into the realm of ketones. And we did see one reaction that allows us to do that. We can make a ketone from an acid halide, using a lithium dialkyl cuprate. So, now we have worked backwards again:

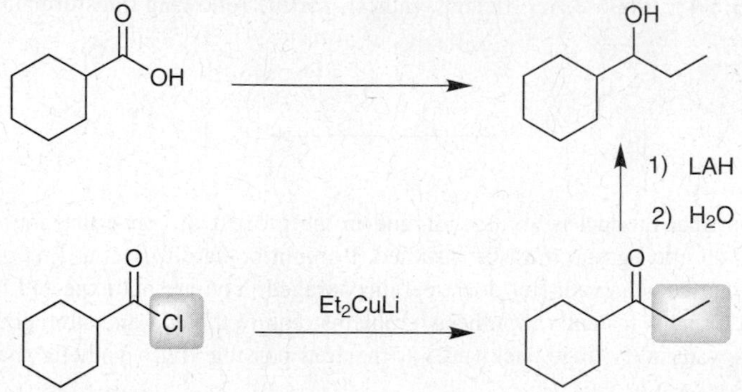

1) LAH

2) H₂O

Et₂CuLi

Finally, we just need to convert the carboxylic acid into an acid halide, and we can do that in one step with thionyl chloride.

So, our answer is:

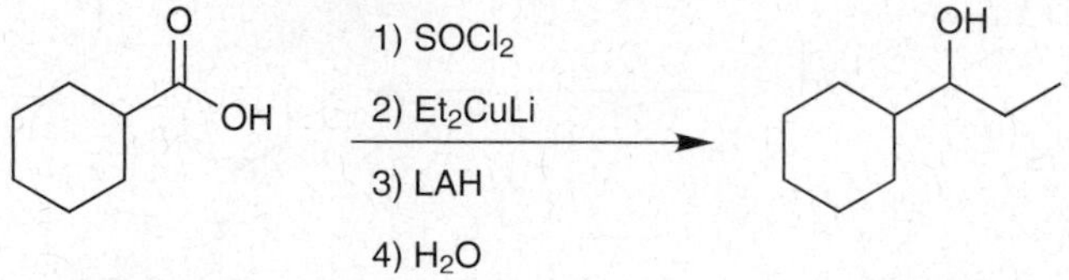

1) SOCl₂

2) Et₂CuLi

3) LAH

4) H₂O

Now let's get some practice with some more "cross-over" problems. In order to do these problems, you will need to review this chapter *and* the previous chapter (ketones and aldehydes)—you will need to have all of the reactions from both chapters at your fingertips.

At first, you might find it difficult to identify the following problems as cross-over problems, but hopefully, you will start to see some trends as you solve these problems. The hope is that you will train your eyes to locate problems that involve "cross-over" reactions.

In solving these problems, make sure that you are familiar with the ways that we have seen for crossing over. We have seen four such reactions so far, all of which are summarized in the following chart:

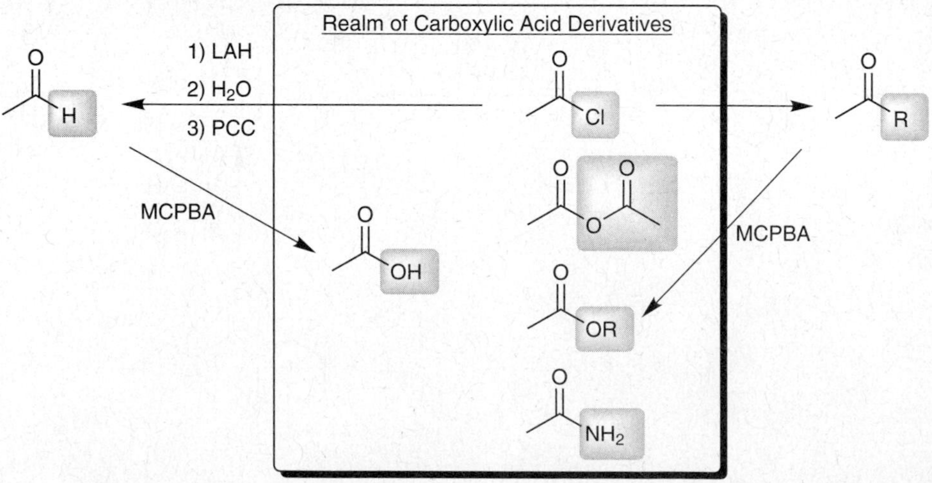

Realm of Carboxylic Acid Derivatives

1) LAH
2) H₂O
3) PCC

MCPBA

MCPBA

Study this chart carefully. To help you remember them, you should notice one thing that all four reactions have in common. They are all reduction-oxidation reactions. This should make sense because carboxylic acid derivatives are at a different oxidation state than ketones and aldehydes.

You will need some time to do the following problems, so don't sit down to do these problems when you only have 5 minutes to study. That would just frustrate you. Make sure that you have some time to spend when you sit down to work through these problems.

PROBLEMS Propose an efficient synthesis for each of the following transformations. In each case, remember to work backwards (retrosynthetic analysis), and try to determine which cross-over reaction to use. When you compare your answers to the answers in the back of the book, keep in mind that there is often more than one way to solve a synthesis problem. If your answer is different from the answer in the back of the book, you should not necessarily conclude that your answer is wrong.

6.55

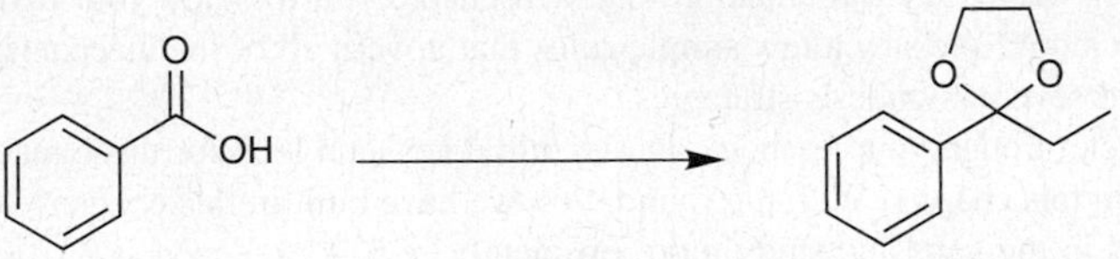

6.56

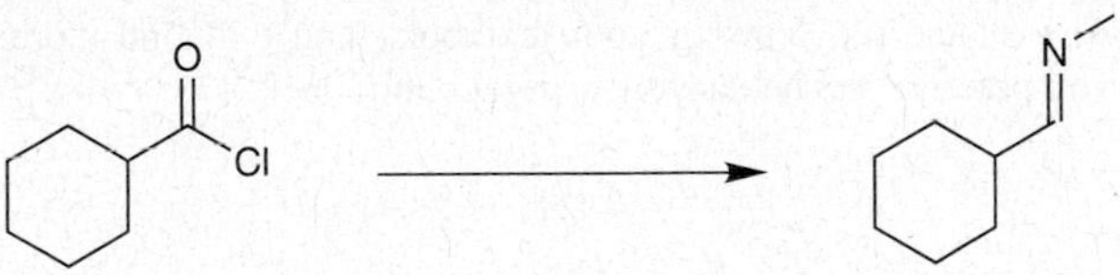

6.57

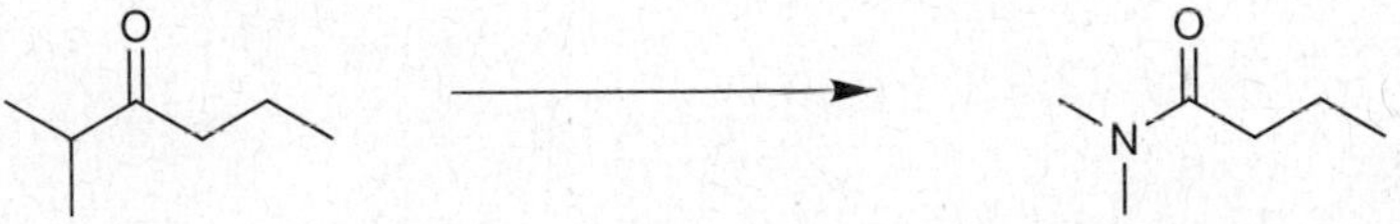

6.58

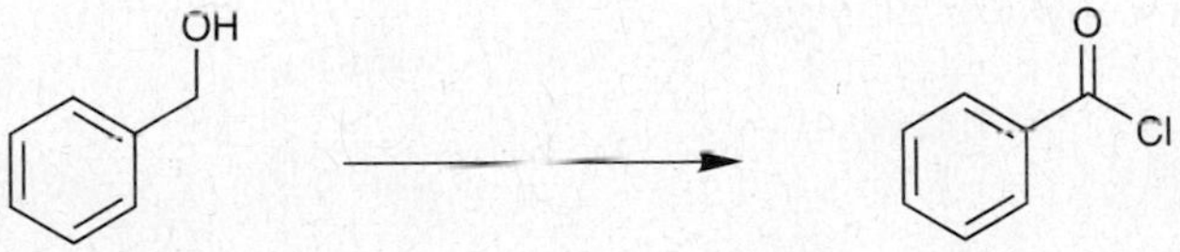

6.59

6.60

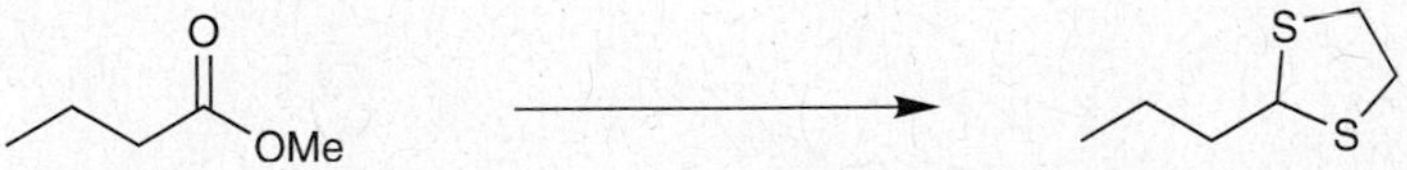

6.61

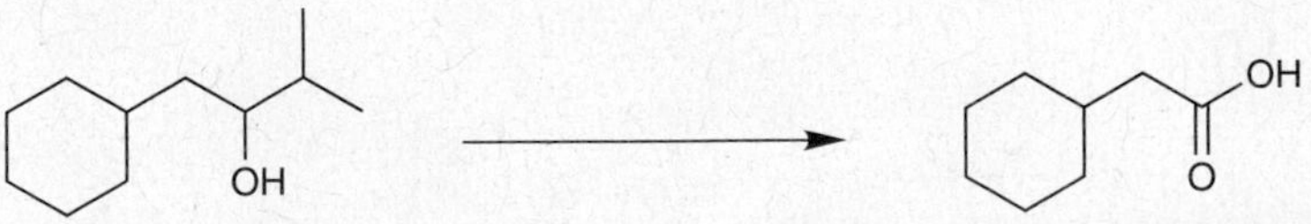

6.62

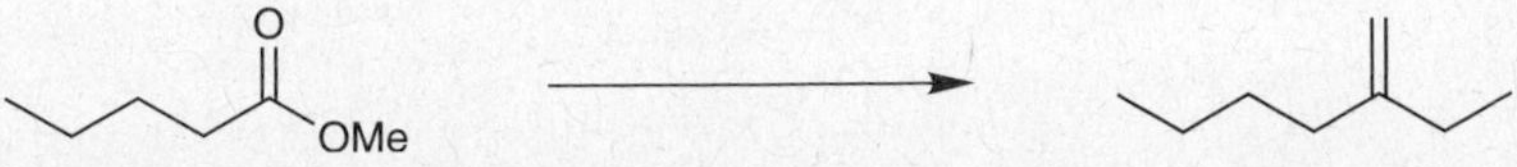

6.63

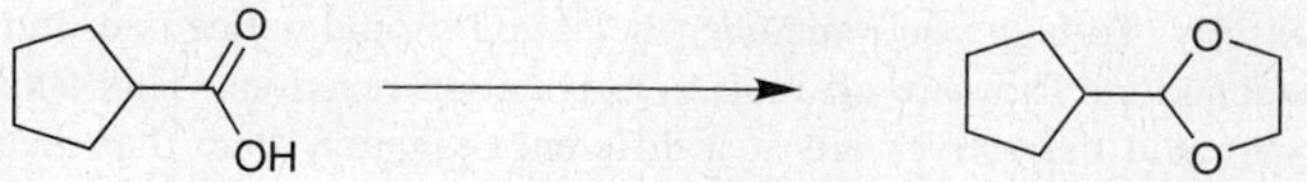

6.64

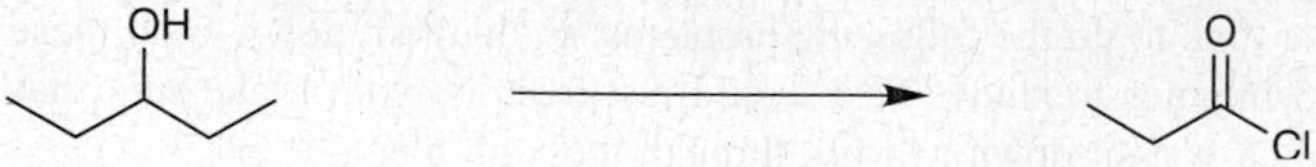

6.65

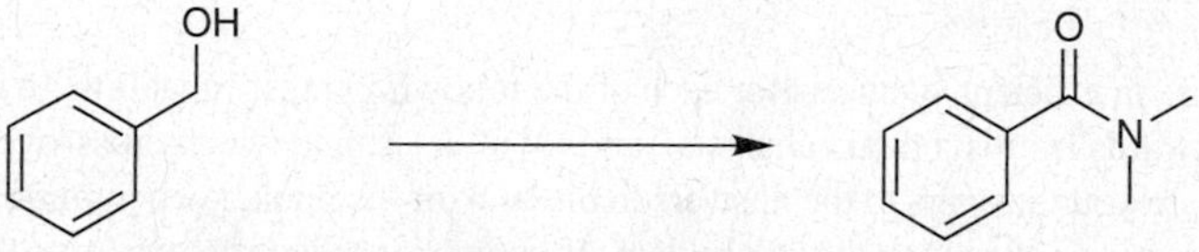

The goal of this chapter was to lay a foundation that will enable you to study your textbook and lecture notes more efficiently. We saw a few simple rules that govern all of the mechanisms in this chapter, and we learned several synthesis strategies.

Now you can go back through your textbook and lecture notes, and look for those reactions that we did not cover here in this chapter. With the foundation we have built in this chapter, you should be in good shape to fill in the gaps and study more efficiently.

And make sure to do ALL of the problems in your textbook. You will find more synthesis problems there. The more you practice, the better you will get. Good luck.

CHAPTER 7

ENOLS AND ENOLATES

7.1 ALPHA PROTONS

In the previous two chapters, we focused on the reactions that can take place when a nucleophile attacks a carbonyl group:

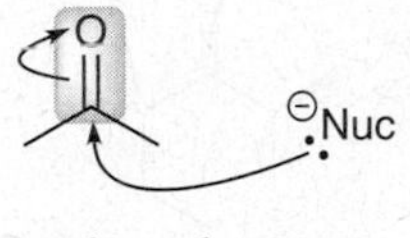

We first learned about nucleophilic attack on ketones and aldehydes (in Chapter 5). Then, in Chapter 6, we explored reactions of carboxylic acid derivatives. Now, we are ready to move away from the carbonyl group, and explore the chemistry that can take place at the alpha (α) carbon:

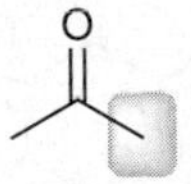

We call this the alpha carbon, because it is the carbon atom directly connected to the carbonyl group. We use the Greek alphabet to label carbon atoms, moving away from the carbonyl group, in either direction:

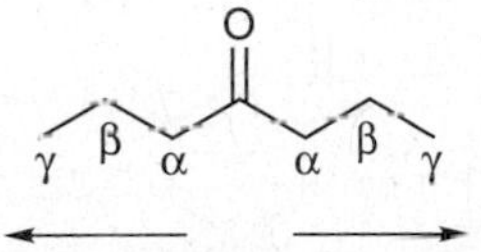

Notice that in this compound, there are two alpha positions. In this chapter, we will focus on the chemistry that can take place at the alpha positions.

Before we get started, we should discuss one more piece of terminology. Any protons connected to an alpha carbon are called alpha protons:

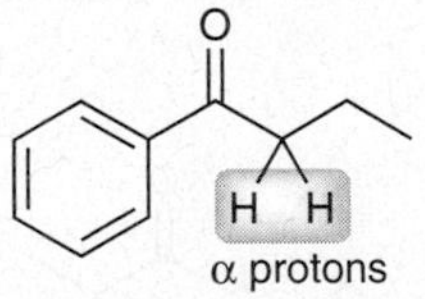

α protons

Not all alpha carbon atoms will have alpha protons. For example, consider the following compound:

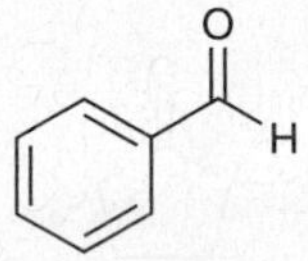

This compound has no alpha protons. If you look just to the right of the carbonyl group, you will see that there is no alpha carbon (it is just an aldehyde). That aldehydic H is NOT an alpha proton because it is not connected to an alpha carbon. And if you look just to the left of the carbonyl group, you will see that there IS an alpha carbon, but this carbon has no protons.

It is important to recognize the presence or absence of alpha protons. We will see a lot of reactions in this chapter, and most of these reactions will be based on the presence of alpha protons. It turns out that alpha protons are somewhat acidic; and the removal of an alpha proton generates an anion that is fairly reactive. We will see this in greater detail very soon. For now, let's just make sure that we can identify alpha protons when we see them.

EXERCISE 7.1   Identify all alpha protons in the following compound:

Answer   To see if there are any α protons, we must first identify the alpha carbon atoms:

The alpha carbon on the right does **not** have any protons. The alpha carbon on the left **does** have a proton:

So, there is just one alpha proton in this compound.

PROBLEMS   For each of the compounds below, identify all alpha protons (some compounds may not have any alpha protons).

7.2

7.3

7.4

7.5

7.6

7.7

7.2 KETO-ENOL TAUTOMERISM

When a ketone has an alpha proton, there is an interesting thing that can happen. In the presence of either acid or base, the ketone exists in equilibrium with another compound:

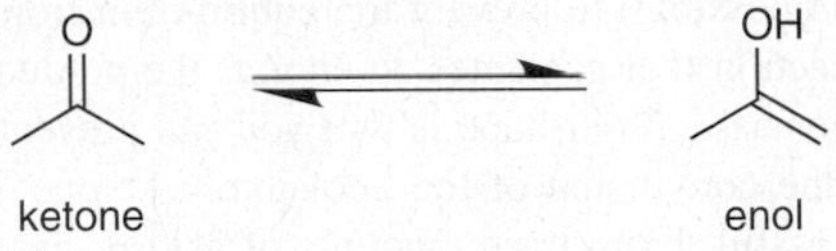

ketone   enol

This other compound is called an **enol**, because it has a C=C bond ("ene") and an OH group ("ol"). The equilibrium shown above is actually very important, because you will see it in many mechanisms. So, let's take a closer look.

If we focus on the connections of atoms, we will find that the two compounds differ from each other in the placement of one proton. The ketone has the proton attached to an alpha carbon, and the enol has the proton connected to oxygen:

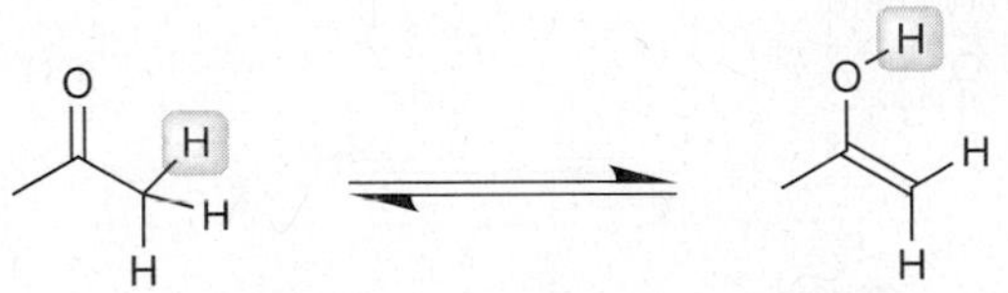

It is true that the π bond is also in a different location. But when we just focus on the atoms (which atoms are connected to which other atoms), we find that the difference is in the placement of just one proton. We have a special name to describe the relationship between compounds that differ from each other in the placement of just one proton. We call them *tautomers*. So, the enol above is said to be the *tautomer* of the ketone, and similarly, the ketone is the *tautomer* of the enol. The equilibrium shown above is called *keto-enol tautomerism*.

Keto-enol tautomerism is **NOT** resonance. The two compounds shown above are NOT two representations of the same compound. They are, in fact, different compounds. These two compounds are in equilibrium with each other.

In most cases, the equilibrium greatly favors the ketone:

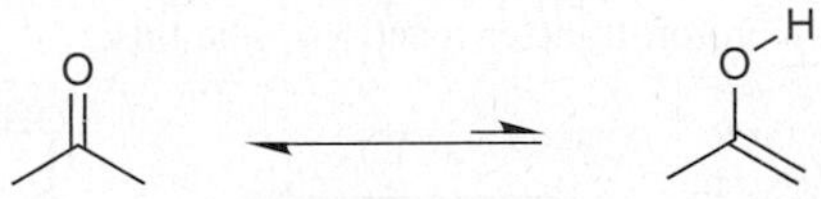

This should make sense, because the last two chapters focused on the formation of C=O bonds as a driving force for reactions. A ketone has a C=O bond, but an enol does not. So we should not be surprised that the equilibrium favors the ketone.

There are some situations where the equilibrium can favor the enol. For example:

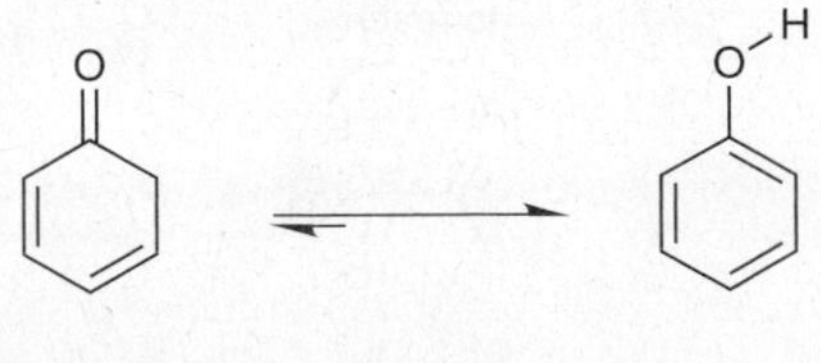

In this case, the enol is an aromatic compound, and it is much more stable than the ketone (which is not aromatic). There are many other situations where an enol can be more stable than its tautomer. You will probably find some of these examples in your textbook (such as 1,3-diketones). But in most cases (other than these few exceptional cases), the equilibrium will favor a ketone over an enol.

It is very hard (close to impossible) to prevent the equilibrium from being established. Imagine that you are performing a reaction that generates an enol as the product, and you take great efforts to remove all traces of acid or base. Your hope is that you can prevent the equilibrium from being established, so as to avoid the conversion of the enol into a ketone. But you will find that your efforts will likely be unsuccessful. Even trace amounts of acid or base adsorbed on the glassware (that you cannot remove) will allow the equilibrium to be established.

We will now explore a mechanism for keto-enol tautomerism. We said that compounds will tautomerize in the presence of either acid or base, so we will need to explore two mechanisms: one under *acidic* conditions, and one under *basic* conditions.

We saw that, by definition, tautomers will differ in the position of one proton. So, the conversion of a ketone into an enol requires two steps: 1) introduce a proton, and 2) remove a proton:

Similarly, the conversion of an enol into a ketone also requires the same two steps: 1) introduce a proton, and 2) remove a proton:

You might wonder why two separate steps are required. Why can't the proton just move over, in one step (in an intramolecular proton-transfer reaction), like this:

This doesn't work, because the oxygen atom is just too far away (in space) from the proton it is trying to remove:

The mechanism requires two separate steps. But these two steps can be in either order: we can either first protonate (acidic conditions) or we can first deprotonated (basic conditions). Here is the mechanism under basic conditions:

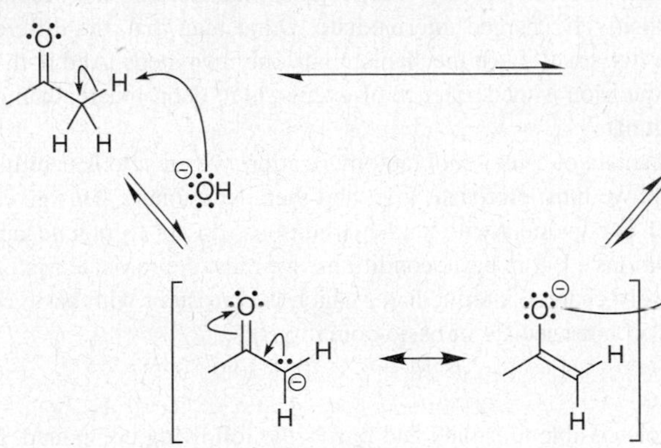

Notice that there is one intermediate (for which we must draw resonance structures), and this intermediate is negatively charged. If you look at the second resonance structure, it looks like an enol that is missing a proton. So, we call this intermediate an **enolate**. Don't be fooled into thinking that the mechanism above has more than two steps. Resonance (of the intermediate) is NOT a step. Our mechanism only has two steps, like this:

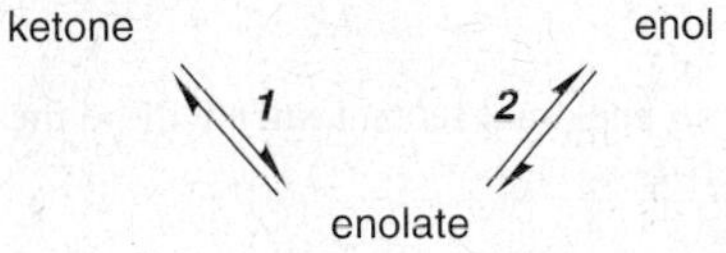

The enolate is very important for the rest of this chapter, because it can function as a nucleophile. We will see many examples in the coming sections. For now, let's finish our discussion of keto-enol tautomerism.

In the mechanism above, we first deprotonated (basic conditions). Under acidic conditions, the first step is to protonate:

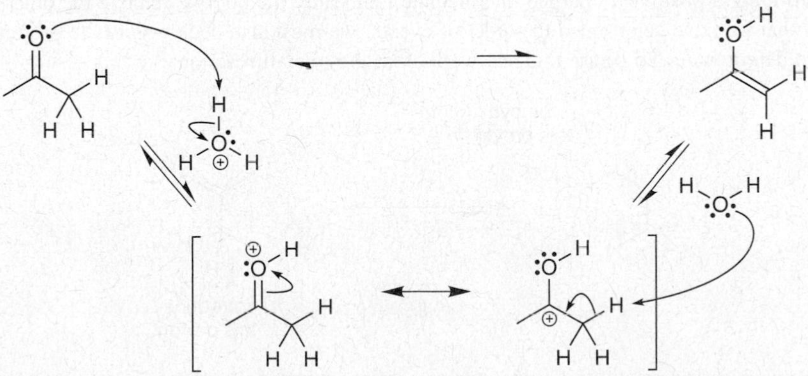

Once again, there are just two steps here. Don't be fooled by the resonance of the intermediate. Resonance is not a step. Resonance is just our way of dealing with the fact that we cannot draw the intermediate with only one drawing. We need two drawings to capture its character. And if you look at these resonance structures, you will see that this intermediate is positively charged.

Notice the difference between these two mechanisms (acidic vs. basic conditions). The first mechanism (basic conditions) has a negatively charged intermediate, and the second mechanism (acidic conditions) has a positively charged intermediate. Other than that, the difference between these two mechanisms is pretty small. Each mechanism has only two steps. And both steps are just proton transfers. The only question is the sequence of events. Is it: deprotonate, then protonate? Or is it: protonate, then deprotonate?

When drawing the mechanism of a keto-enol tautomerization, we must look carefully at the conditions. In acidic conditions, we must protonate first, and then deprotonate. This gives a positively charged intermediate, which is consistent with acidic conditions (do not form a negatively charged intermediate in acidic conditions). But in basic conditions, we must deprotonate first, and then protonate. This gives a negatively charged intermediate, which is consistent with basic conditions (do not form a positively charged intermediate in basic conditions).

EXERCISE 7.8 It is not possible to isolate and purify the following compound, because upon formation, it will rapidly tautomerize to form a ketone. Draw a mechanism for formation of the ketone under acidic conditions.

Answer This compound is an enol, and its tautomer will be the following ketone:

To convert the enol into a ketone, our mechanism will have two steps: protonate and deprotonate. But we must decide what order to use. Do we first protonate? Or do we first deprotonate? To answer this question, we look at the conditions. Since we are in acidic conditions, we should first protonate (forming a positively charged intermediate), and only then do we remove the other proton.

Now that we have determined the order of events, we must also decide **where** to protonate, and **where** to deprotonate. To figure this out, we look at the overall reaction:

When we analyze the reaction like this, it is easy to see **where** to introduce a proton and **where** to remove a proton. This might seem trivial, but it is extremely important because it showed us that we must protonate the double bond (rather than protonating the oxygen atom), like this:

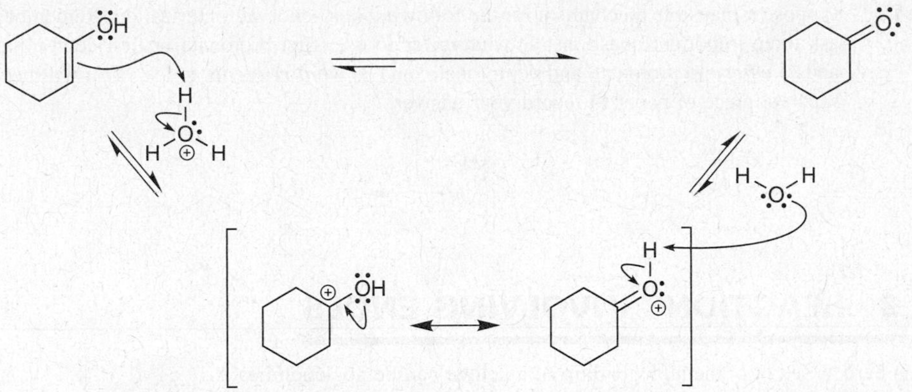

So often, students will start this problem by protonating the OH group. Although that might make sense at first, you will find that this step will NOT lead to formation of the ketone. The first step is to protonate the double bond; *not* the OH group.

When drawing a mechanism, make sure to never use HO⁻ and H_3O^+ in the same mechanism. When acidic conditions are indicated, use H_3O^+ to protonate and use *H_2O* to deprotonate. Don't use hydroxide to remove a proton, because there are not many hydroxide ions present under acidic conditions.

Similarly, when basic conditions are indicated, use HO⁻ to remove the proton and use *H_2O* to protonate. Don't use H_3O^+ to protonate, because we are in basic conditions. Here is the take home message: always stay consistent with your conditions.

So, to recap, there are three things to consider in order to correctly draw the mechanism of a keto-enol tautomerization: 1) what *order* to use (first protonate or first deprotonate), 2) *where to* protonate and where to deprotonate, and 3) *what reagents* to use when drawing the proton transfer steps (stay consistent with the conditions).

PROBLEMS   For each of the following transformations, propose a mechanism that is consistent with the conditions indicated (you will need a separate piece of paper to record your answers):

7.9

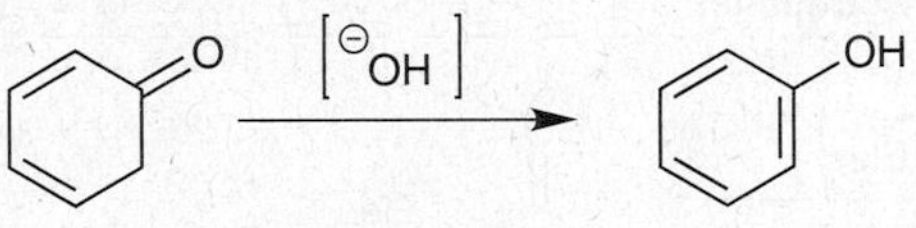

7.10

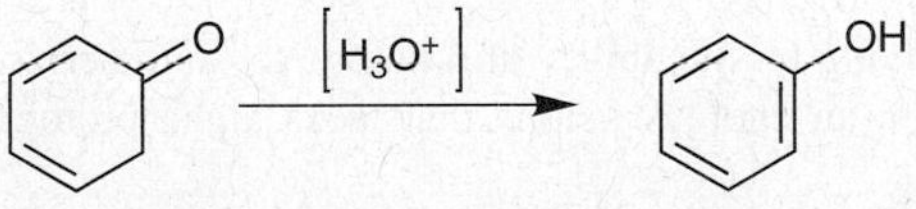

7.11

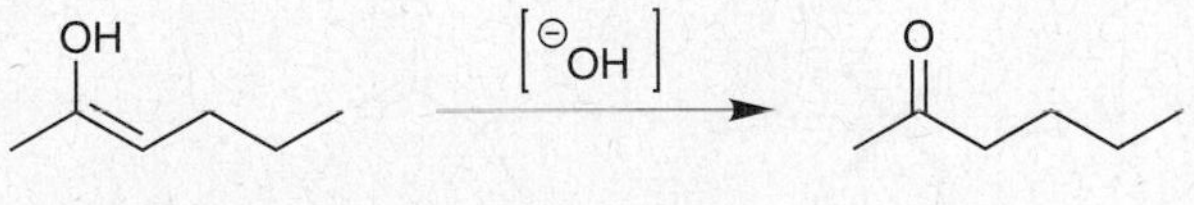

7.12

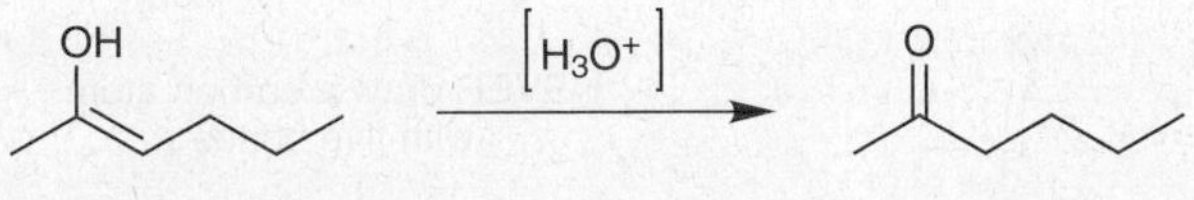

7.13 Propose a plausible mechanism for the following keto-enol tautomerization. Remember to ask three important questions: 1) what *order* to use (first protonate or first deprotonate), and 2) *where* to protonate and deprotonate, and 3) *what reagents* to use. You will need a separate piece of paper to record your answer.

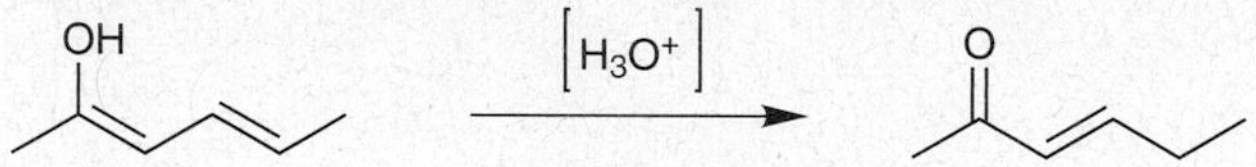

$$\left[H_3O^+\right]$$

7.3 REACTIONS INVOLVING ENOLS

It is hard to see how the alpha carbon of a ketone can be nucleophilic:

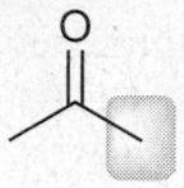

The alpha carbon does ***not*** have a lone pair or a π bond that can function as a nucleophilic center. However, when we examine the structure of the enol (that is in equilibrium with the ketone), we get a different picture:

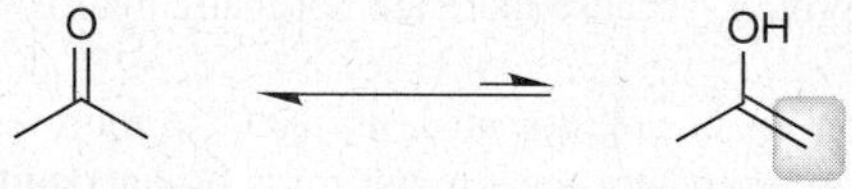

The enol has a π bond on the alpha carbon, which renders it nucleophilic. Also, consider the resonance structure of the enol:

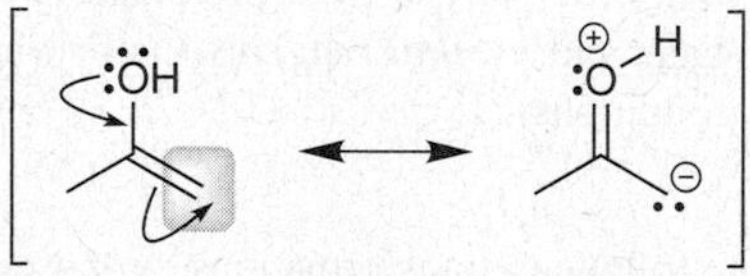

Notice that there is a negative charge on the alpha position, and therefore, the alpha carbon can function as a nucleophile to attack some electrophile:

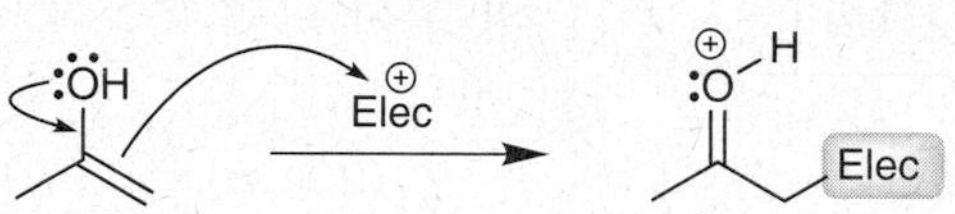

In order for the attack to occur, we are relying on the ability of a ketone to tautomerize. But, not every ketone will exist in equilibrium with an enol. A ketone that lacks alpha protons will ***not*** tautomerize to form an enol:

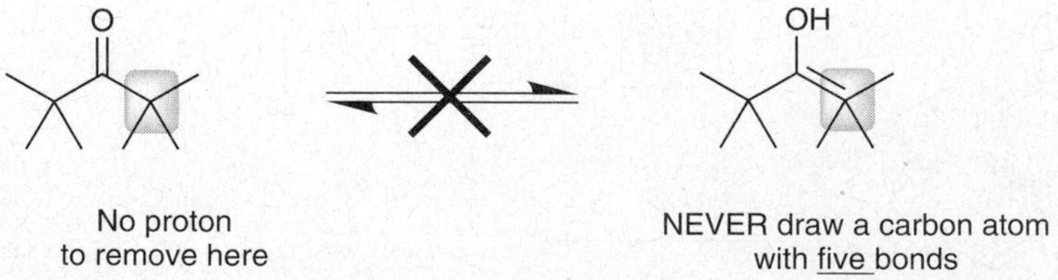

No proton
to remove here

NEVER draw a carbon atom
with <u>five</u> bonds

Most ketones do in fact have alpha protons, and therefore, a typical ketone will exist in equilibrium with an enol. In the previous section, we saw some rare cases where the equilibrium can actually favor the enol, but in general, the equilibrium favors the ketone. Therefore, you will generally only have trace amounts of the enol present in equilibrium with the ketone.

This small amount of enol is able to react as a nucleophile and attack some electrophile. After the enol attacks the electrophile, the keto-enol equilibrium is re-established by producing some more enol (to account for the enol that "disappeared" as a result of the reaction). Slowly but surely, most of the ketone molecules end up converting into enols and reacting with the electrophile. The most common example is alpha-halogenation, which can occur when a ketone is treated with Br_2 in aqueous acid (H_3O^+) to generate an α-halo ketone:

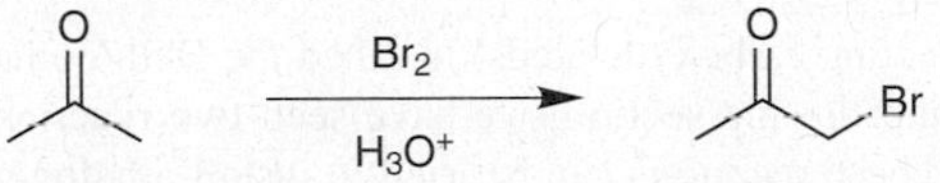

Let's explore how this process occurs. The ketone tautomerizes to generate a small amount of enol. Then comes the critical step: the enol functions as a nucleophile to attack Br_2 (the electrophile):

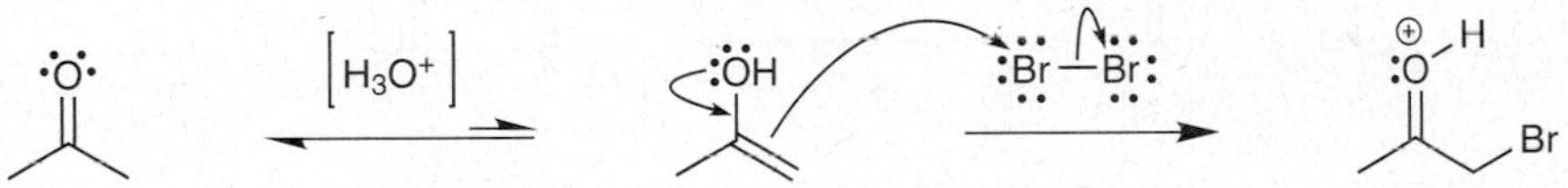

Deprotonation then generates the product:

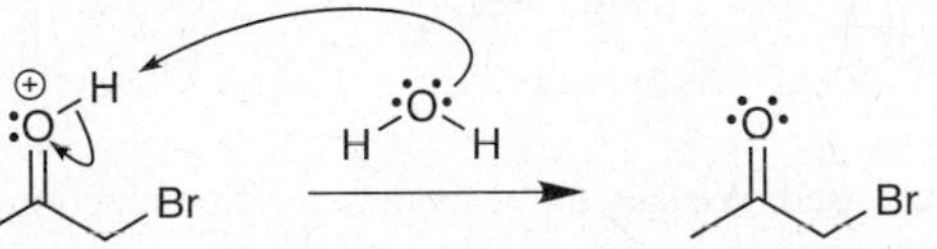

Notice that most of the steps in this mechanism are just proton transfers. Our mechanism represents the following pattern: tautomerize, attack, deprotonate. But "tautomerize" is just a new name for a special combination of two proton transfer steps. There is really only one step where an attack takes place (when the enol attacks the electrophile).

In the end, this provides a method for installing a halogen at the alpha position of a ketone:

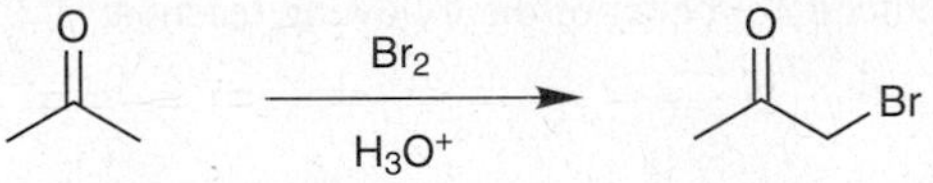

Alternatively, we might see a different reagent other than H_3O^+, for example:

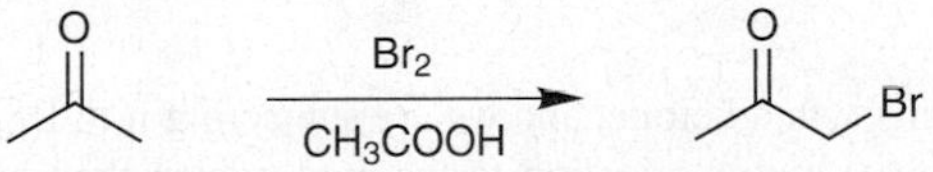

Acetic acid, CH_3COOH, can be used as a mild acid to facilitate the tautomerization. We don't have to worry about this acid undergoing halogenation itself (at its alpha position), like this:

THIS REACTION IS TOO SLOW

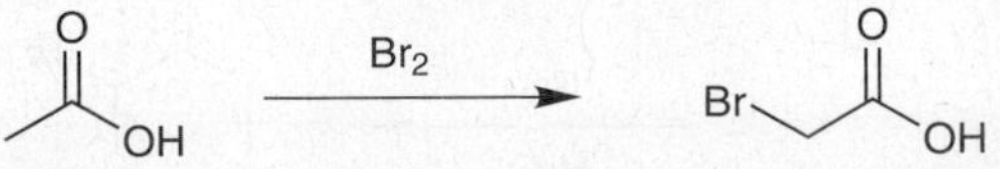

We don't have to worry about this, because carboxylic acids are much slower to react in this kind of reaction.

If we **want** to halogenate the alpha position *of a carboxylic acid*, it is possible, but it will require some extra steps. First, we must convert the carboxylic acid into an acid halide. We do this because the enol of an acid halide will rapidly attack a halogen. Then, in the end, we just convert the acid halide back into a carboxylic acid:

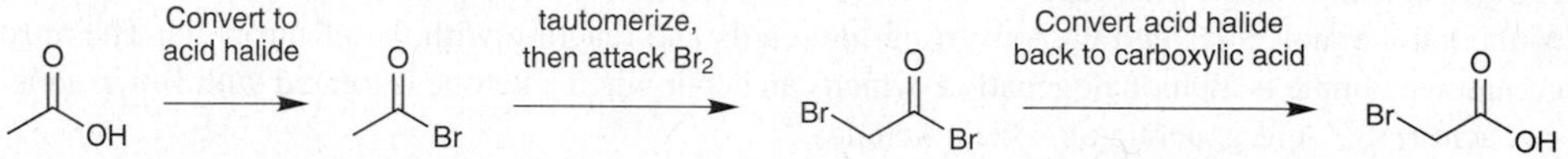

This strategy (for halogenating carboxylic acids) is called the Hell-Volhard-Zelinsky reaction.

Here is the bottom line: in this section, we have seen two reactions that exploit the nucleophilic nature of enols. These reactions can be used to install a halogen at the alpha position of a ketone,

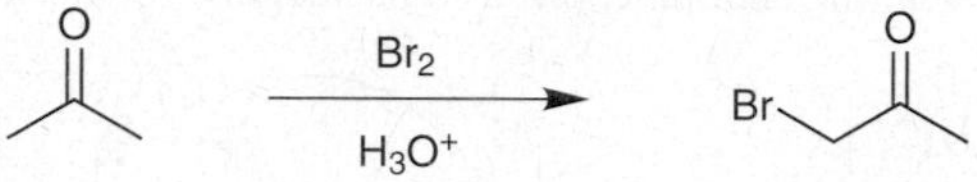

or to install a halogen at the alpha position of a carboxylic acid:

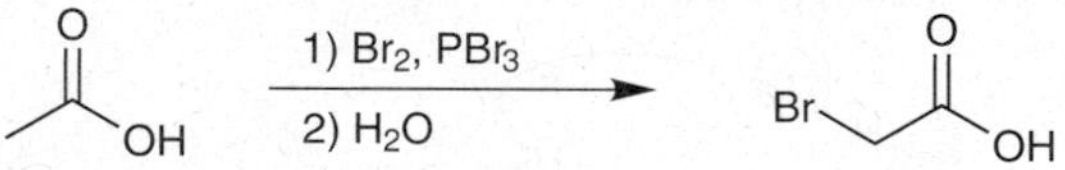

Notice the reagents that we used. We saw the reagent in the first reaction (Br_2 and some mild acid—to halogenate a ketone). But to halogenate a carboxylic acid, we use a different set of reagents. We use Br_2 and PBr_3, followed by H_2O. The function of Br_2 and PBr_3 is to make the acid halide, form the enol, and then have the enol attack Br_2. Then, water is used in the last step to convert the acid halide back into a carboxylic acid.

EXERCISE 7.14 Predict the product of the following reaction:

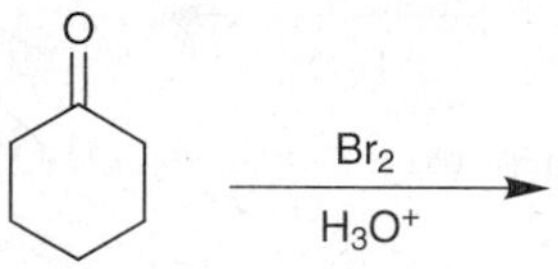

Answer We are starting with a ketone, and we are subjecting it to Br_2 in the presence of a mild acid. The mild acid promotes tautomerization to the enol, which then attacks the Br_2 in an alpha halogenation. So, in the end, our product will have a Br at one of the alpha positions. Either side is the same, so we can just pick a side:

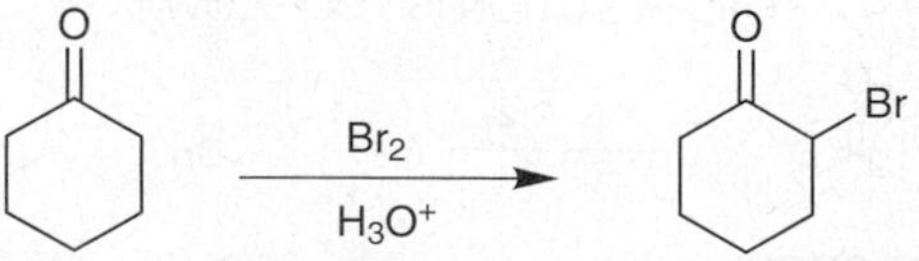

PROBLEMS   Predict the products of each of the following reactions. Remember that you can only halogenate an alpha position that has protons.

7.15

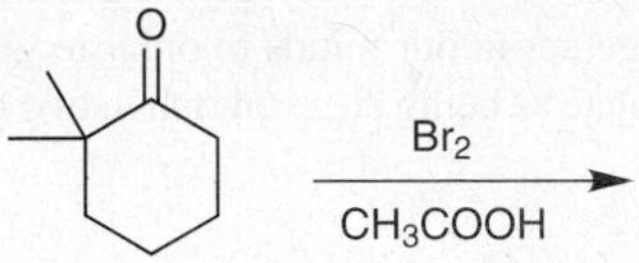

$$\xrightarrow[\text{CH}_3\text{COOH}]{\text{Br}_2}$$

7.16

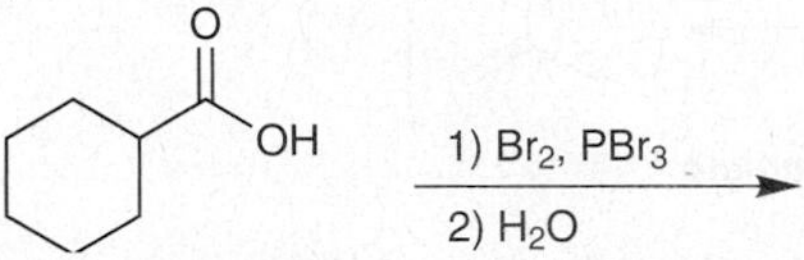

$$\xrightarrow[\text{2) H}_2\text{O}]{\text{1) Br}_2,\ \text{PBr}_3}$$

7.17

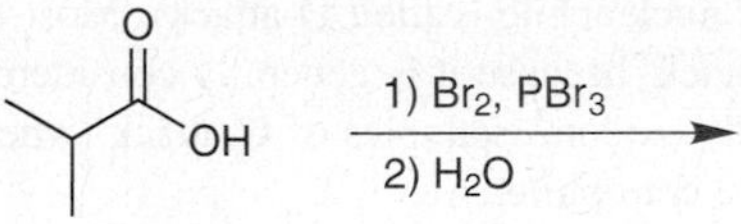

$$\xrightarrow[\text{2) H}_2\text{O}]{\text{1) Br}_2,\ \text{PBr}_3}$$

7.18

$$\xrightarrow[\text{H}_3\text{O}^+]{\text{Br}_2}$$

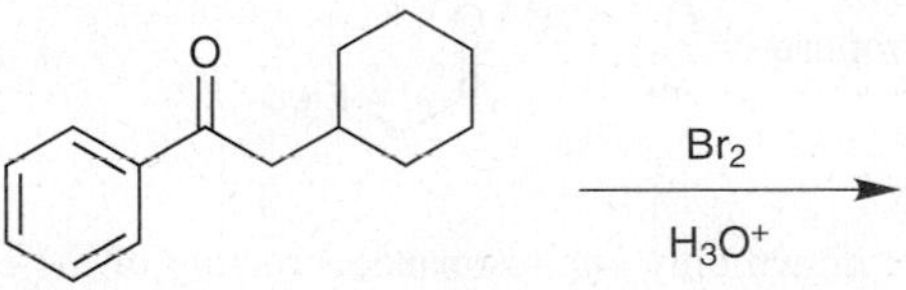

7.4   MAKING ENOLATES

In the previous section, we saw that enols can be nucleophilic. But enols are only mild nucleophiles. So, the question is: how can we make the alpha position even more nucleophilic (so that we can have a broader range of possible reactions)? There is a way to do this. We just need to give the alpha position a negative charge. To see how this can be achieved, let's quickly review the mechanism we saw for tautomerization under basic conditions, and let's focus on the intermediate (highlighted below):

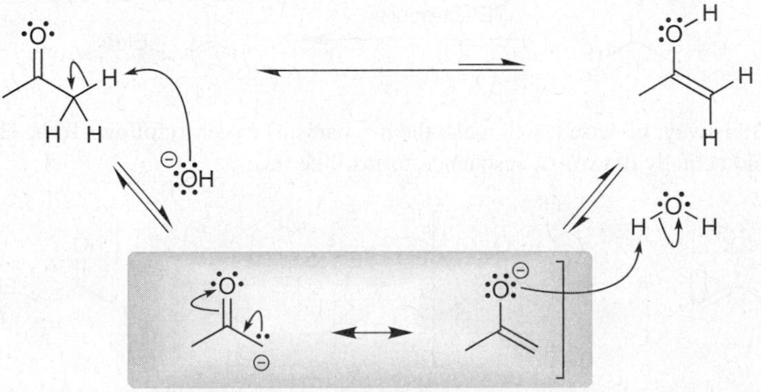

enolate

The intermediate is negatively charged, and we mentioned before that it is called an enolate. In order to capture the essence of the enolate, we must draw resonance structures. Remember what resonance structures represent. We cannot draw this *one* intermediate with any single drawing, so we draw two drawings, and we meld these two images together in our minds in order to get a better picture of this intermediate. And that picture shows the enolate as being electron rich in two locations: the alpha carbon *and* the oxygen atom:

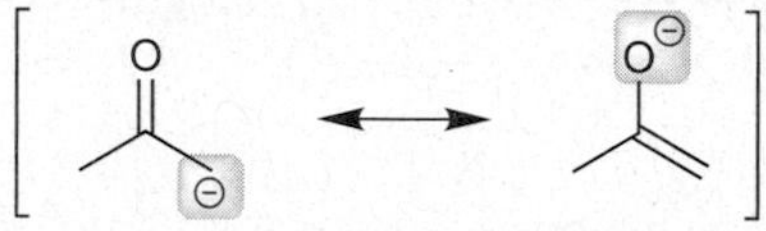

enolate

So, we expect *both* of these locations to be very nucleophilic. Nevertheless, we won't explore any reactions in which the oxygen atom functions as a nucleophile (called O-attack). Most textbooks and instructors do not teach the conditions for O-attack, because it is generally considered to be a more advanced topic. So, from now on, we will only explore examples of C-attack (where the alpha carbon acts as the nucleophile, attacking some electrophile):

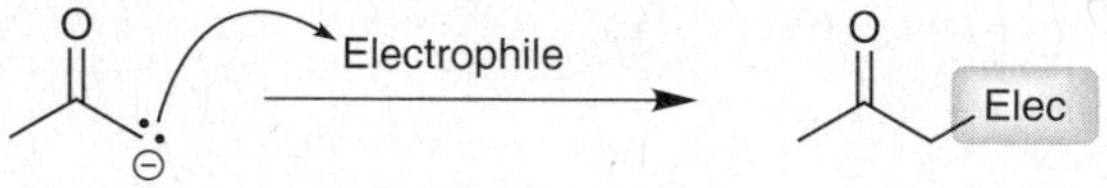

Notice that, in showing the attack, we have drawn only one resonance structure of the enolate. If we had used the other resonance structure, it would have looked like this:

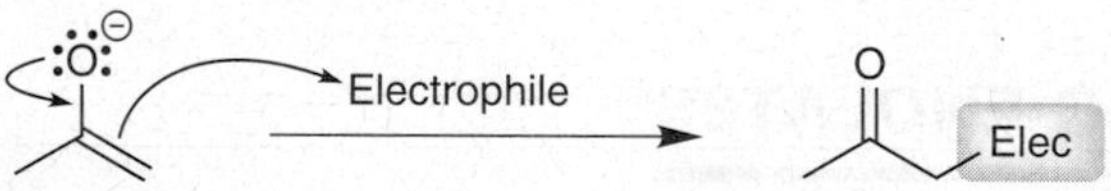

This is just another way of showing the same step. Many textbooks will show it the second way (starting with the resonance form that has the negative charge on oxygen). Perhaps this is more appropriate, because this resonance form is contributing more to the overall character of the enolate. However, in this book, we will use the resonance structure where the negative charge is on carbon:

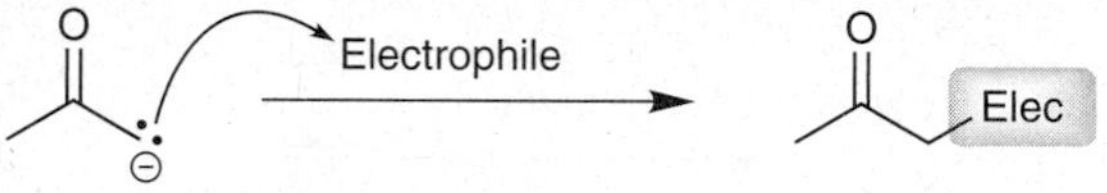

We will do it this way, because it will make the mechanisms easier to follow. To be absolutely correct, we should actually draw *both* resonance forms, like this:

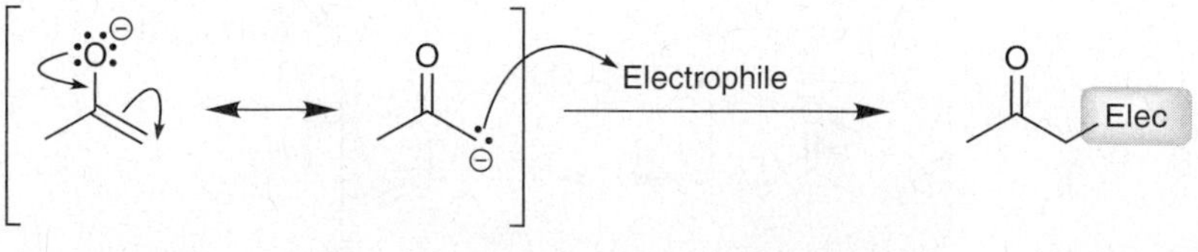

But for simplicity, we will just show one resonance structure for the enolate (in most of the mechanisms that we will see in this chapter).

Now let's think about what kind of base we would need **to make** an enolate. If we use bases such as HO⁻ or RO⁻ (bases with a negative charge on oxygen), we find that these bases are *not* strong enough to completely convert the ketone into an enolate. Rather, an equilibrium is established between the ketone, the enolate, and the enol. This equilibrium only produces very small amounts of the enolate, but that doesn't matter. Once an enolate reacts with an electrophile, the equilibrium produces more enolate to replenish the supply. Over time, all of the ketone can convert into the enolate and then react with some electrophile. This is very similar to the situation we saw with enols. Once again, we are relying on the equilibrium to continuously produce more of the enolate. The major difference here is that enolates are so much more reactive than enols. Therefore, the chemistry of enolates is more robust than the chemistry of enols.

As an example for the richness of enolate chemistry, consider this: some enolates are much more stabilized than other enolates. These "super-stabilized" enolates are more "tame" nucelophiles (more selective in what they react with). For example, a compound with two carbonyl groups (separated by one carbon) can be deprotonated to form an intermediate that is sort of like a *double* enolate:

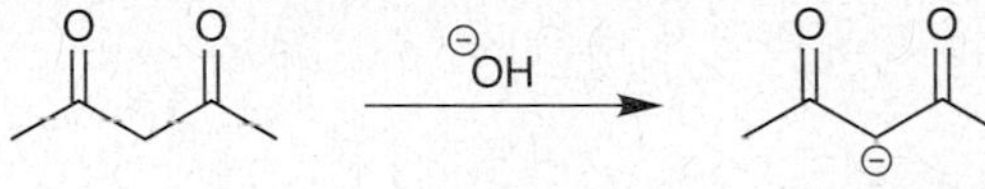

The negative charge in this intermediate is delocalized over both carbonyl groups:

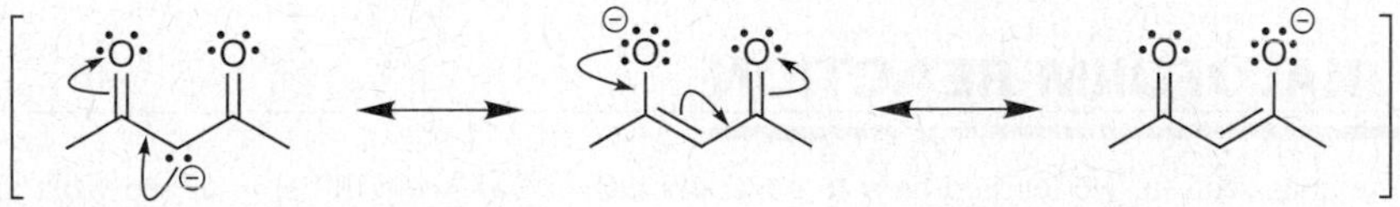

And therefore it is extremely stable. In fact, it is even more stable than HO⁻ or RO⁻. So, when we use bases such as HO⁻ or RO⁻, the equilibrium greatly favors the enolate:

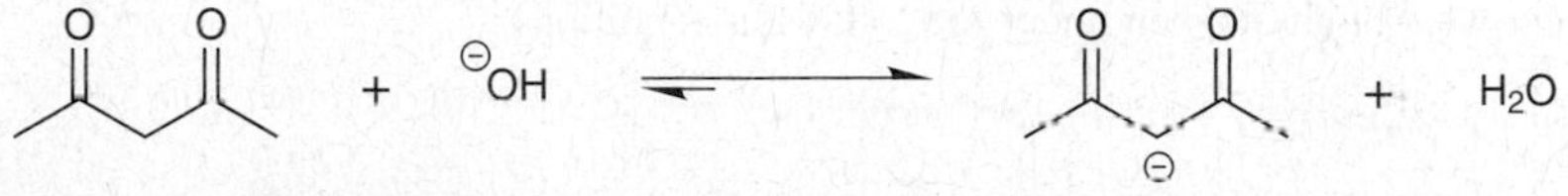

We will soon see that the position of this equilibrium will be a driving force in the Claisen condensation (later in this chapter).

EXERCISE 7.19 Consider the following compound:

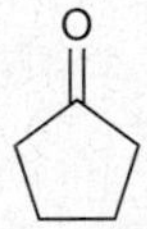

Draw the enolate that is formed when this compound is deprotonated. Make sure to draw all resonance structures.

Answer We just need to identify the alpha proton, and then remove it:

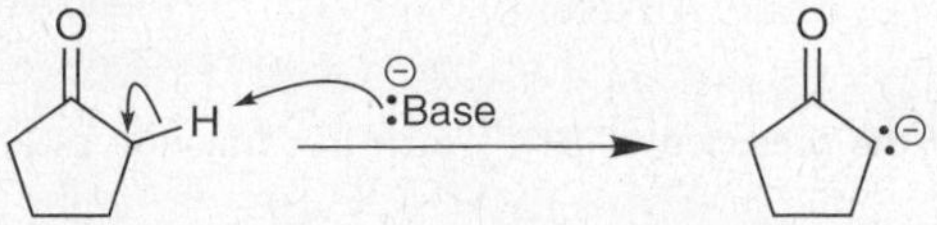

And then we draw the resonance structures:

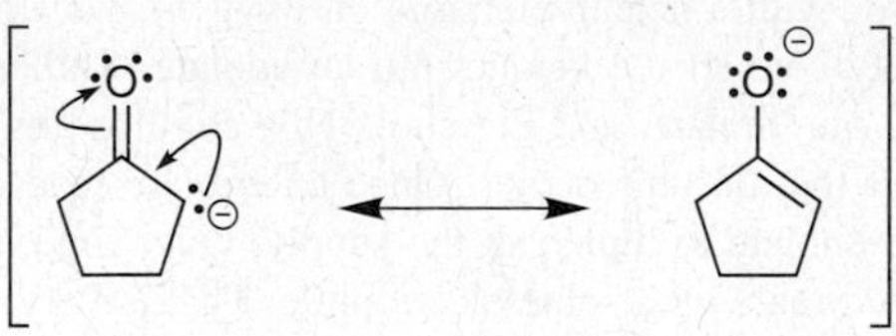

PROBLEMS Draw the enolate that would be generated when each of the following compounds is treated with hydroxide. Make sure to draw all significant resonance structures.

7.20

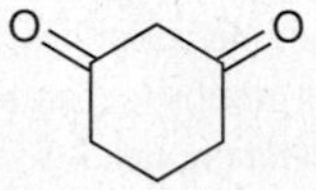

7.21

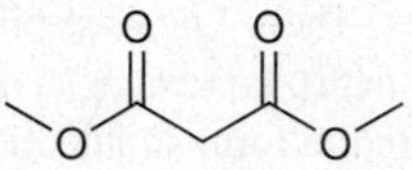

7.22

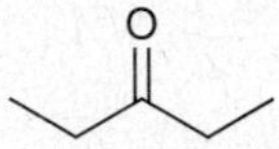

7.23

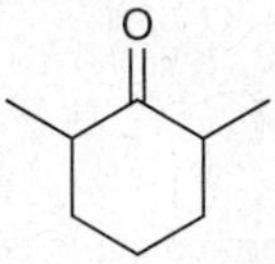

7.5 HALOFORM REACTION

In the previous section, we learned how to make enolates. Now, we will begin to see what an enolate can attack. In this section, we will explore the reaction between an enolate and a halogen (such as Br, Cl, or I). In the following sections, we will explore the reactions between an enolate and other electrophiles.

Consider what might happen under the following conditions:

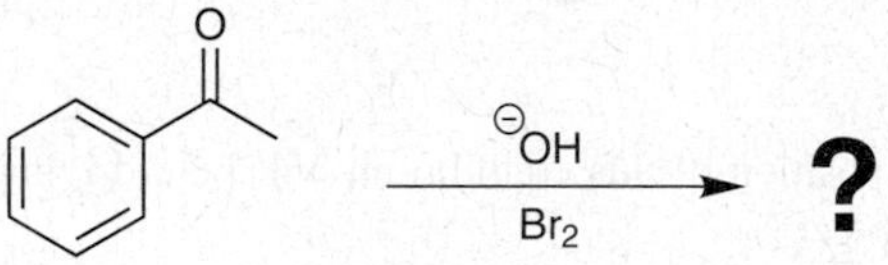

We have a ketone and hydroxide, which means that the equilibrium will involve a small amount of enolate:

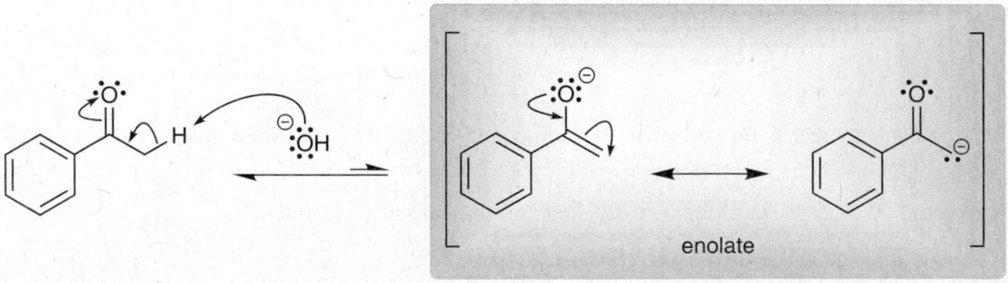

This enolate is formed in the presence of Br$_2$, which can function as an electrophile. The initial product is not surprising:

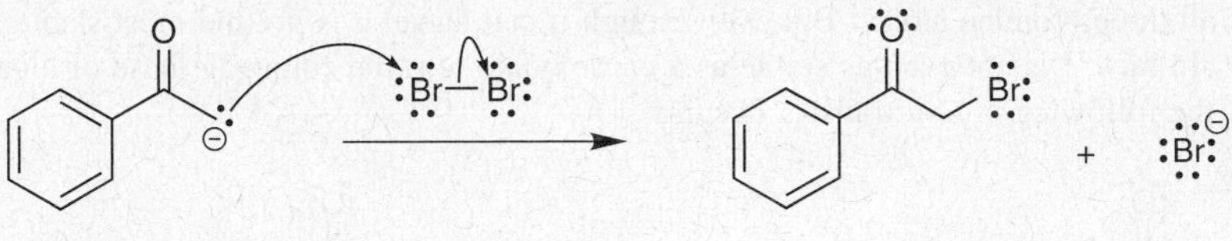

The enolate attacks Br_2 and expels Br^- as a leaving group. The result is that we have installed Br at the alpha position:

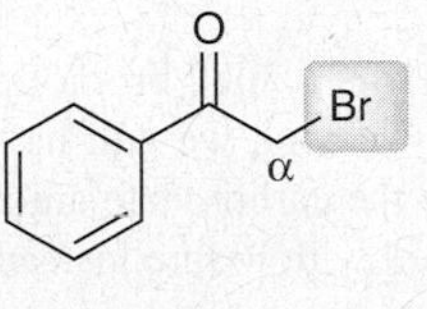

But the reaction doesn't stop there. Remember that the base (hydroxide) is still present in solution. So hydroxide can remove another alpha proton. In fact, it is even easier to remove this proton, because the inductive effect of the bromine atom serves to further stabilize the resulting enolate, which can then attack Br_2 again:

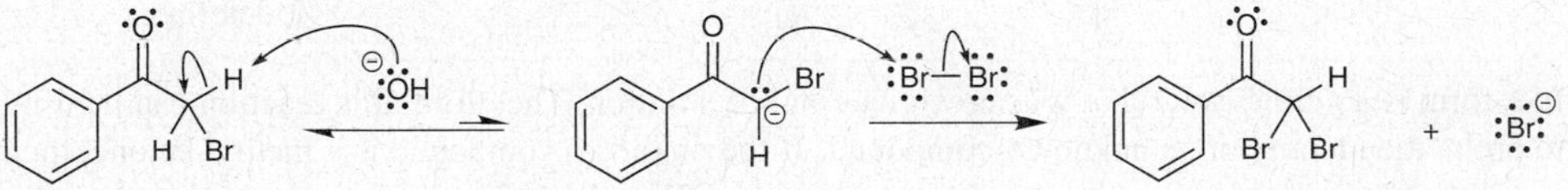

Now we have *two* Br atoms in our compound. And then, it happens again:

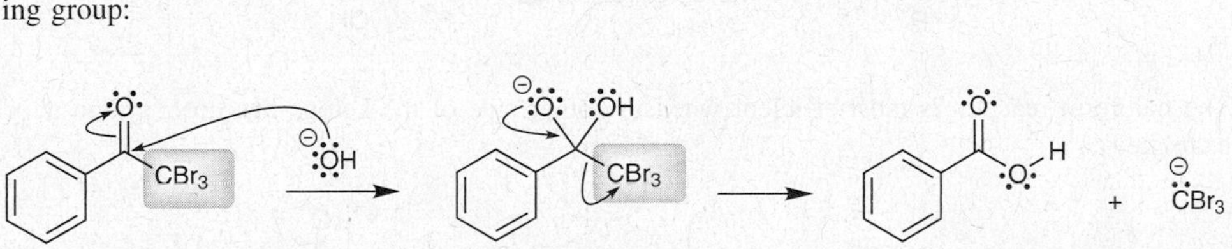

Think about what has happened so far. A methyl group (CH_3) group has been converted into a CBr_3 group. This transformation is very significant, because a CBr_3 group is able to function as a leaving group:

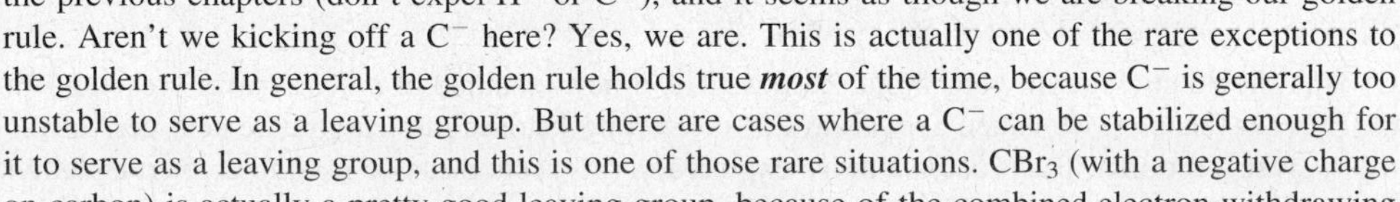

At this point, you should be feeling uncomfortable. You probably remember our golden rule from the previous chapters (don't expel H^- or C^-), and it seems as though we are breaking our golden rule. Aren't we kicking off a C^- here? Yes, we are. This is actually one of the rare exceptions to the golden rule. In general, the golden rule holds true *most* of the time, because C^- is generally too unstable to serve as a leaving group. But there are cases where a C^- can be stabilized enough for it to serve as a leaving group, and this is one of those rare situations. CBr_3 (with a negative charge on carbon) is actually a pretty good leaving group, because of the combined electron-withdrawing effects of all three bromine atoms. But even though it can leave, it is not the most stable anion on the planet. In fact, it is not even as stable as a carboxylate ion (the conjugate base of a carboxylic acid). So, the following proton transfer occurs:

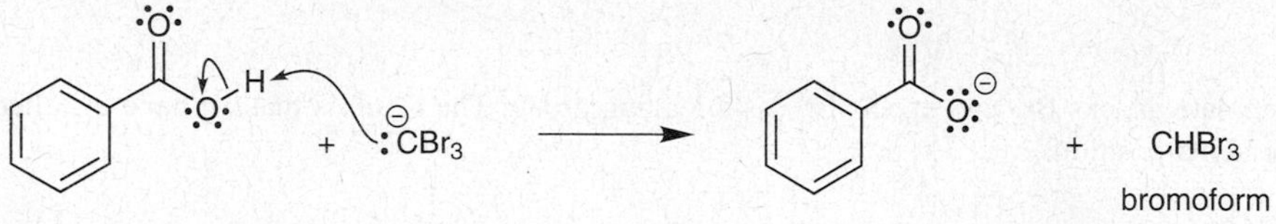

This forms a carboxylate anion and $CHBr_3$ (called bromoform). And this is the end of our mechanism. If we want to isolate the carboxylic acid, we will have to introduce a source of protons into the reaction flask in order to protonate the carboxylate anion.

When the same reaction is performed with iodine instead of bromine, iodoform is obtained as a by-product, instead of bromoform:

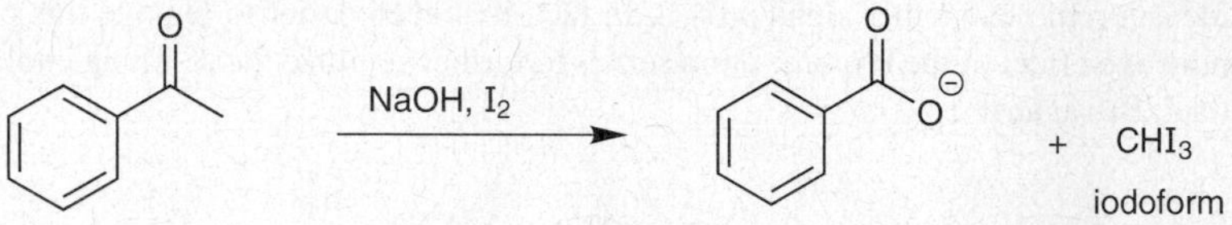

Iodoform is a yellow solid that will precipitate out of solution. Therefore, this reaction can be used to probe the identity of an unknown compound. If the unknown compound is a methyl ketone, then it will produce iodoform under these conditions (NaOH and I_2). This iodoform test is not really used anymore (we now have spectroscopy techniques that give us this information and much, much more). So, this chemical test is really a relic of the past. But for some reason, it is still used in textbook problems. You will usually see it like this: "An unknown compound tests positive for iodoform, and. . . ." The beginning of this problem is telling you that you have a methyl ketone. If you see this in a problem in your textbook, you should know what it means.

But there is a much more important use for this reaction. You can use it when solving synthesis problems. This reaction provides a way to convert a methyl ketone into a carboxylic acid:

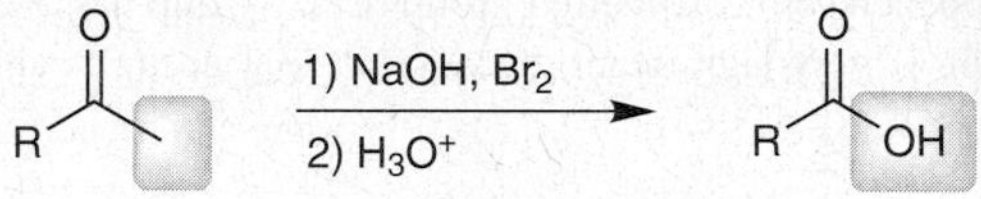

The haloform reaction is most efficient when the other side of the ketone has no α protons, for example.

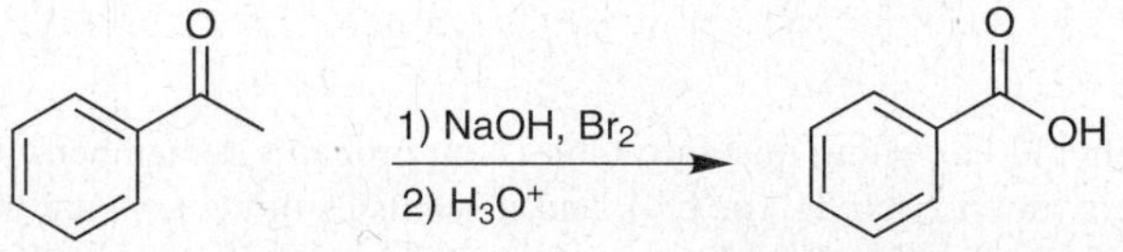

This should stick out in your mind, because it is a new example of a "cross-over" reaction. In the previous chapter, we talked about ways of converting ketones into carboxylic acid derivatives (cross-over reactions). The process that we explored in this section can be used to convert a methyl ketone into a carboxylic acid. You should add this to your synthetic toolbox.

EXERCISE 7.24 Predict the products of the following reaction:

1) NaOH, Br$_2$

2) H$_3$O$^+$

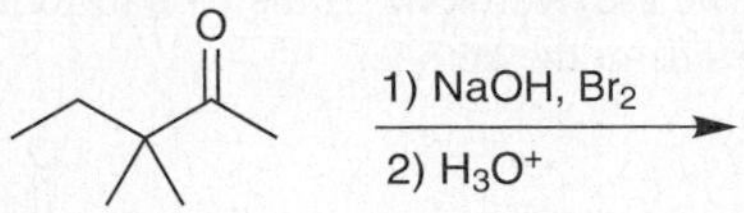

Answer The reactant is a methyl ketone and the reagents will convert a methyl ketone into a carboxylic acid (with a by-product of bromoform):

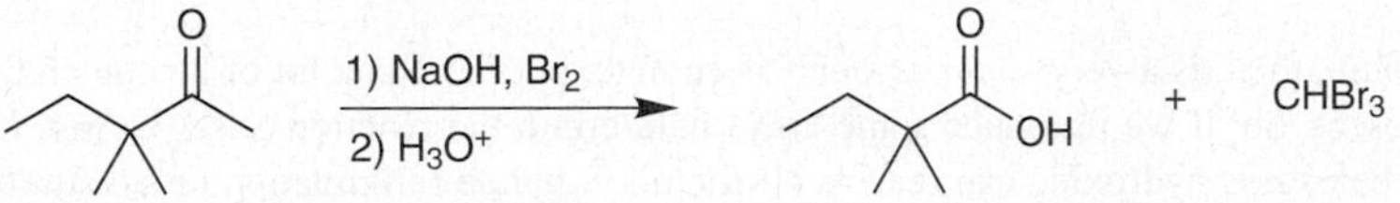

1) NaOH, Br$_2$

2) H$_3$O$^+$

+ CHBr$_3$

PROBLEMS Predict the products for each of the following reactions:

7.25

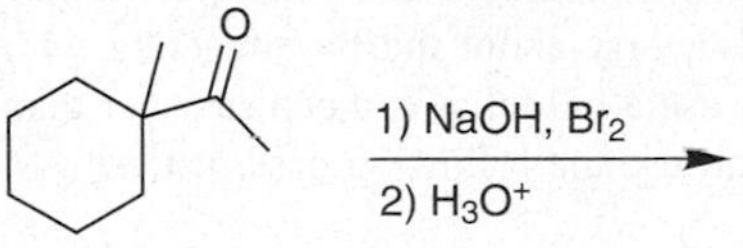

1) NaOH, Br$_2$

2) H$_3$O$^+$

7.26

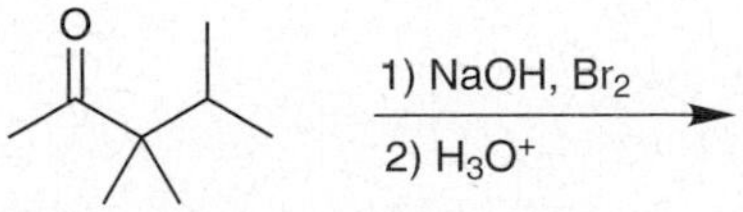

1) NaOH, Br$_2$

2) H$_3$O$^+$

7.27 On a separate piece of paper, draw a mechanism for the transformation in the previous problem (7.26).

7.6 ALKYLATION OF ENOLATES

In this section, we will continue to explore reactions between enolates and electrophiles. Specifically, we will learn how to install an alkyl group at an alpha position:

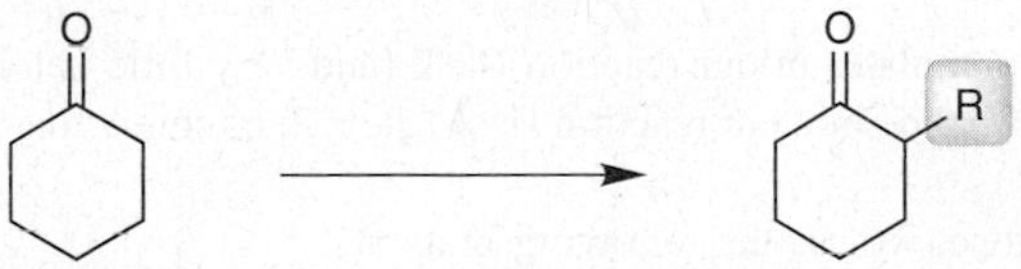

In order to alkylate the alpha position, it makes sense to use an enolate to attack an alkyl halide, for example:

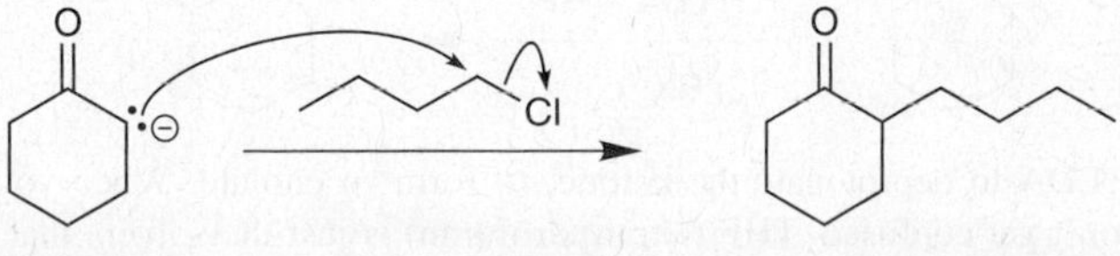

This is just an S$_N$2 reaction, so it will be efficient with *primary* alkyl halides (if a secondary alkyl halide is used, the enolate will function as a base, and elimination will be favored over substitution).

But we run into a major obstacle when we try to make the enolate by treating a ketone with hydroxide. Remember that when we use hydroxide as the base to form our enolate, we find that the equilibrium lies very far to the side of the ketone:

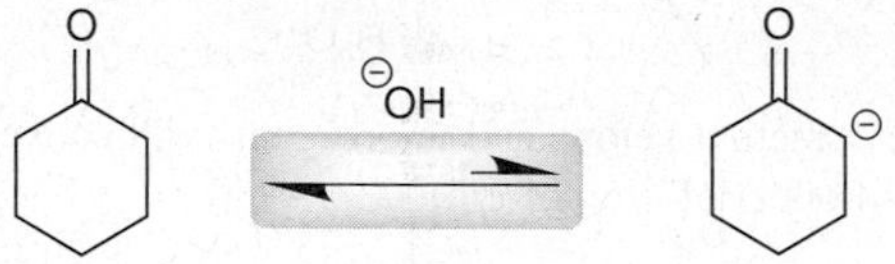

At equilibrium, there is a very small amount of enolate, but there is a lot of ketone and a lot of hydroxide present. So, if we introduce some alkyl halide into the reaction flask, we run into a major obstacle. The excess hydroxide can react with the alkyl halide (elimination or substitution), which creates competing side reactions that generate a mixture of undesired products.

In order to avoid this problem, we will need to form the enolate under conditions where most of the ketone molecules are converted into enolates. If we are able to do this, we will have very little base left over, and therefore, we won't have to worry about the base reacting with the alkyl halide. It is possible to do this, but we will need to use a base that is much stronger than the bases we have been using so far (HO^- and RO^-). We can use the following base instead:

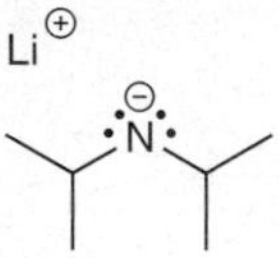

The name of this compound is lithium diisopropylamide, or LDA for short. LDA is a very strong base, because the negative charge is on a nitrogen atom (which is less stable than a negative charge on an oxygen atom). The two isopropyl groups are sterically bulky, so LDA is *not* a good nucleophile. LDA is primarily used as a strong, sterically hindered base, which is exactly what we need in our situation. By using LDA, we can achieve an efficient conversion of the ketone into the enolate. The equilibrium strongly favors the enolate:

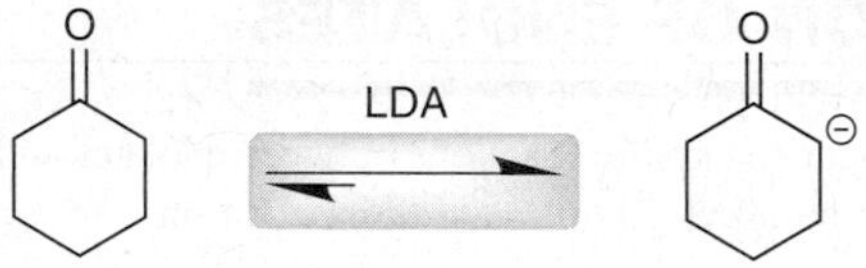

So, we will have mostly enolates in our reaction flask (and very little ketone or base). Now when we introduce some alkyl halide into our reaction flask, the risk of competing side reactions is greatly reduced.

So, to alkylate a ketone, we use the following reagents:

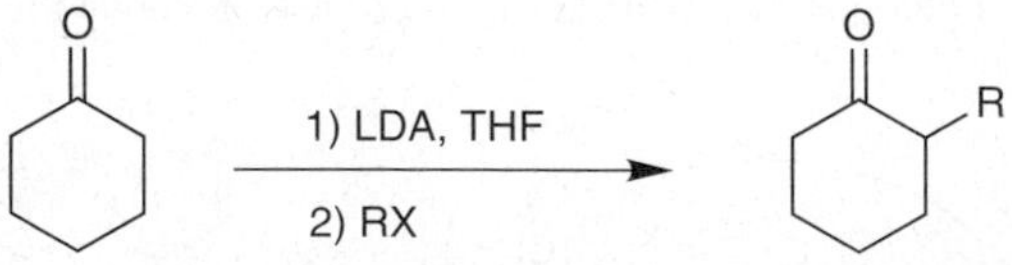

In step 1, we use LDA to deprotonate the ketone, to form an enolate. When you see THF in the reagents above, don't get confused. THF (tetrahydrofuran) is just the solvent that is typically used with LDA. In step 2 above, we use an alkyl halide (RX) to install the alkyl group, where R is some primary alkyl group, and X is a halogen (Cl, Br, or I).

This process works very well when we start with symmetrical ketones, like in the case above with cyclohexanone. But what happens when we start with an unsymmetrical ketone? For example, consider the following situation:

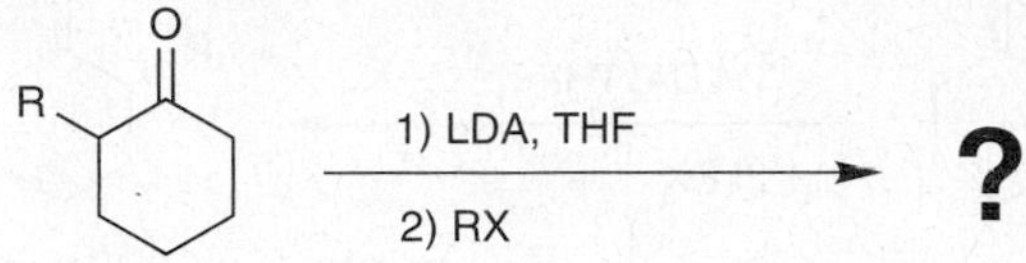

Where will the incoming alkyl group be installed? On the left side, or on the right? In order to answer this question, we need to take a close look at the two possible enolates:

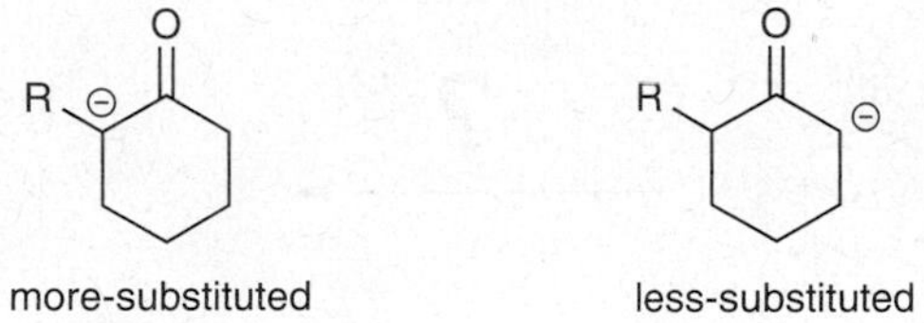

The more-substituted enolate (above left) is the more stable enolate. However, the less-substituted enolate (above right) can form faster, because there are twice as many protons available on the less-substituted side:

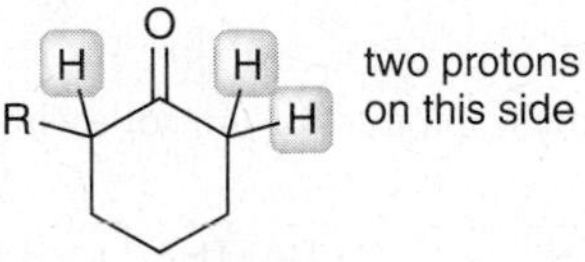

So from a probability point of view, we expect the less-substituted enolate to form more rapidly. Also, we expect the sterically hindered base to have a much easier time removing one of these protons. So, we have two competing arguments:

This is a classic example of thermodynamics vs. kinetics. Thermodynamics is all about stability and energy levels. So, a thermodynamic argument says that we should predominantly form the more stable enolate. However, a kinetic argument tells us to expect the other enolate, simply because it forms faster. Which argument wins? The truth is that a mixture of products is observed. But with LDA at low temperature, there is a clear preference to form the kinetic enolate:

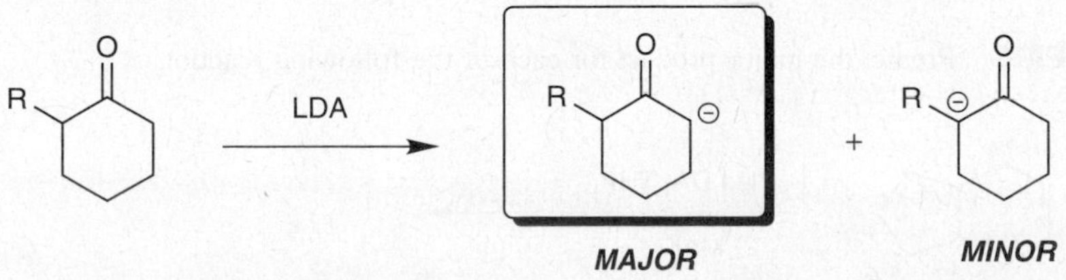

When we introduce the alkyl halide to the reaction flask, alkylation will occur primarily at the less-substituted alpha position:

That works very well if we *want* to install the alkyl group at the less-substituted position. But what if we want to install the alkyl group at the more-substituted position? In other words, what if we want to do this:

There are many different ways to achieve this transformation. Essentially, you need to form the thermodynamic enolate, rather than the kinetic enolate. Some textbooks will teach one or two ways to do this, while other textbooks will skip it altogether. You should look through your textbook and lecture notes to see if you are responsible for knowing how to alkylate the more-substituted side.

EXERCISE 7.28 Predict the major product of the following reaction:

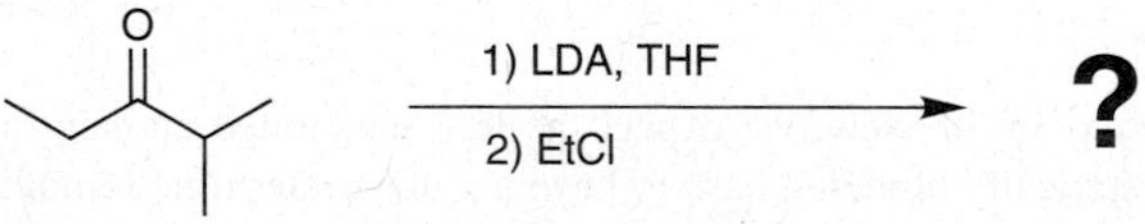

Answer This is an alkylation reaction. In step 1, we are using LDA to form an enolate. And then in step 2, we are using an alkyl halide to alkylate.

Because the alkyl halide is ethyl chloride in this case, we will be installing an ethyl group on an alpha carbon. The only question is: which alpha carbon? The more-substituted carbon or the less-substituted carbon? The use of LDA as our base predominantly gives the kinetic enolate (the less-substituted enolate). Therefore, our major product will have the ethyl group at the less-substituted alpha position:

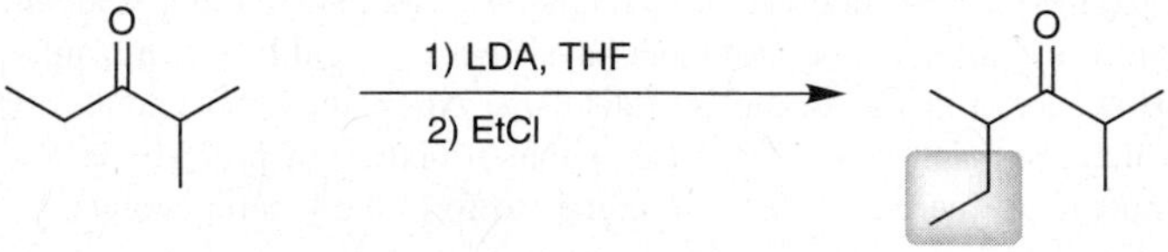

PROBLEMS Predict the major product for each of the following reactions:

7.29

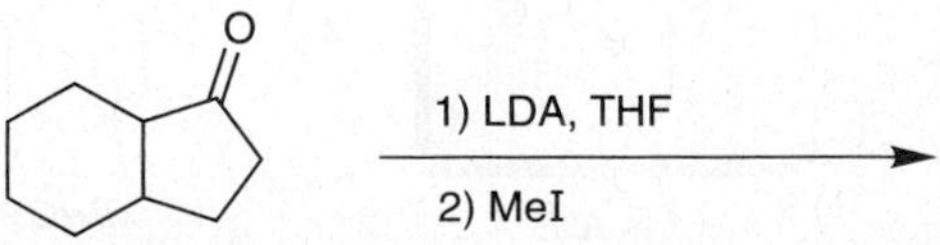

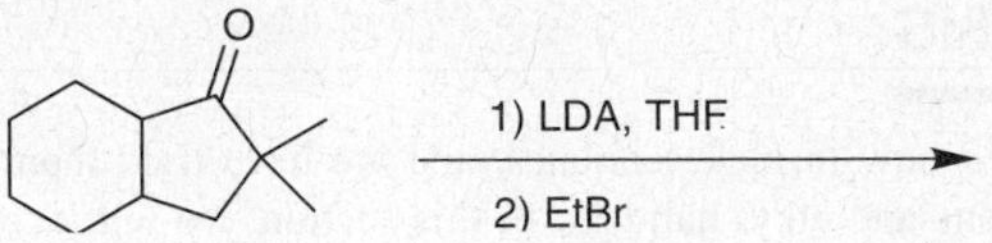

7.30

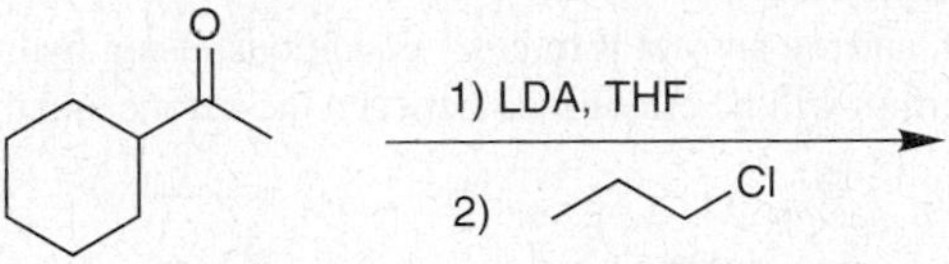

$$\xrightarrow{\text{1) LDA, THF}}$$
2) EtBr

7.31

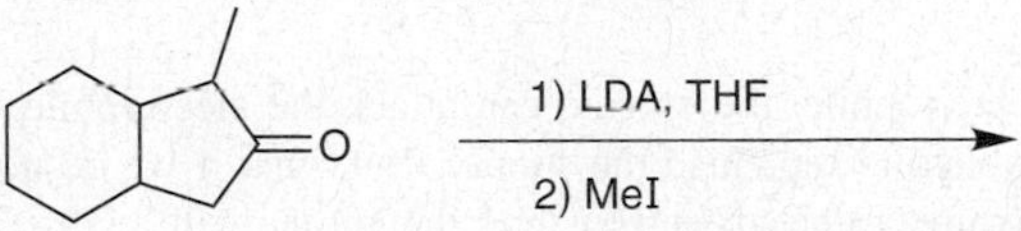

1) LDA, THF

2) ⁀⁀Cl

7.32

1) LDA, THF

2) MeI

EXERCISE 7.33 What reagents would you use to achieve the following transformation:

?

Answer If we look at the difference between the starting material and the product, we will see that there is an extra methyl group that was introduced. This methyl was installed at the less substituted position, which can be achieved with LDA and a methyl halide:

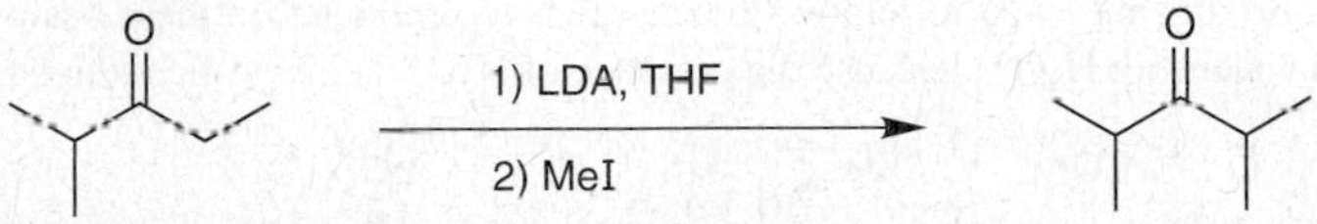

1) LDA, THF

2) MeI

PROBLEMS Identify the reagents you would use to achieve each of the following transformations:

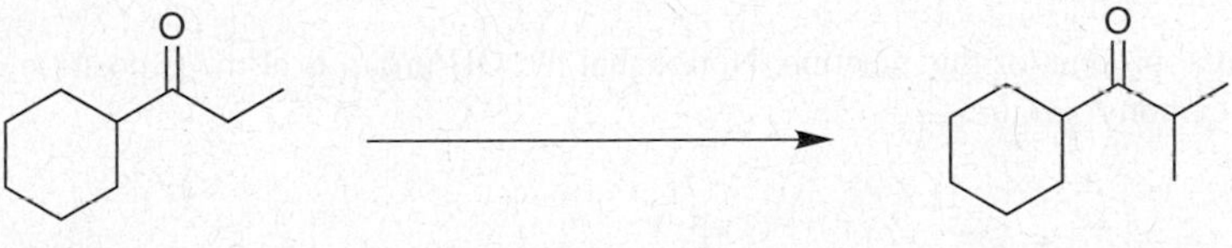

7.34

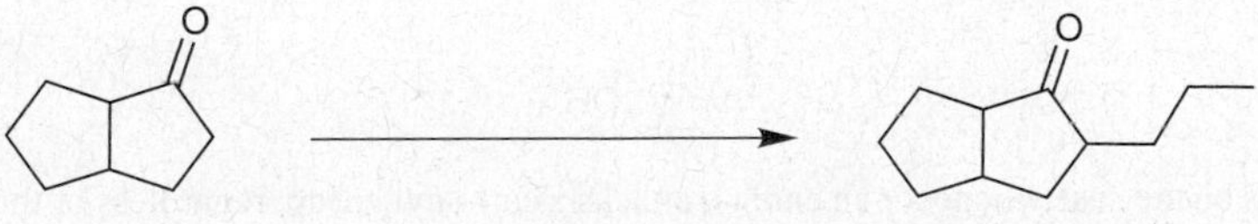

7.35

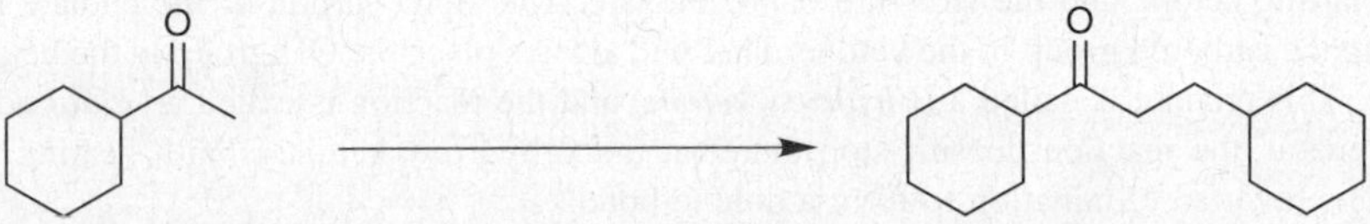

7.36

7.7 ALDOL REACTIONS

So far in this chapter, we have learned how to make enolates, and we have used them to attack various electrophiles (including halogens and alkyl halides). In this section, we will explore what happens when an enolate attacks a ketone or aldehyde.

Suppose we start with a simple ketone, and we subject it to basic conditions, using hydroxide as a base. We have already seen that an equilibrium will be established between the ketone and the enolate:

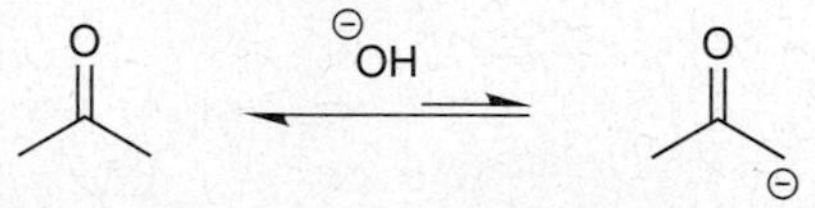

If we do this in the presence of an electrophile, the enolate can attack the electrophile. And then the equilibrium will produce more enolate to replenish the supply. But what if we do not add any other electrophiles to the reaction mixture? What if we just treat the ketone with hydroxide?

It turns out that there actually is an electrophile present. We said that the enolate is in equilibrium with the ketone (and there is a lot of ketone present). Well, ketones are electrophilic, aren't they? We devoted an entire chapter to the reactions that take place when ketones get attacked. So, what happens when an enolate attacks a ketone?

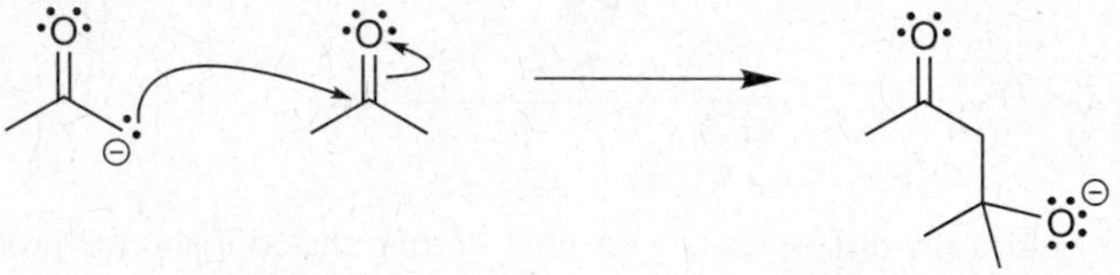

The enolate attacks the ketone, forming a tetrahedral intermediate. Now, our golden rule tells us to try and re-form the carbonyl, but don't expel H^- or C^-. In this case, we have no leaving groups that can be expelled. So, the only way to remove the charge is to protonate. In these basic conditions, the proton source is water (not H_3O^+, because the presence of H_3O^+ is insignificant in basic conditions):

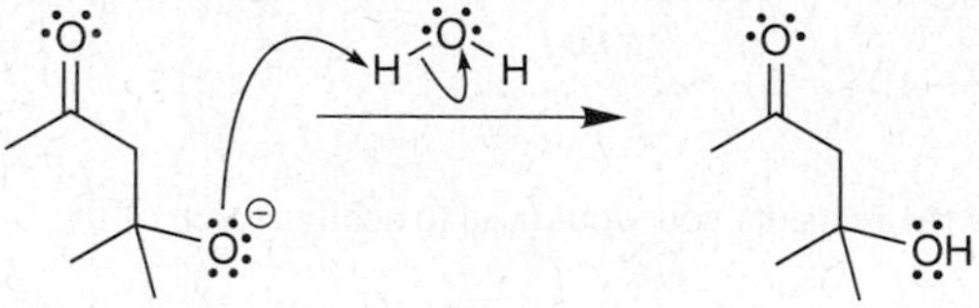

This is the initial product of this reaction. Notice that the OH group is at the β position relative to the surviving carbonyl group:

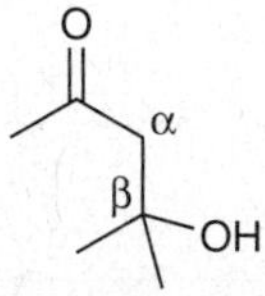

This will always be the case whenever an enolate attacks a carbonyl group, regardless of the structure of the starting ketone and the structure of the enolate. The alpha carbon of the enolate is directly attacking the carbonyl group of the ketone. That will always place the OH group in the beta position. Always. This product is called a *β-hydroxy ketone*, and the reaction is called an ***aldol*** addition.

In general, the reaction doesn't stop there (at the β-hydroxy ketone). With heating, the basic conditions favor an elimination to form a double bond:

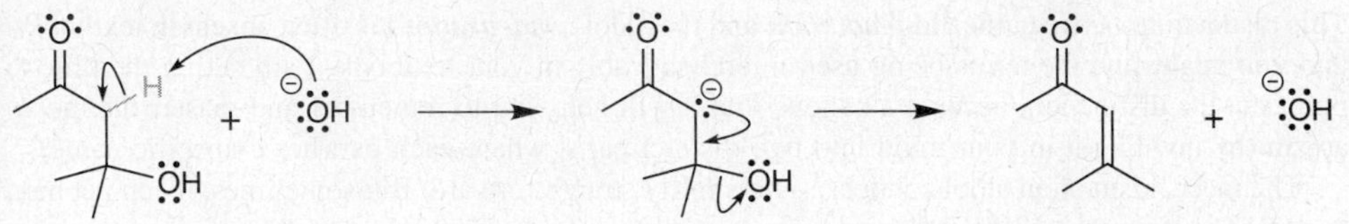

This product has a double bond in conjugation with the carbonyl group. The double bond is located between the α and β positions. So, the product is called an *α,β-unsaturated ketone*.

In the laboratory, we can often control how far the reaction goes. By carefully controlling the conditions of the reaction (temperature, concentrations, etc.), we can usually control whether the reaction stops at a *β-hydroxy ketone*, or whether it continues to form an *α,β-unsaturated ketone*. So, you can use an aldol reaction to form either product.

But you should be familiar with the proper terminology. When we go all the way to the *α,β-unsaturated ketone*, we call the reaction an aldol ***condensation***. By definition, a condensation is any reaction where two molecules come together, and in the process, a small molecule is liberated. The small molecule can be N_2 or CO_2 or H_2O, etc. In this case, we have two molecules of ketone coming together, and in the process, a molecule of water is liberated:

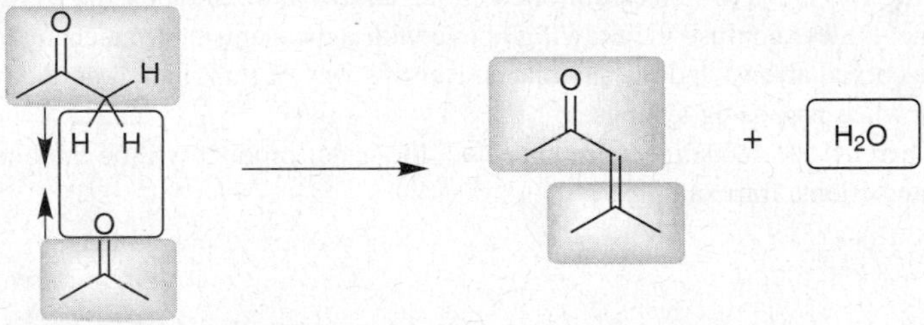

Therefore, we call this reaction an aldol ***condensation***. But what if we control the reaction conditions so that we stop at the *β-hydroxy ketone*?

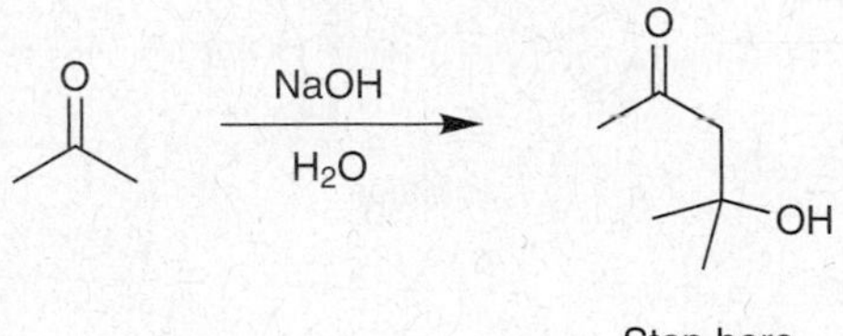

Stop here

If we stop here, then we cannot call it a condensation reaction anymore, because a water molecule was not lost in the process. So, instead, we call it an aldol addition. The difference between an aldol ***condensation*** and an aldol ***addition*** is how far we go in the process:

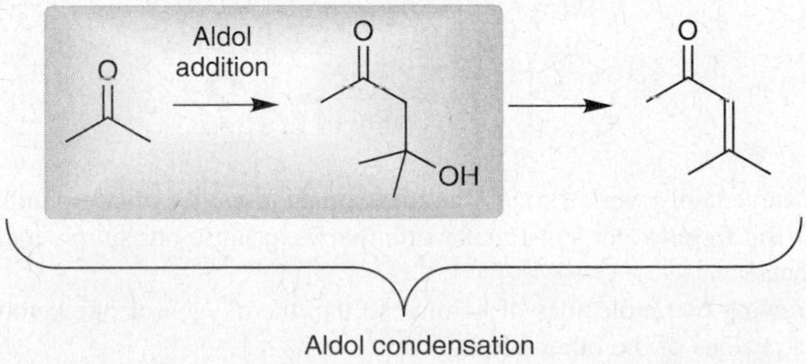

Aldol condensation

This distinction (between the aldol *addition* and the aldol *condensation*) is often absent in textbooks, and you might find the terms being used interchangeably in your textbook. I am taking the time to point out the distinction, because I believe that it will help you to remember and master the mechanism (by dividing it in your mind into two distinct parts, where each part has a specific name).

The mechanism of an aldol condensation is fairly straightforward. But sometimes, it can get hard to see what reagents to use when proposing a synthesis. So try to think of it the way we showed it just a few moments ago:

We are removing two alpha protons from one ketone, and we are removing the oxygen atom from the other ketone. Do *not* confuse the drawing above with a mechanism. A mechanism is when you show all of the curved arrows and intermediates. But this way of thinking about the reaction might come in handy when proposing syntheses.

In a case where two stereoisomers are expected, the major product will be the one that exhibits fewer steric interactions, for example:

Let's get some practice:

EXERCISE 7.37 Draw the aldol condensation product that is obtained when the following compound is heated in the presence of hydroxide ions:

Answer We can certainly work through the mechanism to get the answer, and we will soon get practice with that. But for now, let's just make sure that we can use our simple method for drawing the expected product.

We start by drawing two molecules of ketone, so that the oxygen of one ketone is pointing directly at the alpha protons of the other ketone:

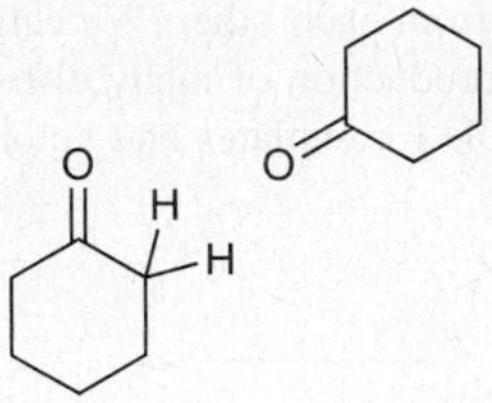

Then, we erase the two alpha protons and the oxygen atom, and we push the fragments together (connecting them with a double bond):

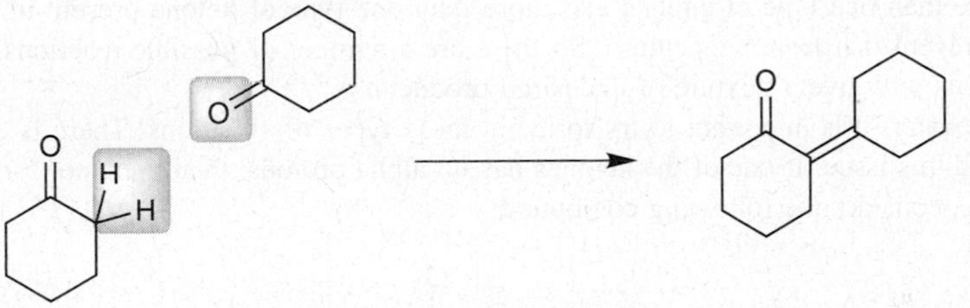

That's all there is to it. It is a simple, but powerful, way of thinking about this reaction.

PROBLEMS Predict the major product for each of the following reactions. In each case, assume that a condensation takes place, and draw the α,β-unsaturated ketone that is produced.

7.38

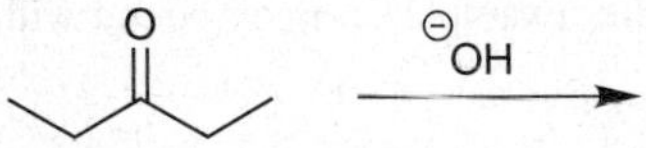

7.39

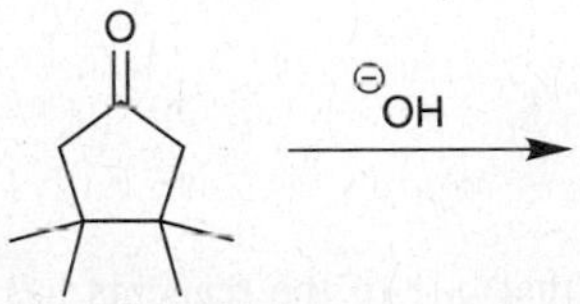

7.40

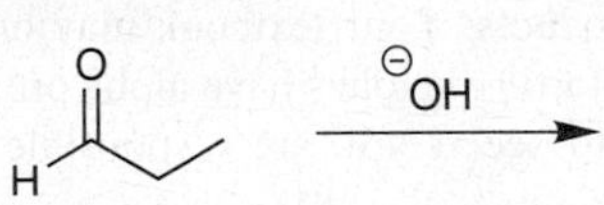

In all of the aldol condensations that we have seen so far, two molecules *of the same ketone* were reacting with each other. One molecule of ketone was deprotonated to give an enolate, which then attacked another molecule of the same ketone. But what if we had used two different ketones? For example, what if we try to do this:

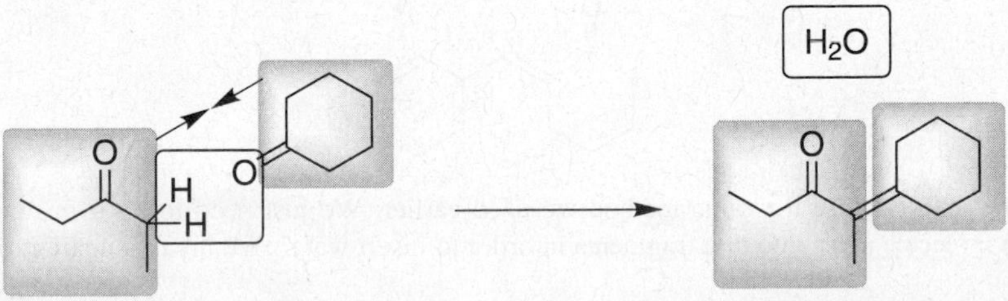

Notice that the ketones are different from each other. We call this a ***crossed***-aldol. This can work, but care must be taken to avoid the production of many different products. To see why this is the case, we must realize that it is possible for enolates and ketones to exchange protons:

So, you can't really control which ketone will be converted into the enolate. This means that there will be more than one type of enolate and more than one type of ketone present in solution (and you can't prevent that from happening). So there are a number of possible reactions that can take place, and this will give a mixture of undesired products.

So, practically, it is important to try to avoid these types of situations. There is one very easy way to avoid this issue. If one of the ketones has no alpha protons, then it cannot form an enolate. For example, consider the following compound:

This compound, called benzaldehyde, has no alpha protons. Therefore, it cannot be converted into an enolate. It will just wait to be attacked. Here is another example of a compound with no alpha protons:

So, one way to achieve a crossed aldol is to make sure that one of the reagents has no alpha protons. That will minimize the number of potential products. Your textbook may or may not show methods for achieving a crossed aldol where both starting ketones have alpha protons. You should look through your textbook and lecture notes to see if you are responsible for such methods.

EXERFCISE 7.41 What reagents would you use to make the following compound, using an aldol condensation:

Answer We can use the same method we used earlier. We just need to do it in reverse. We break the molecule apart into two fragments in order to insert water. We break it apart at the C═C bond:

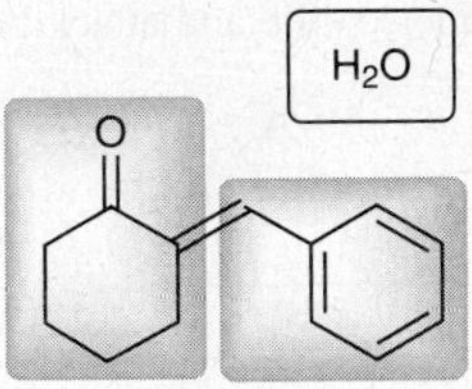

And we just have to decide which fragment gets the oxygen atom and which fragment gets the protons. The fragment on the left already has a carbonyl group, so that fragment must get the two alpha protons. The fragment on the right will get a carbonyl group in place of the C=C bond:

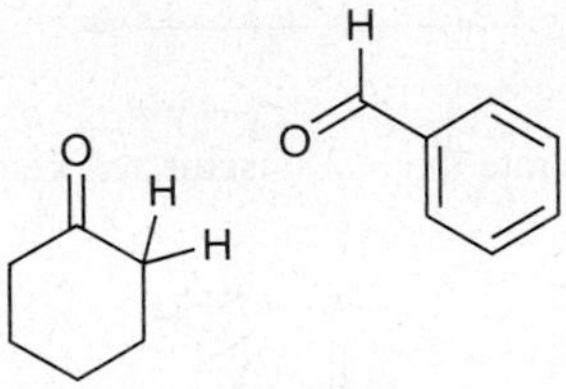

PROBLEMS Identify what reagents you would use to make each of the following compounds (using an aldol condensation):

7.42

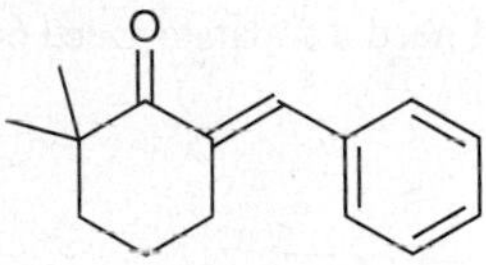

7.43

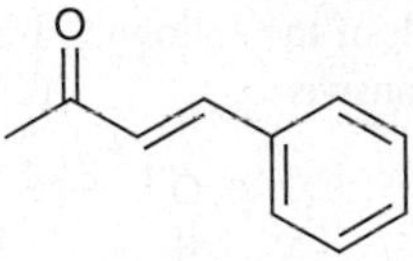

7.44

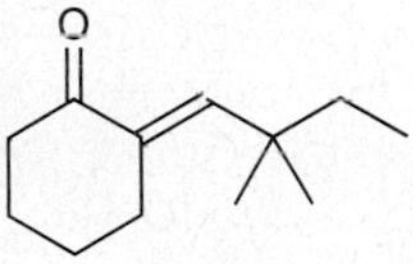

7.45

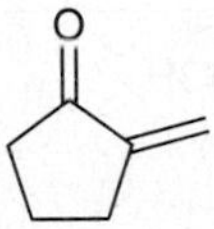

EXERCISE 7.46 Propose a plausible mechanism for the following transformation:

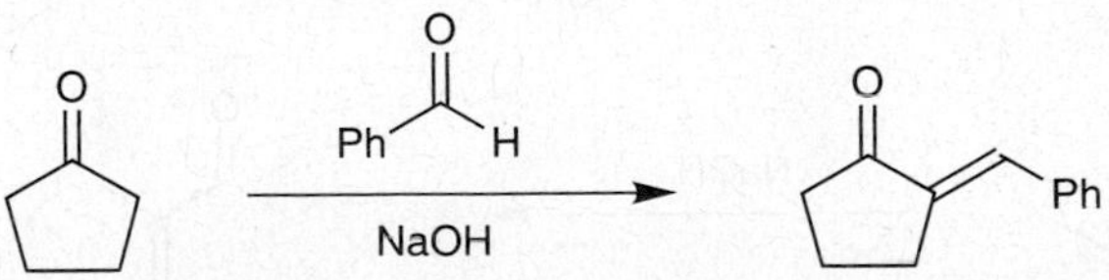

Answer In the first step, hydroxide functions as a base and removes a proton to generate an enolate:

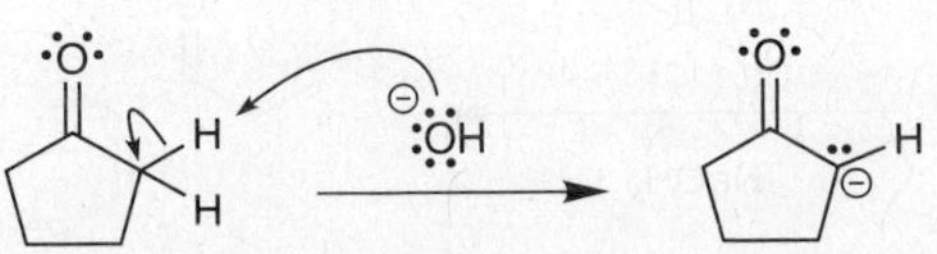

Then, this enolate can function as a nucleophile and attack benzaldehyde:

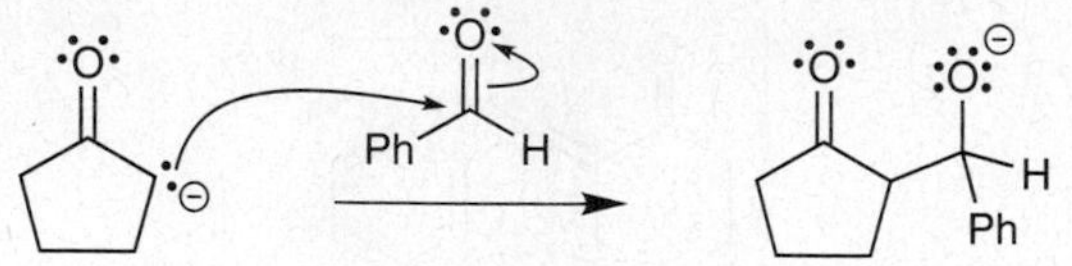

Water then functions as a proton source to generate the β-hydroxy ketone:

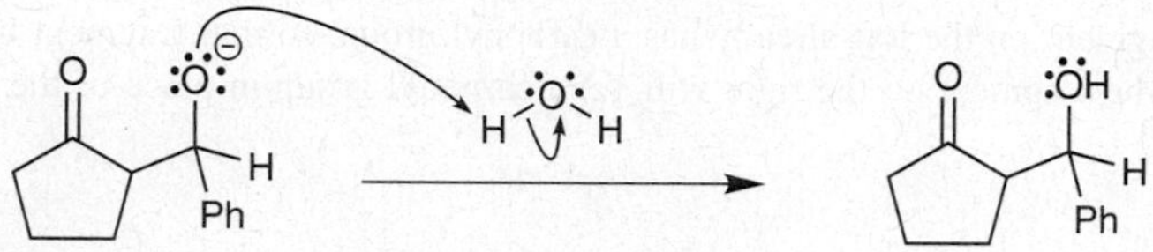

Finally, water is eliminated to generate the α,β-unsaturated ketone:

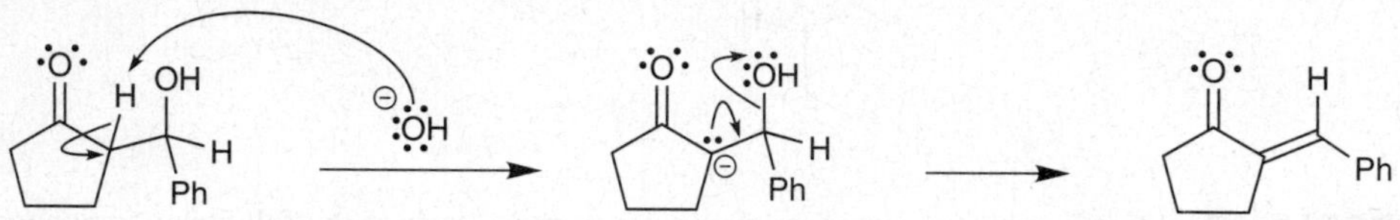

PROBLEMS Now let's get some practice drawing mechanisms for aldol condensations. Draw a mechanism for each of the following transformations. You will need a separate piece of paper to record each of your answers:

7.47

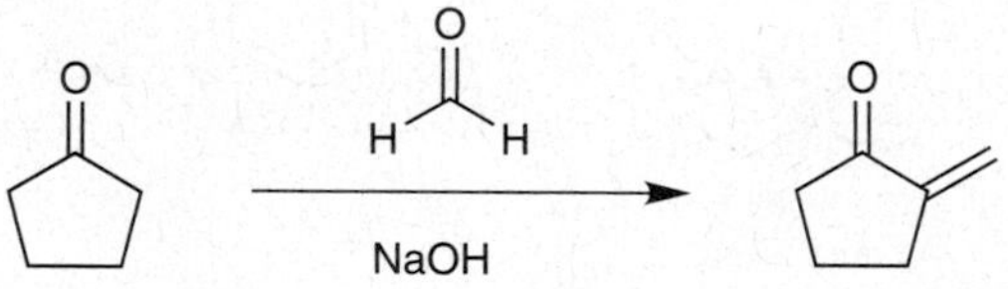

7.48

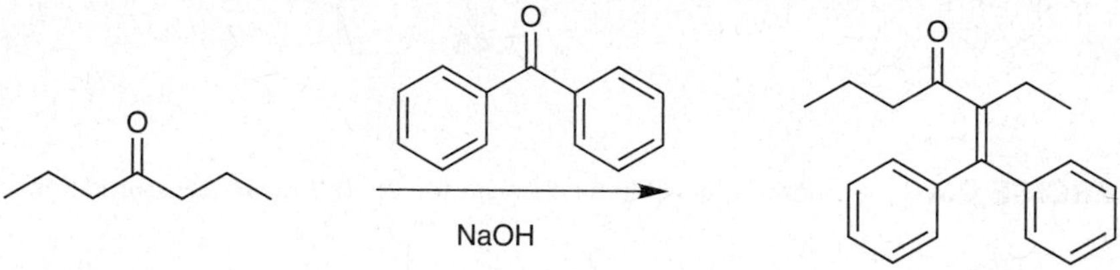

7.49

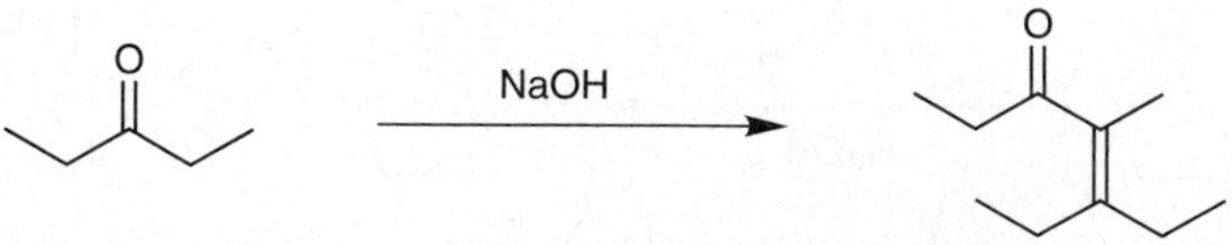

7.50

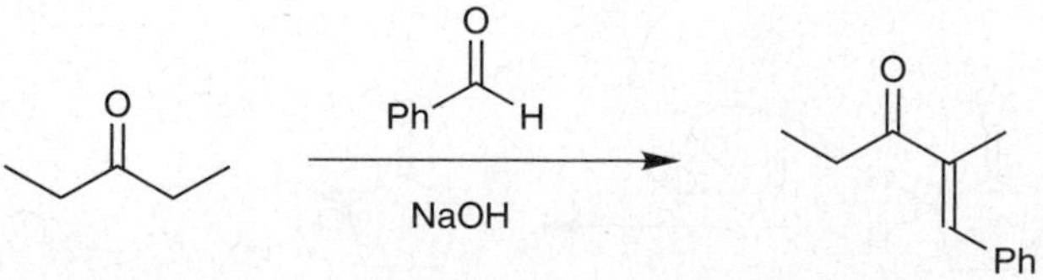

7.8 CLAISEN CONDENSATION

In the previous section, we saw that an enolate can attack a ketone:

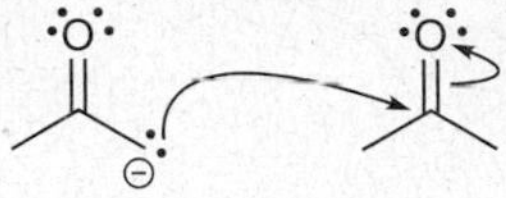

In this section, we will explore what happens when an *ester enolate* attacks an ester:

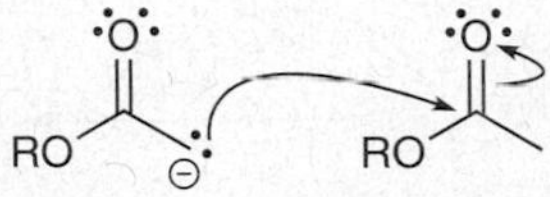

An ester enolate is similar to a regular enolate: an ester enolate is nucleophilic, and it will also attack a carbonyl group. When an ester enolate attacks an ester (shown above), the reaction that takes place is called a Claisen condensation. Here is the overall transformation:

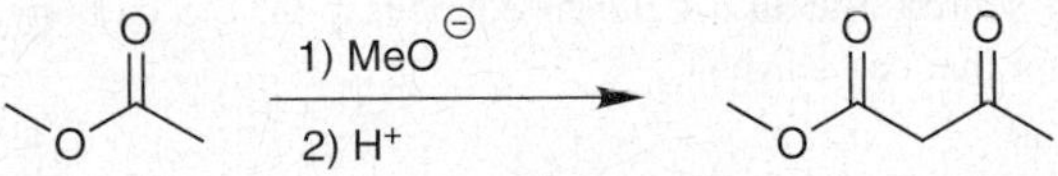

The product is called a β-keto ester:

The ester gets priority over the other carbonyl, so we label the carbon atoms (α, β, γ, etc.) moving away *from the ester*

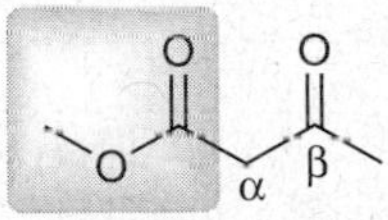

The "keto" group is located at the *beta* position

At first glance, this product seems very different from the α,β-unsaturated ketones obtained from aldol condensations. But when we explore the mechanism, we will see the parallel between the aldol and Claisen condensations.

Let's start with the first step: preparing the enolate:

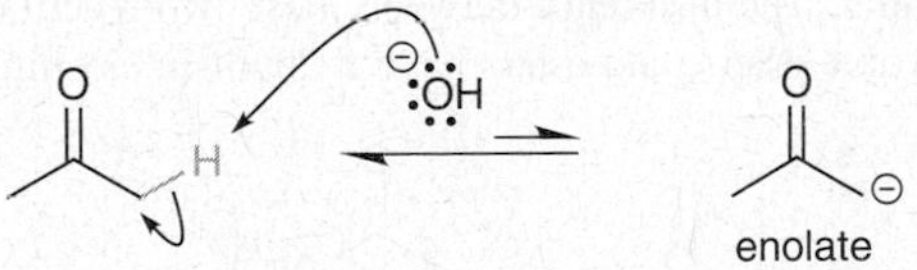

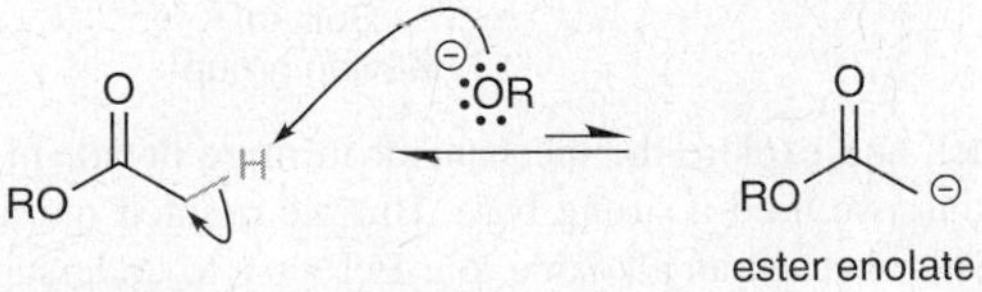

So far, both mechanisms are almost identical. The only difference is the choice of base (the aldol condensation uses hydroxide, and the Claisen condensation uses an alkoxide), and we will discuss the reason for this shortly. For now, let's continue comparing the mechanisms. In the next step, the enolate attacks:

Aldol

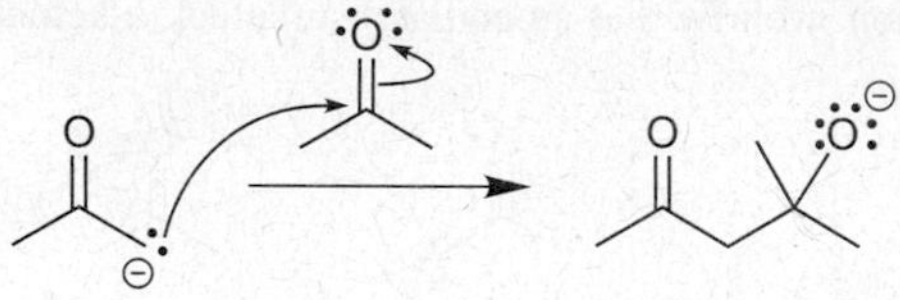

Claisen

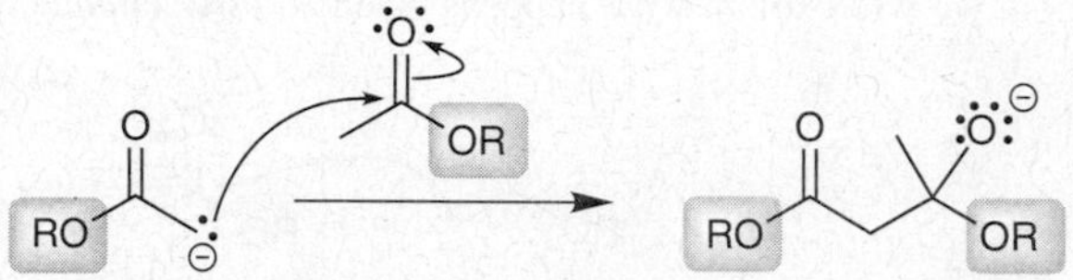

Once again, we see that both mechanisms are essentially identical. In the Claisen condensation, the alkoxy groups seem to just come along for the ride.

But now the two reactions take different routes. And we can use our golden rule to understand why. In the aldol reaction, the carbonyl group cannot re-form, so the oxygen atom must be protonated with a suitable proton source. But in a Claisen condensation, the carbonyl group CAN re-form, because there is a group that can leave:

Aldol

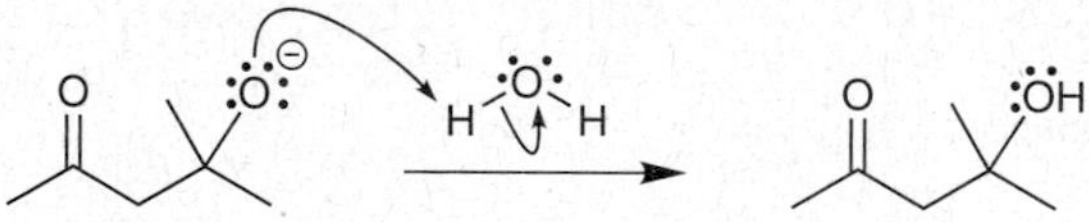

Claisen

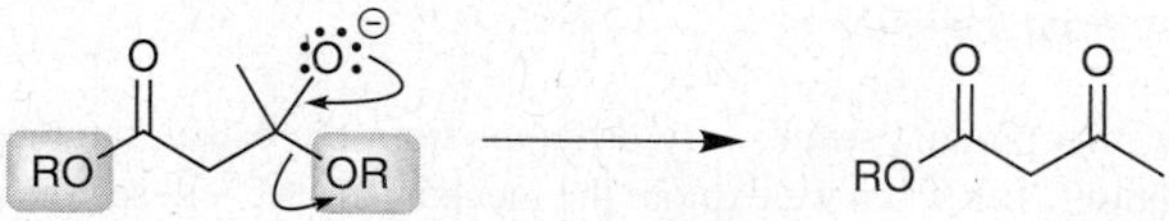

And this is why the product of a Claisen condensation looks very different from the product of an aldol condensation. But when you understand the mechanisms, you can appreciate that these reactions are very similar. The difference between these two reactions stems from the fact that Claisen condensations involve esters; and esters have a "built-in" leaving group:

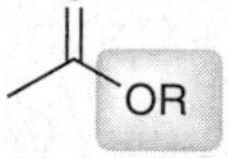

Built-in
leaving group

Now let's go back and explore the mechanism in more detail. In the first step, we made an ester enolate. To do this, we used a strong base. But we pointed out at the time that we did NOT use hydroxide. Instead, we used an alkoxide ion. Let's try to understand why.

If we had used hydroxide, then we might have observed a competing reaction. Instead of hydroxide acting as a base to remove a proton, it is possible for hydroxide to function as a nucleophile, attacking the carbonyl group of the ester:

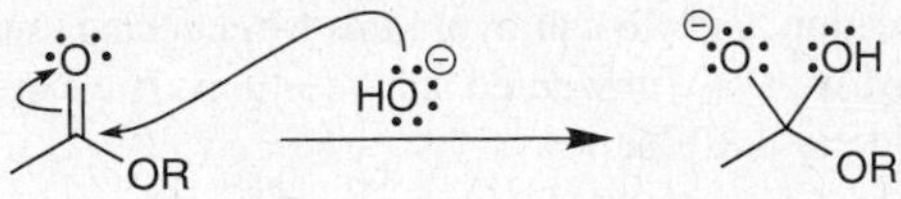

After the initial attack, the carbonyl group could re-form by expelling the alkoxy group. This unwanted side reaction would hydrolyze the ester and produce a carboxylate ion (we saw this reaction in the previous chapter):

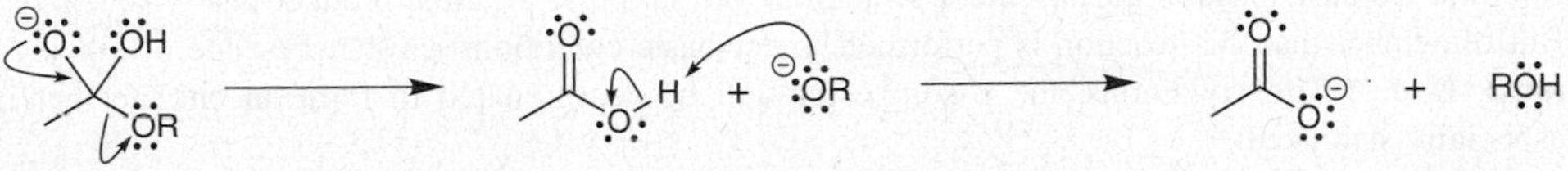

In order to avoid this, we use an alkoxide as our base. It is true that alkoxides can *also* function as nucleophiles, but think about what happens if the alkoxide ion functions as a nucleophile and attacks:

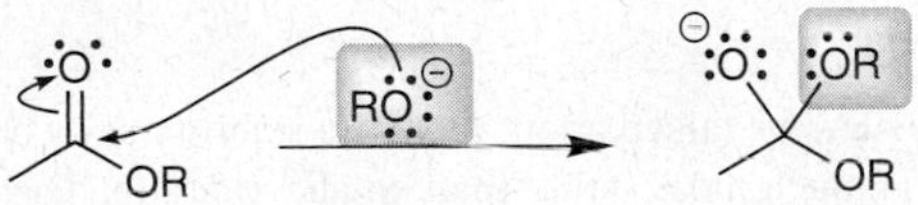

When the carbonyl group re-forms, it doesn't matter which alkoxy group gets expelled. Either way, the original ester will be regenerated:

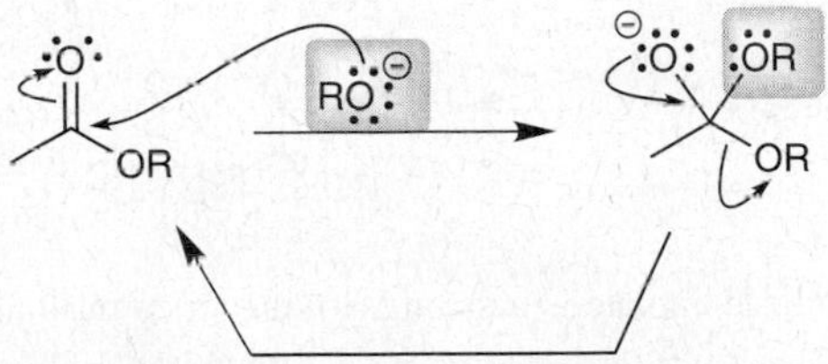

Although the alkoxide *can* attack the carbonyl group, we don't have to worry about it, because it does not actually lead to new products. So, we can avoid unwanted side reactions by using an alkoxide ion as the base for a Claisen condensation.

Be careful, though. We cannot just use *any* alkoxide ion. We must choose the alkoxide ion carefully. If we are dealing with a methyl ester, then we would use methoxide:

The reason for this is simple. Suppose we used ethoxide in this case. This would actually change some of our ester:

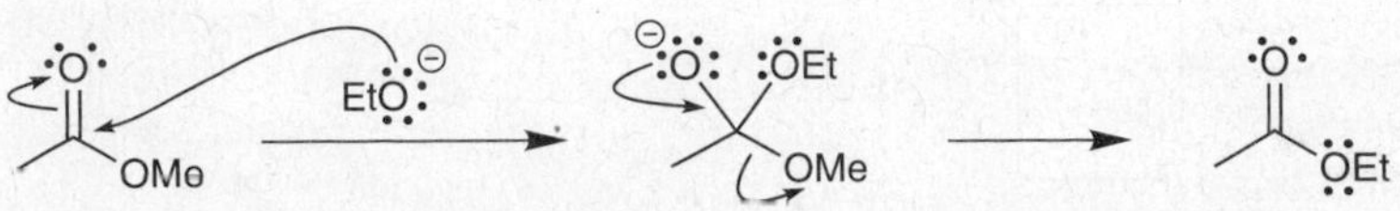

This is called *trans*-esterification, and we can avoid this by choosing our base to match the alkoxy group of the ester. That way, we avoid unwanted side reactions. If we are dealing with an ethyl ester, then we just use ethoxide as our base:

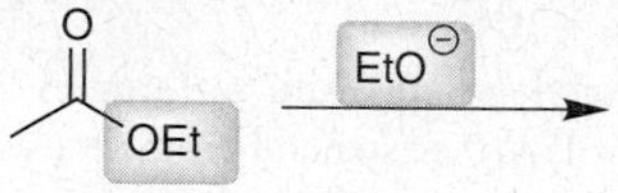

Now that we know what base to choose for a Claisen condensation, let's talk about another special role that the base plays in a Claisen condensation. We said that the final product is a β-keto ester. But remember that this reaction is performed under basic conditions (in the presence of alkoxide ions). Under these conditions, the β-keto ester will be deprotonated to form an enolate that is especially stabilized:

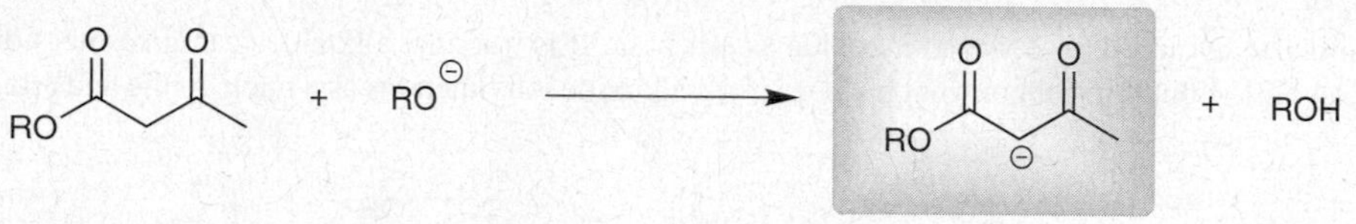

In the beginning of this chapter, we talked about the extra stability associated with this kind of "double" enolate. This enolate is much more stable than an alkoxide ion. That is an important point, because it means that the reaction will favor formation of the product. Why? Because the reaction is converting alkoxide ions into enolate ions (which are more stable):

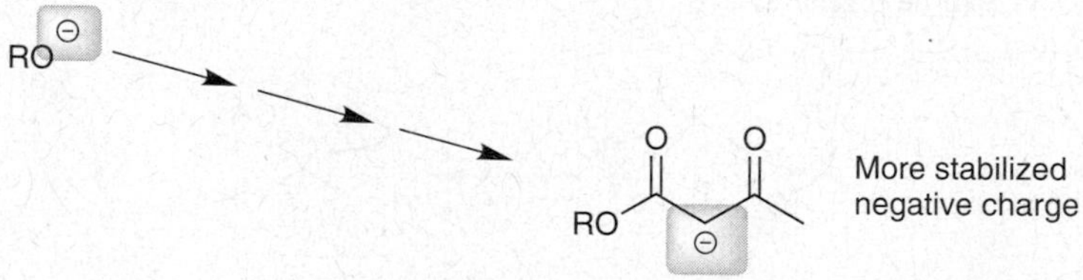

More stabilized negative charge

The formation of this stabilized enolate is a strong driving force pushing this reaction toward the formation of products.

So, when the reaction is finished, a proton source must be introduced into the reaction flask in order to protonate the enolate and obtain the product:

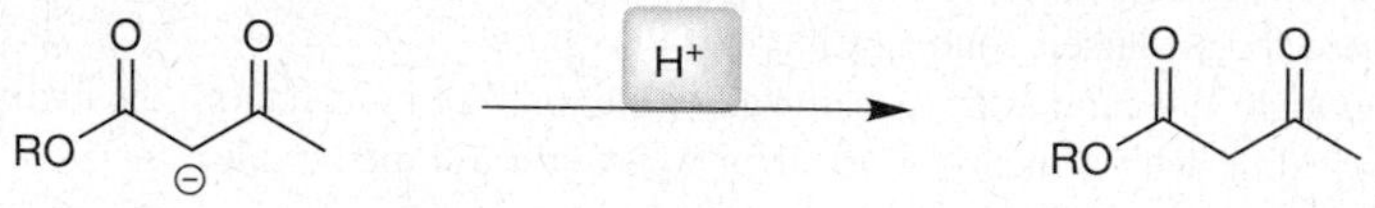

The Claisen condensation is important because it gives us a way to make β-keto esters. And we will soon see that there is a clever synthetic trick that you can perform with β-keto esters. So, let's make sure that we have mastered the Claisen condensation.

Overall, here is what is happening:

ROH

We are removing an alpha proton from one ester, and we are removing an alkoxy group from the other ester. The remaining fragments are then joined together. Notice that a small molecule is liberated in the process (ROH). That is why we call this reaction a Claisen *condensation*.

Now let's practice predicting products. We will soon come back and master the mechanism. But for now, let's make sure that you train your eyes to see the products of a Claisen condensation in an instant. You will need that skill for proposing syntheses.

EXERCISE 7.51   Predict the products of the following reaction:

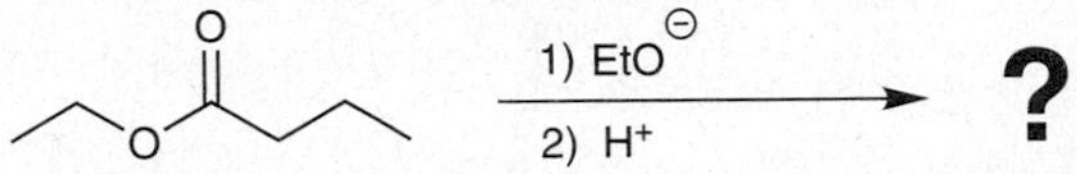

Answer   The following two fragments are adjoined, while EtOH is expelled, like this:

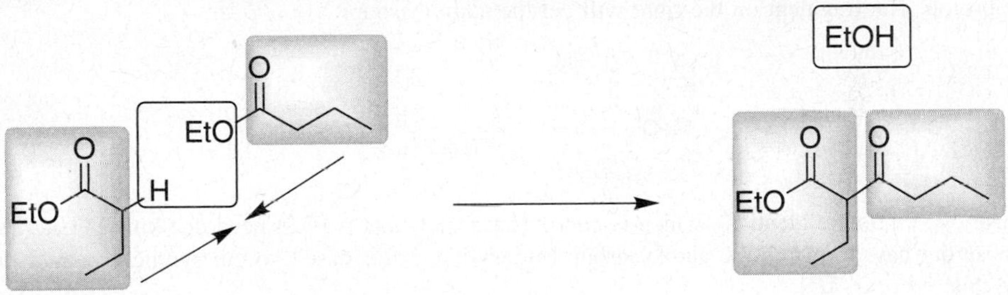

PROBLEMS   Predict the products of the following reactions:

7.52

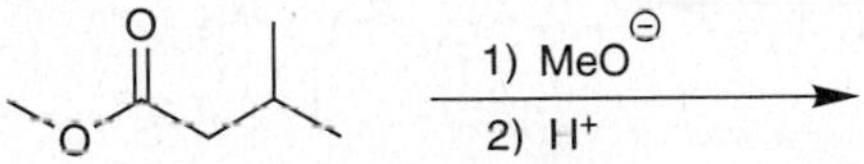

7.53

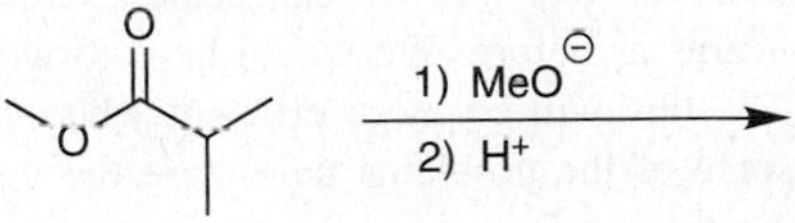

7.54

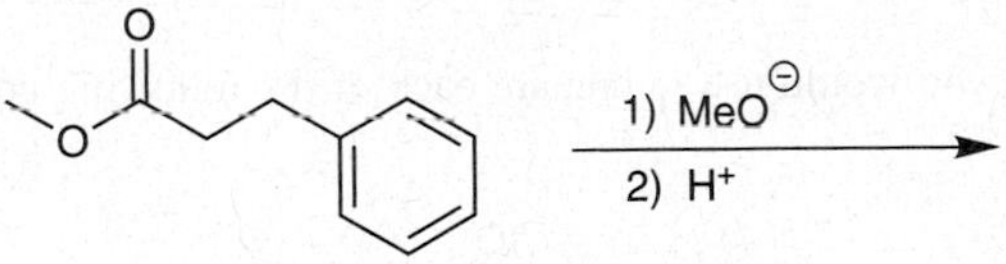

7.55

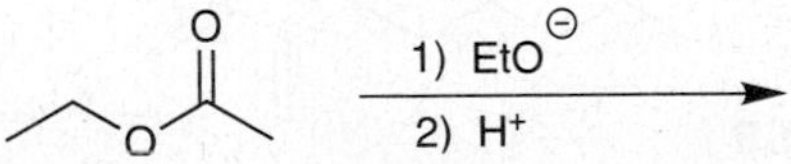

EXERCISE 7.56   Identify the reagents you would use to prepare the following compound using a Claisen condensation:

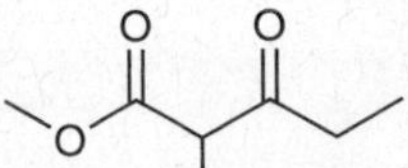

Answer We break the molecule apart into two fragments in order to insert MeOH. We break it apart between the α and β positions:

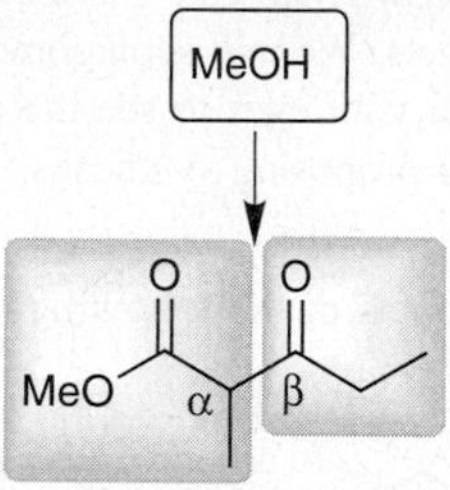

And we just have to decide which fragment gets the methoxy group and which fragment gets the proton. The fragment on the left already has its alkoxy group, so that must be the fragment that gets the proton. The fragment on the right will get the alkoxy group:

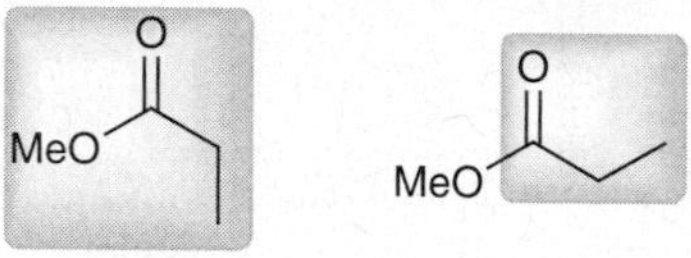

These two esters are identical, which is good. That means that we just need one kind of ester. We choose our base to match the alkoxy group (methoxide in this case), so our synthesis would look like this:

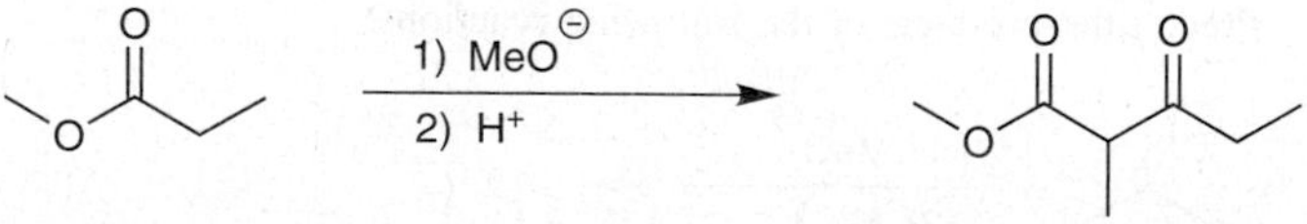

It is possible to achieve crossed Claisen condensations (just like we can achieve crossed aldol condensations), but we would have the same concerns as before. We would have to worry about potential side reactions. A crossed Claisen condensation will be more efficient when one of the esters has no alpha protons. You will see that some of the problems below are the products of crossed Claisen condensations. Keep an eye out for them.

PROBLEMS Identify the reagents you would use to prepare each of the following compounds using a Claisen condensation:

7.57

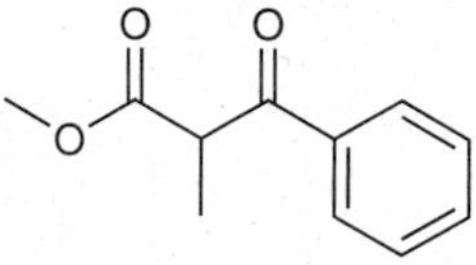

7.58

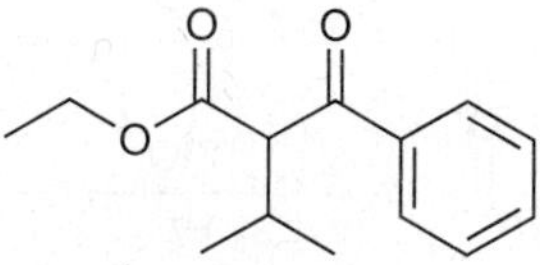

7.59

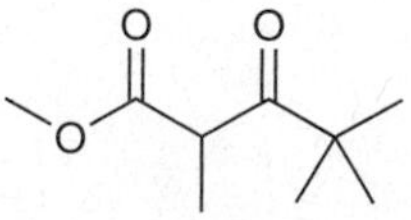

7.60

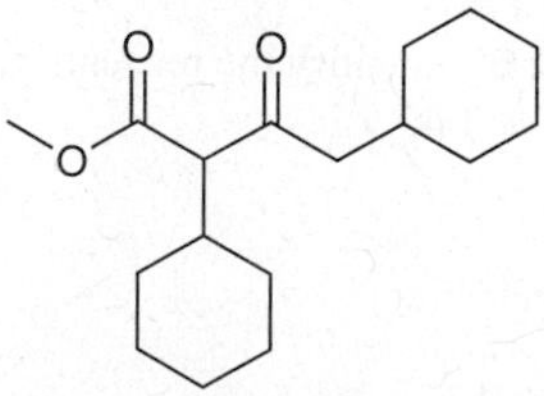

EXERCISE 7.61   Propose a mechanism for the following transformation:

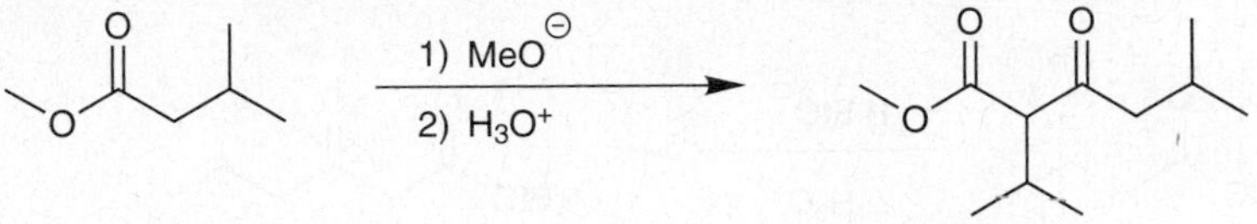

1) MeO⊖
2) H₃O⁺

Answer   First, methoxide functions as a base and deprotonates the ester:

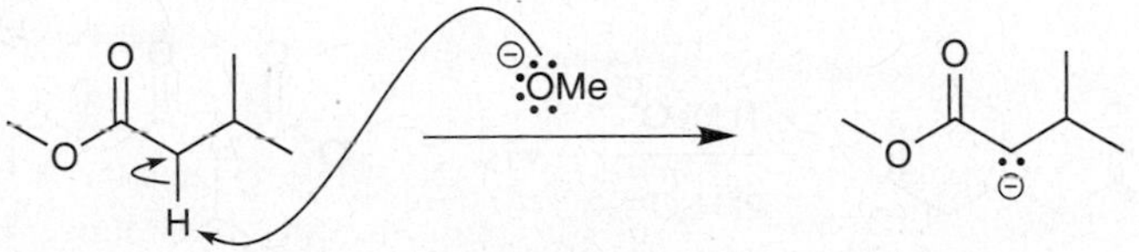

This ester enolate then attacks an ester to give the following intermediate:

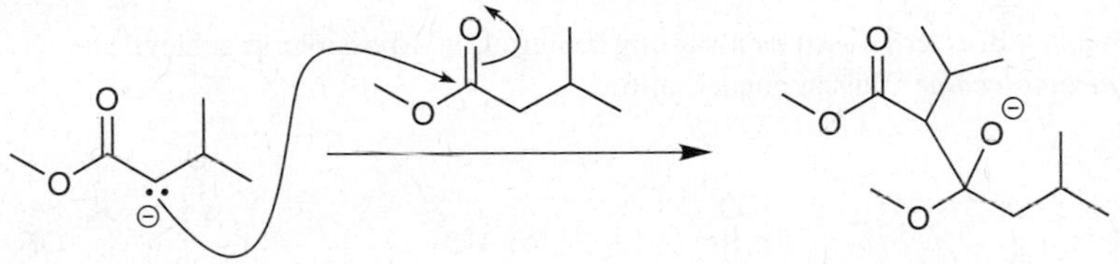

This intermediate then can re-form the carbonyl group by expelling methoxide:

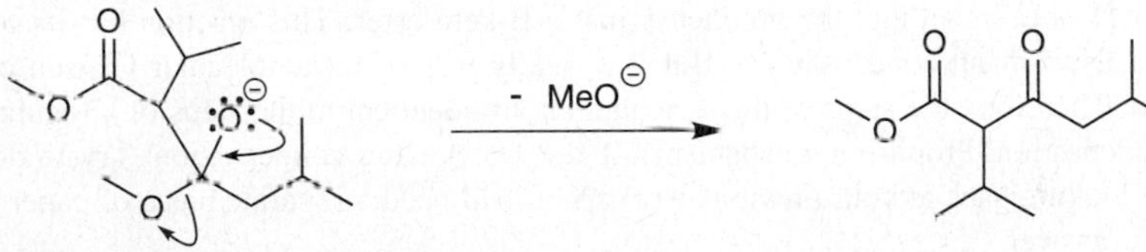

Under these basic conditions (methoxide), the β-keto ester is deprotonated:

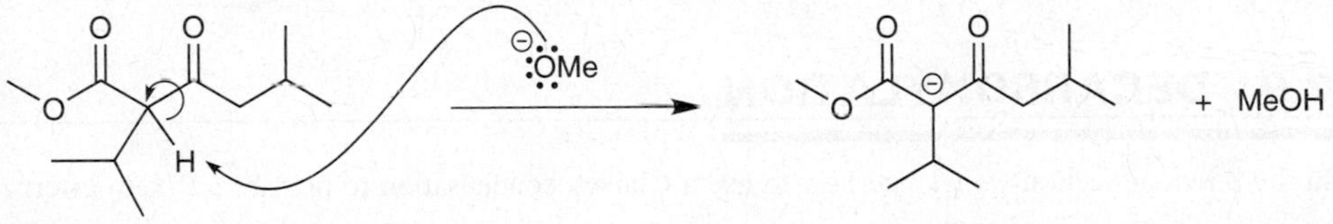

+   MeOH

This deprotonation step is important, because the formation of this stabilized anion is the driving force for the reaction. That is why we must show this step. That explains why the reagents indicate that acid is added to the flask at the end of the reaction. A source of protons is required to regenerate the final product:

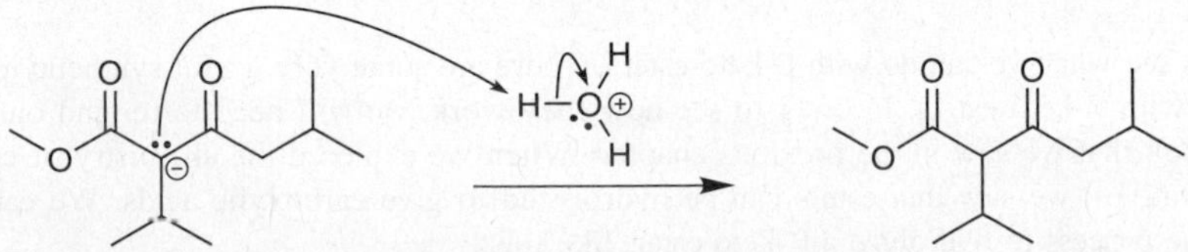

PROBLEMS Propose a mechanism for each of the following reactions. You will need a separate piece of paper to record your answers:

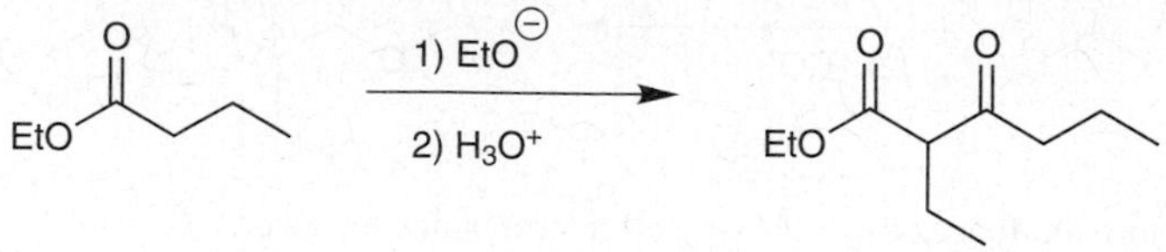

7.62

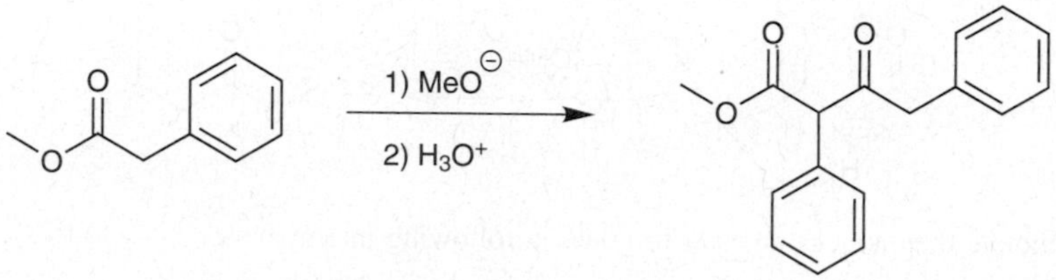

1) EtO$^{\ominus}$

2) H$_3$O$^+$

7.63

1) MeO$^{\ominus}$

2) H$_3$O$^+$

7.64 When a di-ester is used as a starting material, it is possible to achieve an *intramolecular* Claisen condensation:

1) RO$^{\ominus}$

2) H$_3$O$^+$

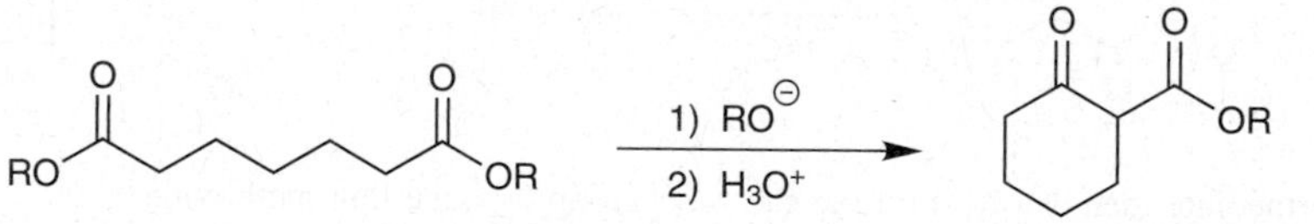

Notice once again that the product is just a β-keto ester. This reaction has its own name (the Dieckmann condensation). But it is really just an intramolecular Claisen condensation. Therefore, the steps of this mechanism are identical to the steps of a regular Claisen condensation. Propose a mechanism for the Dieckmann condensation. Try to do it without looking back at your previous work. You will need a separate piece of paper to record your answer.

7.9 DECARBOXYLATION

In the previous section we learned how to use a Claisen condensation to prepare a β-keto ester:

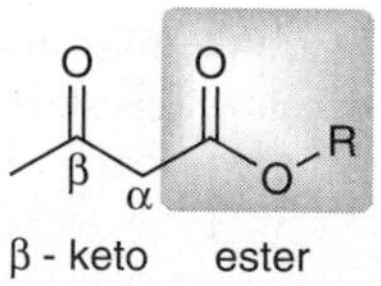

β - keto ester

Now let's see what we can do with β-keto esters. There are some very useful synthetic techniques that start with β-keto esters. In order to see how they work, we will need to remind ourselves of one reaction that we saw in the previous chapter. When we explored the chemistry of carboxylic acid derivatives, we saw that esters can be hydrolyzed to give carboxylic acids. We can use the exact same process to hydrolyze a β-keto ester, like this:

β - keto ester → H_3O^+ → β - keto acid

And the product is a β-keto acid, which will undergo a unique reaction when heated. Specifically, the carboxyl group is completely expelled:

This carboxyl group is expelled → heat → + CO_2

We call this process a *decarboxylation*. This reaction is the basis for the synthetic techniques we will learn in this section, so let's make sure we understand how a decarboxylation occurs. The process begins with a pericyclic reaction that liberates CO_2 as a gas:

heat → +

Pericyclic reactions are characterized by a ring of electrons moving around in a circle. There are many kinds of pericyclic reactions (the Diels-Alder reaction, which you probably explored already last semester, is a type of pericyclic reaction). Pericyclic reactions truly deserve their own chapter, and unfortunately, many textbooks do not devote an entire chapter to pericyclic reactions (they are just scattered throughout the various chapters). Perhaps your instructor will spend some time on pericyclic reactions. We will not cover them right now, as we must continue with the topic at hand.

In the reaction above, CO_2 gas is liberated (that's how the carboxyl group is expelled), generating an enol. And we know that enols will quickly tautomerize to give ketones:

So, when a β-keto acid is heated, the carboxyl group is expelled, and we end up with a ketone.

Now consider what we have just done. We took a β-keto ester (which is the product of a Claisen condensation), and we hydrolyzed it to produce a β-keto *acid*. Then, we heated this compound, and we expelled the carboxyl group:

β - keto ester → H_3O^+ → β - keto acid → heat / - CO_2 → ketone

In the end, the product is a ketone. To see why this is so useful, we need to add just one more step at the very beginning of the overall process. Imagine that we first alkylate the β-keto ester:

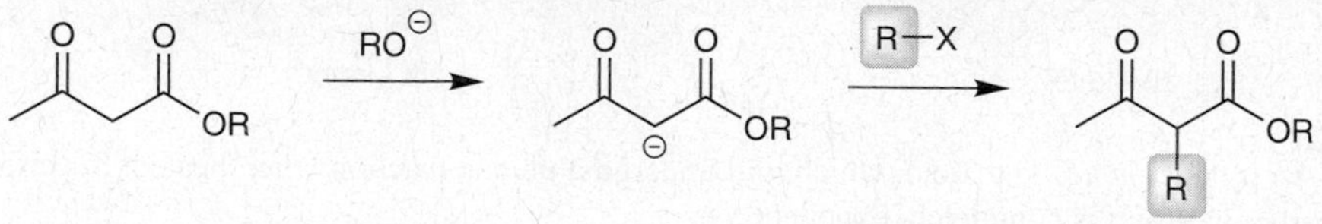

We have already seen this kind of reaction before (Section 7.5). It is just an alkylation. We used an alkoxide ion to produce a doubly stabilized enolate, which then attacks the alkyl halide in an S_N2 reaction. If we then continue with the rest of the strategy (hydrolysis, followed by decarboxylation), the following product is obtained:

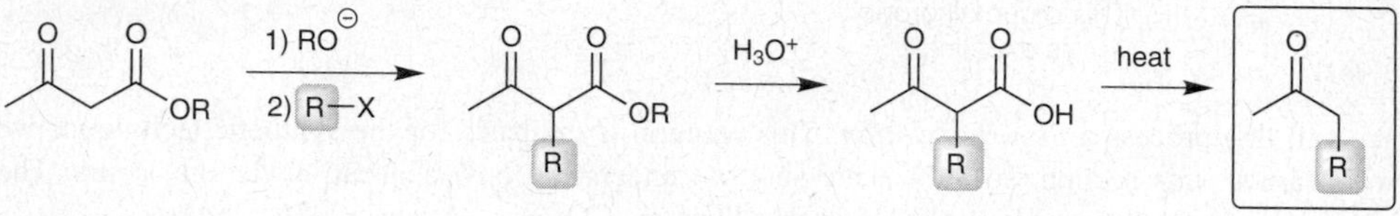

Take a close look at the product. This compound is a substituted derivative of acetone:

acetone   a substituted
          derivative of acetone

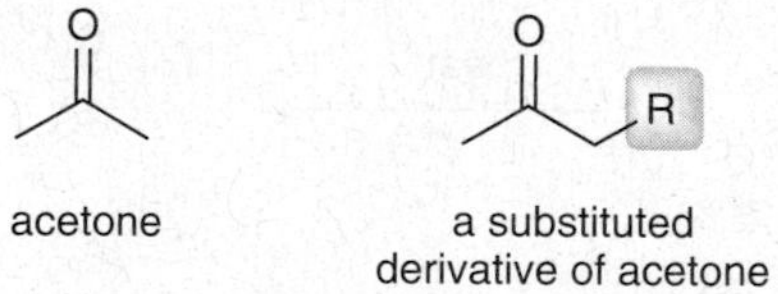

This provides a method to make a wide variety of substituted derivatives of acetone.

This is useful, because we would encounter an obstacle if we tried to alkylate acetone directly:

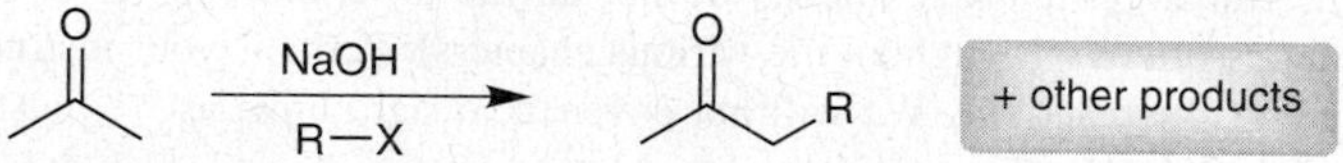

The desired product would be produced together with many other undesired products (from polyalkylation and from elimination reactions). So the strategy we have learned provides a clean way to make substituted derivatives of acetone. But be careful—remember that the alkylation step is an S_N2 process, so primary alkyl halides will be more efficient. In other words, you *could* use this strategy to prepare the following compound:

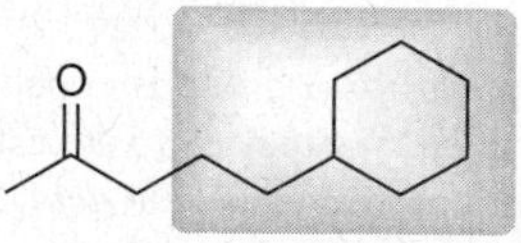

But you could *not* use this synthetic strategy to make this compound:

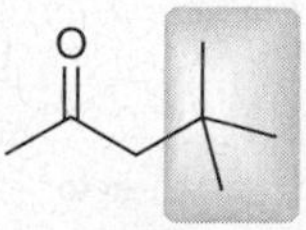

because that would have required an alkylation step involving a tertiary alkyl halide, which cannot function as a substrate in an S$_N$2 process.

In order to use this synthetic strategy, we will always have to start with the following compound:

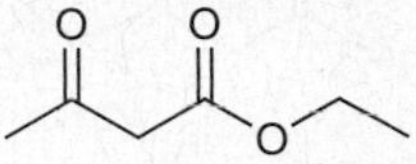

This compound is called ethyl acetoacetate. This compound belongs to a class of compounds called acetoacetic esters. Therefore, we call our strategy the *acetoacetic ester synthesis*.

To summarize what we have seen, the acetoacetic ester synthesis has three main steps: alkylate, hydrolyze, and then decarboxylate. Say that ten times real fast.

Now let's get some practice using this synthetic strategy:

EXERCISE 7.65   Starting with ethyl acetoacetate, show how you would prepare the following compound:

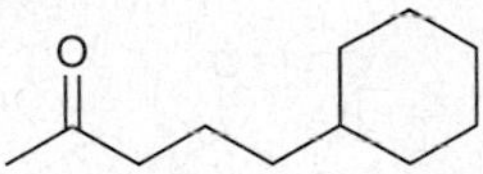

Answer   Remember that the acetoacetic ester synthesis has the following steps: alkylate, hydrolyze, and then decarboxylate. So, in order to solve this problem, we must identify the alkyl group that should be installed:

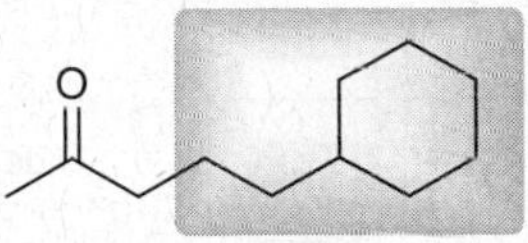

So, we will need the following alkyl halide:

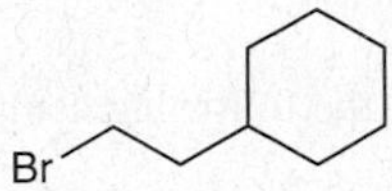

Now that we have determined what alkyl halide to use, we are ready to propose our synthesis:

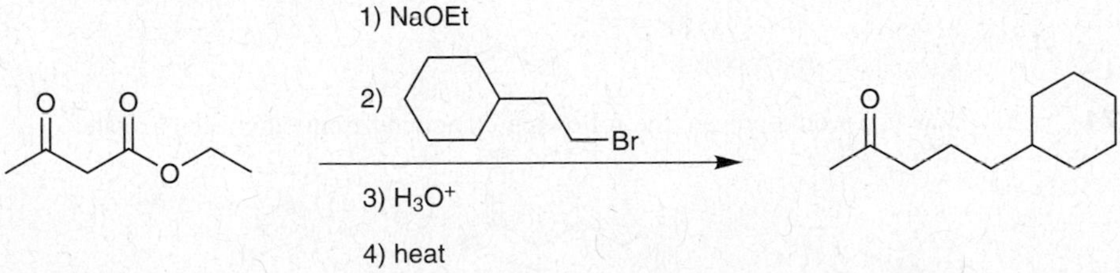

1) NaOEt

2) [cyclohexylethyl bromide structure]

3) H$_3$O$^+$

4) heat

PROBLEMS   Show you would prepare each of the following compounds from ethyl acetoacetate:

7.66

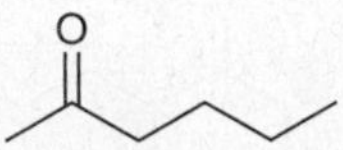

7.67

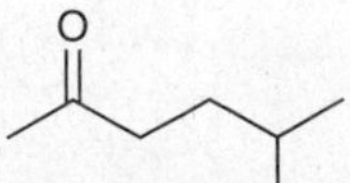

7.68

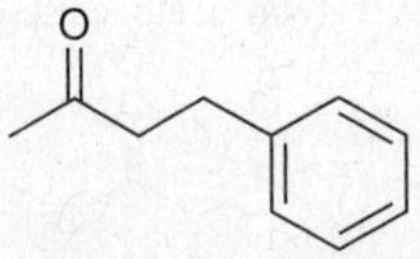

7.69 Explain why the following compound *cannot* be made with an acetoacetic ester synthesis.

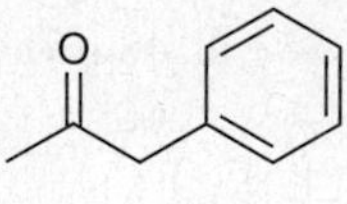

In all of the problems we have done so far, we have focused on alkylating *once*. But it is also possible to alkylate twice, which would give a product with two alkyl groups:

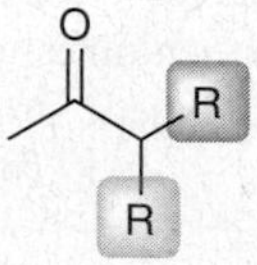

And the R groups don't even have to be the same. In the following sequence, you should notice that two different alkyl groups are installed:

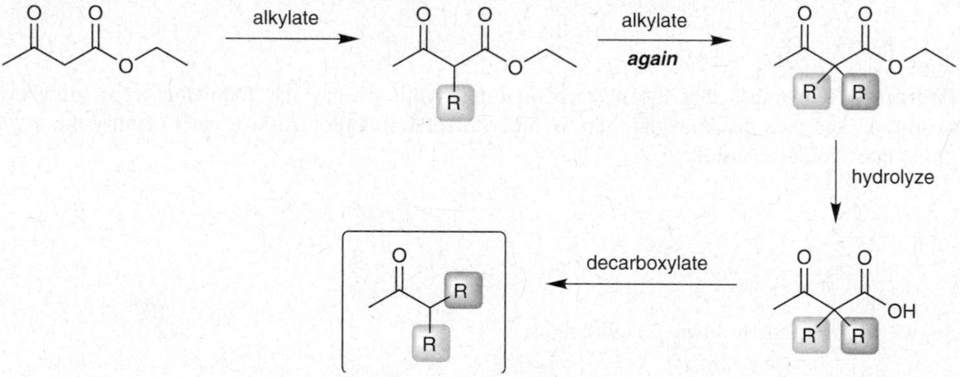

alkylate → alkylate *again* →

hydrolyze

decarboxylate ←

7.70 Show how you would prepare the following compound from ethyl acetoacetate:

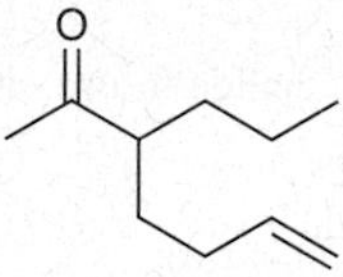

7.71 Show how you would prepare the following compound from ethyl acetoacetate:

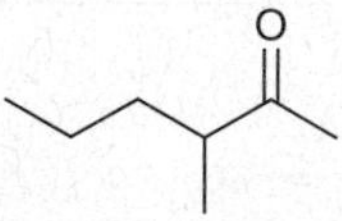

7.72 Propose a synthesis for the following transformation. (*Hint:* This reaction is similar to an acetoacetic ester synthesis, but we are just starting with a different β-keto ester):

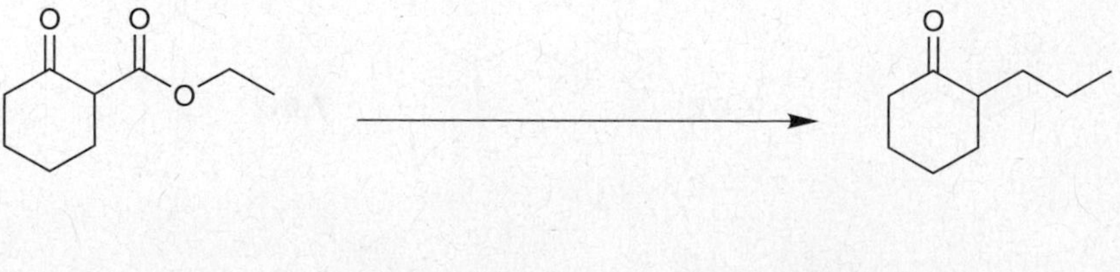

There is another common synthetic strategy that utilizes the same concepts as the acetoacetic ester synthesis. So, let's now focus on this other strategy. It is called the *malonic ester synthesis*, because the starting material is a malonic ester (called diethyl malonate):

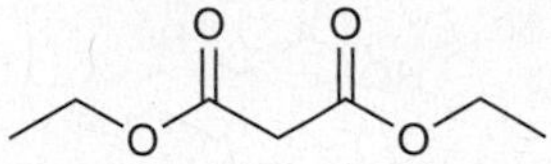

We follow the same three steps that we followed in our previous strategy: alkylate, hydrolyze, and then decarboxylate. The only difference is that we start with a slightly different starting material (malonic ester, instead of acetoacetic ester), and therefore, our product will be slightly different. Compare the structures of ethyl acetoacetate and diethyl malonate:

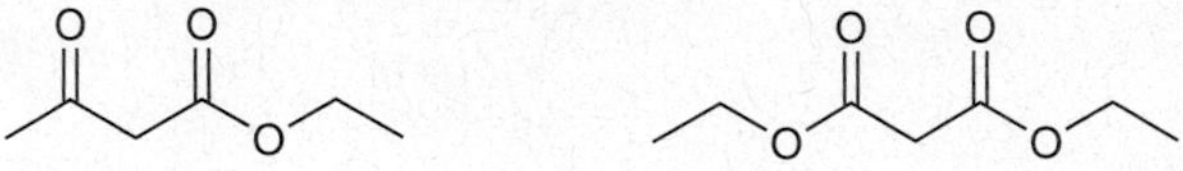

Notice that diethyl malonate has *two* carboxyl groups (as opposed to ethyl acetoacetate, which has one carboxyl group and one carbonyl group). To see how this extra carboxyl group affects the structure of our end product, let's go through the three steps: alkylate, hydrolyze, and then decarboxylate.

We start with an alkylation:

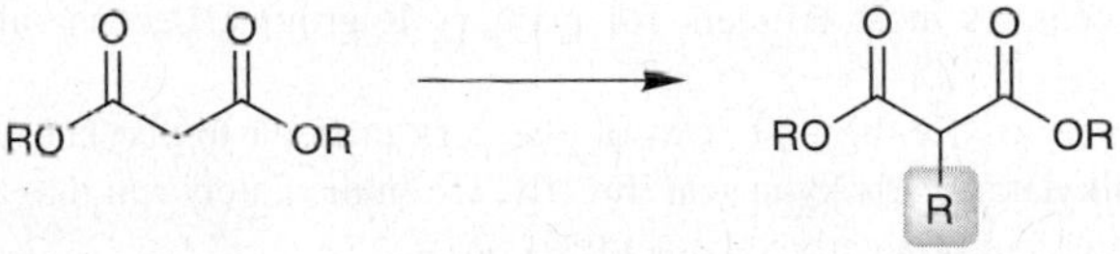

Then, we hydrolyze:

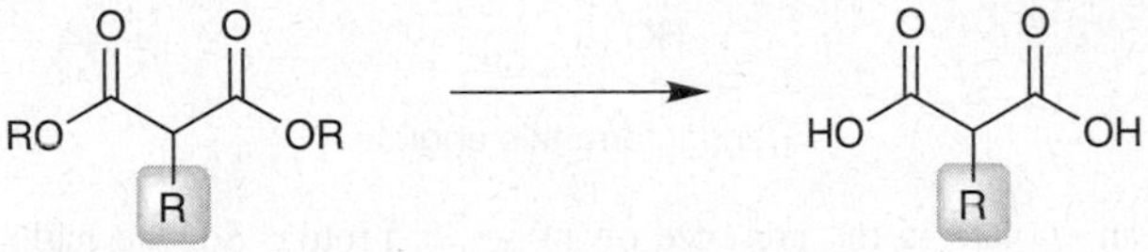

Notice that *both* sides get hydrolyzed.

Then, finally, we decarboxylate:

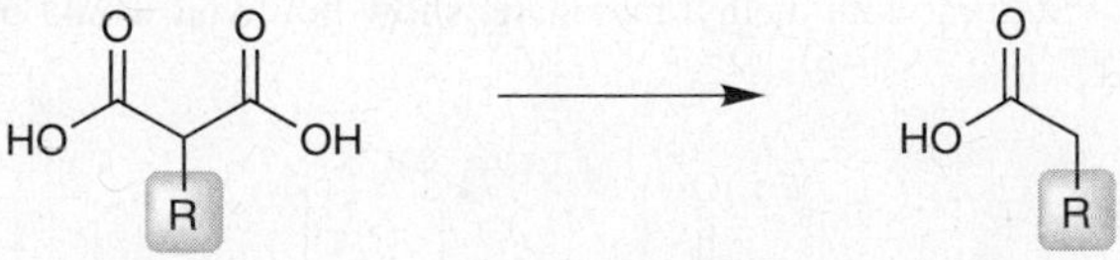

Only one side undergoes decarboxylation. Why? Remember how a decarboxylation works. It is a pericyclic reaction that can occur when a C=O bond is α,β to a carboxylic acid moiety. After the first carboxyl group is expelled, there is no longer a C=O bond that is β to the remaining carboxylic acid group. Try to draw a mechanism for the second carboxyl group leaving, and you should find that you can't do it.

Notice that the product is now a substituted carboxylic acid. This is the power of the malonic ester synthesis. It provides a method for making a wide variety of substituted carboxylic acids:

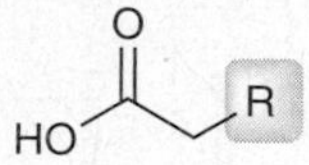

This synthesis can also be used to install *two* alkyl groups (just like we did with the acetoacetic ester synthesis). We would just alkylate twice at the beginning of our procedure:

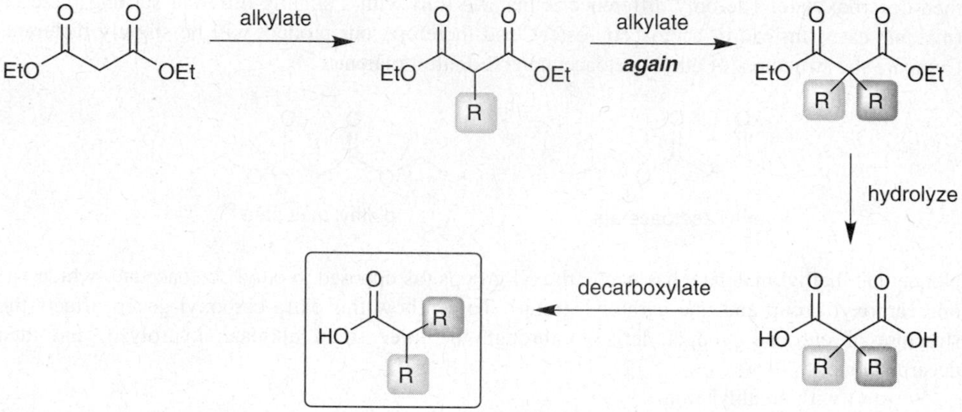

Once again, the process is most efficient for primary R groups, because alkylation is an S_N2 process.

This strategy is very useful, because it would be very difficult to alkylate a carboxylic acid directly. If we try to alkylate a carboxylic acid directly, we immediately run into an obstacle, because we cannot form an enolate of a carboxylic acid:

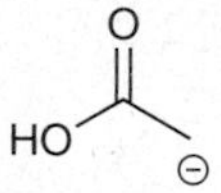

Cannot form this enolate

You cannot form an enolate in the presence of an acidic proton. So, the malonic ester synthesis gives us a way around this obstacle. It provides a method for making substituted carboxylic acids. Let's get some practice with this:

EXERCISE 7.73 Starting with diethyl malonate, show how you would prepare the following compound:

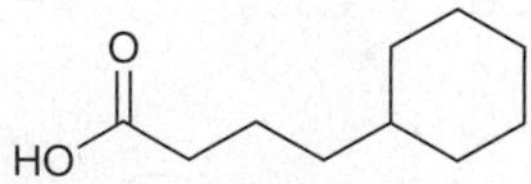

Answer Remember that the malonic ester synthesis has the following steps: alkylate, hydrolyze, and then decarboxylate. So, in order to solve this problem, we must identify the alkyl group that should be installed.

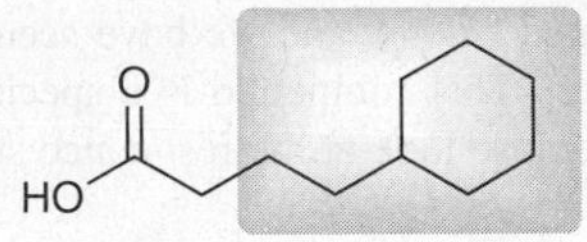

So, we will need the following alkyl halide:

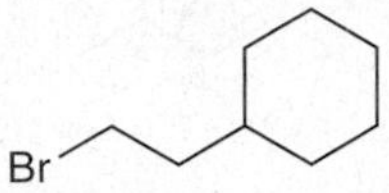

Now that we have determined what alkyl halide to use, we are ready to propose our synthesis:

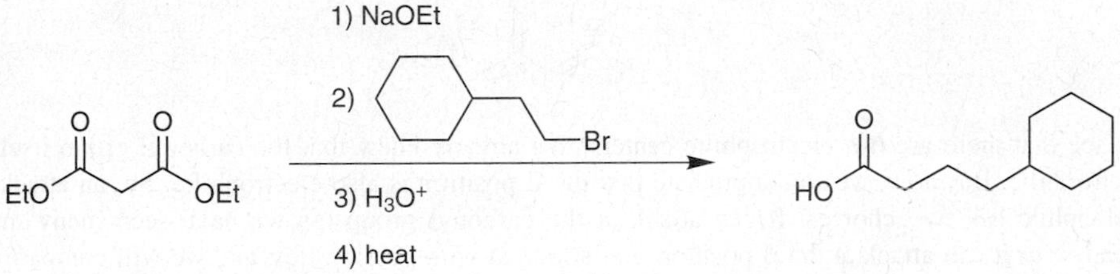

PROBLEMS Identify what reagents you would use to achieve each of the following transformations:

7.74

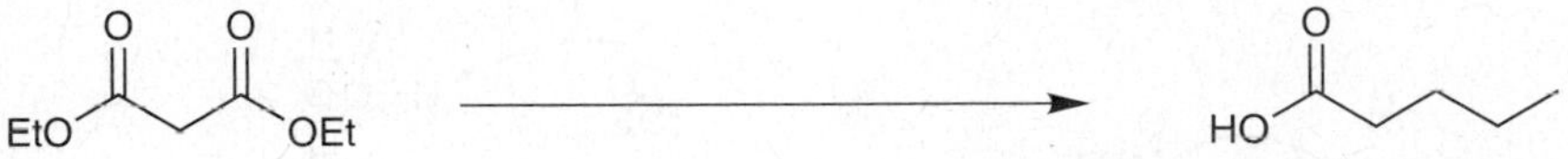

7.75

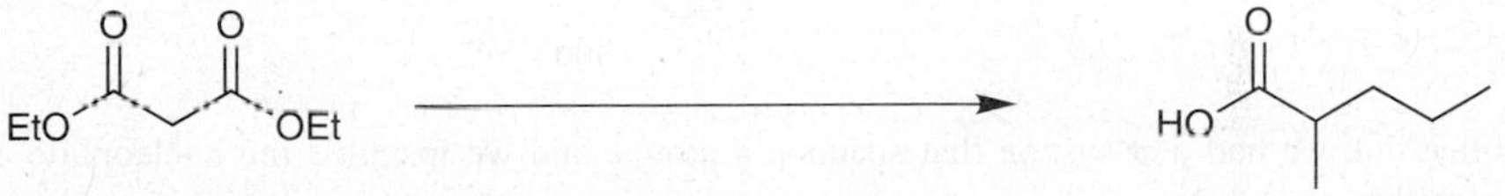

7.76

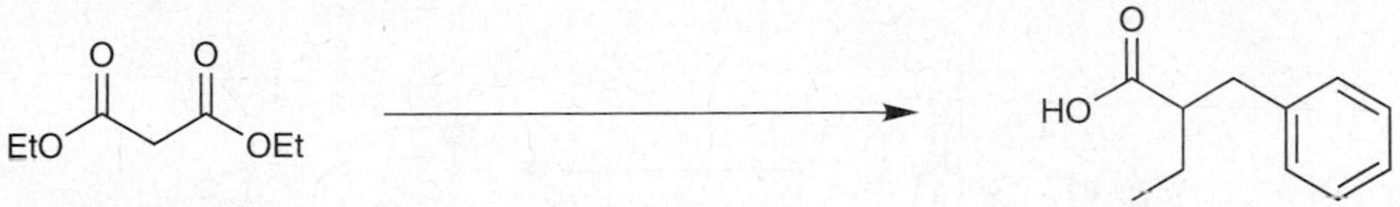

7.10 MICHAEL REACTIONS

In this chapter, we have seen that enolates can attack a wide variety of electrophiles. We started the chapter with the reaction between enolates and halogens. Then we looked at the reaction between enolates and alkyl halides. We also saw that enolates can attack ketones or esters. In this section, we will conclude our discussion of enolates by looking at a special kind of electrophile that can be attacked by an enolate. Consider the following compound:

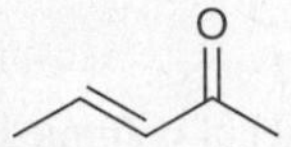

This compound is an α,β-unsaturated ketone, and we have seen that compounds of this type can be made with an aldol condensation. This compound is a special kind of electrophile. To understand why it is special, let's take a close look at the resonance structures:

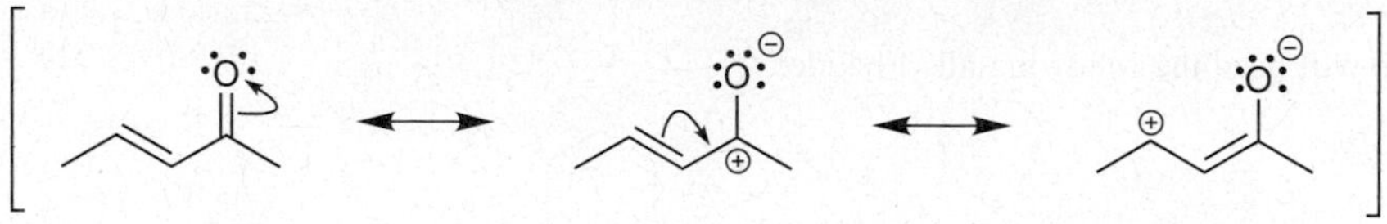

These resonance structures paint the following picture:

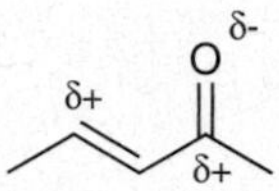

We see that there are *two* electrophilic centers. We already knew that the carbonyl group itself is electrophilic. But now, we can appreciate that the β position is also electrophilic. So, an attacking nucleophile has two choices. It can attack at the carbonyl group (as we have seen many times already), *or* it can attack at the β position. Let's look at both possibilities, and we will compare the products.

Consider what happens if the nucleophile attacks the carbonyl group, and the resulting intermediate is then protonated:

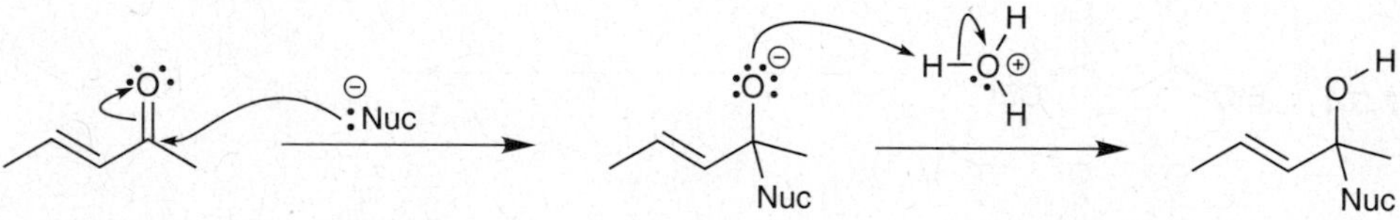

Notice that we had a π system that spanned 4 atoms, and we installed the nucleophile and the H in positions 1 and 2:

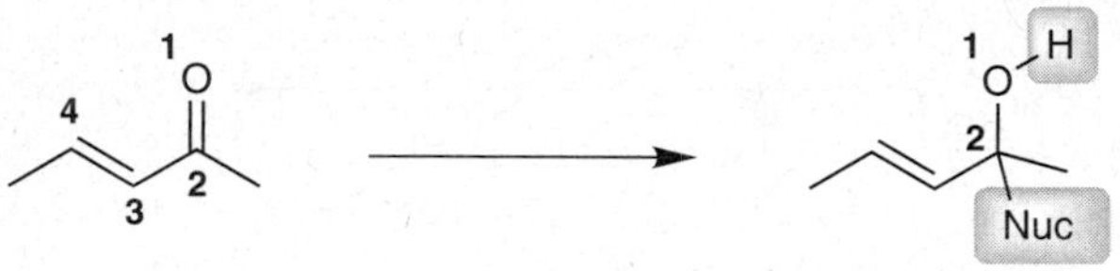

Therefore, we call this a *1,2-addition*.

Now consider what happens if the nucleophile attacks the β position, rather than the carbonyl group. The initial intermediate is an enolate:

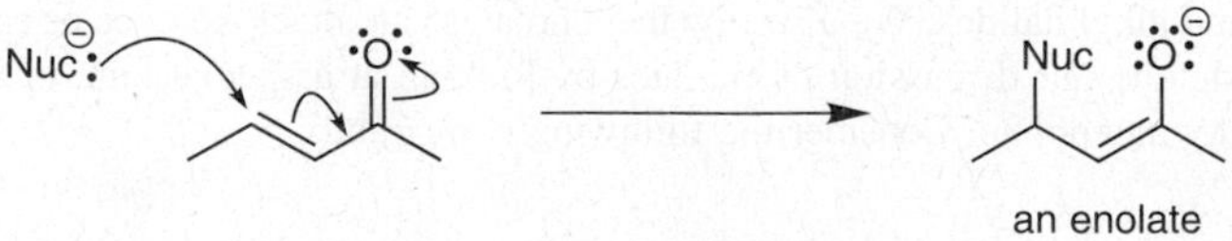

an enolate

Then, when this enolate is protonated, an enol is formed:

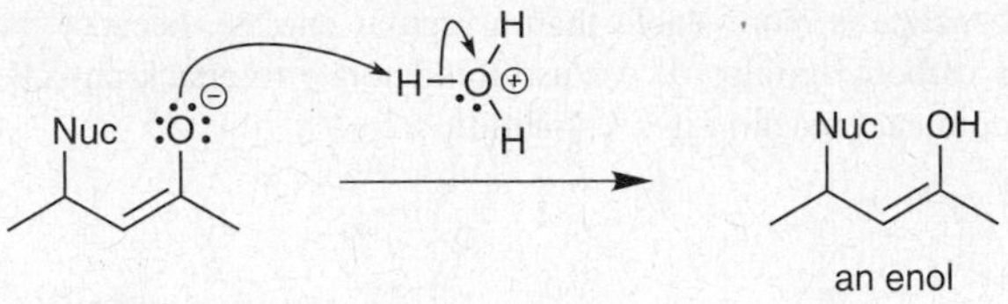

Once again, we have added the nucleophile and H across the π system. But this time, we have added them across the *ends* of this system:

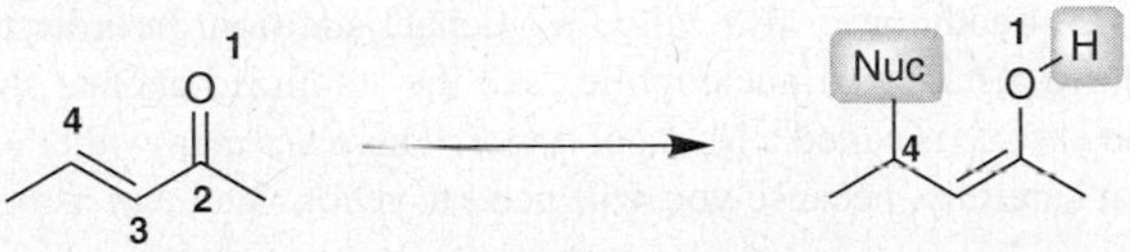

So, we call this a *1,4-addition*. Chemists have given this reaction other names as well. A 1,4-addition is often called a **conjugate addition**, or a **Michael addition**.

We know that the product of a 1,4-addition is not going to stay in the form of an enol, because an enol will tautomerize to form a ketone:

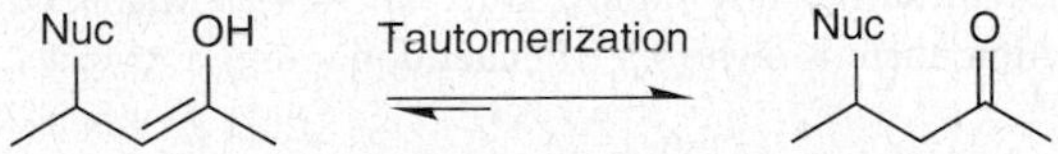

When you look at this ketone, it is hard to see why we call it a 1,4-addition. After all, it looks like the nucleophile and the H have added across the C=C bond:

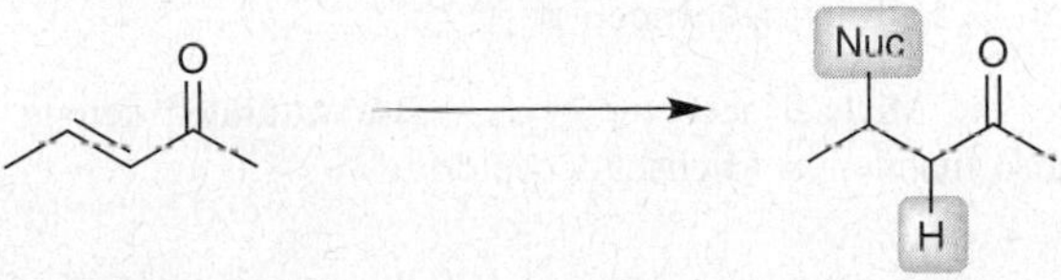

You need to draw the entire mechanism (as we did just a moment ago) in order to see why we call it a 1,4-addition.

Now that we know the difference between a 1,2 addition and a 1,4-addition, let's take a look at what happens when our attacking nucleophile is an enolate.

If an α,β-unsaturated ketone is treated with an enolate, a mixture of products is obtained. Not only do we observe both possibilities (the enolate attacking the carbonyl group, or the enolate attacking the β position), but it gets even more complicated. The product of the 1,4 addition is a ketone, which can be attacked again by an enolate. You can get crossed aldol condensations, and all sorts of unwanted products. So, we can't use an enolate to attack an α,β-unsaturated ketone. The enolate is simply too reactive, and we observe a mixture of undesired products.

The way around this problem is to create an enolate that is more stabilized. A more stable enolate will be less reactive, and therefore, it will be more selective in what it reacts with. But how do we make a more stabilized enolate? We have actually already seen such an example in this chapter. Consider the following enolate:

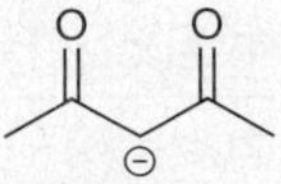

We argued that this enolate is more stable than a regular enolate, because the negative charge is delocalized over ***two*** carbonyl groups. If we use this enolate to attack an α,β-unsaturated ketone, we find that the predominant reaction is a 1,4-addition:

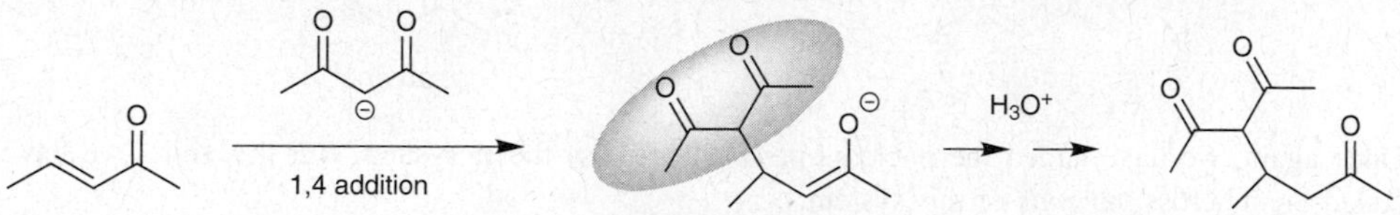

We said earlier that a 1,4-addition is also called a Michael addition. In order to get a Michael addition, you need to have a stabilized nucleophile, like the stabilized enolate shown in the reaction above. This stabilized enolate is called a Michael donor. There are many other examples of ***Michael donors***. Look at them carefully, because you will need to recognize them as being Michael donors when you see them:

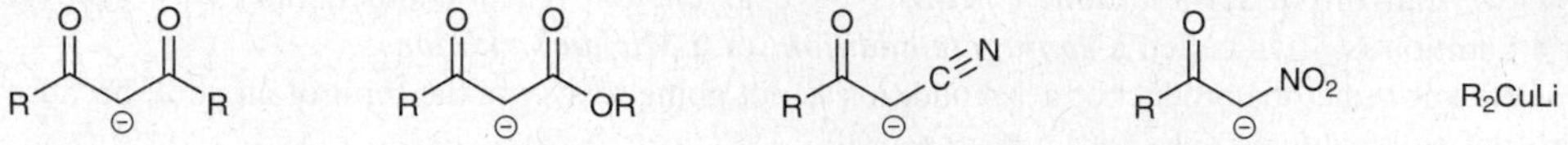

All of these nucleophiles are sufficiently stabilized to function as Michael donors.

In any Michael reaction, there is always a Michael donor ***and*** a Michael acceptor:

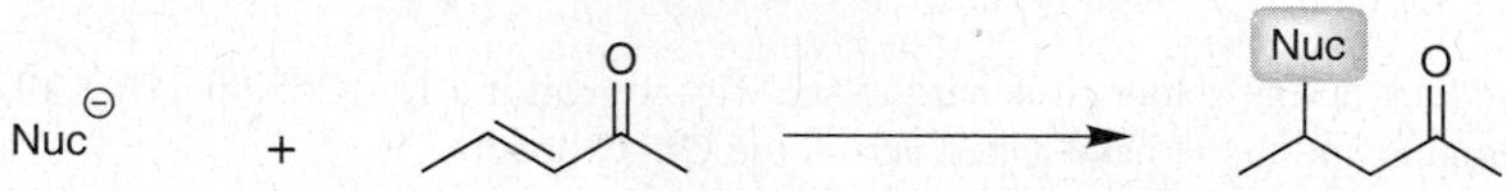

Michael Donor Michael Acceptor

In the reaction above, the Michael acceptor is an α,β-unsaturated ketone. But there are other compounds that can also function as Michael acceptors:

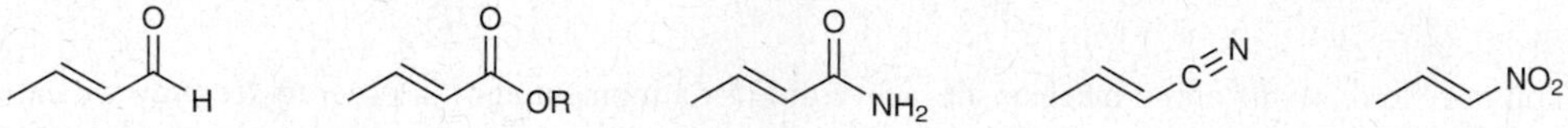

You can use any Michael donor to attack any Michael acceptor. For example, the following reaction would also be called a Michael reaction:

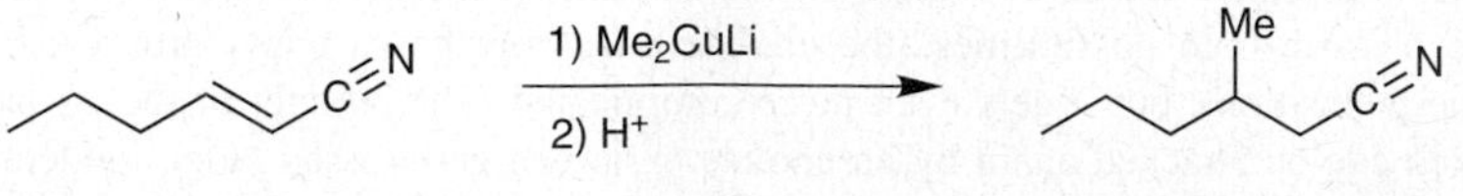

In order to do this next set of problems, you will need to review the lists of Michael donors and Michael acceptors.

EXERCISE 7.77 Draw the expected product of the following Michael reaction:

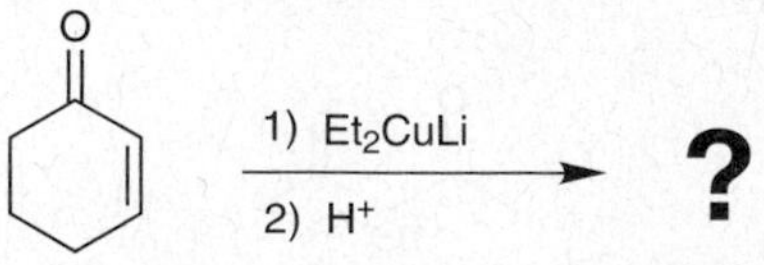

Answer First identify the Michael donor and the Michael acceptor. The α,β-unsaturated ketone is a Michael acceptor, and the lithium dialkyl cuprate is a Michael donor. So we expect the following Michael reaction:

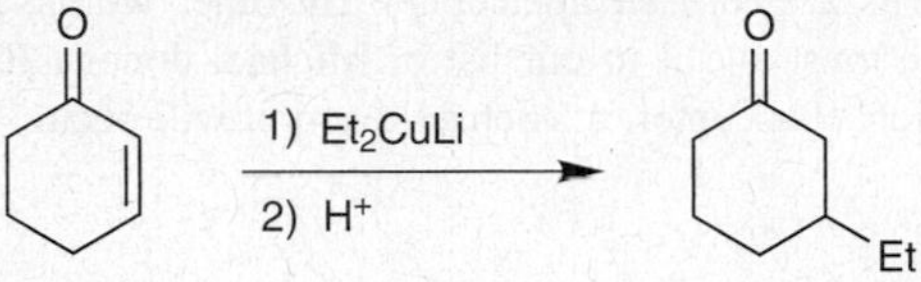

Stare at this transformation for a few moments. It might be helpful to you on an exam.

PROBLEMS For each of the reactions below, determine whether you expect an efficient Michael reaction. If so, then draw the product you expect. If you do not expect an efficient Michael reaction, then simply indicate that a mixture of undesired products will be obtained.

7.78

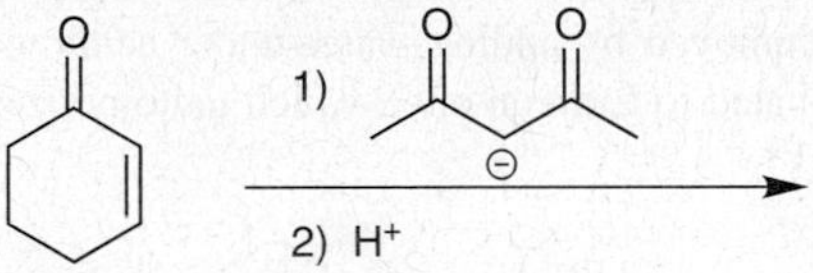

7.79

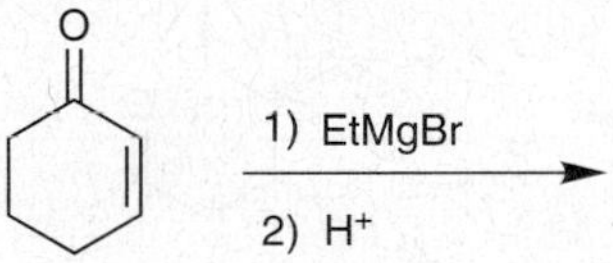

7.80

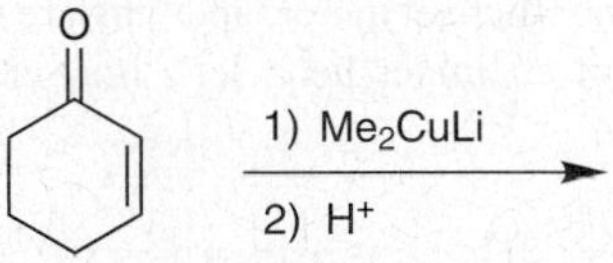

There is one more Michael donor that requires special mention. Enamines are very special Michael donors, because they provide us with a useful synthetic strategy.

When we learned about ketones and aldehydes (Chapter 5), we saw that you can prepare an enamine by treating a ketone with a secondary amine, under the following conditions:

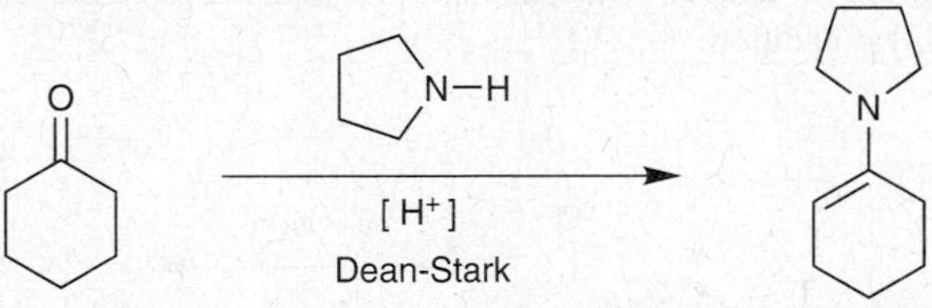

This was our way of converting a ketone into an enamine. To understand how an enamine can function as a Michael donor, let's take a close look at the resonance structures of an enamine:

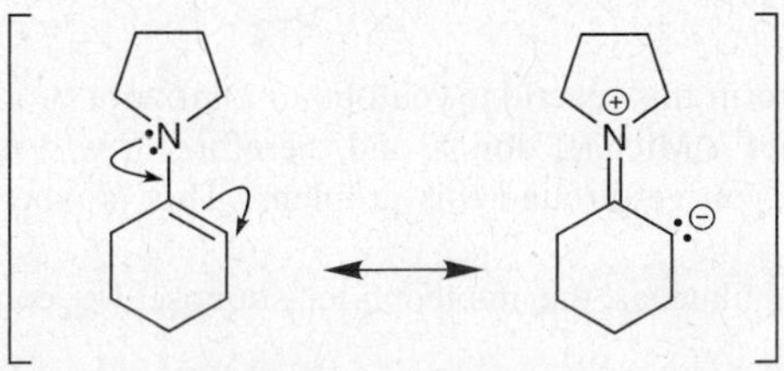

When we meld these two images together in our minds, we see that the carbon atom is nucleophilic (it has some partial negative character). But, it is a fairly weak nucleophile, because the compound does not have a *full* negative charge. Rather, the carbon atom only has *partial* negative character. Therefore, this compound is a *stabilized* nucleophile (in other words, it will be selective in its reactivity). This means we must add it to our list of Michael donors. If we use an enamine as a nucleophile to attack a Michael acceptor, a Michael reaction will occur:

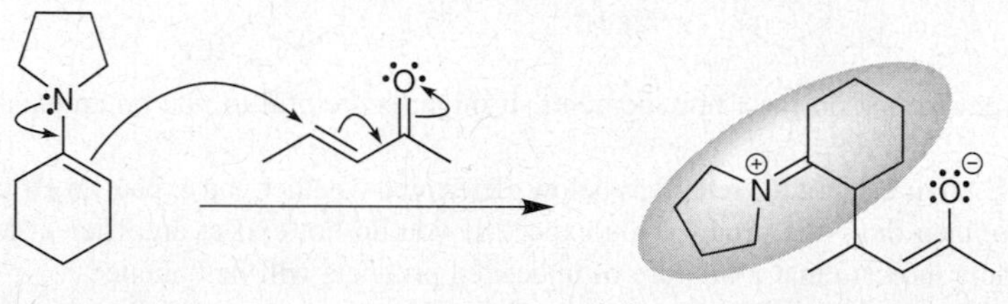

And then, the resulting iminium group can be removed by adding water under acidic conditions (and under these conditions, the enolate is protonated to form an enol, which tautomerizes to form a ketone):

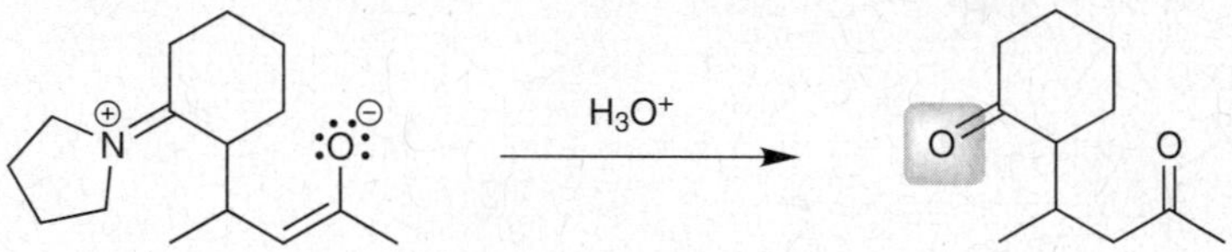

But why is this so important? Why am I singling out this one Michael donor, and why are we learning about enamines again? To understand the usefulness of enamines here, let's imagine that we wanted to achieve the following transformation:

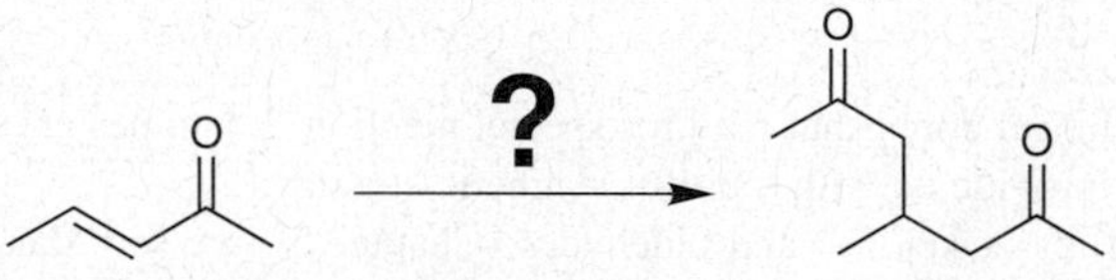

You decide that it should be simple. Your plan is to use a nucleophile that can attack an α,β-unsaturated ketone in a 1,4 addition:

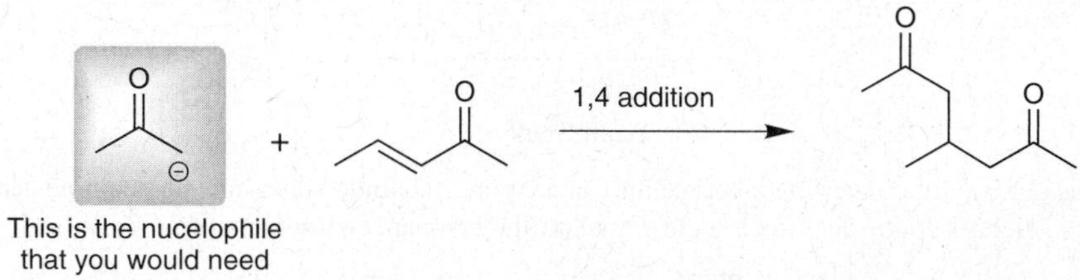

This is the nucelophile
that you would need

But, when you try to perform this reaction, you obtain a mixture of undesired products. Why? Because this enolate is **not** a Michael donor, and therefore, it will not efficiently attack in a 1,4-addition. So, how do you get around this problem? This is where our enamine comes in handy.

Instead of using an enolate as the nucleophile, suppose we convert the ketone into an enamine:

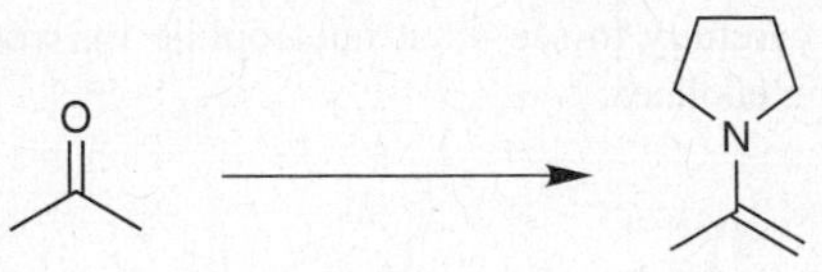

This enamine *is* a Michael donor, and it *will* attack cleanly in a 1,4-addition:

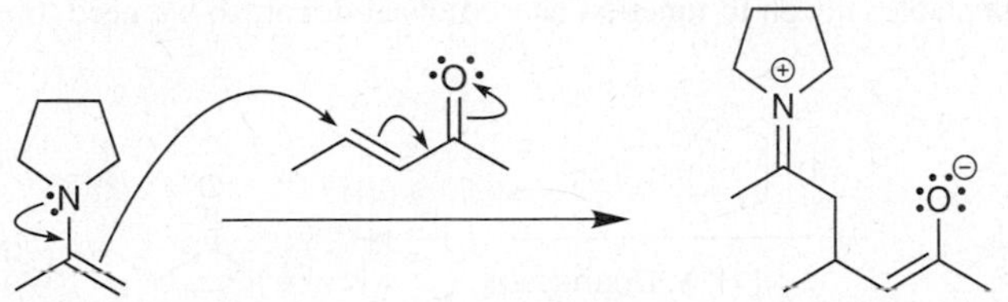

Finally, we use H_3O^+ to remove the iminium group and protonate the enolate (which then turns into a ketone):

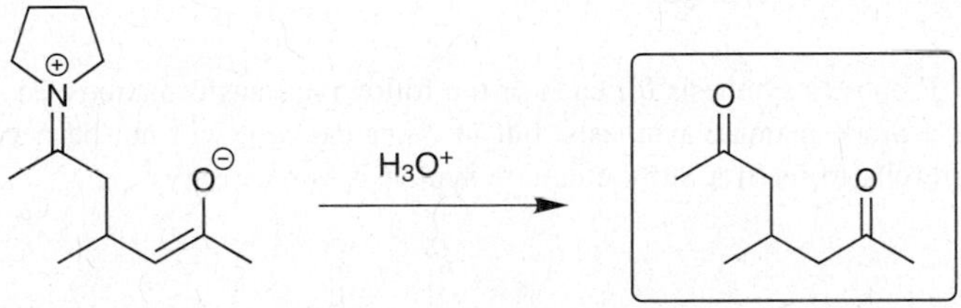

In the end, we prepared the desired product. This synthetic strategy is called the *Stork enamine synthesis*, and it can come in handy when you are proposing syntheses. Whenever you are trying to propose a synthesis, and you decide that you need an enolate to attack as a nucleophile in a 1,4-addition, you will have a problem. Regular enolates are not stable enough to be Michael donors. But, you can convert it into an enamine, which *is* stable enough to be a Michael donor. Then, you can remove the enamine in the end. The enamine serves as a way of temporarily modifying the reactivity of the enolate so that we can achieve the desired result. It is very clever when you really think about it.

EXERCISE 7.81 Propose a plausible synthesis for the following transformation:

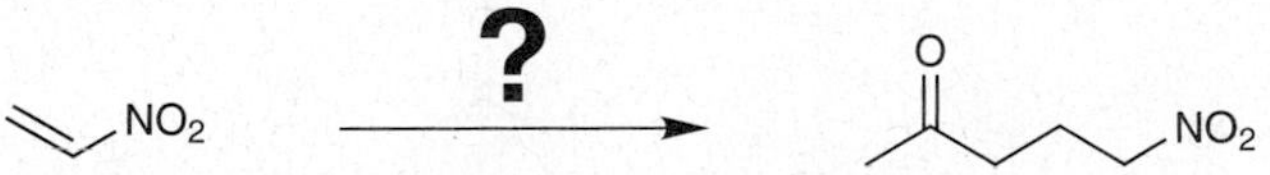

Answer When we inspect this transformation, we see that we need to install the following fragment:

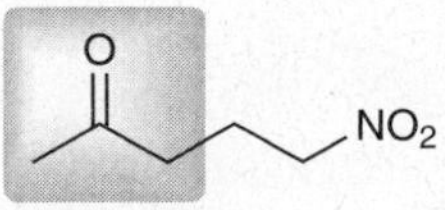

and at the same time, we need to remove the double bond that was in the starting material. We can accomplish both of these at the same time by performing a 1,4 addition with the appropriate nucleophile. When we look carefully to see what nucleophile we would need, we realize that we will need to use the following enolate:

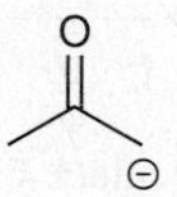

This enolate is **not** stable enough to function as a Michael donor, so we need to use a Stork enamine synthesis:

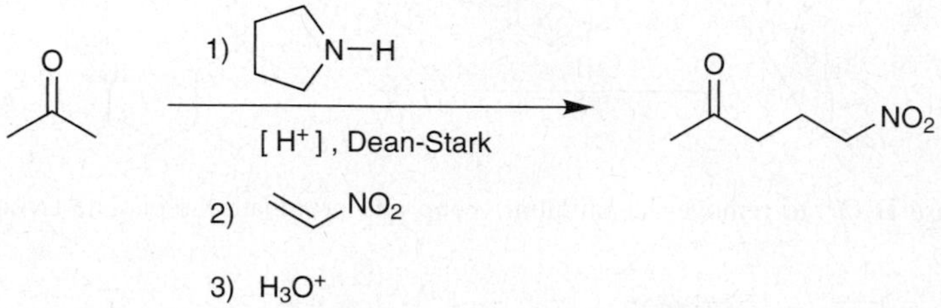

PROBLEMS Propose a synthesis for each of the following transformations. In some cases, you will need to use a Stork enamine synthesis, but in other cases, it will not be necessary. Analyze each problem carefully to see if a Stork enamine synthesis is necessary.

7.82

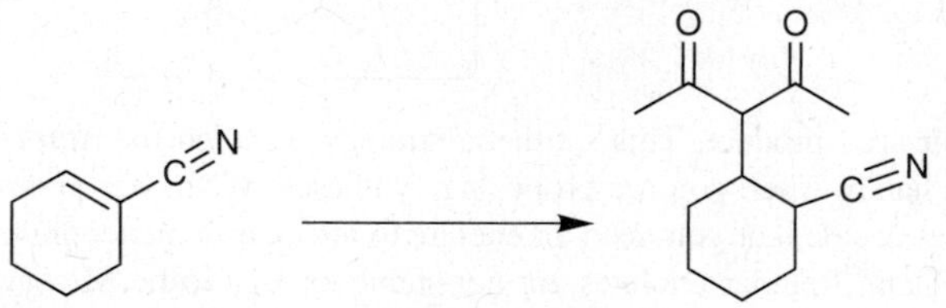

7.83

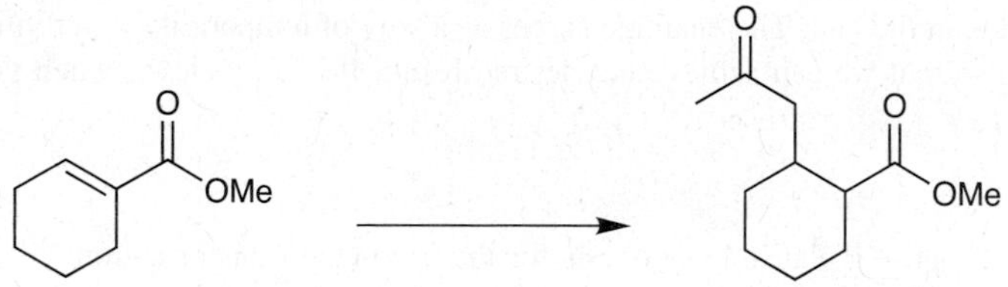

7.84

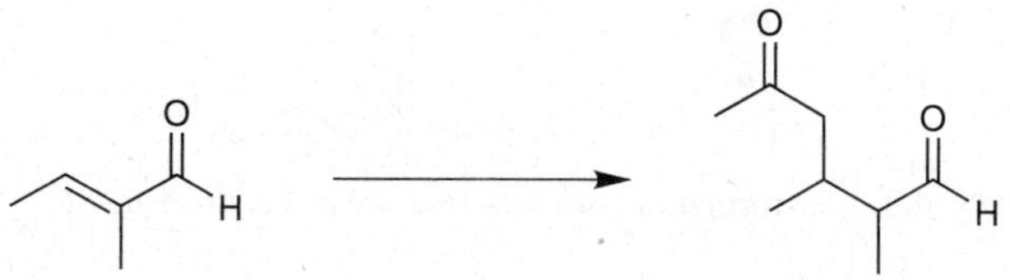

7.85

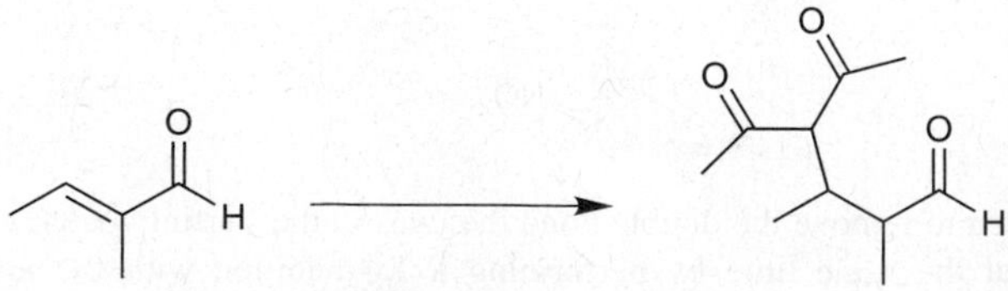

CHAPTER 8

AMINES

8.1 NUCLEOPHILICITY AND BASICITY OF AMINES

Amines are classified based on the number of alkyl groups attached to the central nitrogen atom:

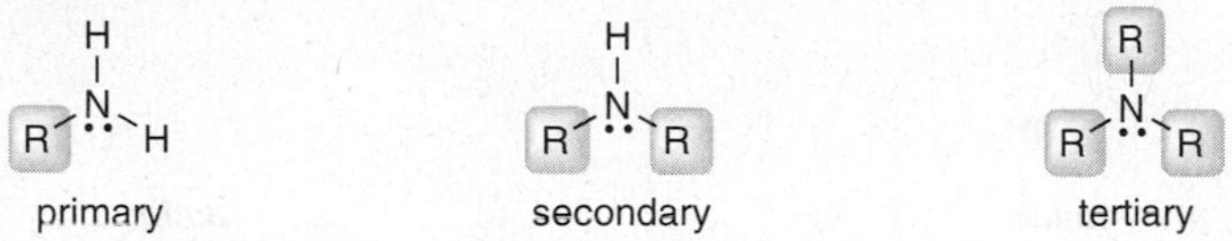

primary   secondary   tertiary

The reactivity of all amines derives from the presence of a lone pair on the nitrogen atom. All amines have this lone pair:

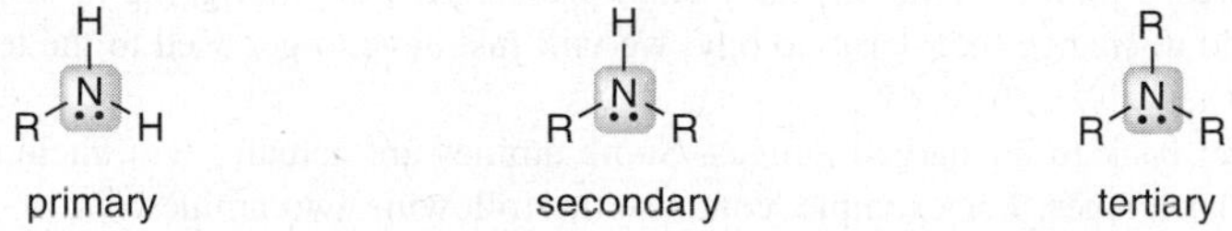

primary   secondary   tertiary

This lone pair can function as a nucleophile (attacking an electrophile):

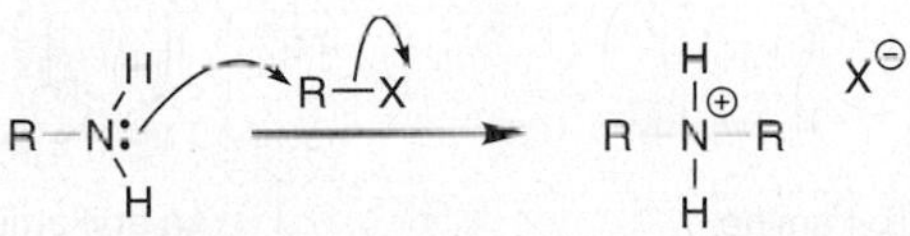

or it can function as a base (receiving a proton):

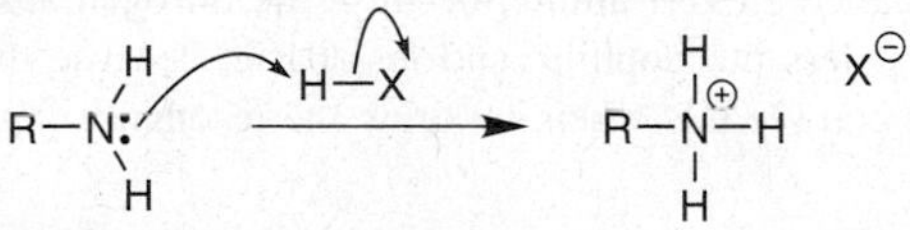

By focusing on this lone pair, we can understand why amines are good nucleophiles *and* good bases. When an amine participates in a reaction, the first step will always be one of these two possibilities: either the lone pair will receive a proton or the lone pair will attack an electrophile.

Of course, if we had a negative charge on the nitrogen atom, that would be even better. To get a negative charge on the nitrogen atom, we must deprotonate the amine. Tertiary amines don't have a proton that can be removed, but primary and secondary amines can be deprotonated to give the following anions:

These anions are stronger nucleophiles and stronger bases than uncharged amines. We call these structures *amides*. Here are two examples of amides that we have seen so far in this course:

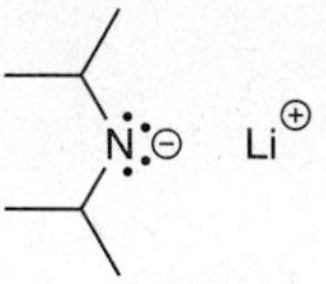

Lithium diisopropyl*amide*
(LDA)

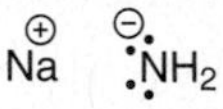

Sodium *amide*

The term *amide* is actually a terrible name, because we have already used this term to describe a type of carboxylic acid derivative:

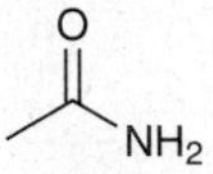

an *amide*

amide ion

Perhaps chemists could have assigned different names to these structures (so that it would be less confusing to students). But historically, chemists have used the term *amide* to refer to both of these types of structures. And old habits die hard. Since we are probably not going to convince chemists around the world to change their terminology, we will just have to get used to the terminology that is in use. Don't let this confuse you.

Now let's get back to uncharged amines. Some amines are actually less nucleophilic and less basic than regular amines. For example, compare the following two amines:

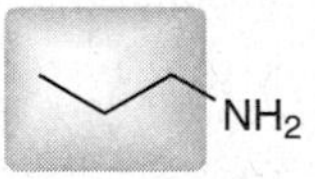

An alkyl amine

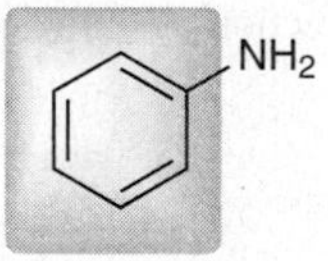

An aryl amine

The first amine is called an *alkyl* amine, because the nitrogen atom is connected to an alkyl group. The second compound is called an *aryl* amine, because the nitrogen atom is connected to an aromatic ring. Aryl amines are less nucleophilic and less basic, because the lone pair is delocalized into the aromatic ring. We can see this when we draw the resonance structures:

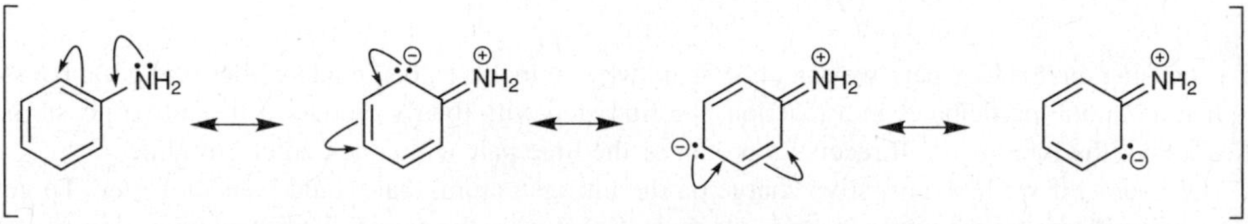

Since the lone pair is delocalized, it will be less available to function as a nucleophile or as a base. That does not mean that an aryl amine can't attack something. In fact, we will soon see a reaction where an aryl amine *is* used as a nucleophile. It *can* function as a nucleophile—but it is just *less* nucleophilic than an alkyl amine.

Now that we have had an introduction to amines, let's focus on a few ways to make amines.

8.2 PREPARATION OF AMINES THROUGH S$_N$2 REACTIONS

Suppose you wanted to make the following primary amine:

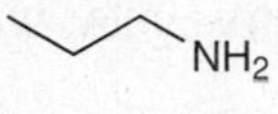

It might be tempting to suggest the following synthesis:

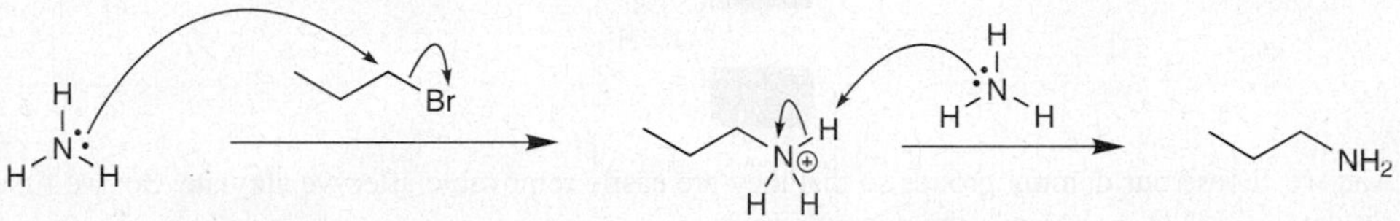

This is just an S$_N$2 reaction, followed by a deprotonation. This approach does work, BUT it is difficult to get the reaction to stop after monoalkylation. The product is also a nucleophile and it competes with NH$_3$ for the alkyl halide. In fact, it is an even better nucleophile than NH$_3$, because the alkyl group is electron donating. Therefore, it will attack again:

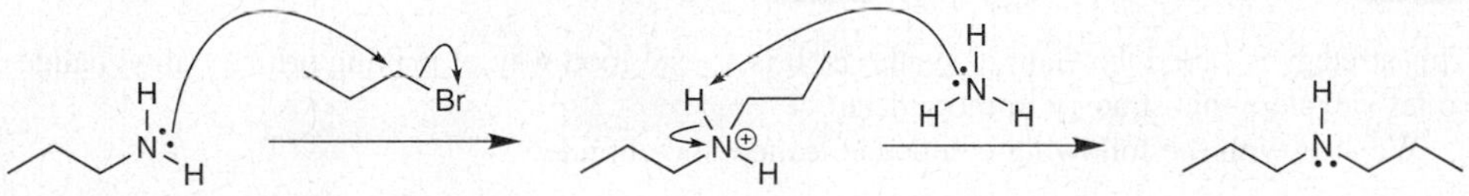

And then it attacks again:

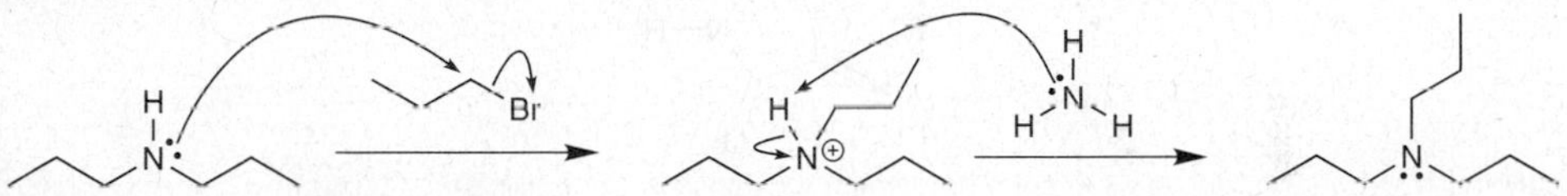

And then one last time:

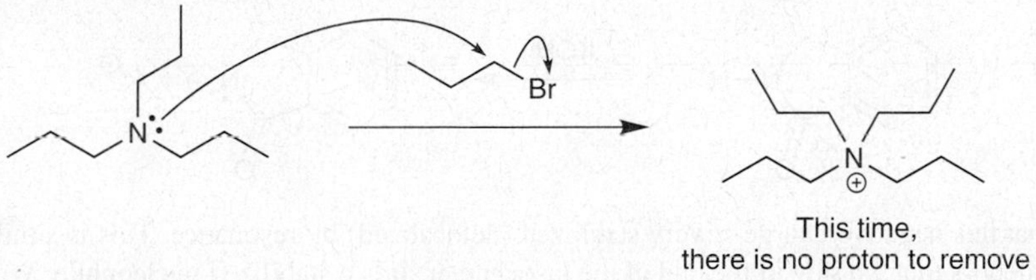

This time,
there is no proton to remove

The final product has four alkyl groups (which is referred to as *quaternary*), and the nitrogen has a positive charge (called an *ammonium* ion). So, we call this a *quaternary ammonium salt*. This structure does not have a lone pair, so it is neither nucleophilic nor basic.

If our intention is to make a quaternary ammonium salt, then the synthesis above is a good approach. But what if we want to make a primary amine? We cannot simply alkylate ammonia:

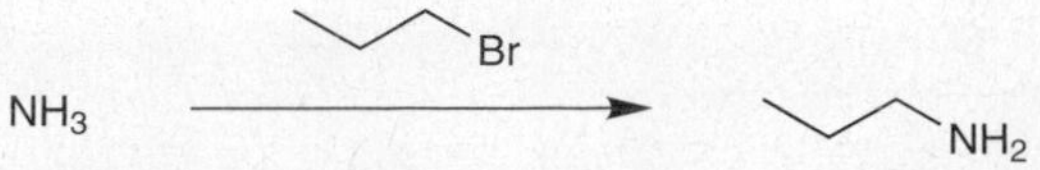

NH$_3$

because it is too difficult to stop the reaction at this stage. The alkyl halide will react again with the primary amine. Even if we try to use only one mole of alkyl halide and one mole of ammonia, we will still get a mixture of undesired products. We will get some polyalkylated products, and we will get some ammonia that could not find any alkyl halide to react with. So what do we do?

To get around this problem, we use a clever trick. We use a starting amine that already has two dummy groups on it:

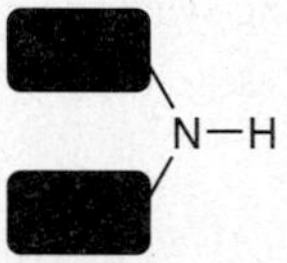

And we choose our dummy groups so that they are easily removable after we alkylate. So, we first alkylate, and then we remove the dummy groups:

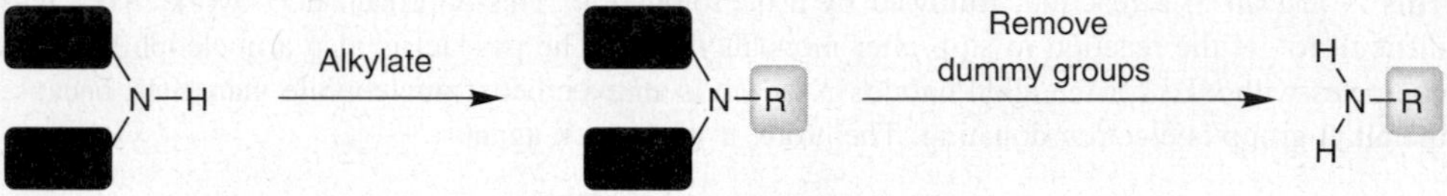

This strategy is called the Gabriel synthesis. It is a very good way of making primary alkyl halides, so let's explore this strategy in more detail.

We start with the following compound, called phthalimide:

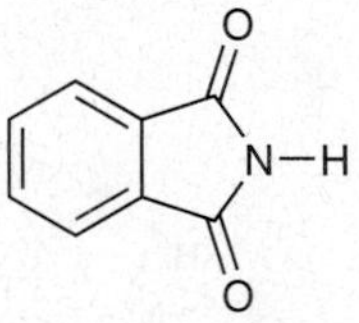

We use a base (KOH) to remove the proton, which gives the following anion:

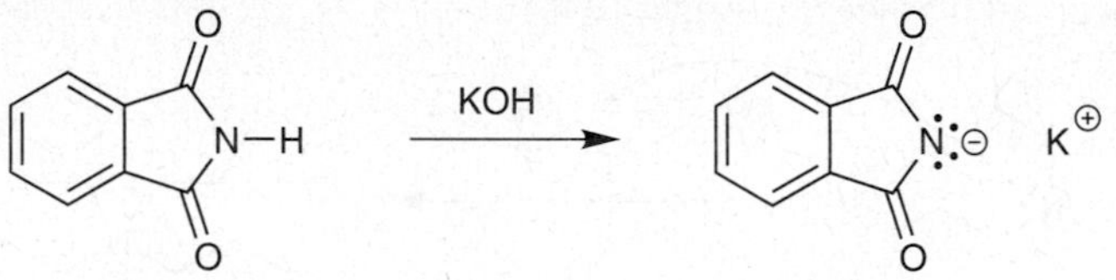

Notice that this negative charge is very stabilized (delocalized) by resonance. This is similar to the Michael donors that we saw in the end of the last chapter. It is a stabilized nucleophile. We use this nucleophile to attack an alkyl halide:

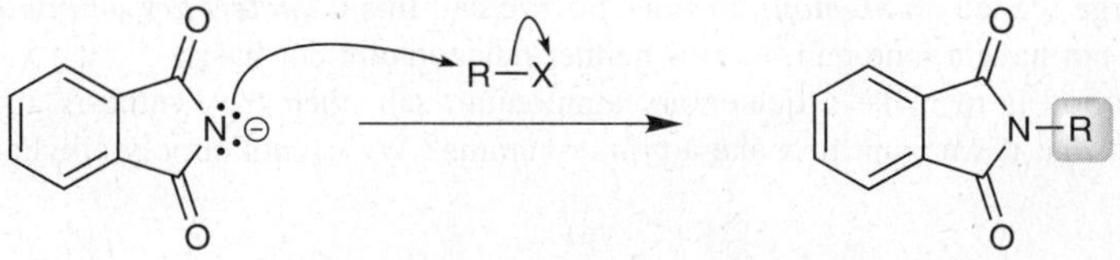

This reaction proceeds via an S$_N$2 process, so tertiary alkyl halides cannot be used. Acid-catalyzed or base-catalyzed hydrolysis is then performed to release the amine. Acidic conditions are more common than basic conditions.

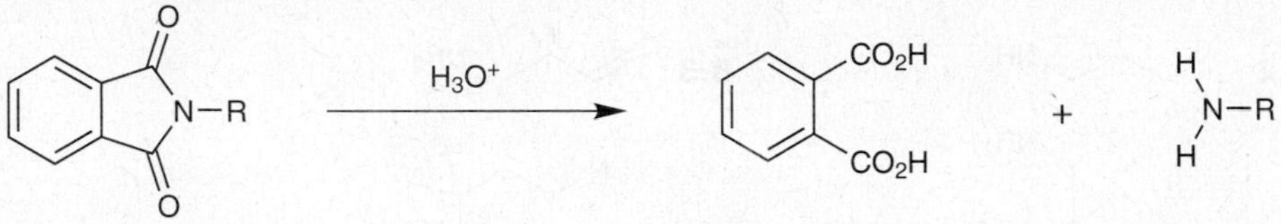

The mechanism of hydrolysis is directly analogous to the hydrolysis of amides, as seen in Section 6.6. The hydrolysis step is slow, and many alternative approaches have been developed. One such alternative employs hydrazine to release the amine:

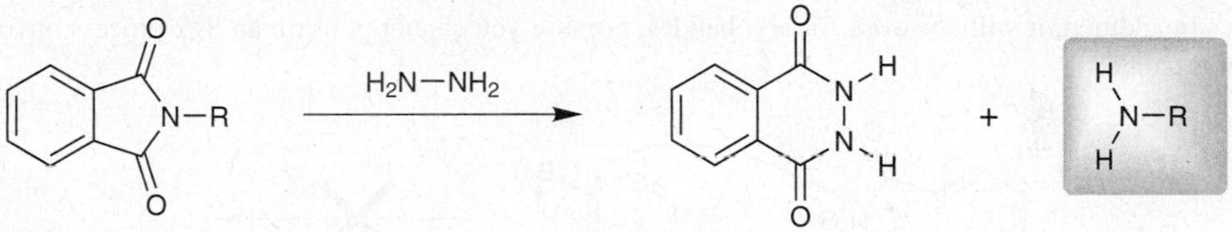

In this step, the dummy groups are removed, generating the desired product.

The Gabriel synthesis is very useful for making primary amines from primary alkyl halides:

EXERCISE 8.1 Identify how we would use a Gabriel synthesis to prepare the following amine:

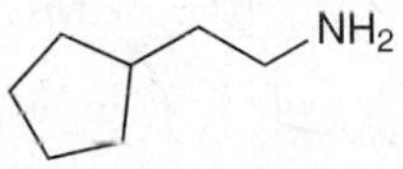

Answer In order to perform a Gabriel synthesis, there is only one thing we need to determine. We must identify the alkyl halide that is necessary. To do this, we just draw a halogen instead of the NH$_2$, like this:

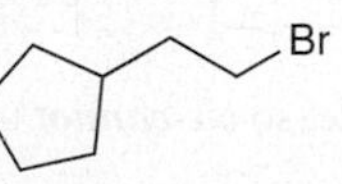

Now that we know what alkyl halide to use, we are ready to propose our synthesis:

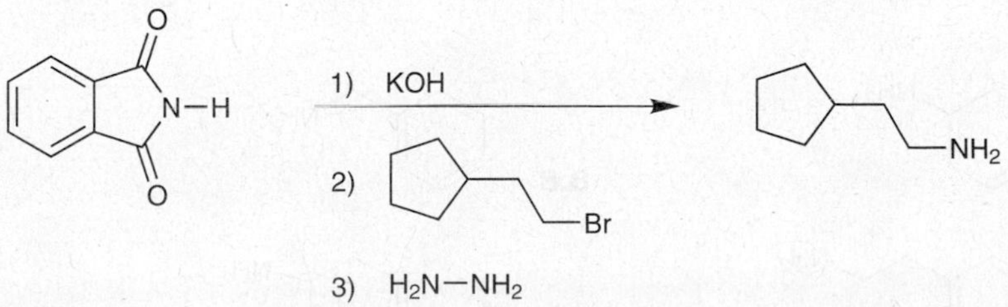

1) KOH

2)

3) H$_2$N—NH$_2$

PROBLEMS Identify how you would use a Gabriel synthesis to prepare each of the following compounds.

8.2

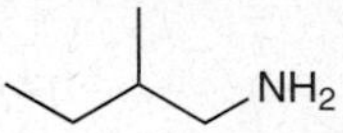

8.3

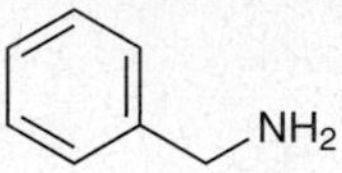

8.4

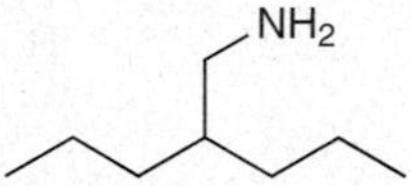

8.5

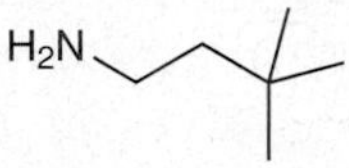

The Gabriel synthesis has its limitations, though. Since it relies on an S_N2 reaction, it will work well with **_primary_** alkyl halides. But it is not so great for secondary alkyl halides, and it will not work at all for tertiary alkyl halides.

In addition, it will not work for aryl halides, because you cannot perform an S_N2 process on an aryl halide:

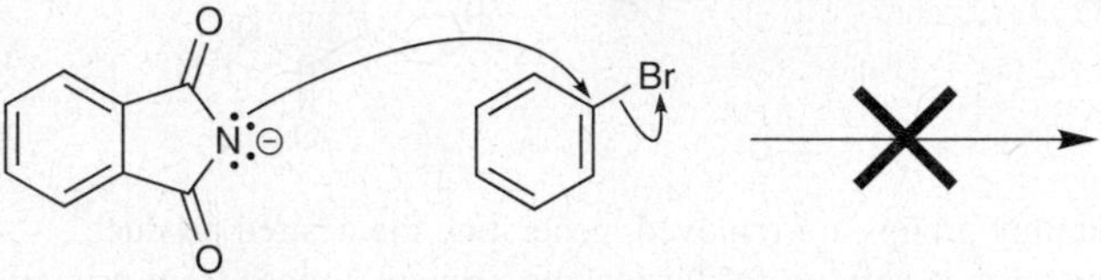

EXERCISE 8.6 Is it possible to prepare the following compound with a Gabriel synthesis?

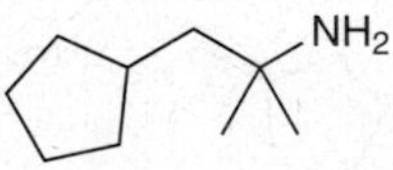

Answer We draw the alkyl halide that would be necessary,

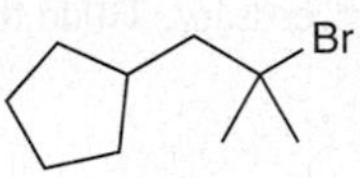

and we see that it is a tertiary alkyl halide, so we **_cannot_** use a Gabriel synthesis to make the desired product in this case.

PROBLEMS Identify whether each of the following compounds could be made with a Gabriel synthesis:

8.7

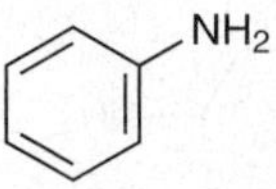

8.8

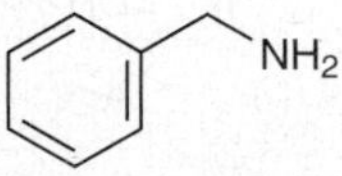

8.9

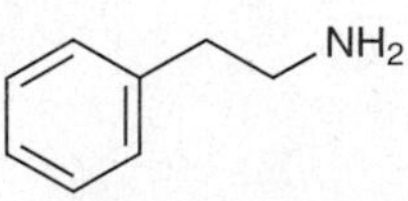

8.10

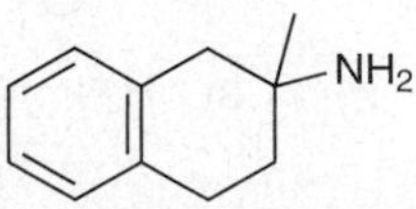

8.3 PREPARATION OF AMINES THROUGH REDUCTIVE AMINATION

In the previous section, we learned how to make amines via an S_N2 process. That method was best for making primary amines. In this section, we will learn a way to make secondary amines, using a two-step synthesis (where the first step is a reaction that we have already seen). When we learned about ketones and aldehydes, we saw how to make *imines*:

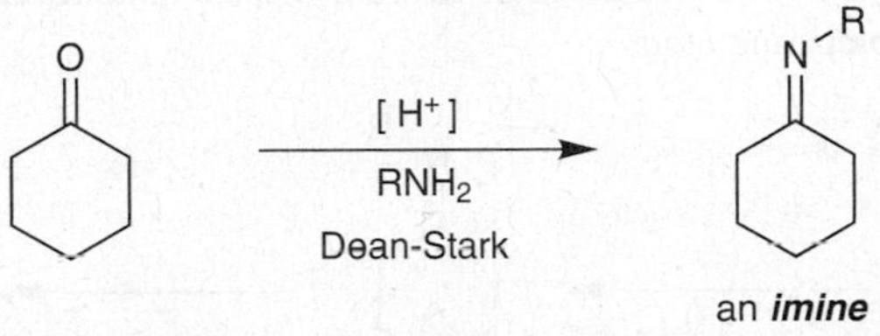

an *imine*

We saw that a primary amine can react with a ketone (under acidic conditions) to give an imine. Now, we will use this reaction to make amines.

When we form an imine, we are forming the essential C-N connection that you need in order to have an amine:

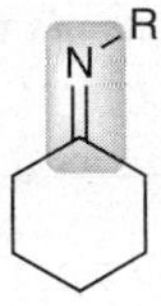

But, the oxidation state is not correct. In order to get an amine, we need to do the following conversion:

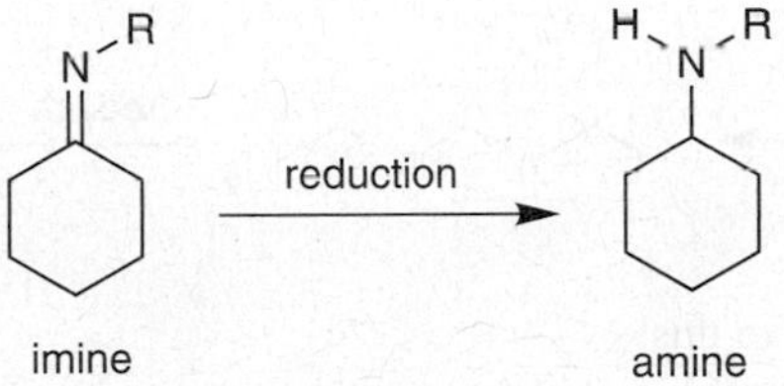

imine   amine

In order to convert an *i*mine into an *a*mine, we will need to perform a reduction. One way to do so is to reduce the imine in much the same way that a ketone is reduced, using LAH:

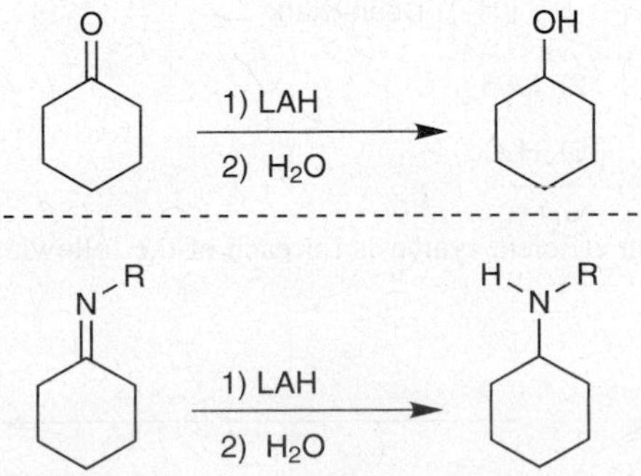

Alternatively, the C=N bond can undergo hydrogenation (in the presence of a catalyst):

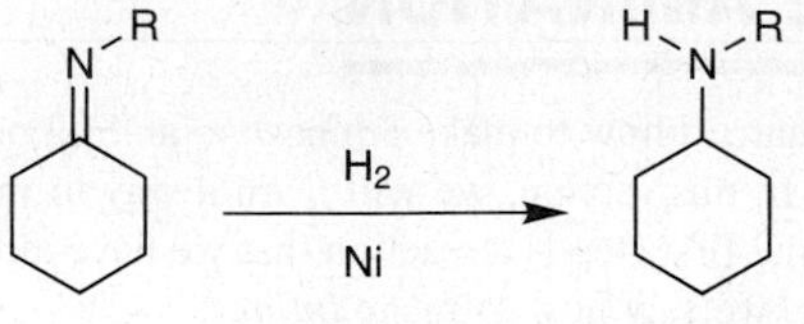

There are many other ways to reduce an imine as well. But the bottom line is that we now have a two-step synthesis for preparing amines:

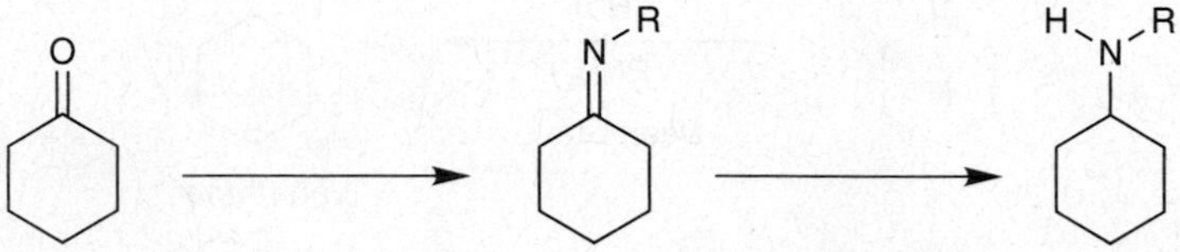

This process is called **reductive amination**, because we are forming an amine (a process called *amination*) via a *reduction*.

EXERCISE 8.11 Suggest an efficient synthesis for the following transformation:

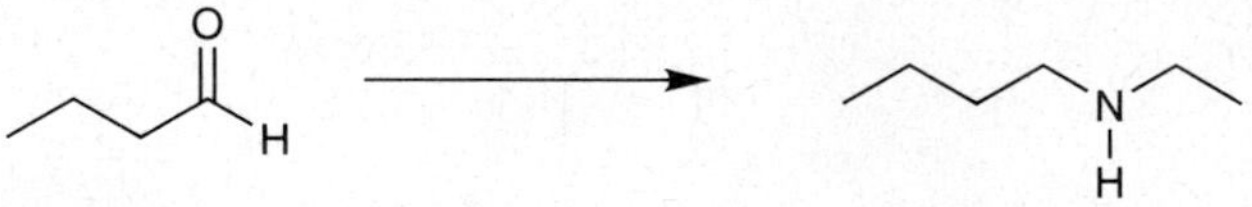

Answer This problem involves the conversion of an aldehyde into an amine. This should alert us to the possibility of a reductive amination. If we used a reductive amination, our strategy would go like this:

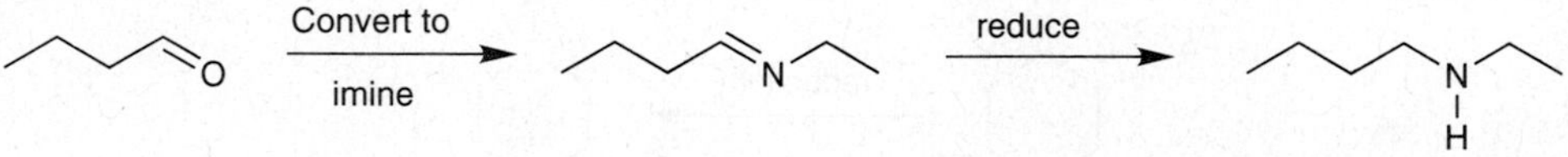

So, our synthesis will look like this:

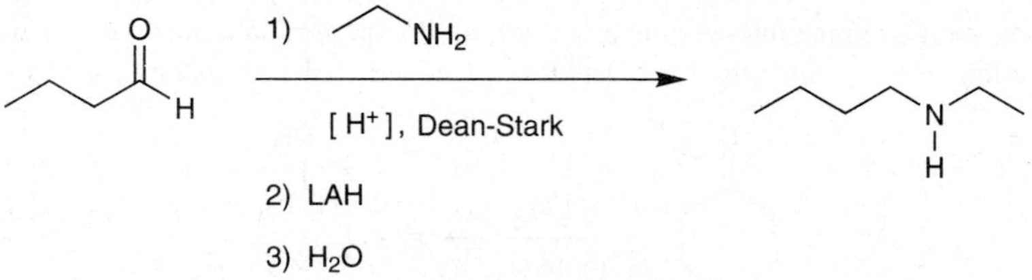

1) ⌃⌃NH₂

[H⁺], Dean-Stark

2) LAH

3) H₂O

PROBLEMS Suggest an efficient synthesis for each of the following transformations.

8.12

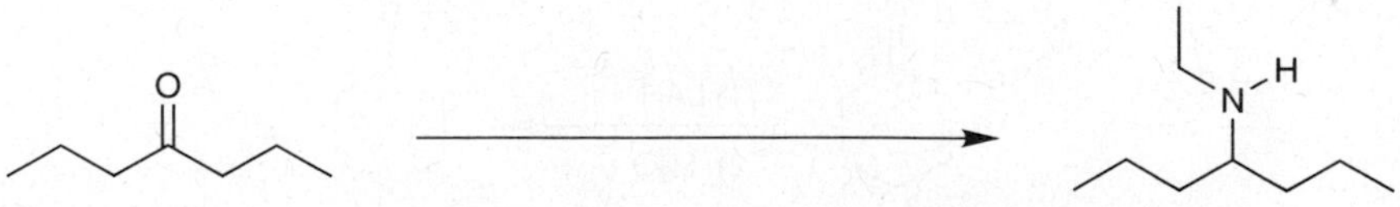

8.13

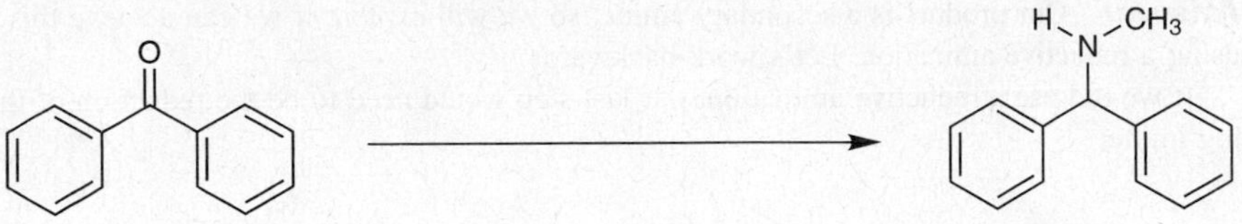

8.14

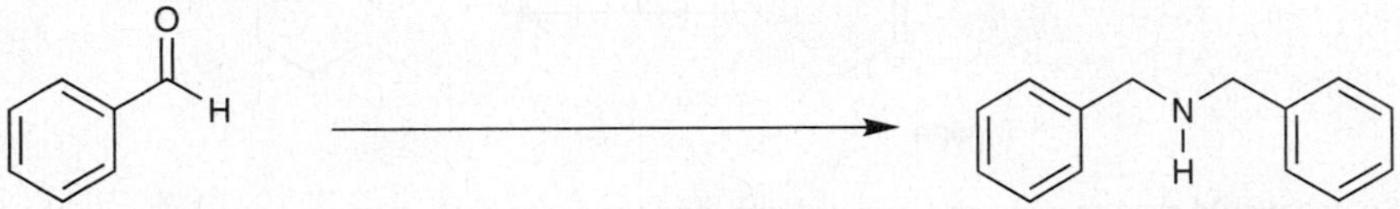

8.15

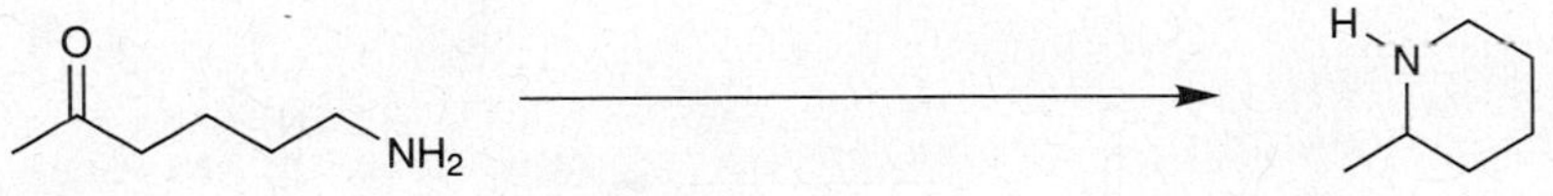

8.16

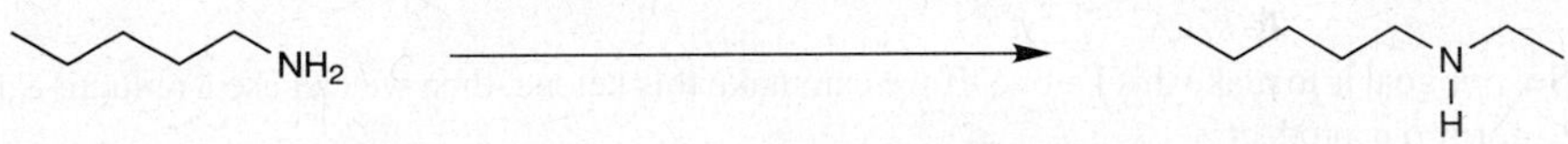

Reductive amination is a useful technique, because the starting material is a ketone or aldehyde. And we have seen many ways to make ketones. This gives us a way to make amines from a variety of compounds:

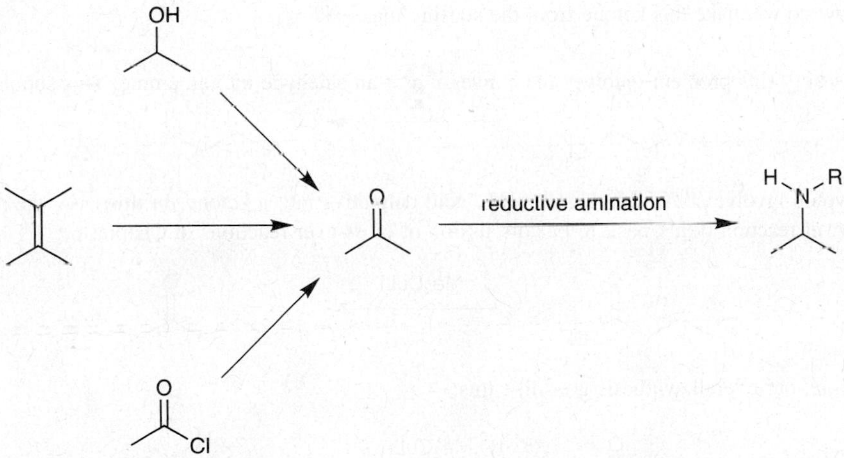

These reactions were covered in Chapter 5. Students often have difficulty combining reactions from different chapters to propose a synthesis, so let's get some practice:

EXERCISE 8.17 Suggest an efficient synthesis for the following transformation:

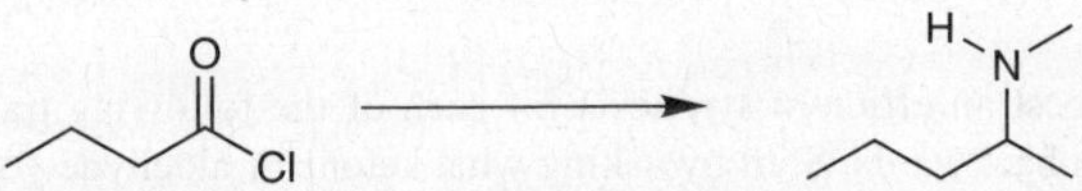

Answer Our product is a secondary amine, so we will explore if we can achieve this synthesis using a reductive amination. Let's work backwards.

If we did use a reductive amination, our last step would need to be the reduction of the following imine:

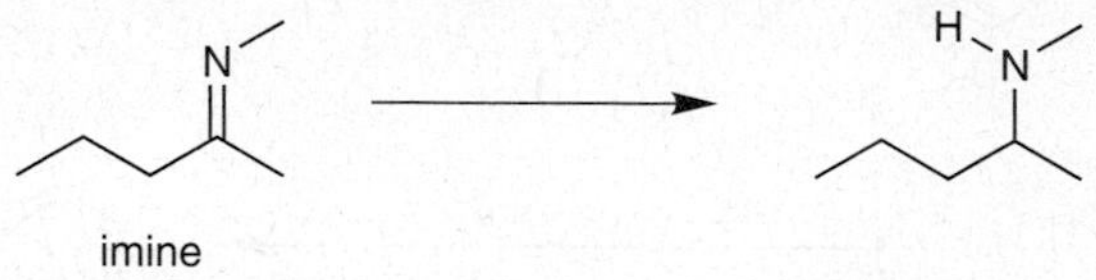

imine

So, we would need to make the imine above, and we could have done that starting with the following ketone:

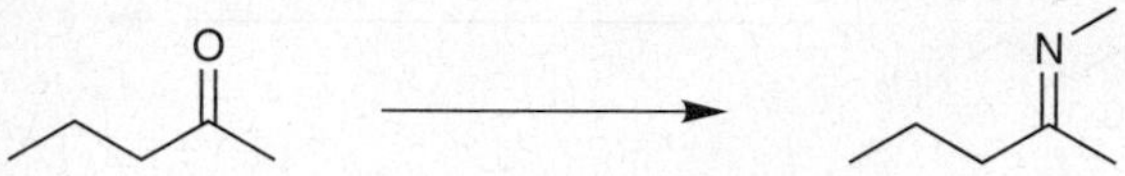

So, our goal is to make this ketone. If we can make this ketone, then we can use a reductive amination to form our product:

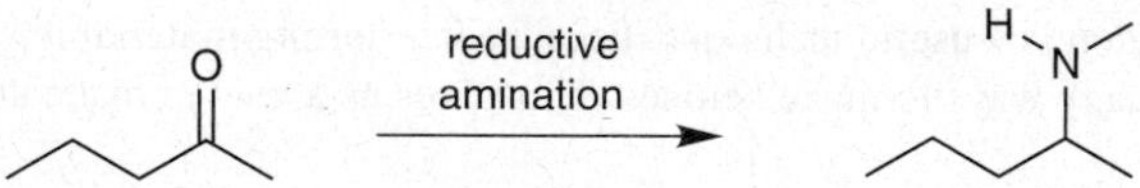

reductive
amination

But how do we make this ketone from the starting material?

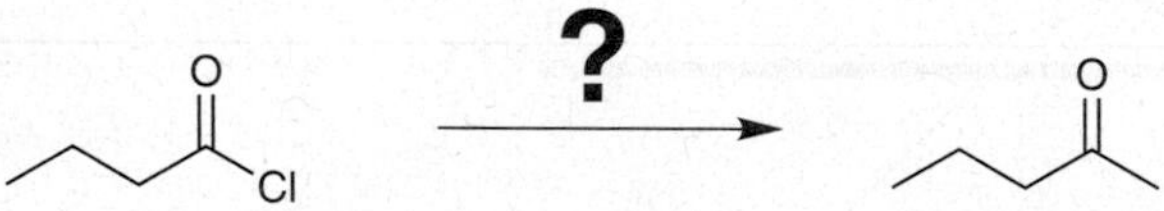

?

This would involve converting a carboxylic acid derivative into a ketone. In this case, we need a cross-over reaction (think back to our discussion of cross-over reactions in Chapter 6):

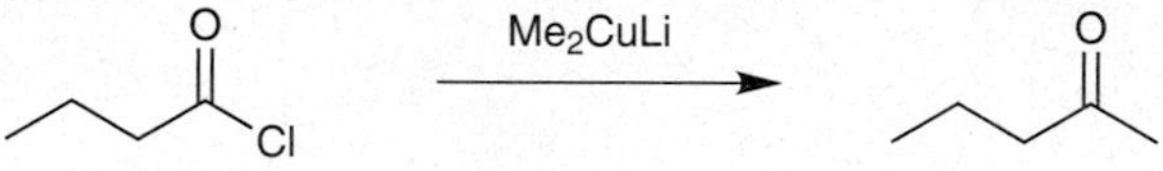

Me₂CuLi

Therefore, our overall synthesis goes like this:

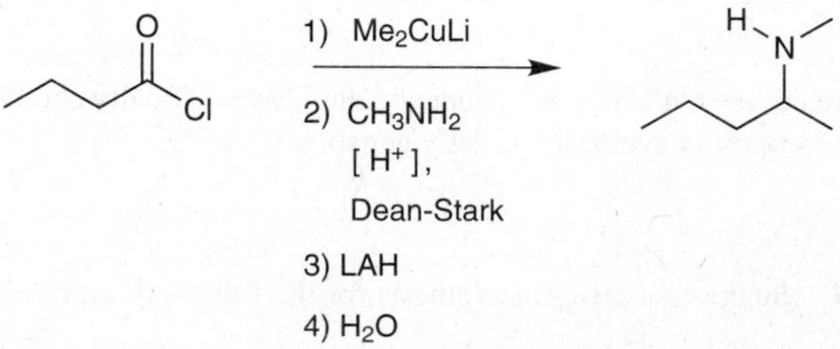

1) Me₂CuLi
2) CH₃NH₂
   [H⁺],
   Dean-Stark
3) LAH
4) H₂O

PROBLEMS Suggest an efficient synthesis for each of the following transformations. In each case, you should work backwards. Start by asking what ketone or aldehyde you would need in order to make the desired product via a reductive amination. Then, ask yourself how you could make that ketone from the starting material.

8.18

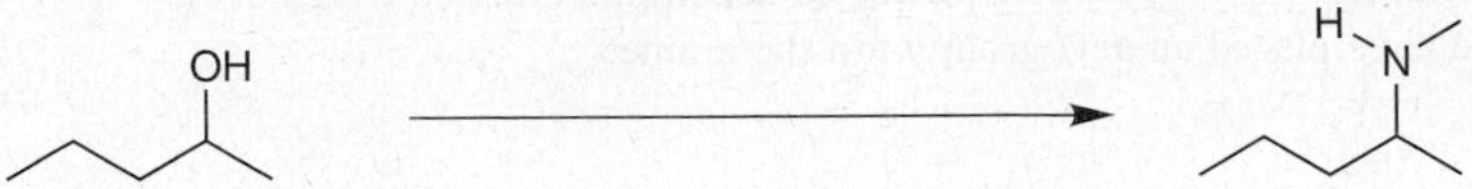

8.19

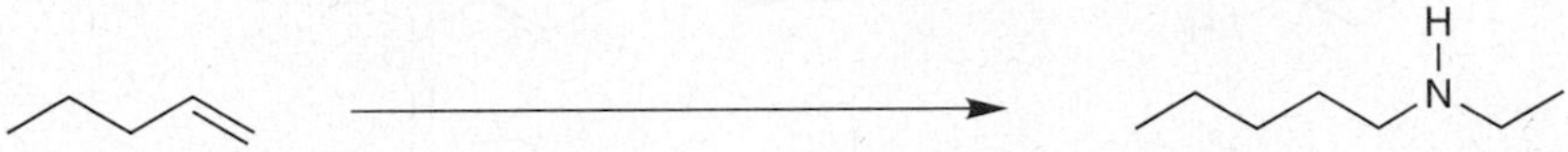

8.20

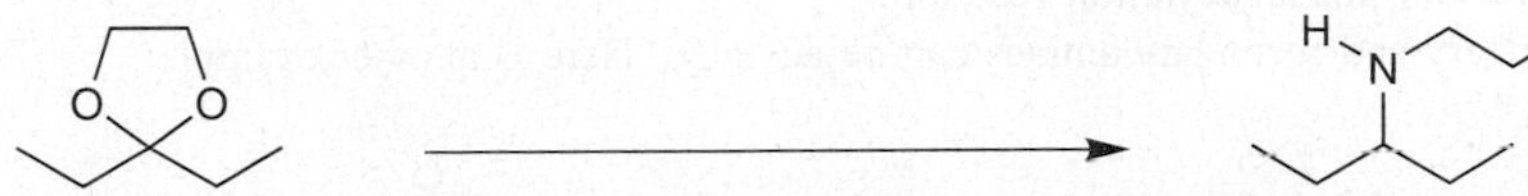

8.21

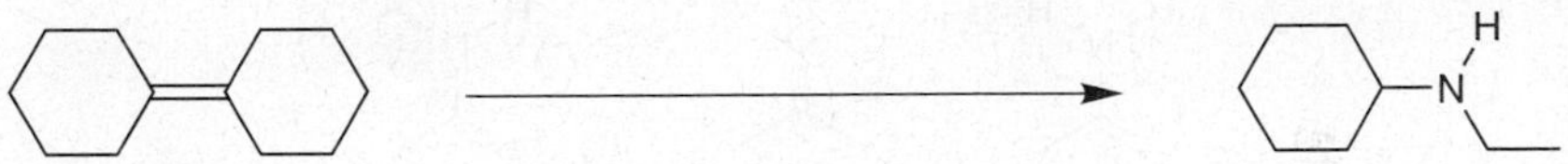

8.22

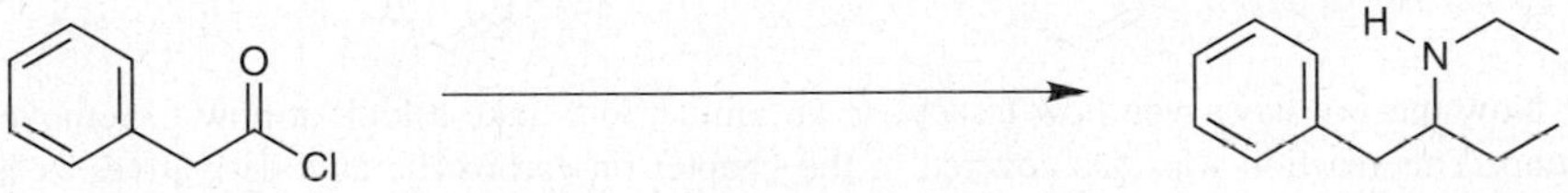

8.4 ACYLATION OF AMINES

So far in this chapter, we have focused on ways of *making* amines. For the rest of the chapter, we will shift our focus. We will now explore reactions of amines.

We will begin our survey with a reaction that we have actually already seen in a previous chapter. When we learned about carboxylic acid derivatives (Chapter 6), we saw that you can convert an acid halide into an amide. For example:

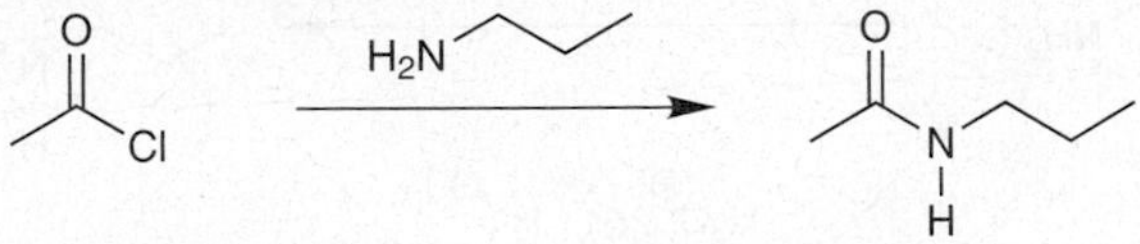

When we draw it this way (with the amine over the reaction arrow), the focus is on what happens to the acid halide (it is converted into an amide). But what if we choose to focus on the amine instead? In other words, let's rewrite the same reaction a bit differently. Let's put the acid halide on top of the reaction arrow, like this:

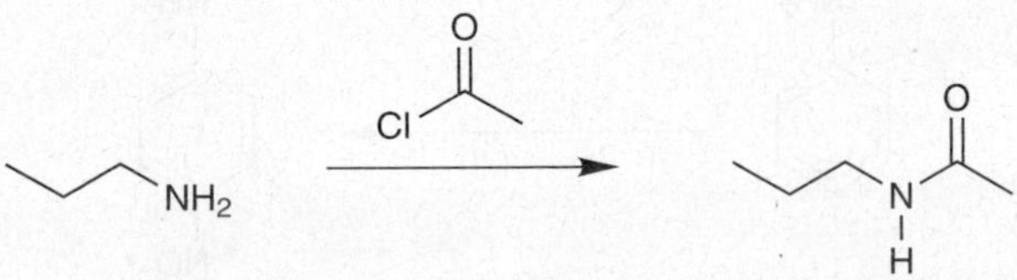

We have not changed the reaction at all. It is still the same reaction (an amine reacting with an acid halide). But when we draw it like this, our attention focuses on converting the *amine* into an *amide*. The acid halide is just the reagent that we use to accomplish this conversion.

Overall, we have placed an *acyl* group onto the amine:

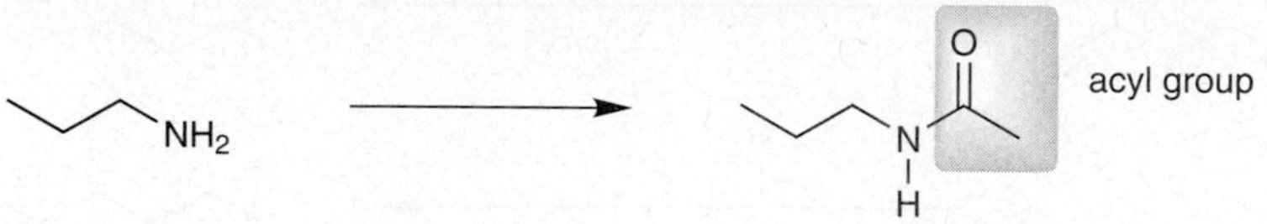

Therefore, we call this an *acylation* reaction.

Most primary and secondary amines can be acylated. Here is another example:

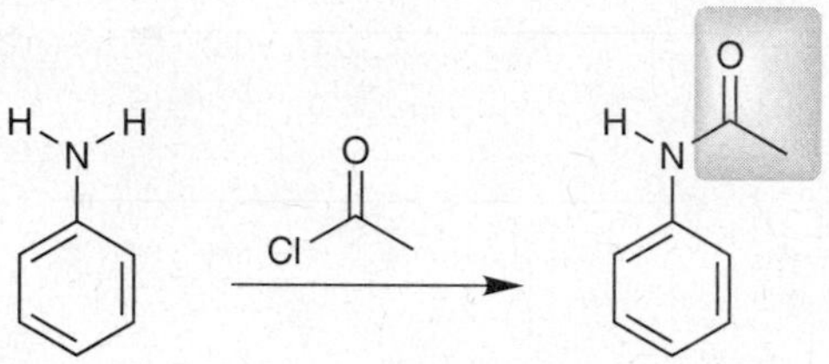

Now that we have seen how to acylate an amine, let's take a look at how to remove the acyl group. This reaction was also covered in the chapter on carboxylic acid derivatives. It is just the hydrolysis of an amide:

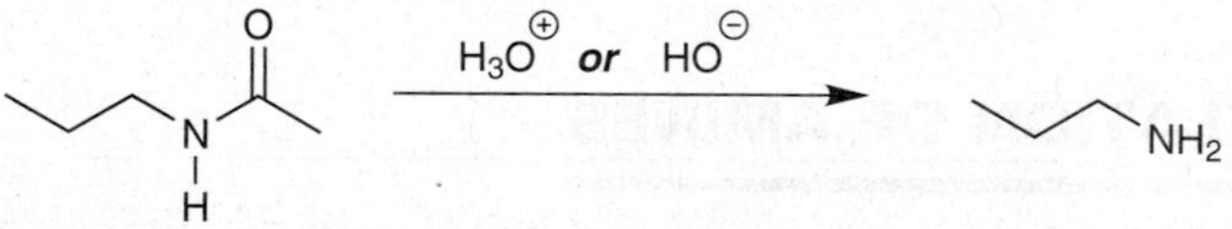

Notice that we have removed the acyl group to regenerate the amine. So, now we know how to install an acyl group, and we know how to remove it:

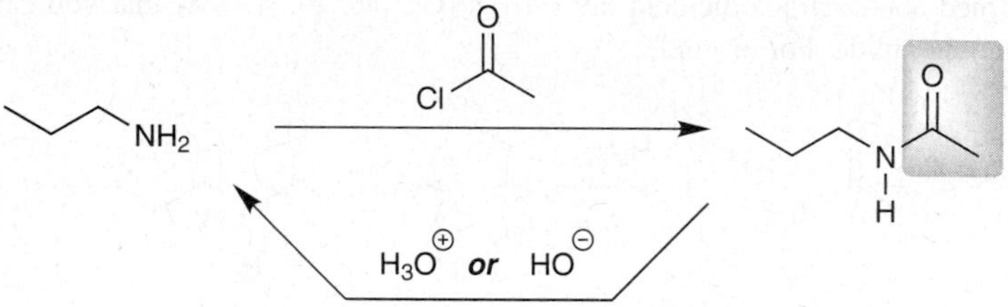

But the obvious question is: why would we want to do that? Why would we ever install a group, just to remove it later? The answer to this question is very important, because it illustrates a common strategy that organic chemists use. Let's try to answer this question through a specific example.

Imagine that we want to achieve the following transformation:

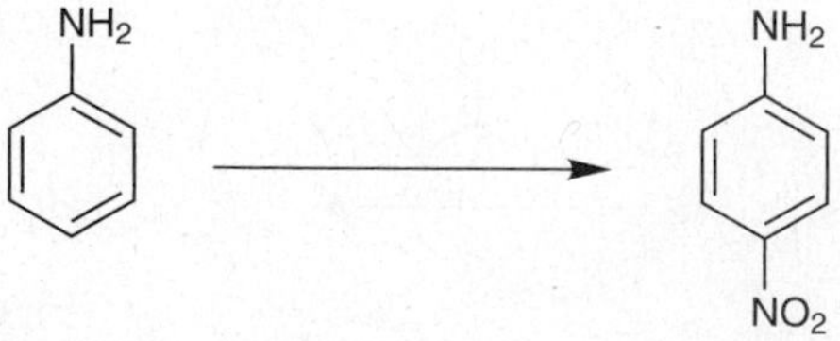

This seems easy to do. Do you remember how to install a nitro group on an aromatic ring (Chapter 3). We just used a mixture of nitric acid and sulfuric acid. The amino group is an activator, so it will direct to the *ortho* and *para* positions, with a preference for *para* substitution (for steric reasons). So, we propose the following:

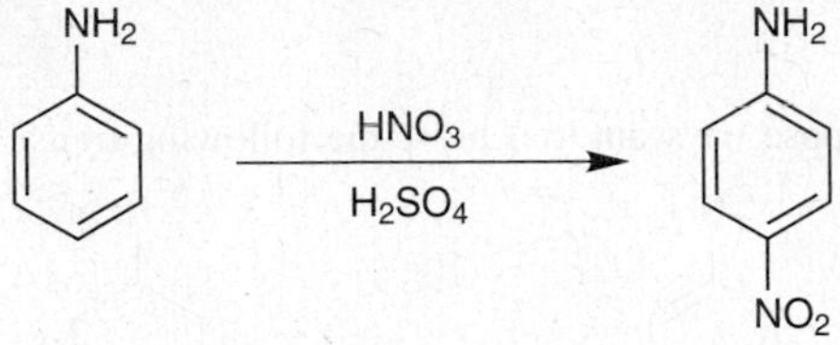

But when we try to perform this reaction, we find that it does not work. It is true that the amino group is a strong activator. The problem is that it is *too strong* of an activator. In this case, a highly activated ring is being exposed to a very strong oxidizing agent. The ring is so highly activated that the mixture of nitric acid and sulfuric acid will produce undesired oxidation reactions. The ring will be oxidized, destroying aromaticity. This is certainly not a good thing if you just want to install a nitro group on the ring. So, how can we generate the desired product?

The way to do it is actually very clever. We first acylate the amino group:

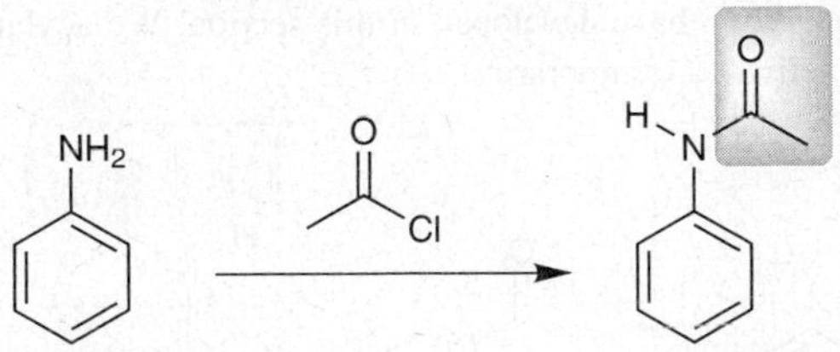

This converts the amino group (which is a *very strong* activator) into a *moderate activator*. Now that it is a moderate activator, we no longer observe the undesired oxidation reactions. The ring will undergo nitration without any problems:

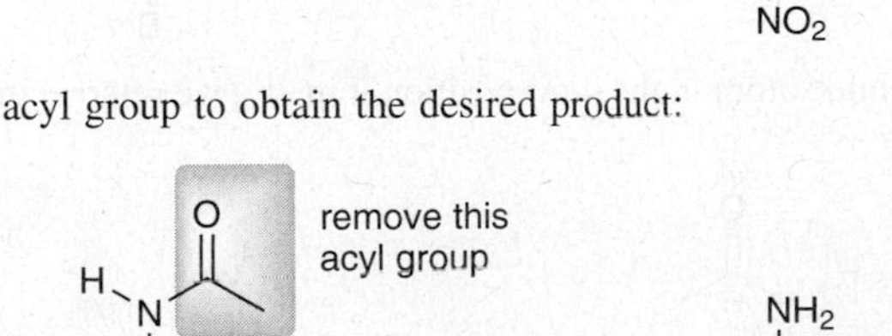

Then, we remove the acyl group to obtain the desired product:

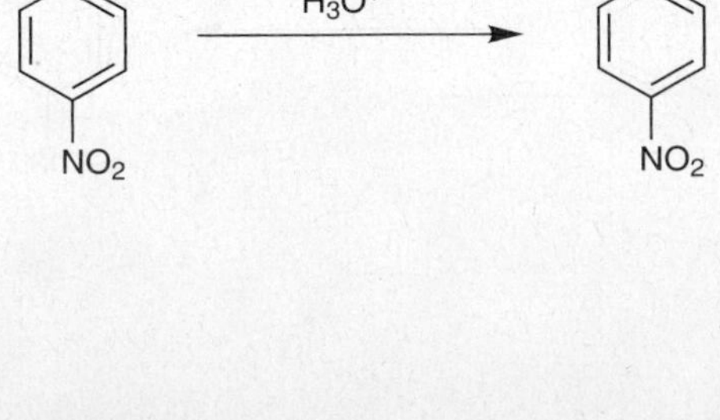

Think about what we just did. We used an acylation process as a way of **temporarily modifying** the electronics of the amino group, so that it would not interfere with the desired reaction. This strategy is used all of the time by organic chemists. This idea of temporarily modifying a functional group (and then converting it back later) is an idea that is used in many other situations as well (not just for acylation of amines).

EXERCISE 8.23 Suppose we want to achieve the following transformation:

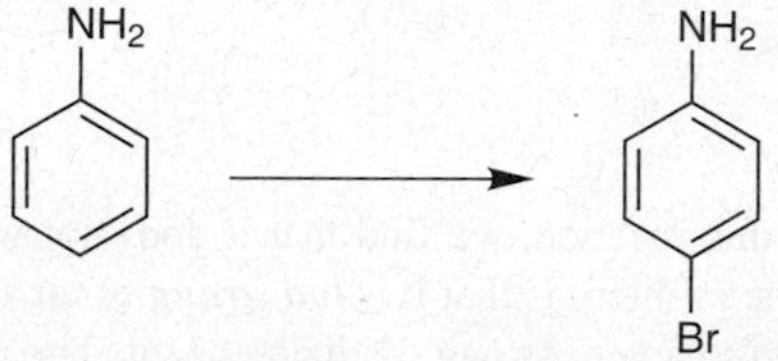

We try to perform this transformation using Br_2, but we find that aniline is too reactive, and we get a mixture of mono-, di-, and tri-brominated products. What can we do to obtain the desired product and avoid polybromination?

Answer The problem is the amino group. It is too strongly activating. To circumvent this obstacle, we use the strategy we have developed in this section. We acylate the amino group, and that makes the ring less activated (temporarily):

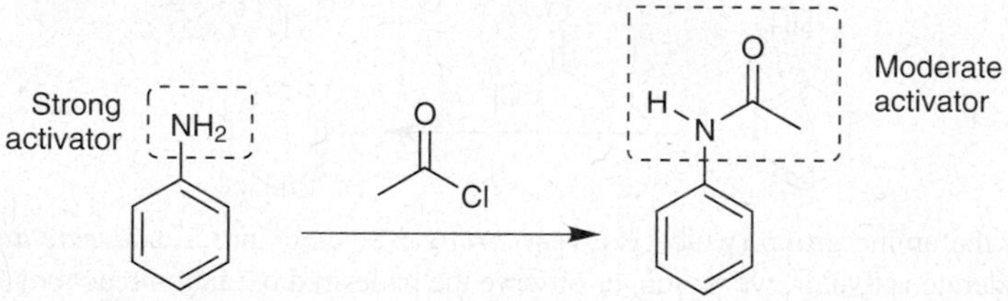

Now we are able to brominate:

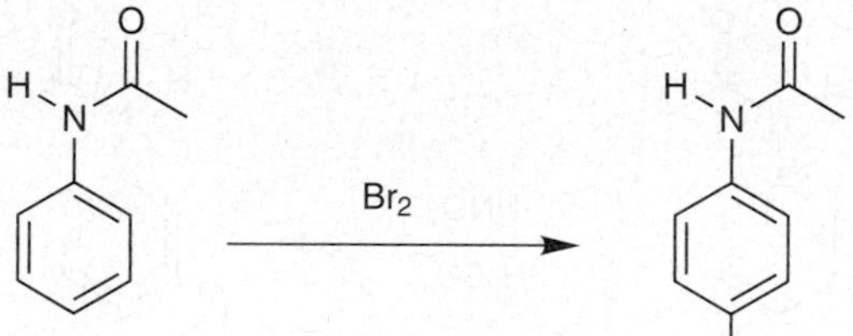

and that installs one bromine atom in the *para* position. Finally, we remove the acyl group to obtain the desired product:

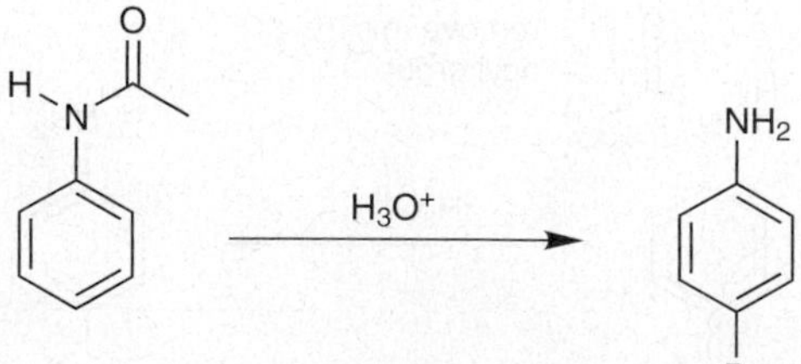

PROBLEMS   Propose an efficient synthesis for each of the following transformations:

8.24

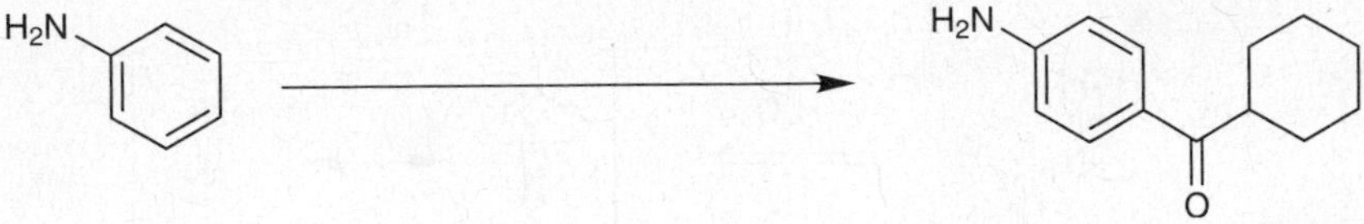

H₂N

⟶

H₂N

8.25

H₂N

⟶

H₂N

Cl

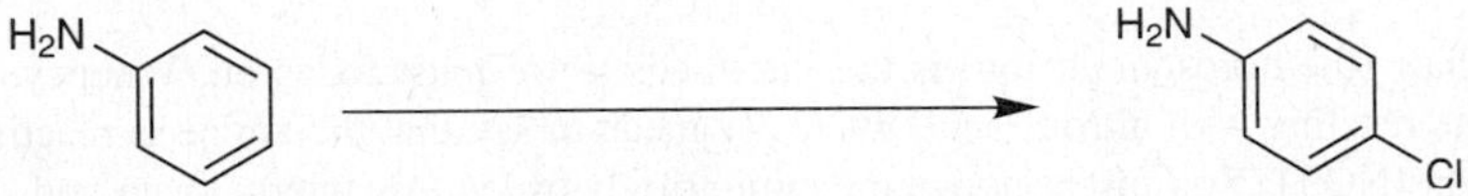

8.5   REACTIONS OF AMINES WITH NITROUS ACID

In this section, we will begin to explore the reactions that take place between amines and nitrous acid. Compare the structures of nitr*ous* acid and nitr*ic* acid:

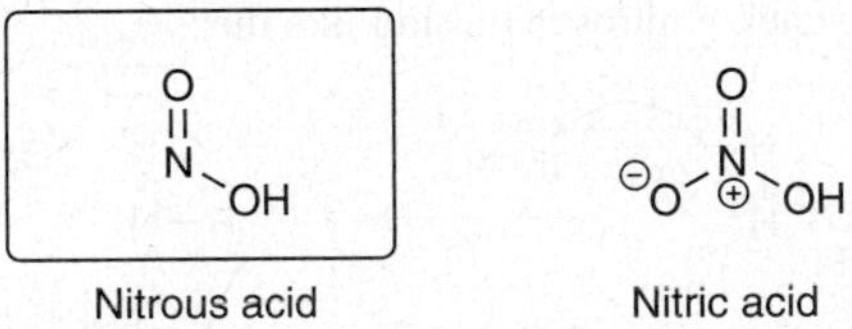

Nitrous acid

Nitric acid

When nitrous acid reacts with amines, the products are very useful. We will soon see that these products can be used in a large number of synthetic transformations. So, let's make sure that you are comfortable with the reactions between amines and nitrous acid.

Let's start by looking at the source of nitrous acid. It turns out that nitrous acid is fairly unstable and, therefore, we cannot just purchase it. You won't find it stored in a bottle. Rather, we have to make nitrous acid in the reaction flask. To do this, we use sodium nitrite ($NaNO_2$) and HCl:

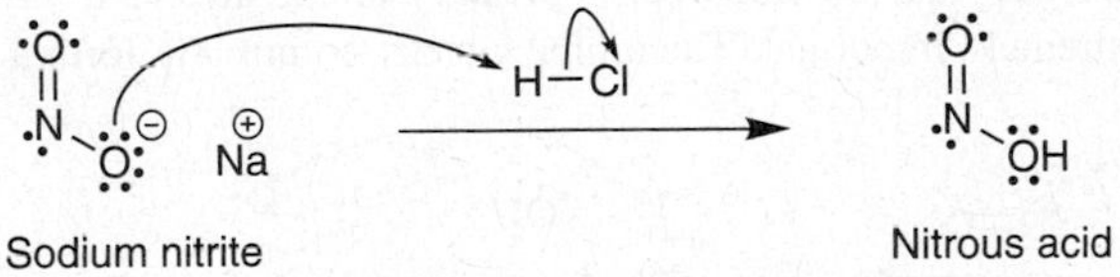

Sodium nitrite

Nitrous acid

Under these (acidic) conditions, nitrous acid is protonated again, to produce a positively charged intermediate:

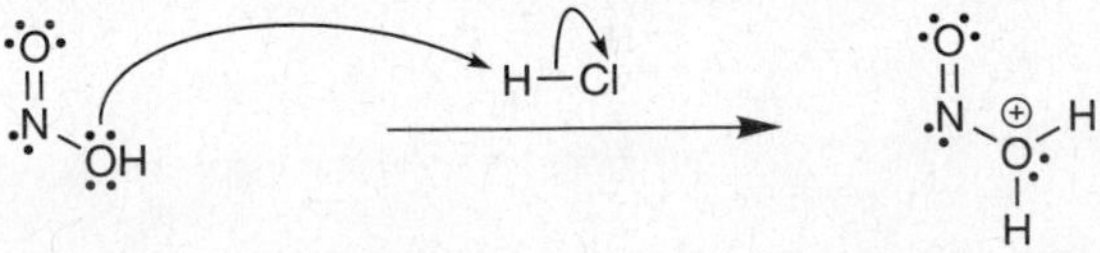

This intermediate can then lose water to give a highly reactive intermediate, called a nitrosonium ion:

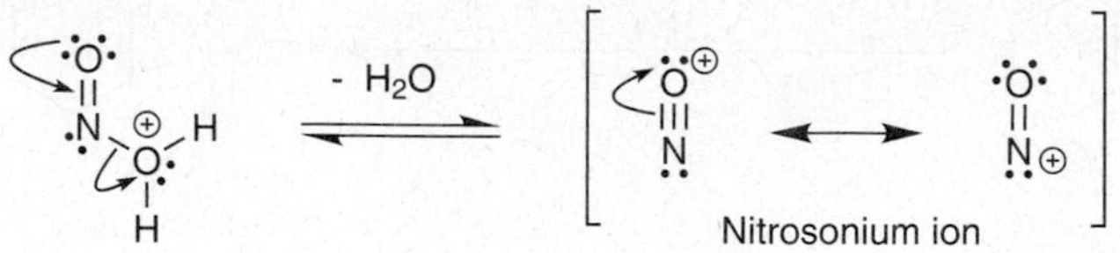

This intermediate (the nitrosonium ion) is the intermediate we must focus on. Whenever we talk about an amine reacting with nitrous acid, we really mean to say that the amine is reacting with a nitrosonium ion (NO^+). You might notice the similarity between this intermediate and the NO_2^+ intermediate (that we used in nitration reactions). Do not confuse these two intermediates. NO^+ and NO_2^+ are different intermediates. In this section, we are only talking about the reactions of amines with the nitrosonium ion (NO^+).

As we said a few moments ago, nitrosonium ions cannot be stored in a bottle. Instead, we must make them *in the presence of an amine*. That way, as soon as the nitrosonium ion is formed, it will immediately react with the amine before it has a chance to do anything else. This is called an *in situ* preparation.

So, now the question is: what happens when an amine reacts with a nitrosonium ion? Let's begin by exploring secondary amines (and then we will explore primary amines).

A secondary amine can attack a nitrosonium ion like this:

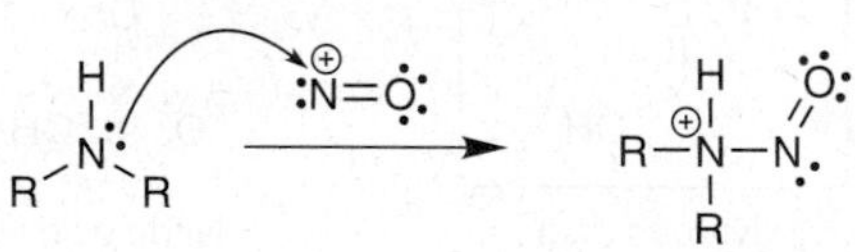

Deprotonation then generates the product:

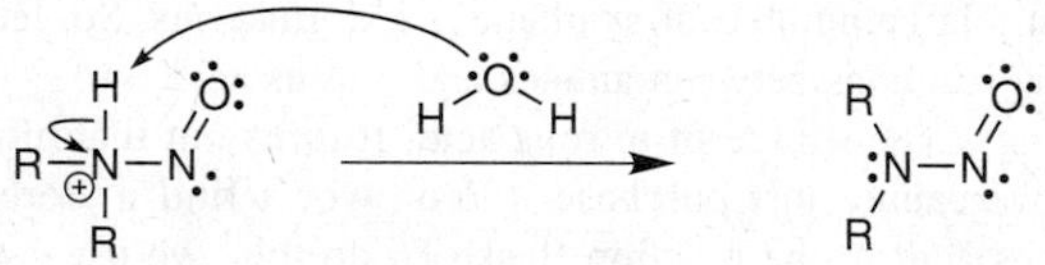

This product is called an *N-nitroso amine*. For short, chemists often call it a ***nitrosamine***.

This reaction is not very useful. But when a ***primary*** amine attacks a nitrosonium ion, the resulting reaction is extremely important. The amine attacks, to initially form a nitrosamine:

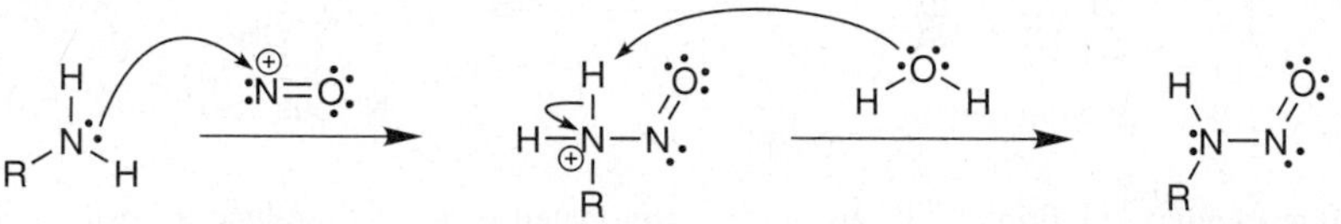

Since we started with a primary amine, we notice that we have a proton in our nitrosamine:

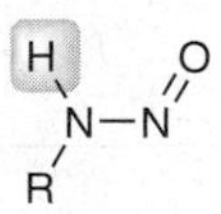

Because of this proton, the nitrosamine continues to react in the following way:

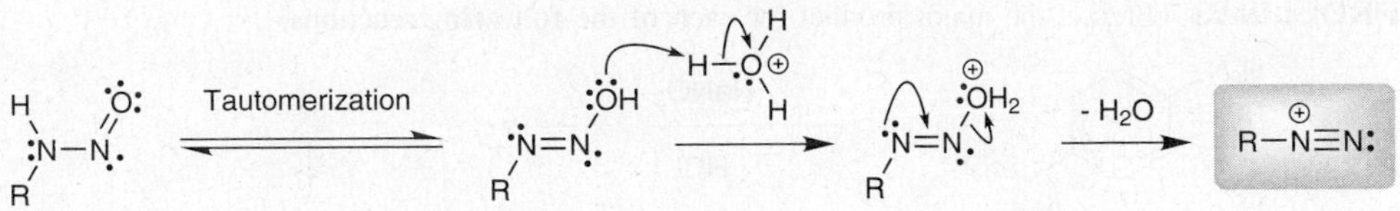

This product is called a *diazonium* ion. The term *azo* means nitrogen, so *diazo* means two nitrogen atoms. And, of course, *onium* means a positive charge. That is what we have here: two nitrogen atoms connected to each other, and a positive charge; thus, the name *diazonium*.

Primary *alkyl* amines will give *alkyl* diazonium salts, and primary *aryl* amines will give *aryl* diazonium salts:

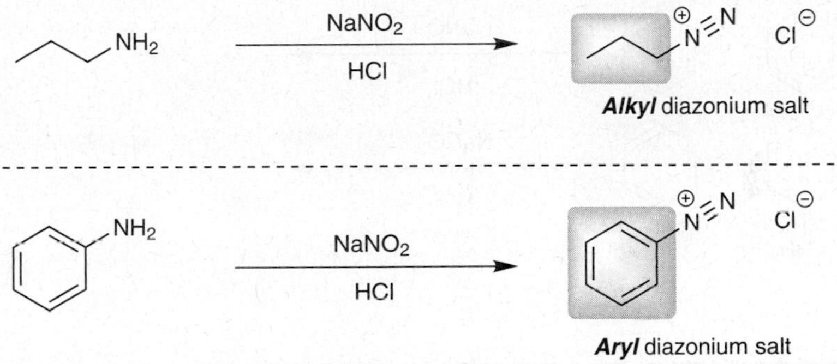

Alkyl diazonium salt

Aryl diazonium salt

Alkyl diazonium salts are not terribly useful. They are very explosive, and as a result, they are very dangerous to prepare. But *aryl* diazonium salts are much more stable, and they are incredibly useful, as we will see in the upcoming section. For now, let's just make sure that we know how to make diazonium salts:

EXERCISE 8.26 Predict the product of the following reaction:

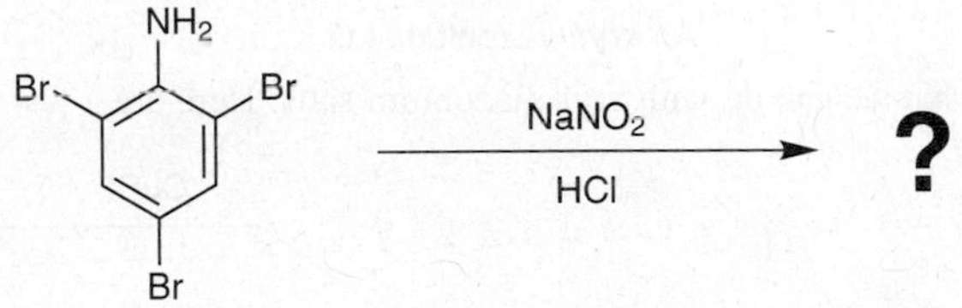

Answer Our starting material is a primary amine. These reagents (sodium nitrite and HCl) are used to form nitrous acid, which then forms a nitrosonium ion. Primary amines react with a nitrosonium ion to give a diazonium salt. So, the product of this reaction is:

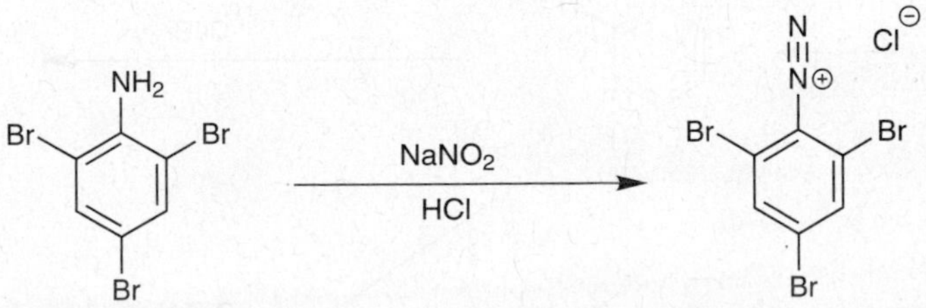

PROBLEMS   Predict the major product for each of the following reactions:

8.27

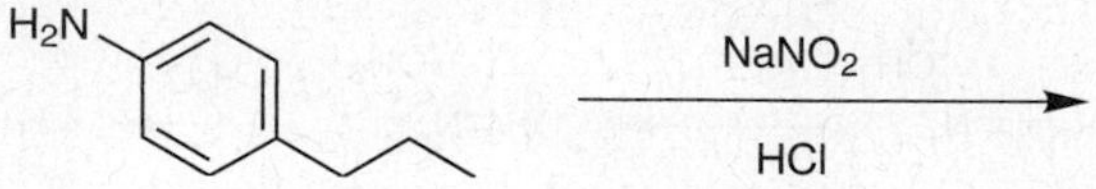

H₂N — [benzene ring with propyl group]   →   NaNO₂ / HCl

8.28

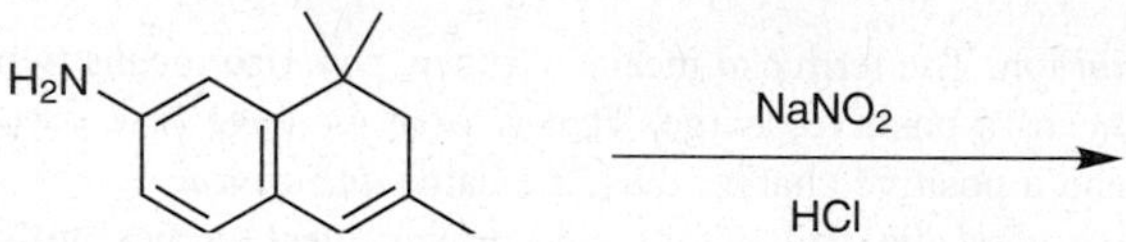

H₂N — [fused bicyclic structure with methyl groups]   →   NaNO₂ / HCl

8.29

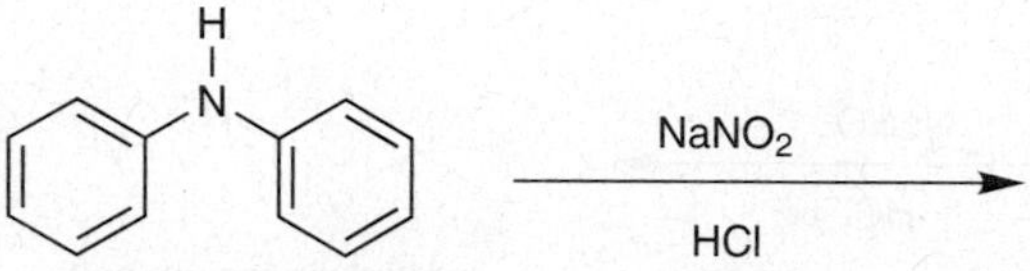

[diphenylamine structure with N–H]   →   NaNO₂ / HCl

8.30

[2-benzylaniline structure with NH₂]   →   NaNO₂ / HCl

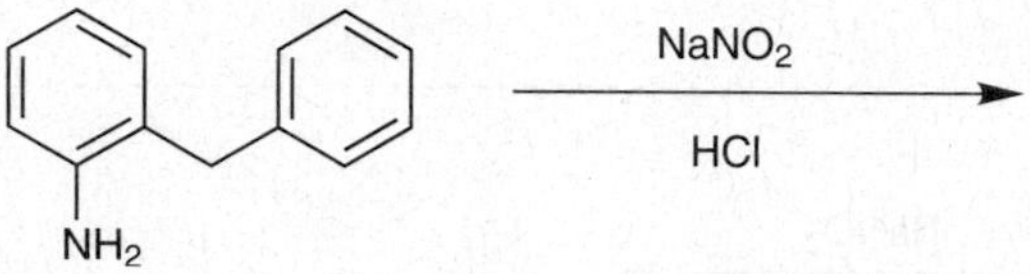

8.6   AROMATIC DIAZONIUM SALTS

In the previous section, we learned how to make aryl diazonium salts:

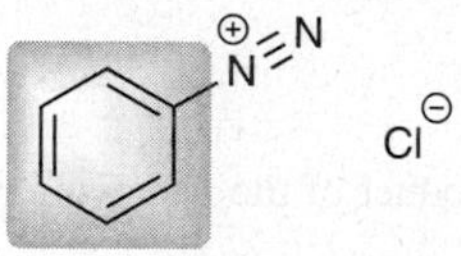

[structure: benzene ring with N⁺≡N group and Cl⁻]

An *aryl* diazonium salt

Now we will learn what we can do with aryl diazonium salts. Here are a just few reactions:

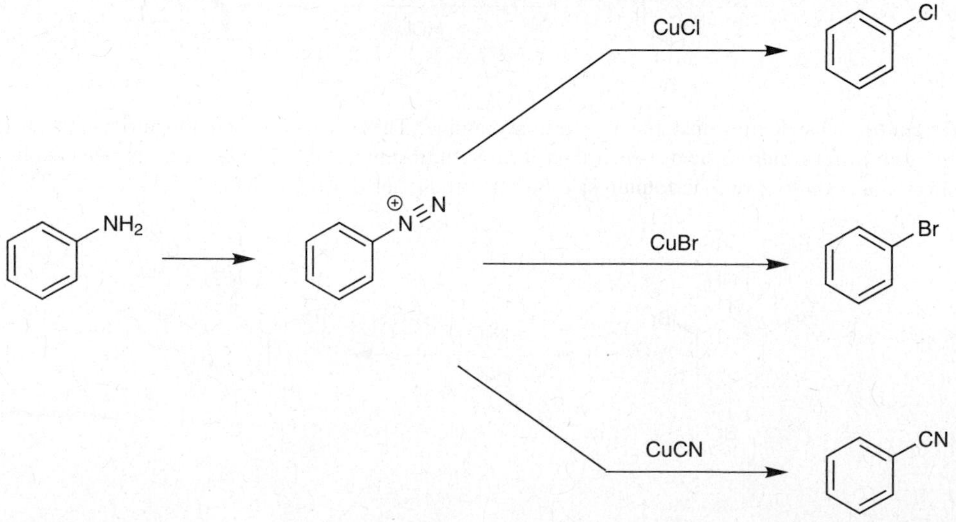

[reaction scheme: aniline (NH₂) → aryl diazonium salt (N⁺≡N), which branches to:
 - CuCl → chlorobenzene (Cl)
 - CuBr → bromobenzene (Br)
 - CuCN → benzonitrile (CN)]

In all of these reactions, we are using copper salts as the reagents. These reactions are called *Sandmeyer reactions*. They are useful, because they allow us to achieve transformations that we could not otherwise achieve with the chemistry that we learned in Chapters 3 and 4 (electrophilic and nucleophilic aromatic substitution). As a case in point, we did not see how to install a cyano group on an aromatic ring. This is our first way to do this.

EXERCISE 8.31 What reagents would you use to achieve the following transformation:

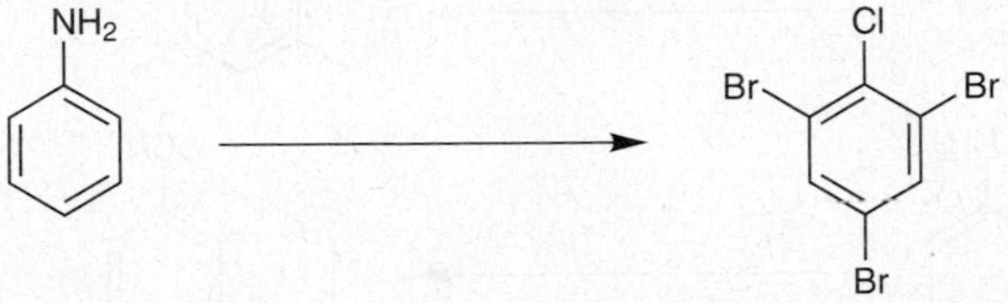

Answer If we brominate aniline, we will find that the amino group is so activated that we will obtain a tribrominated product:

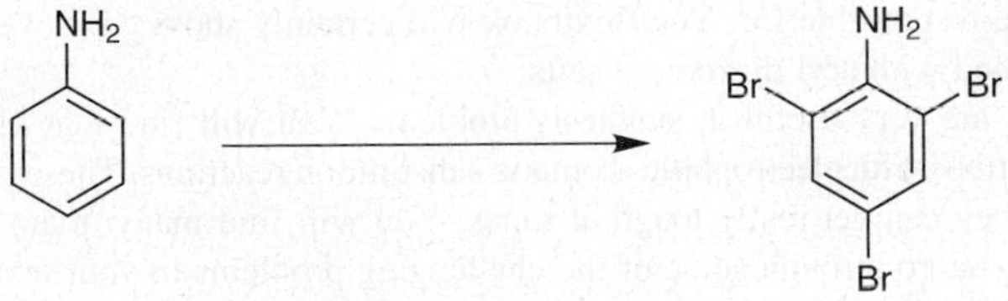

Then, we can convert the amino group into a chloro group. To do this, we make a diazonium salt, followed by a Sandmeyer reaction:

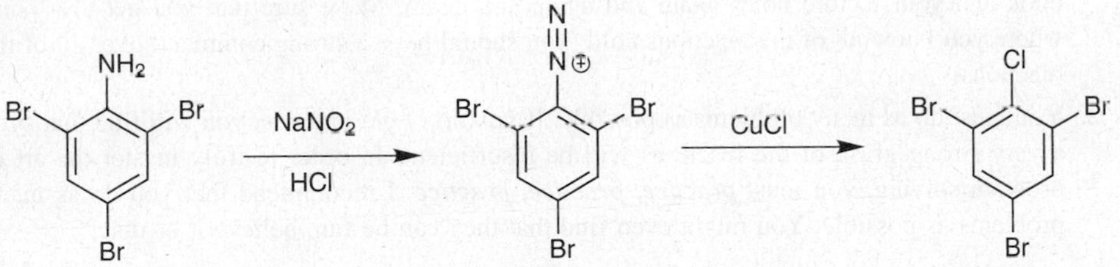

PROBLEMS What reagents would you use to achieve each of the following transformations:

8.32

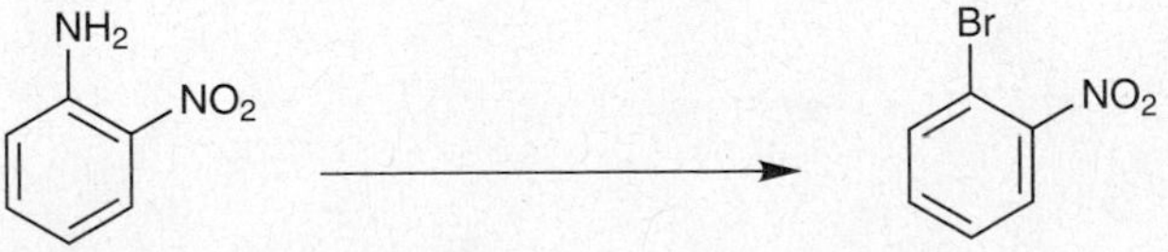

8.33

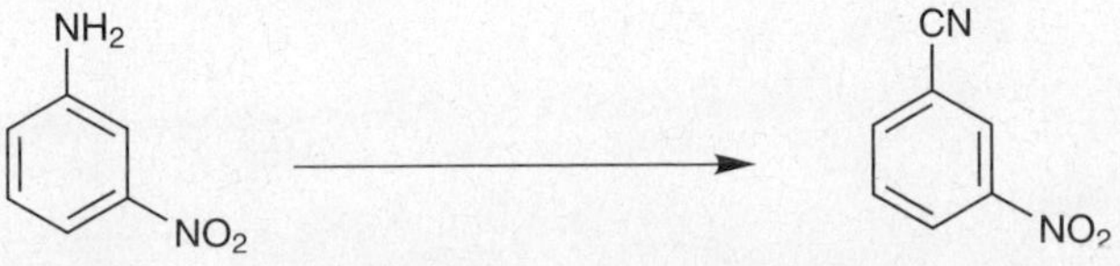

8.34

8.35

8.36

So far, we have seen a few reactions that you can perform with aryl diazonium salts. Many instructors will cover additional reactions of aryl diazonium salts. You should look through your lecture notes to see what you are responsible for. Your textbook will certainly show you several more reactions that can be performed with aryl diazonium salts.

These reactions are very useful in synthesis problems. You will find that some problems will combine these reactions with electrophilic aromatic substitution reactions. These problems can range in difficulty, and they can get really tough at times. You will find many, many such problems in your textbook. As you go through some of the challenging problems in your textbook, I will leave you with a bit of last-minute advice:

There are two important activities that you must do in order to master these types of problems.

1. You must review the reactions and principles from the entire course. Go through your text-book and your lecture notes again and again and again. Make sure that you get to a point where you know all of the reactions cold (you should have a strong command over all of the reactions).

2. You must do as many problems as possible. If you don't get practice, you will find that even a very strong grasp of the reactions will be insufficient. In order to truly master the art of problem solving, you must **practice, practice, practice**. I recommend that you do as many problems as possible. You might even find that they can be fun, believe it or not.

This book was meant to serve as a launching pad for your study efforts. This book did NOT cover everything that you need to know. My intention was to provide you with the **skills** and **understanding** that you need to study more efficiently. Good luck.

ANSWERS

CHAPTER 1
1.1)

1.2) No   **1.3)** No
1.4) Yes   **1.5)** Yes
1.6)

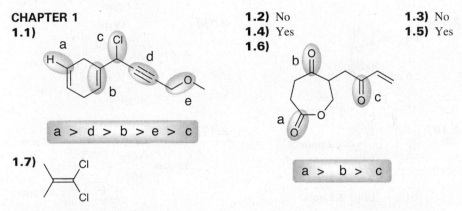

a > d > b > e > c

a > b > c

1.7)

1.8) Consider the third resonance structure, shown below. This resonance structure has a positive charge, indicating that the highlighted carbon atom is electron-poor. As a result, the C=C bond has an unusually strong dipole moment, leading to an unusually strong signal in IR spectroscopy.

1.9) Alcohol   **1.10)** Neither   **1.11)** Carboxylic acid   **1.12)** Alcohol
1.13) Carboxylic acid   **1.14)** Neither   **1.15)** Alcohol
1.16) Carboxylic acid   **1.17)** Primary amine
1.18) Ketone (the little blip at 3400 can be ignored. See exercise 1.21 for an explanation)
1.19) Secondary amine   **1.20)** Alcohol
1.21)

A   F   B

E   C   D

CHAPTER 2

2.2) Two signals   **2.3)** Three signals   **2.4)** One signal   **2.5)** Two signals

2.6) Five signals   **2.7)** Three signals   **2.8)** Six signals   **2.9)** Four signals

2.10) Four signals

2.11)

2.13)

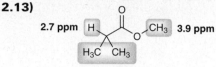

2.14)

2.3 ppm

2.0 ppm

4.2 ppm

2.15)

3.9 ppm

3.4 ppm   H₃C

1.9 ppm

2.7 ppm

2.16)

4.2 ppm

5.9 ppm

2.17)

4.7 ppm

1.4 ppm

2.2 ppm

1.9 ppm

2.18)

1.0 ppm

1.4 ppm

2.2 ppm

3.7 ppm

3.7 ppm

2.19)

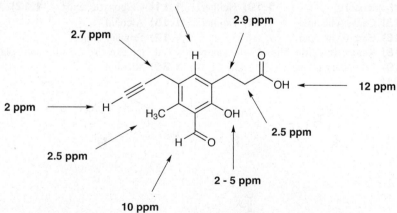

2.21) 5:2:2:1   **2.22)** 2:12   **2.23)** 2:2:2

2.25)

septet H — C(=O) — CH₃ singlet
        |
     H₃C CH₃
      doublet

2.26)

singlet H₃C, H₃C — (C=O) O ring
triplet H, H
           triplet

2.27)

singlet H₃C, H₃C, H₃C — C — O — CH₂ (triplet) — CH₂ (triplet) — C(=O) — CH₃ singlet

2.28)

H H singlet
H C — C H
    O   O
   H₃C CH₃ singlet

2.29)

singlet H₃C, H₃C — ring — O — CH₂ (singlet) — C=O
triplet H H, H H triplet

2.30)

quartet
H H   CH₃ doublet
H — C   CH₃
triplet H₃C   O   H septet

2.31) Isopropyl **2.32)** Ethyl
2.33) Isopropyl **2.34)** Ethyl
2.36) Two degrees of unsaturation
2.37) Three degrees of unsaturation
2.38) Two degrees of unsaturation
2.39) No degrees of unsaturation
2.40) No degrees of unsaturation
2.41) Four degrees of unsaturation

2.43)

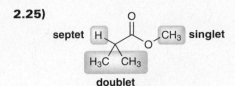

2.44)

2.45)

2.46)

2.47)

2.48)

2.50) Total = Six signals (two signals between 0–50 ppm, and four signals between 100–150 ppm)

2.51) Total = Nine signals (one signal between 0–50 ppm, one signal between 50–100 ppm, six signals between 100–150 ppm, and one signal between 150–220 ppm)

2.52) Total = Five signals (three signals between 0–50 ppm, and two signals between 100–150 ppm)

2.53) Total = Five signals (three signals between 0–50 ppm, and two signals between 50–100 ppm)

2.54) Total = Three signals (two signals between 0–50 ppm, and one signal between 50–100 ppm)

2.55) Total = Five signals (three signals between 0–50 ppm, and two signals between 50–100 ppm)

CHAPTER 3

3.2)

Cl—Al(Cl)(Cl) + :Cl—Cl: → Cl—Al⁻(Cl)(Cl)—Cl⁺—Cl:

3.3)

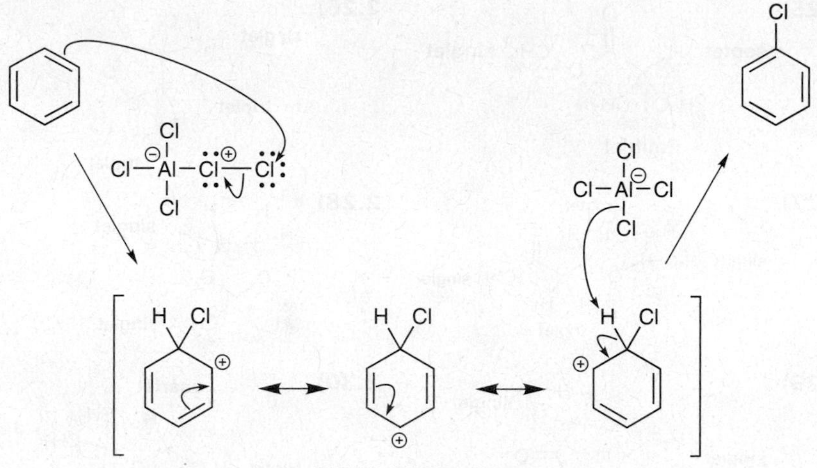
Cl
Cl—Al—Cl—Cl
Cl
Cl—Al—Cl
H Cl
SIGMA COMPLEX

3.4)

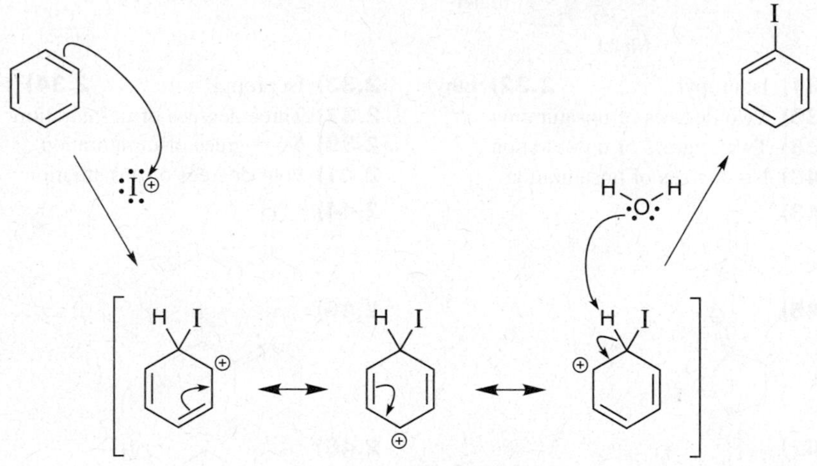
I
H O H
H I
SIGMA COMPLEX

3.5)

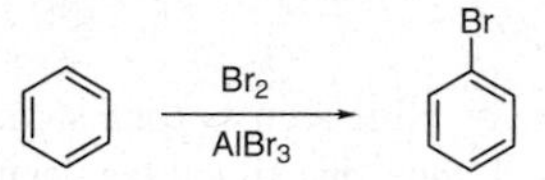
Br
Br_2
$AlBr_3$

3.6)

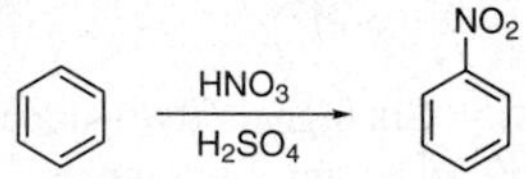
NO_2
HNO_3
H_2SO_4

3.7)

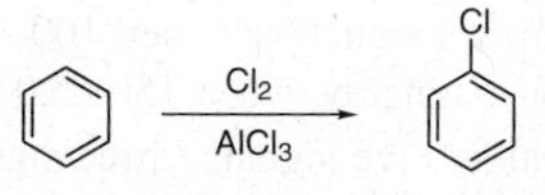
Cl
Cl_2
$AlCl_3$

3.10)

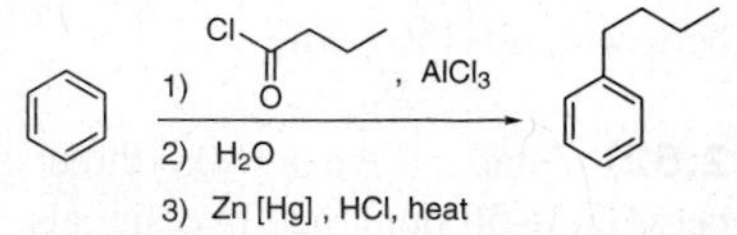
1) Cl O , $AlCl_3$
2) H_2O
3) Zn [Hg] , HCl, heat

3.11)

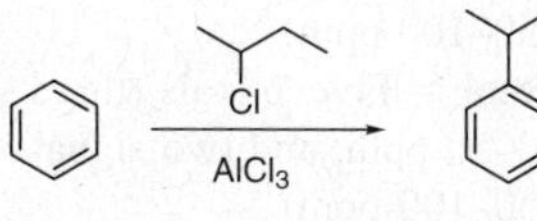
Cl
$AlCl_3$

3.12)

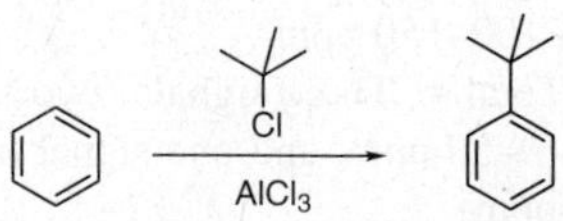
Cl
$AlCl_3$

3.13)

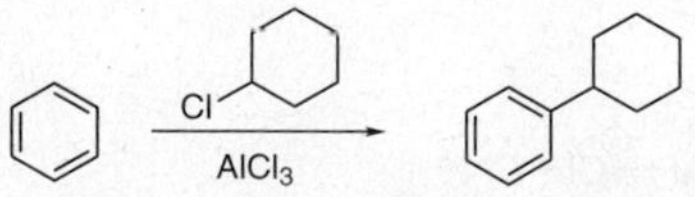
Cl
$AlCl_3$

3.14)

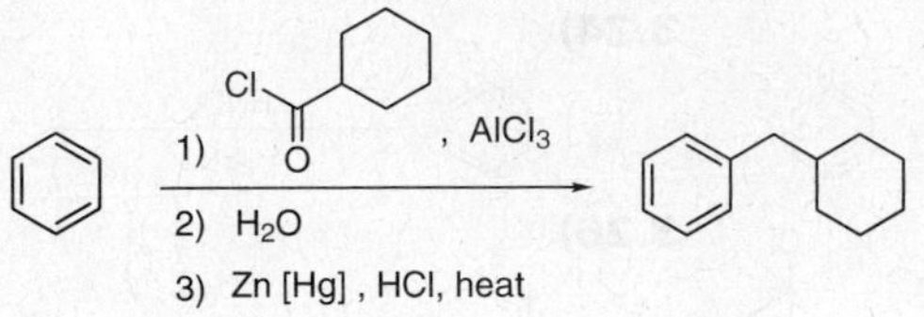
Cl
O
1) , AlCl$_3$
2) H$_2$O
3) Zn [Hg] , HCl, heat

3.15)

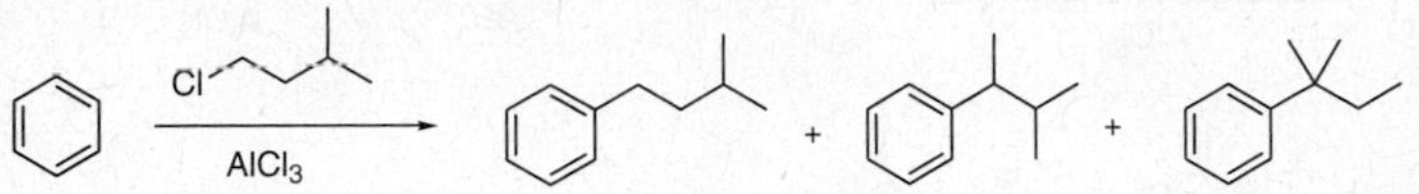
Cl
AlCl$_3$
+
+

3.17)

Formation of acylium ion:

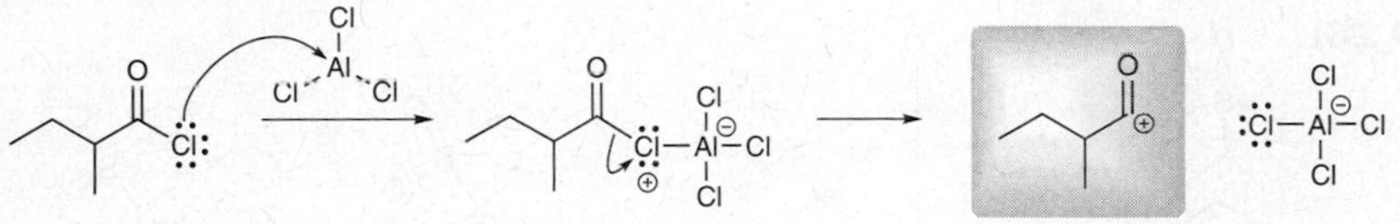
O
Cl
Al
Cl
Cl

Electrophilic aromatic substitution:

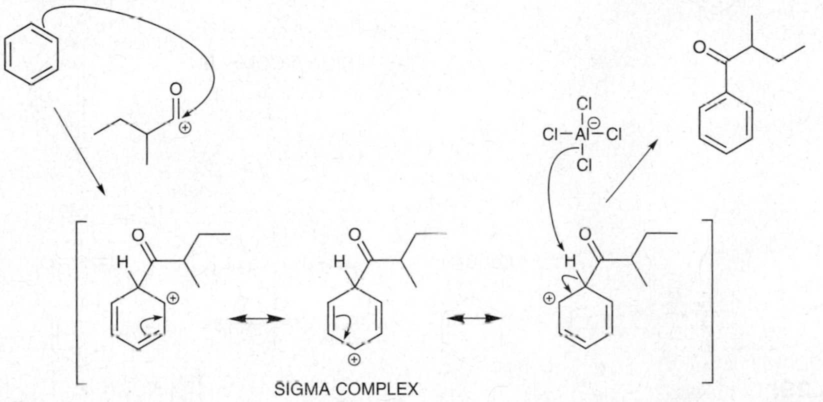
SIGMA COMPLEX

3.19)

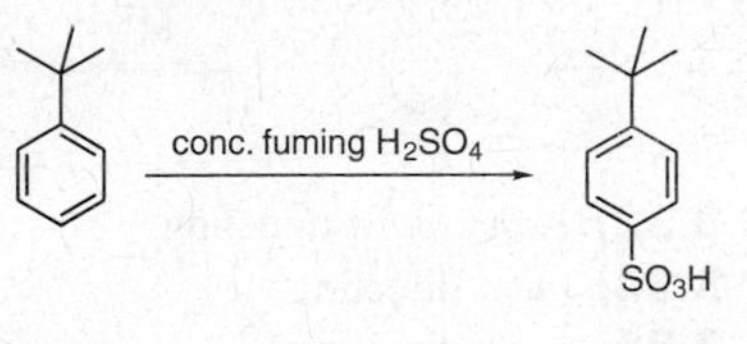
conc. fuming H$_2$SO$_4$
SO$_3$H

3.20)

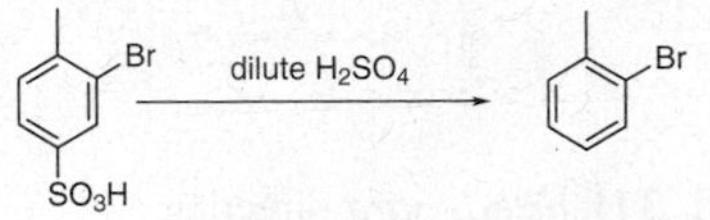
Br
dilute H$_2$SO$_4$
SO$_3$H
Br

3.21)

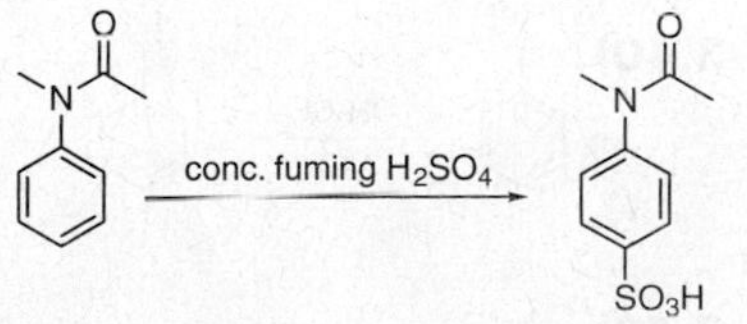
O
N
conc. fuming H$_2$SO$_4$
SO$_3$H

3.22)

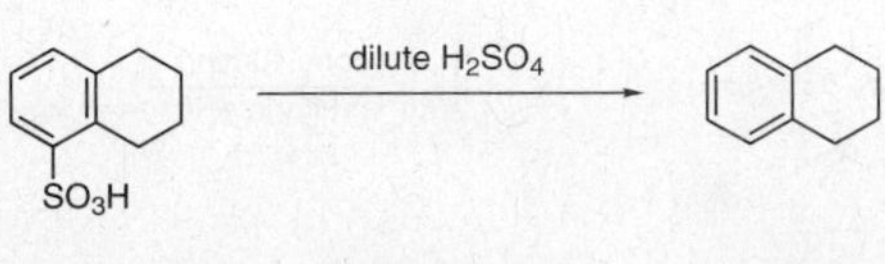
dilute H$_2$SO$_4$
SO$_3$H

3.23)

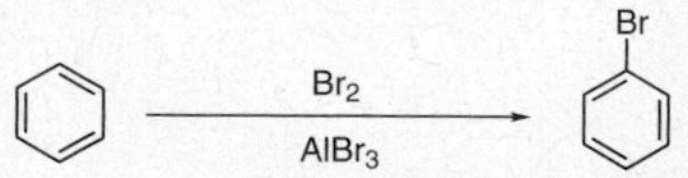

3.24)

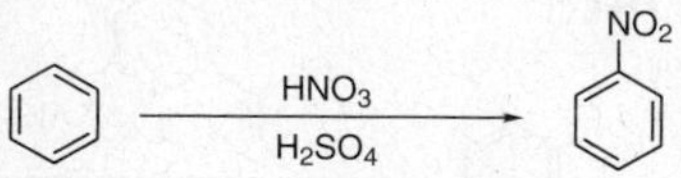

3.25)

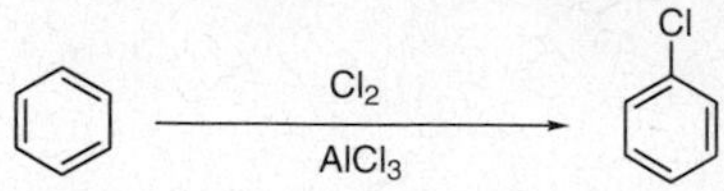

3.26)

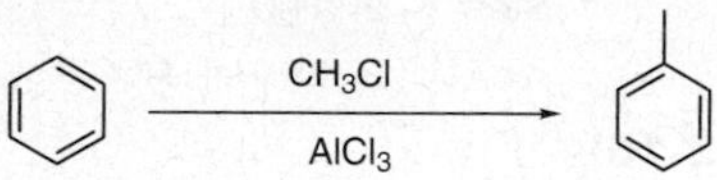

3.27)

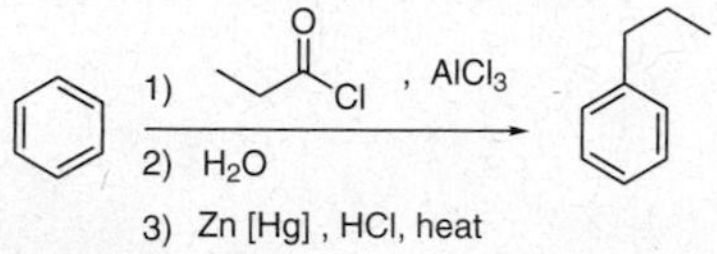

1) (acyl chloride), AlCl₃

2) H₂O

3) Zn [Hg] , HCl, heat

3.28)

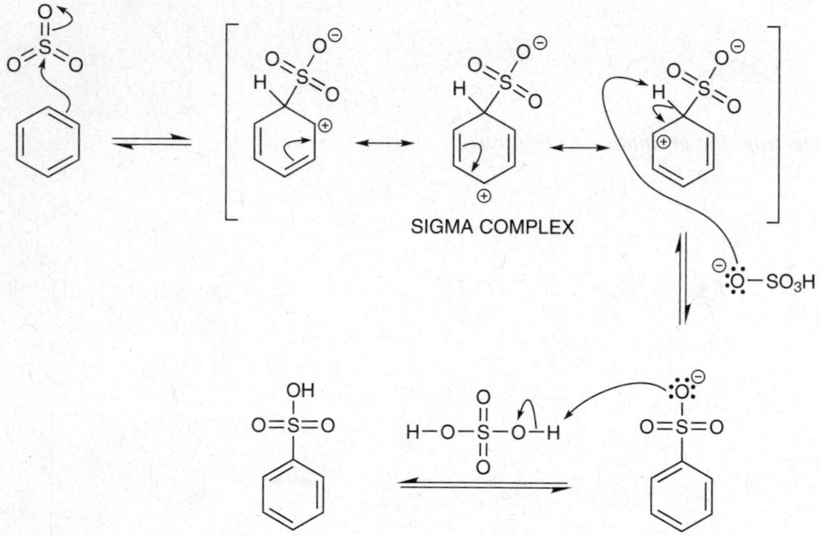

SIGMA COMPLEX

3.29)

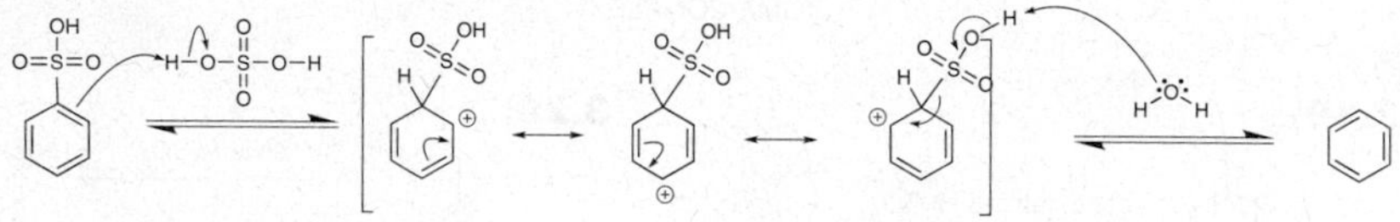

3.31) *ortho, para* directing

3.33) *meta* directing

3.35) *ortho, para* directing

3.37) *ortho, para* directing

3.32) *ortho, para* directing

3.34) *meta* directing

3.36) *meta* directing

3.39)

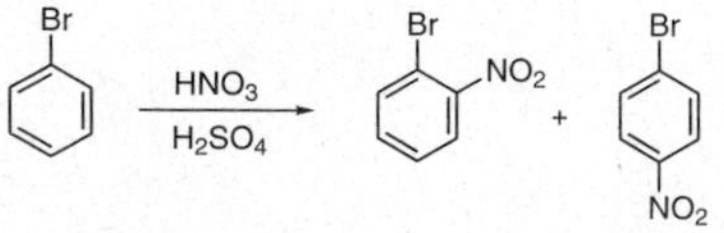

3.40)

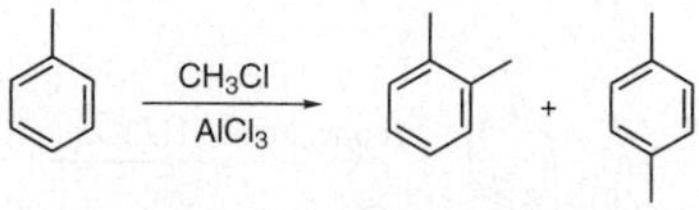

3.41)

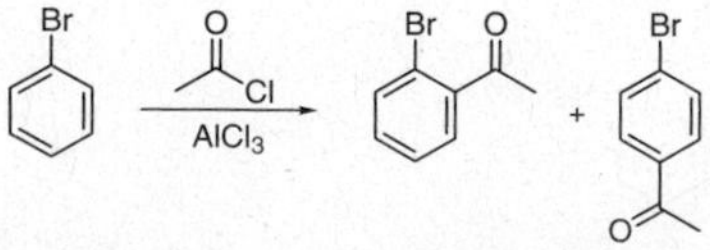

3.42)

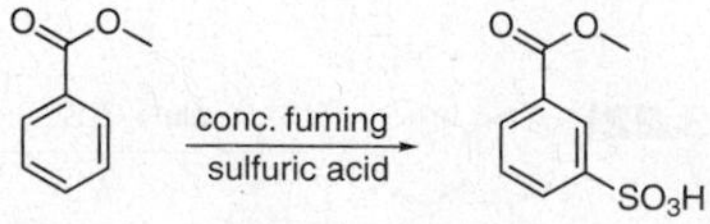

3.43)

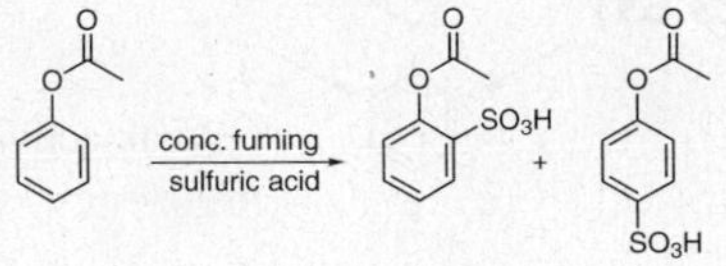
conc. fuming
sulfuric acid
SO3H
+
SO3H

3.44)

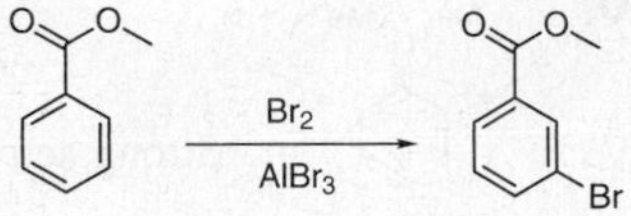
Br2
AlBr3
Br

3.45)

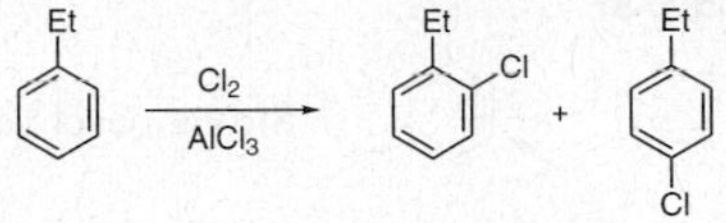
Et
Cl2
AlCl3
Et
Cl
+
Et
Cl

3.47)

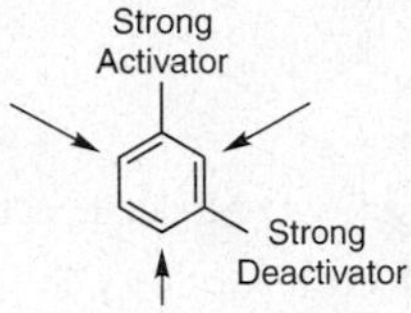
Strong
Activator
Strong
Deactivator

3.48)

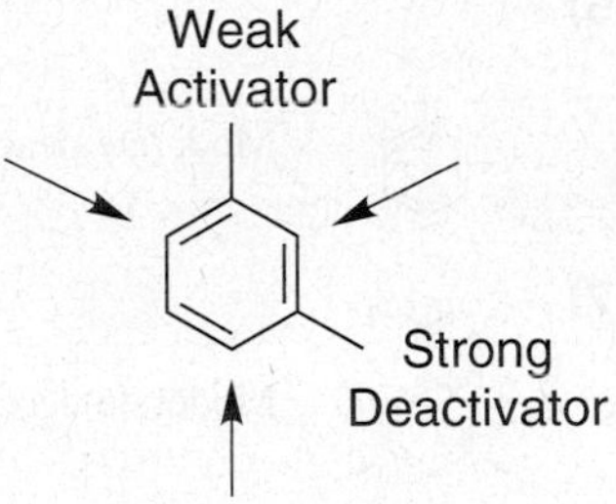
Weak
Activator
Strong
Deactivator

3.49)

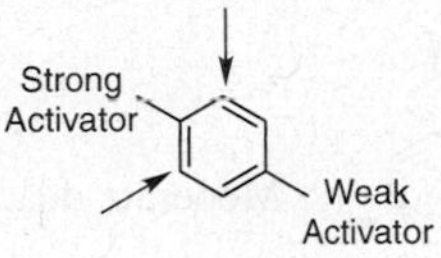
Strong
Activator
Weak
Activator

3.50)

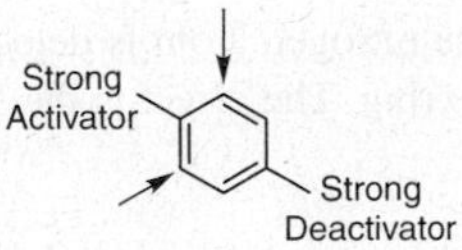
Strong
Activator
Strong
Deactivator

3.51)

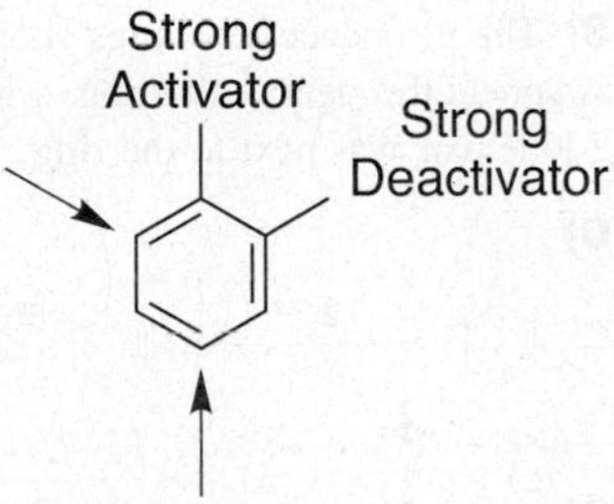
Strong
Activator
Strong
Deactivator

3.52)

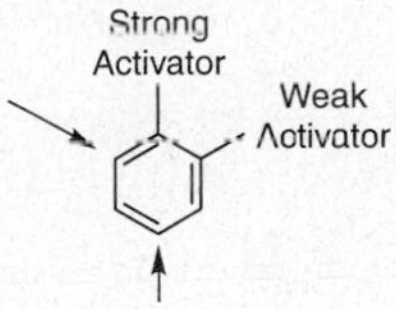
Strong
Activator
Weak
Activator

3.53)

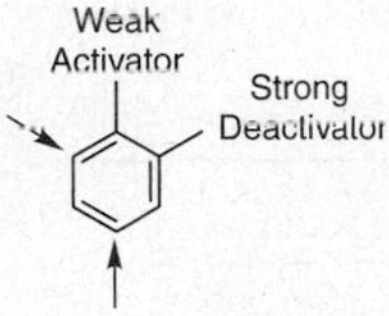
Weak
Activator
Strong
Deactivator

3.54)

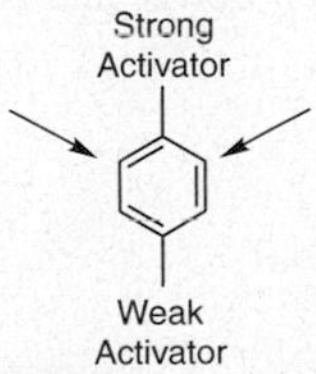
Strong
Activator
Weak
Activator

3.55)

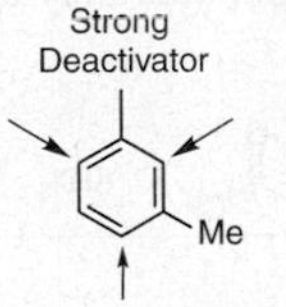
Strong
Deactivator
Me

3.56)

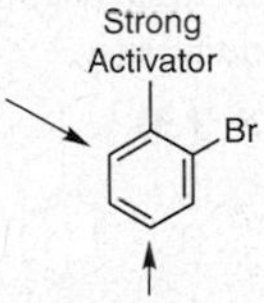
Strong
Activator
Br

3.58)

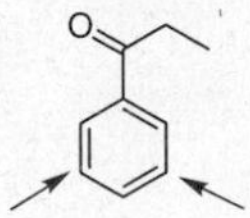

Moderate deactivator

3.59)

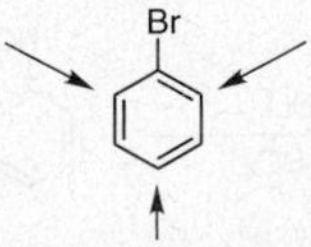
Br

Weak deactivator

3.60) Strong activator

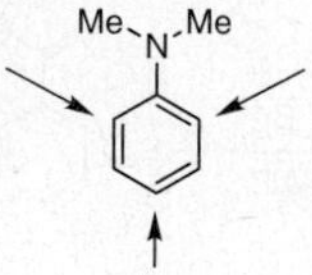

3.61) Weak activator

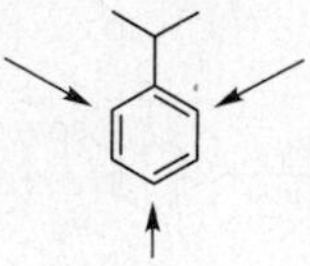

3.62) Moderate activator

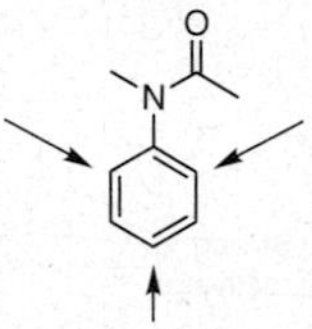

3.63) Strong deactivator

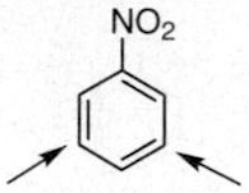

3.64) Moderate deactivator

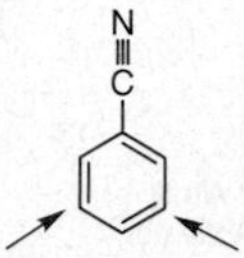

3.65) Moderate deactivator

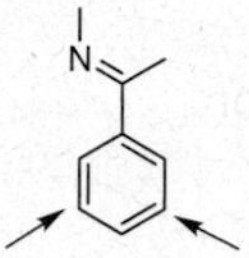

3.66) Strong deactivator

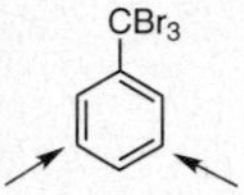

3.67) Moderate deactivator

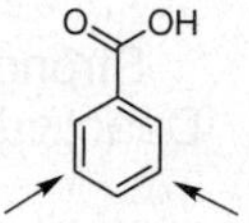

3.68) The resonance structures show that the lone pair on the nitrogen atom is delocalized, and is spread throughout the ring, which strongly activates the ring The effect is the same as if the lone pair was next to the ring.

3.70)

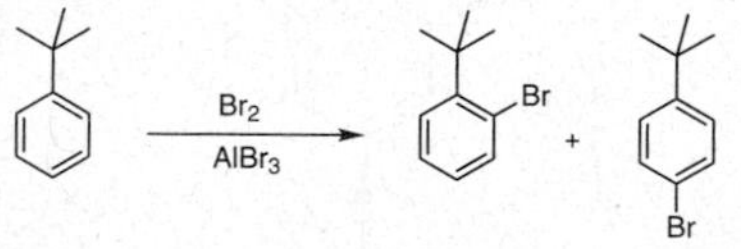

3.71)

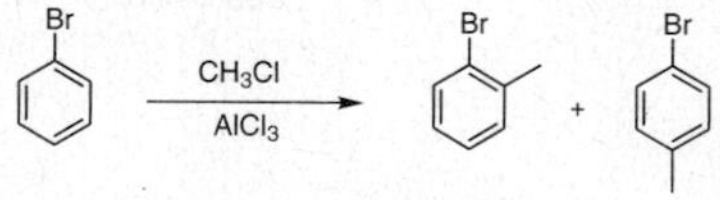

3.72)

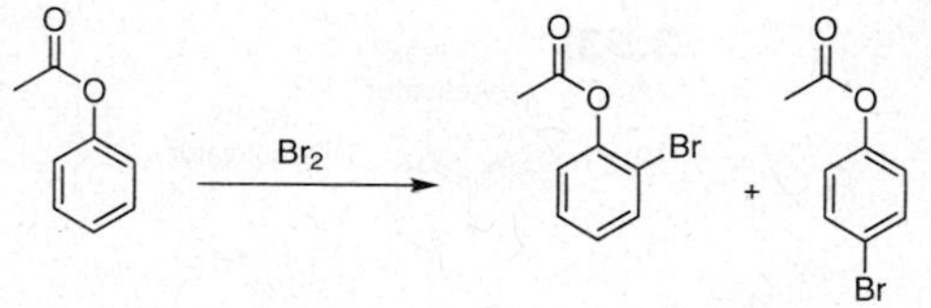

In this case, the ring is moderately activated toward bromination, so a Lewis acid is not necessary.

3.73)

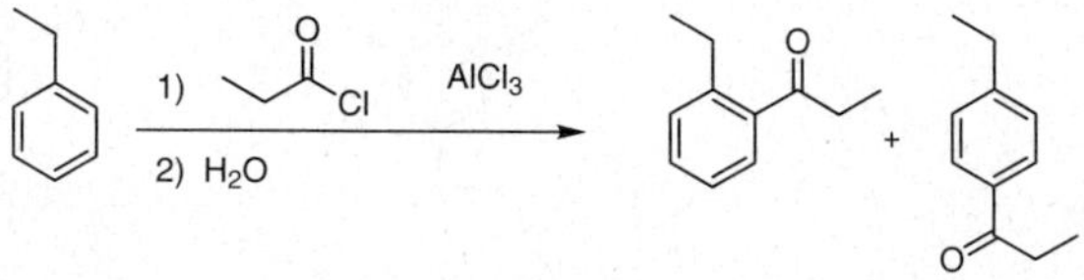

3.74)

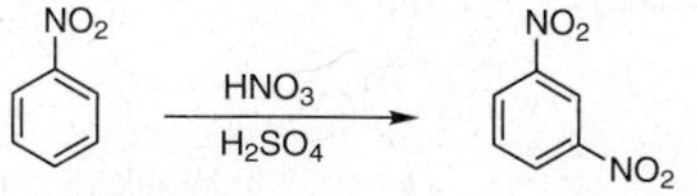

3.76)

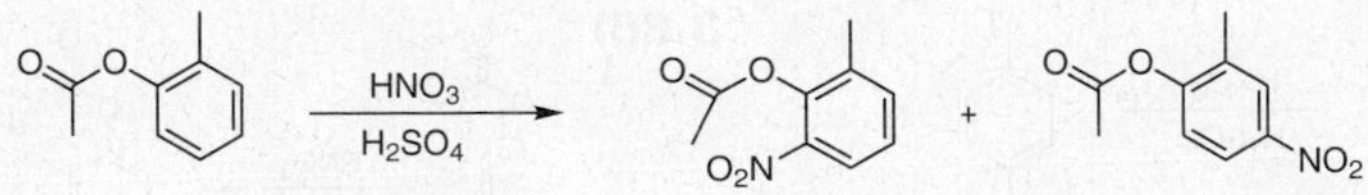
HNO3
H2SO4
O2N
NO2
+

3.77)

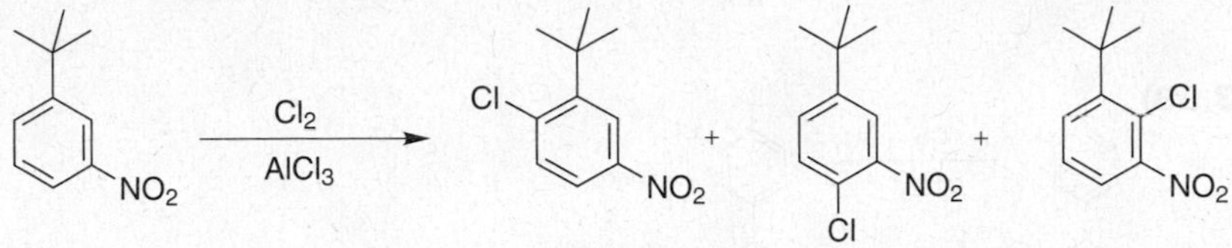
NO2
Cl2
AlCl3
Cl
NO2
+
NO2
Cl
+
Cl
NO2

3.78)

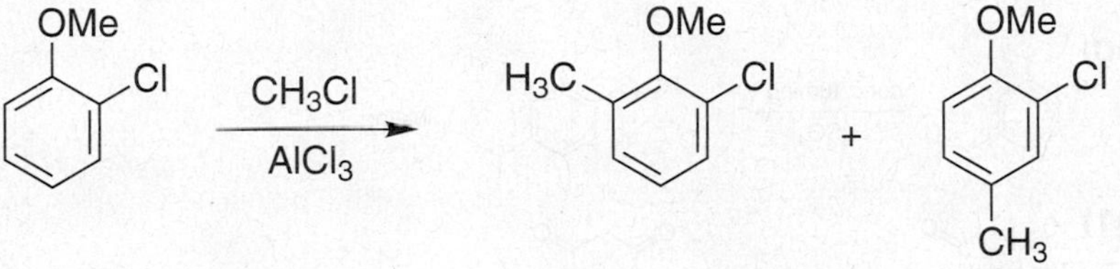
OMe
Cl
CH3Cl
AlCl3
OMe
H3C
Cl
+
OMe
Cl
CH3

3.79)

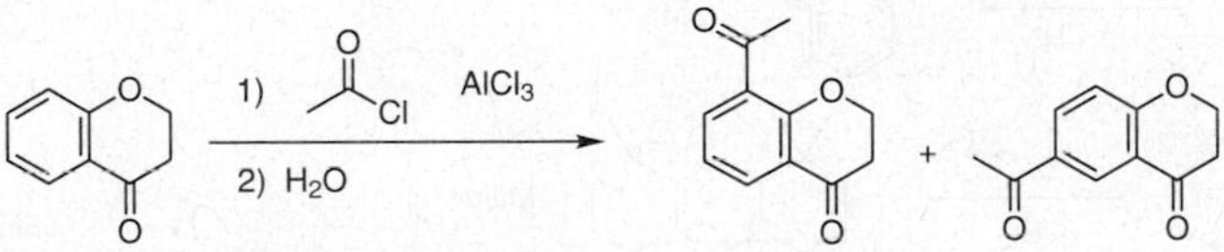
1)
Cl
AlCl3
2) H2O
+

3.81)

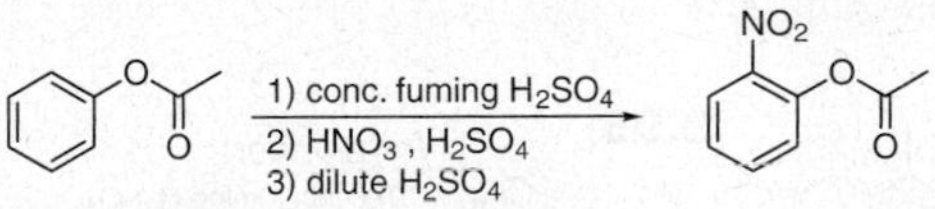
1) conc. fuming H2SO4
2) HNO3 , H2SO4
3) dilute H2SO4
NO2

3.82)

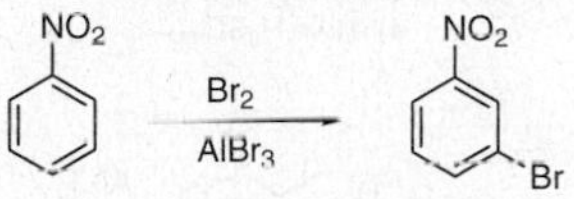
NO2
Br2
AlBr3
NO2
Br

3.83)

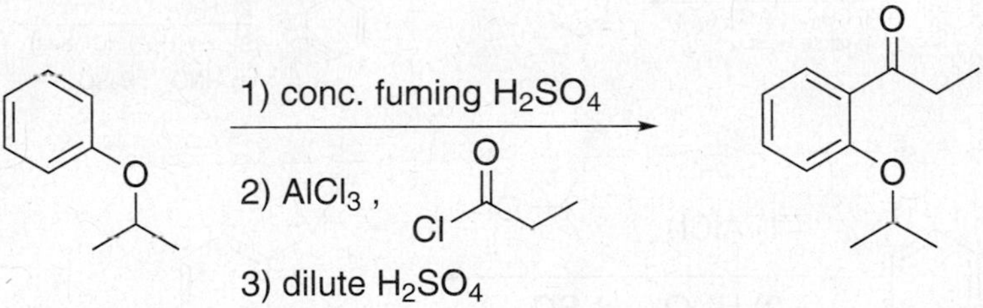
1) conc. fuming H2SO4
2) AlCl3 ,
Cl
3) dilute H2SO4

3.84)

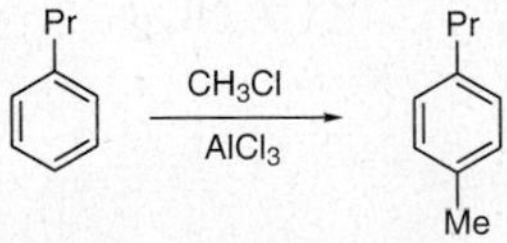
Pr
CH3Cl
AlCl3
Pr
Me

3.85)

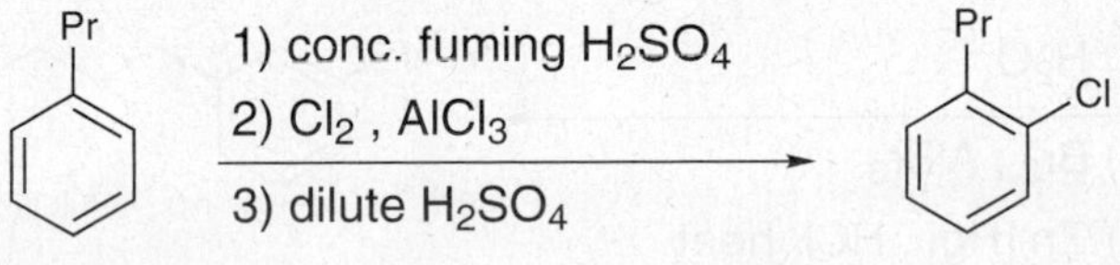
Pr
1) conc. fuming H2SO4
2) Cl2 , AlCl3
3) dilute H2SO4
Pr
Cl

3.87)

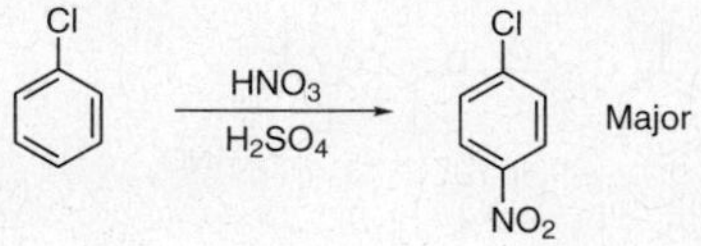

Cl

$\xrightarrow[\text{H}_2\text{SO}_4]{\text{HNO}_3}$

Cl, NO$_2$ Major

3.88)

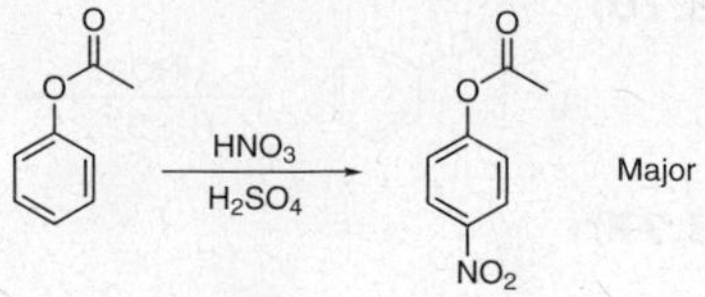

$\xrightarrow[\text{H}_2\text{SO}_4]{\text{HNO}_3}$

NO$_2$ Major

3.89)

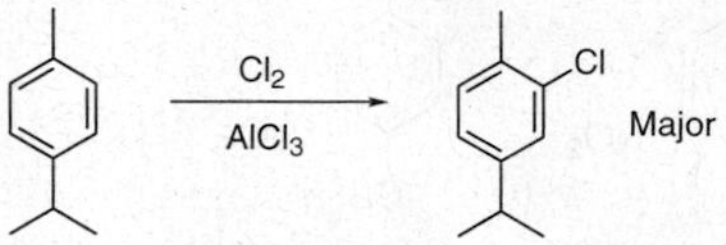

$\xrightarrow[\text{AlCl}_3]{\text{Cl}_2}$

Cl   Major

3.90)

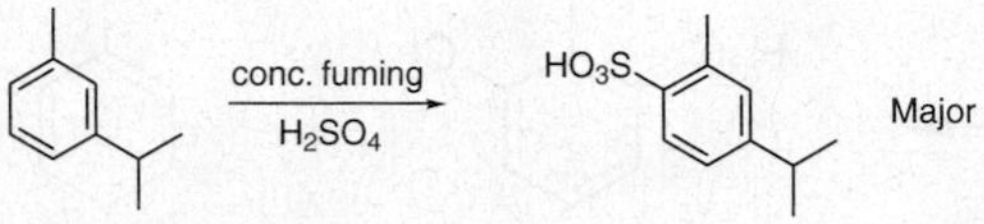

$\xrightarrow[\text{H}_2\text{SO}_4]{\text{conc. fuming}}$

HO$_3$S   Major

3.91)

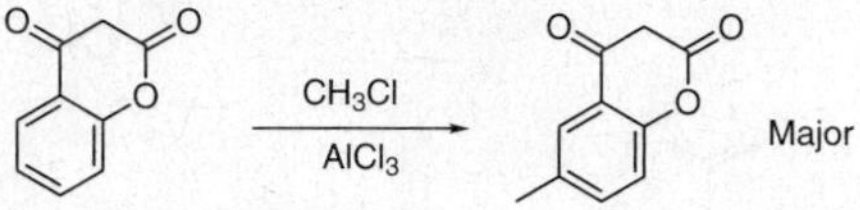

$\xrightarrow[\text{AlCl}_3]{\text{CH}_3\text{Cl}}$

Major

3.92)

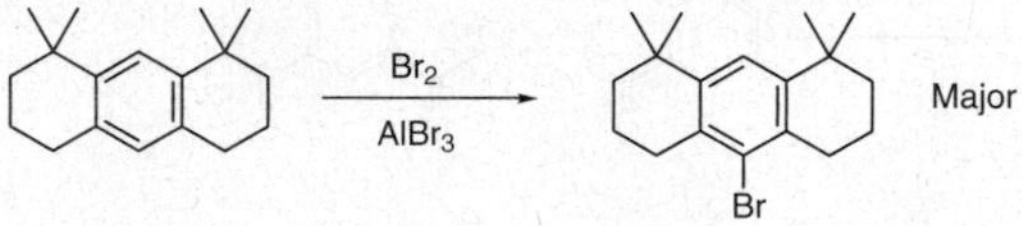

$\xrightarrow[\text{AlBr}_3]{\text{Br}_2}$

Br   Major

3.94)

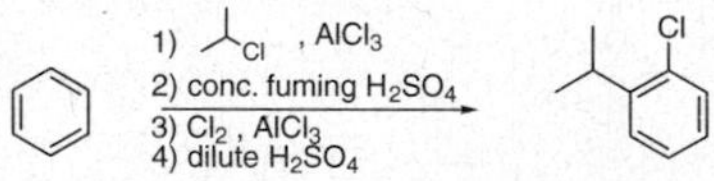

1) $\diagup\!\!\diagdown$Cl , AlCl$_3$
2) conc. fuming H$_2$SO$_4$
3) Cl$_2$, AlCl$_3$
4) dilute H$_2$SO$_4$

Cl

3.95)

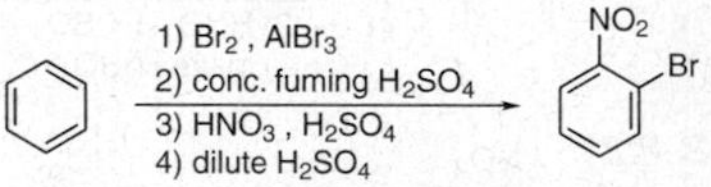

1) Br$_2$, AlBr$_3$
2) conc. fuming H$_2$SO$_4$
3) HNO$_3$, H$_2$SO$_4$
4) dilute H$_2$SO$_4$

NO$_2$, Br

3.96)

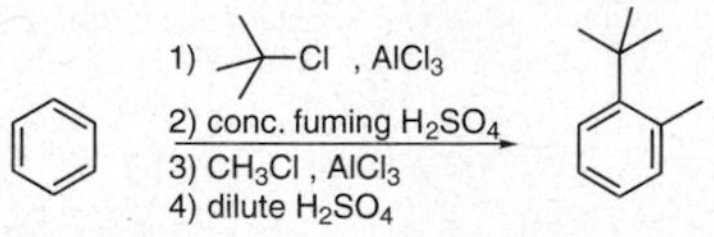

1) $\diagup\!\!\diagdown$Cl , AlCl$_3$
2) conc. fuming H$_2$SO$_4$
3) CH$_3$Cl , AlCl$_3$
4) dilute H$_2$SO$_4$

3.97)

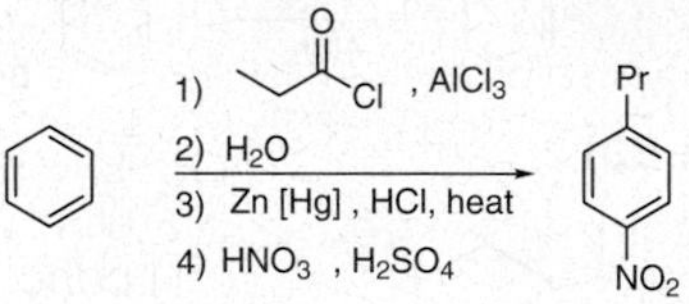

1) $\diagup\!\!\diagdown$C(=O)Cl , AlCl$_3$
2) H$_2$O
3) Zn [Hg] , HCl, heat
4) HNO$_3$, H$_2$SO$_4$

Pr, NO$_2$

3.98)

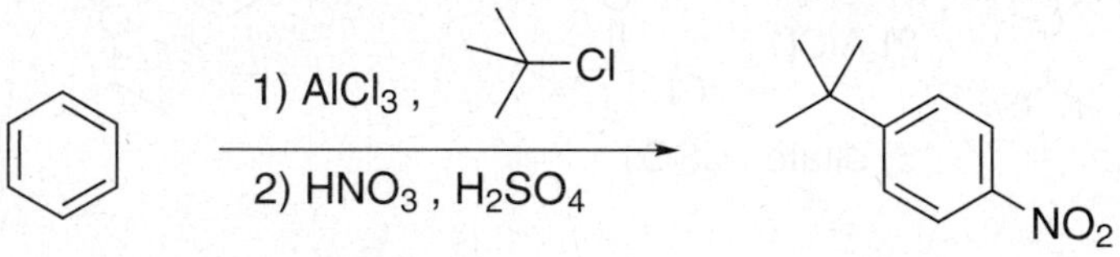

1) AlCl$_3$, $\diagup\!\!\diagdown$Cl

2) HNO$_3$, H$_2$SO$_4$

NO$_2$

3.99)

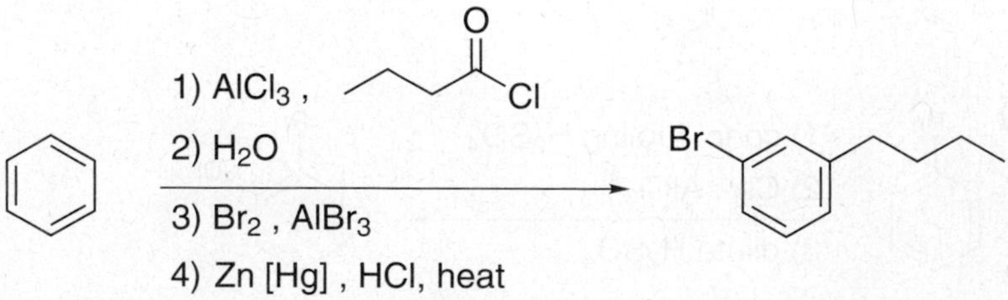

1) AlCl$_3$, (acid chloride)

2) H$_2$O

3) Br$_2$, AlBr$_3$

4) Zn [Hg] , HCl, heat

Br

3.100)

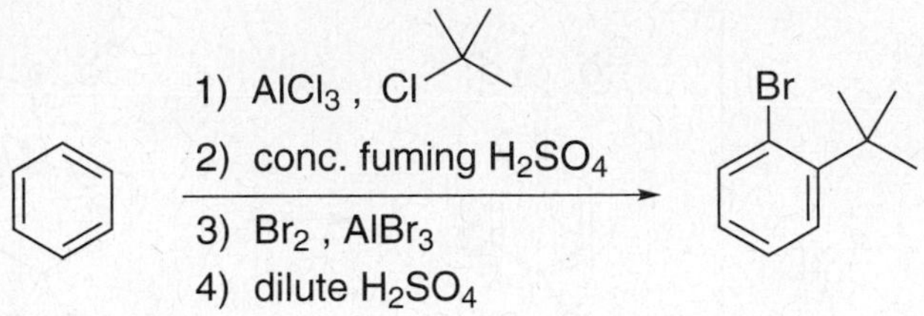
1) $AlCl_3$, Cl
2) conc. fuming H_2SO_4
3) Br_2, $AlBr_3$
4) dilute H_2SO_4
Br

3.101)

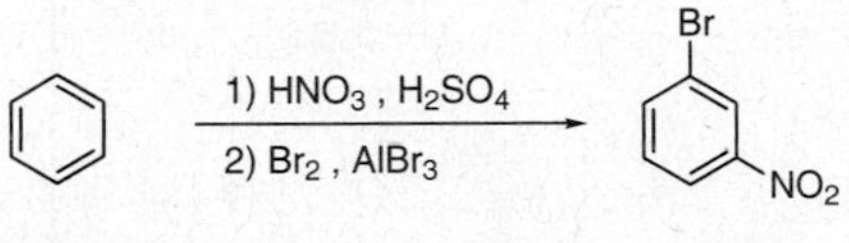
1) HNO_3, H_2SO_4
2) Br_2, $AlBr_3$
Br
NO_2

3.102)

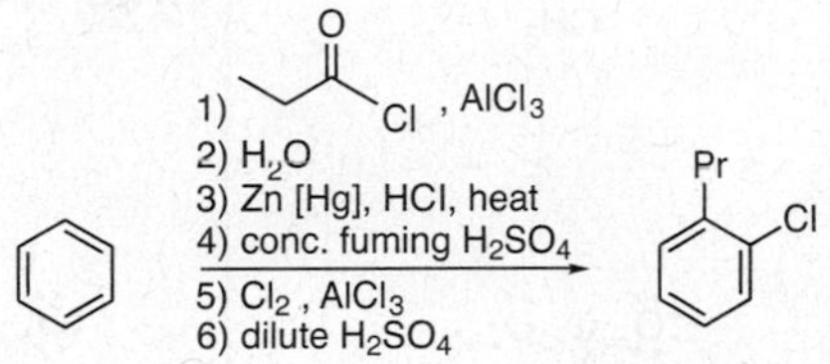
O
1) Cl, $AlCl_3$
2) H_2O
3) Zn [Hg], HCl, heat
4) conc. fuming H_2SO_4
5) Cl_2, $AlCl_3$
6) dilute H_2SO_4
Pr
Cl

CHAPTER 4

4.2) Yes **4.3)** No **4.4)** No **4.5)** Yes **4.6)** No **4.7)** No

4.9)

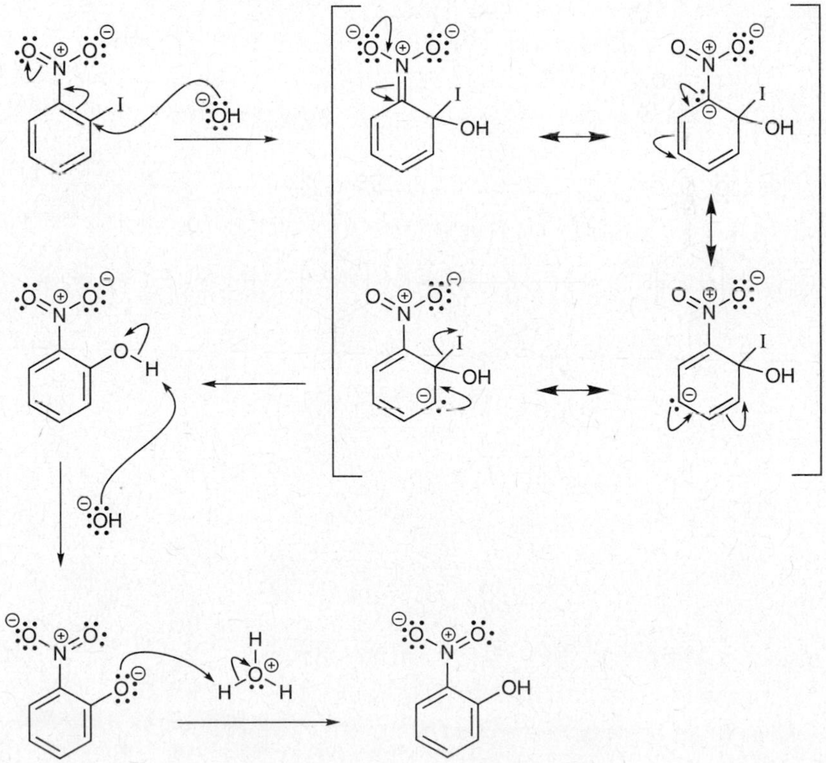

4.10)

4.11)

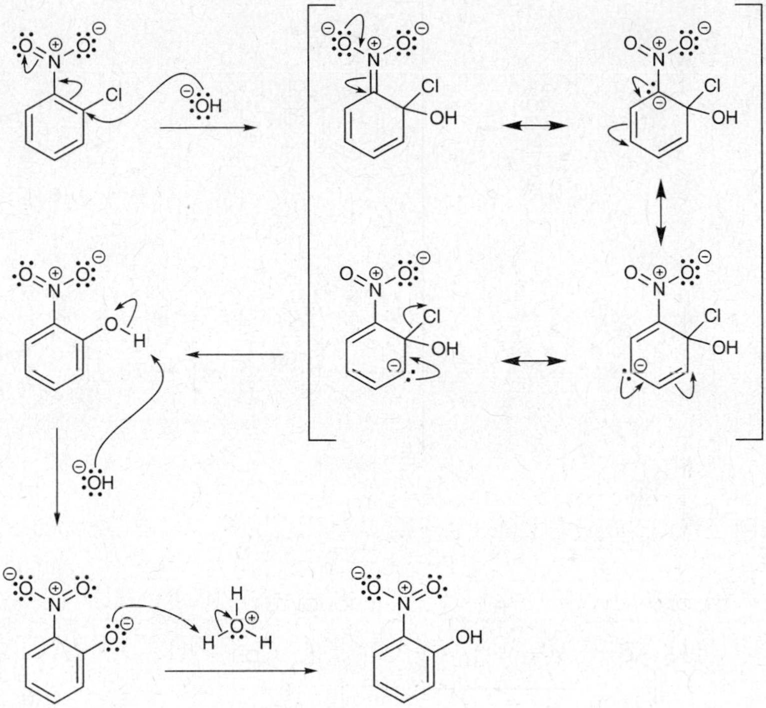

4.13)

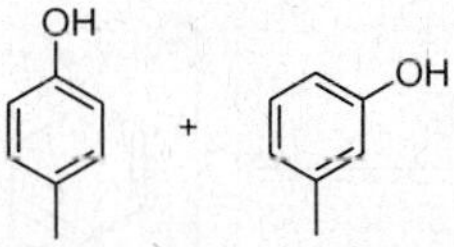

4.14)

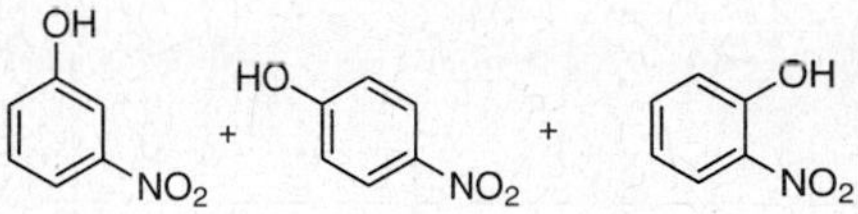

4.15)

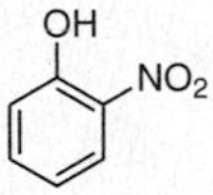

4.16)

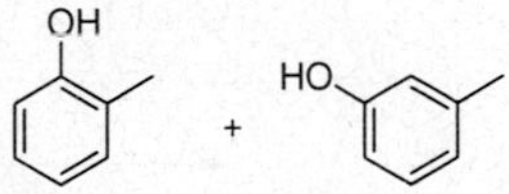

4.17)

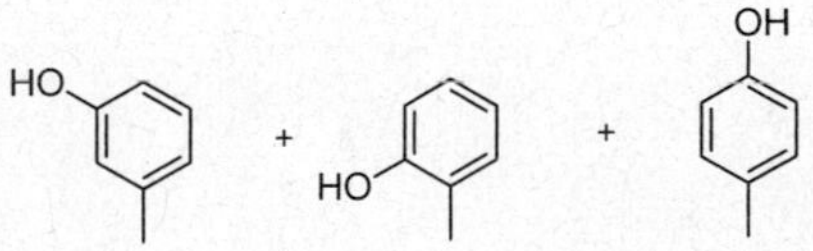

4.18)

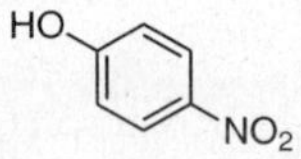

4.20)

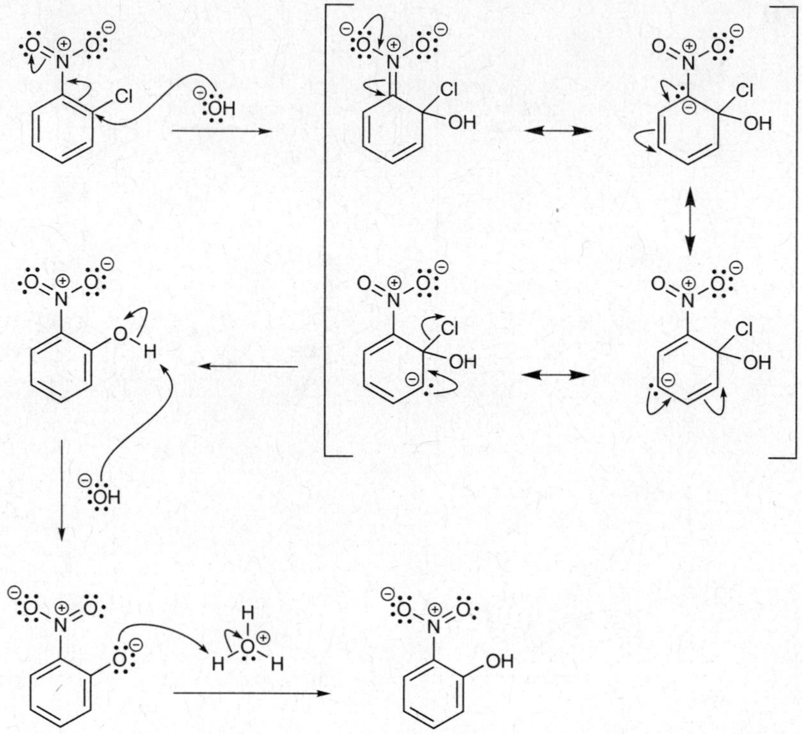

4.21)

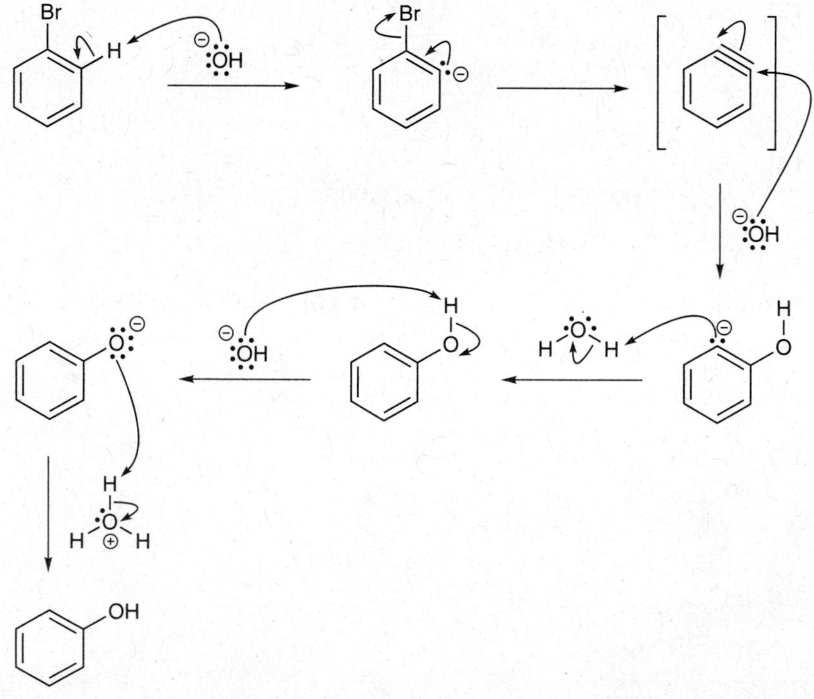

4.22)

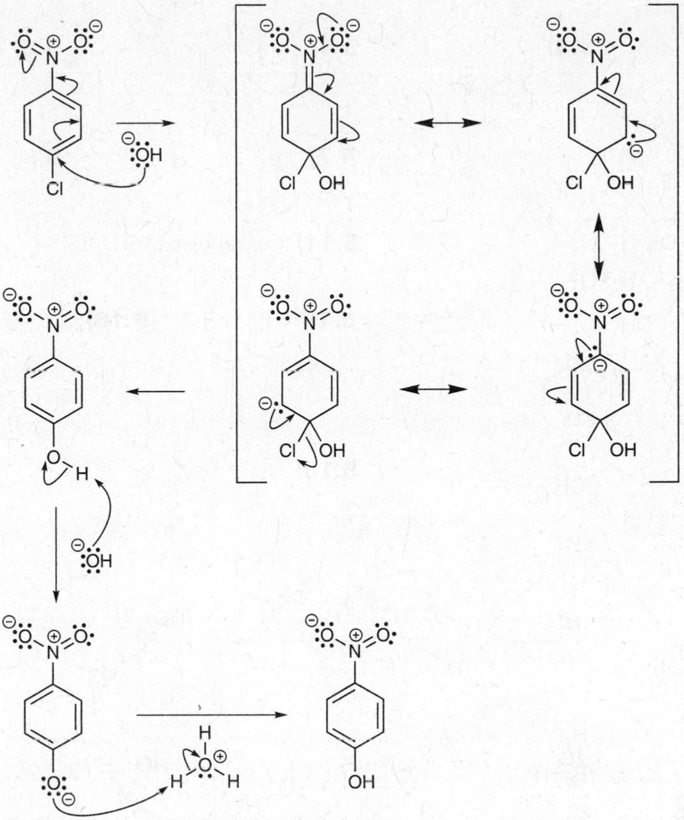

4.23)

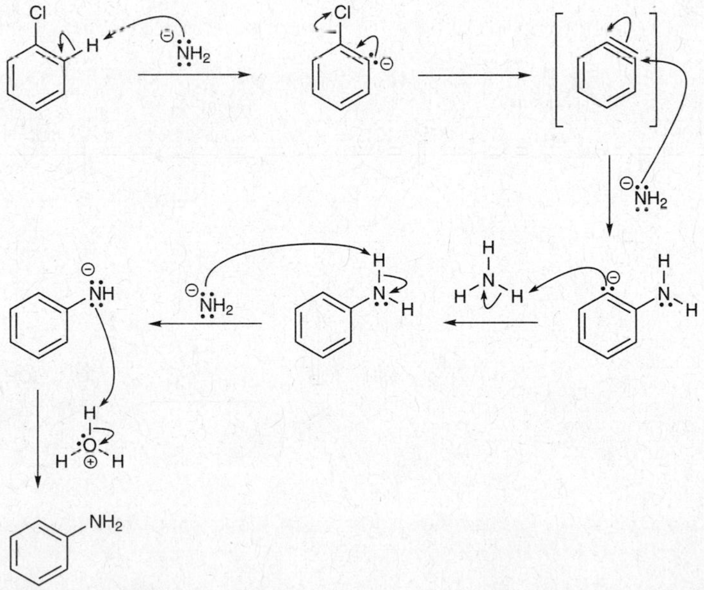

CHAPTER 5

5.2) 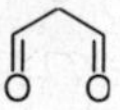
5.3) 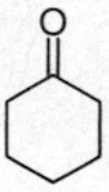
5.4) 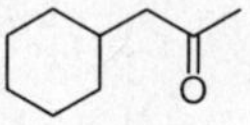
5.5)

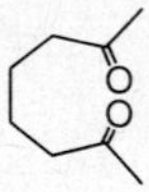

5.6) 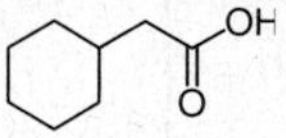
5.7) 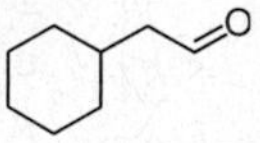
5.9) PCC

5.10) $Na_2Cr_2O_7$, H_2SO_4
5.11) O_3, followed by DMS

5.12) $Na_2Cr_2O_7$, H_2SO_4

5.14) 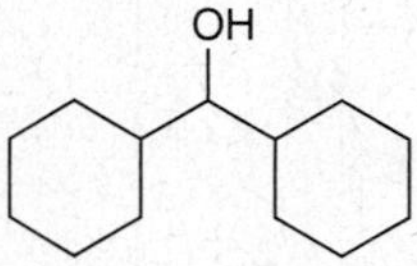
5.15) 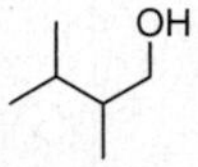
5.16)

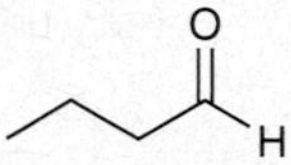

5.17) 
5.18)

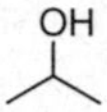

5.20)

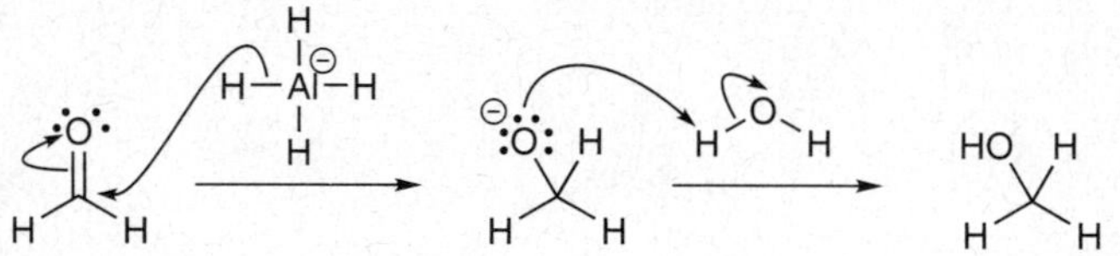

5.21)

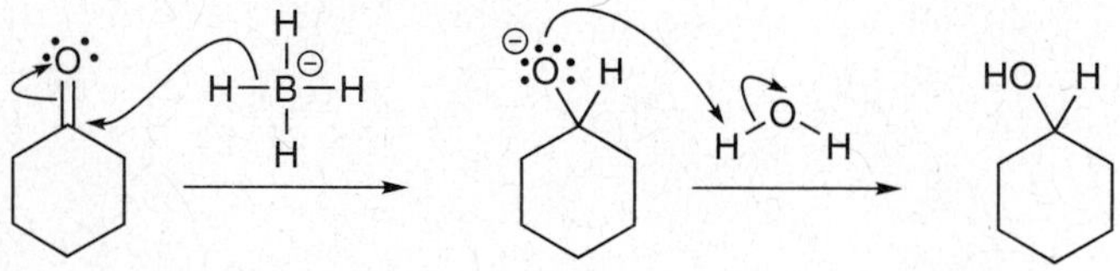

5.22)

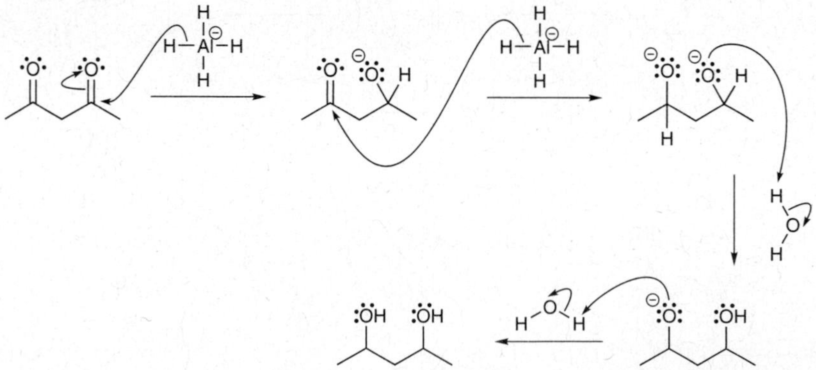

5.24)

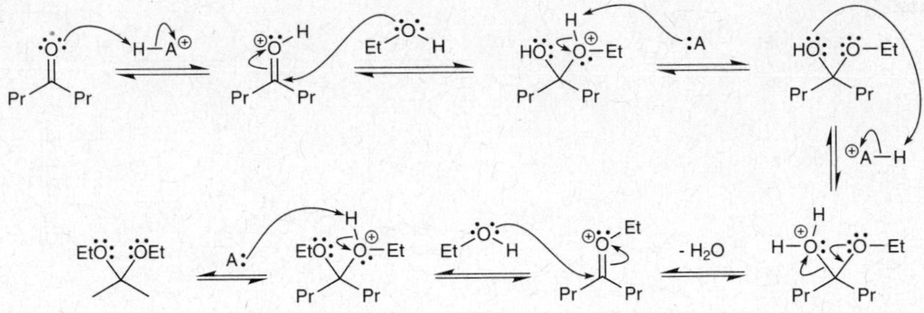

5.25)

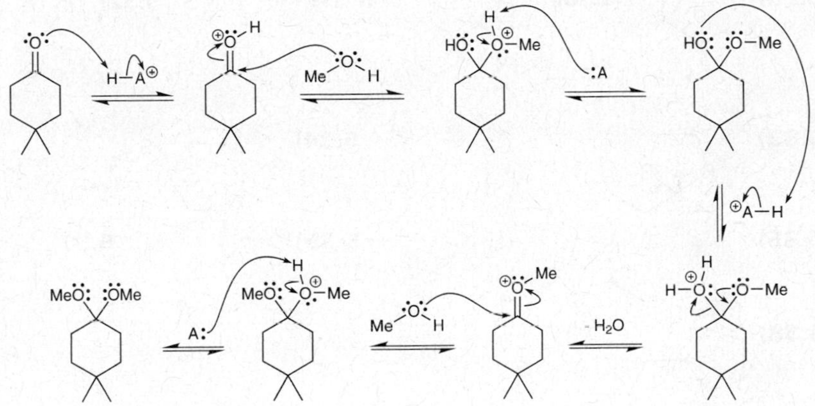

5.26)

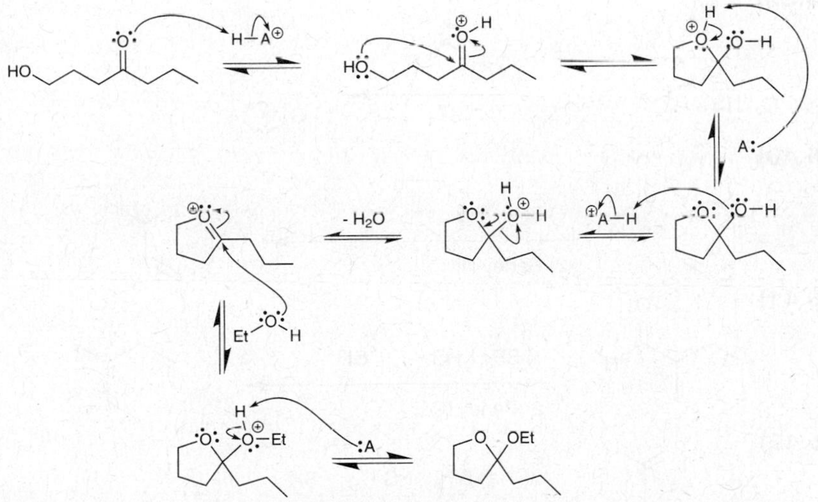

5.27)

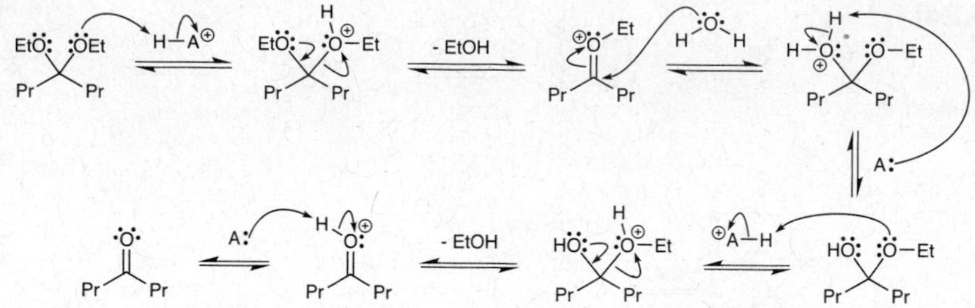

5.29) 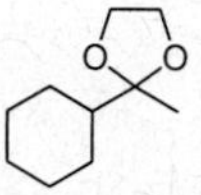 **5.30)** 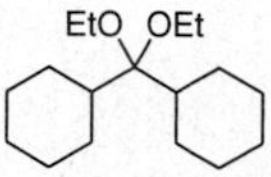 **5.31)** 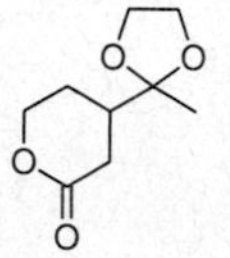 **5.32)**

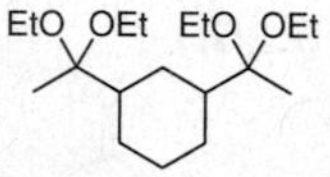

5.33) 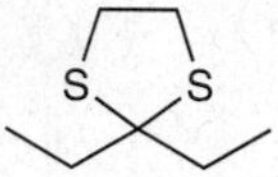 **5.34)**

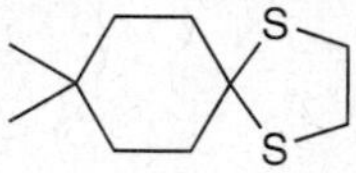

5.35) 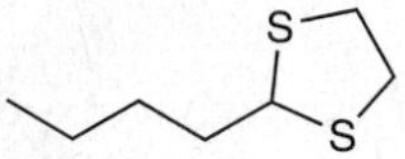 **5.36)** 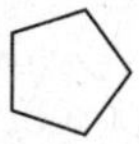 **5.37)**

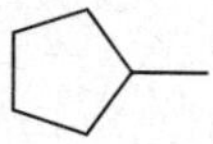

5.38)

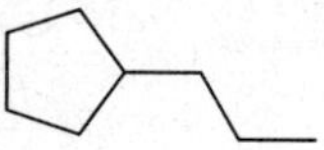

5.39)

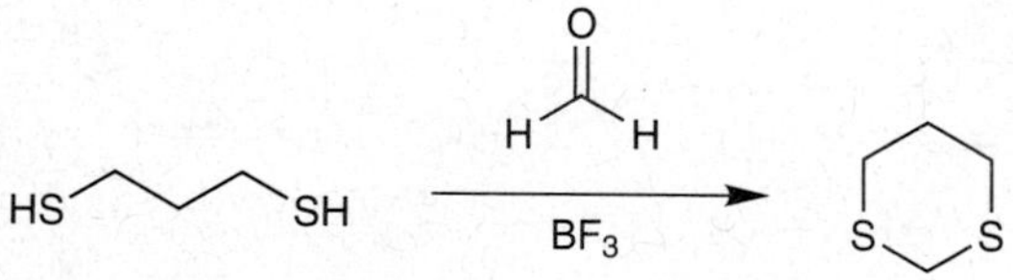

5.40)

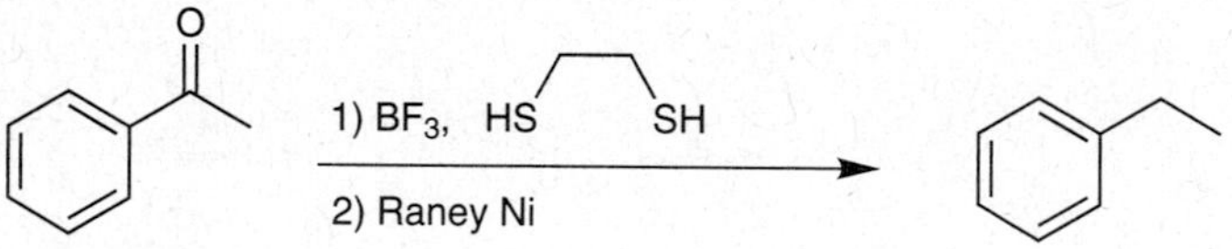

5.41)

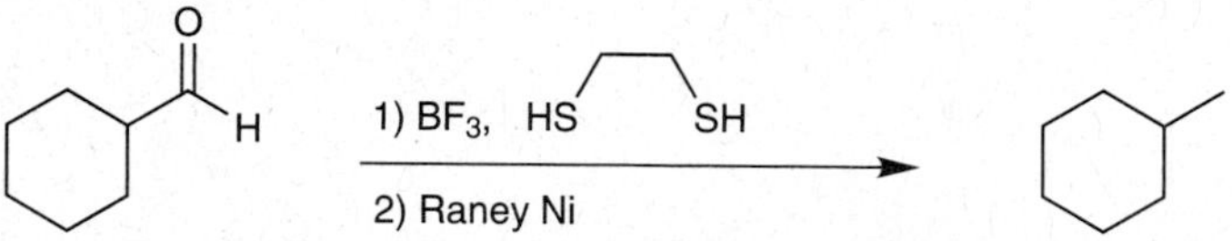

5.42)

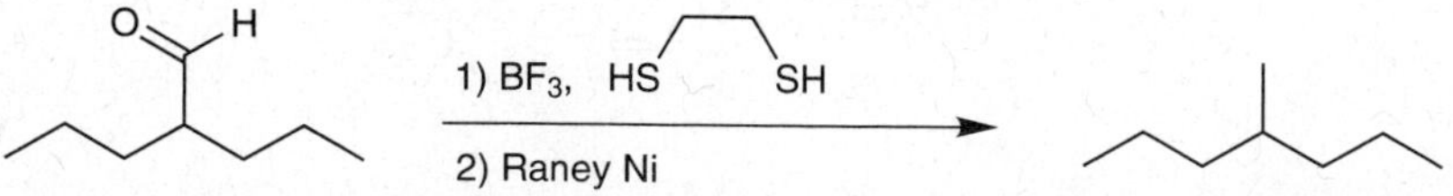

5.43)

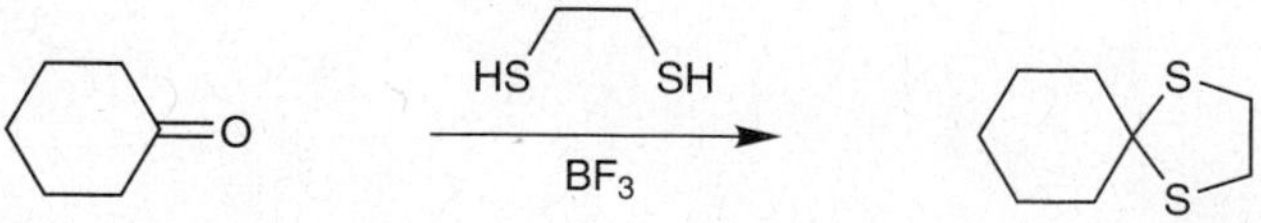

5.44)

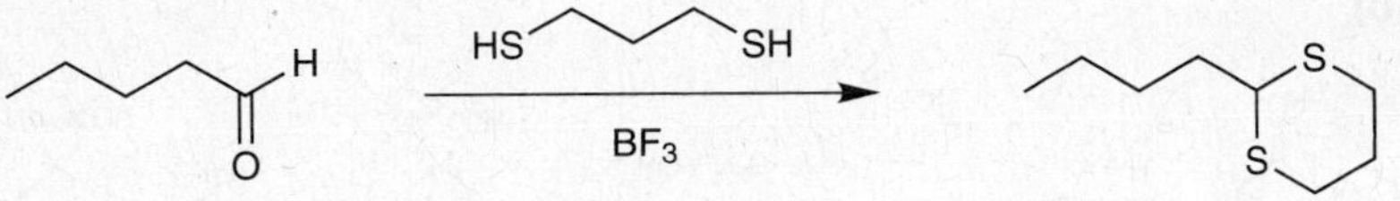

5.45)

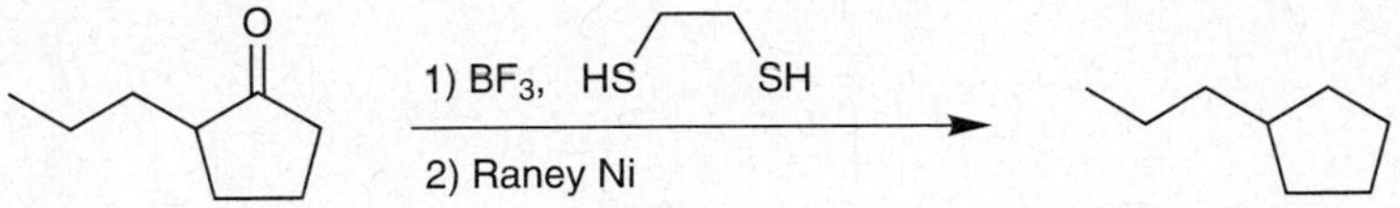

5.46)

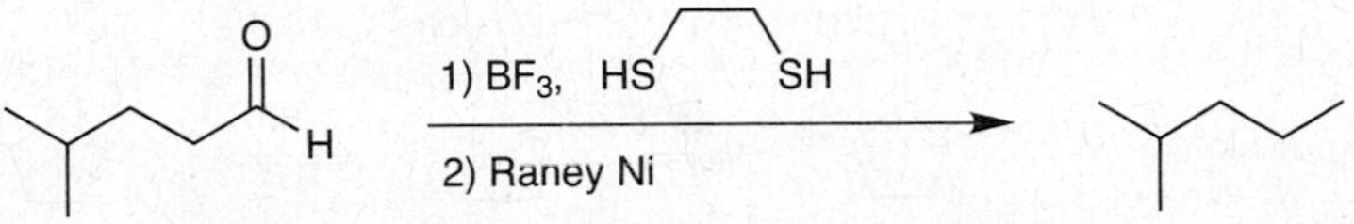

5.48)

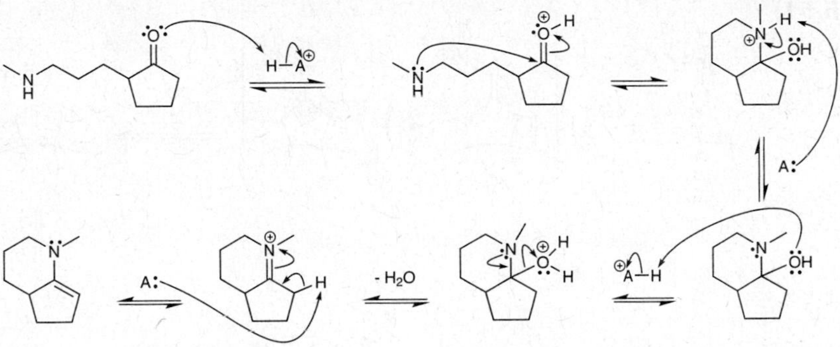

5.49)

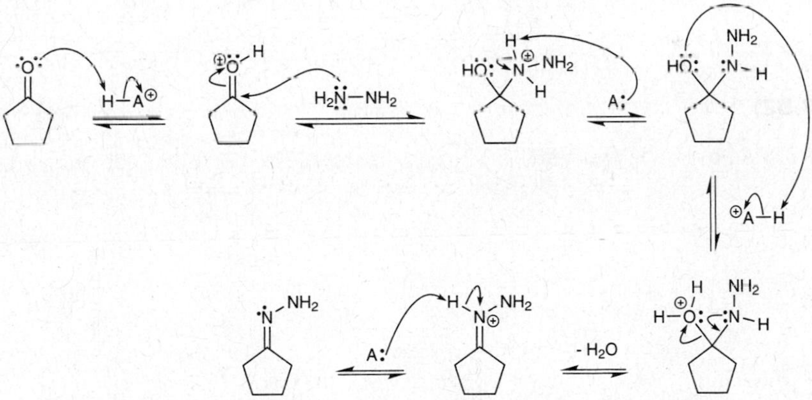

5.50)

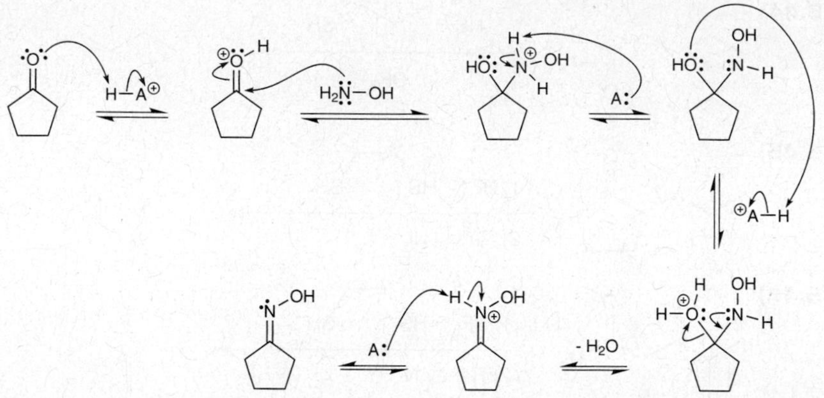

5.51)

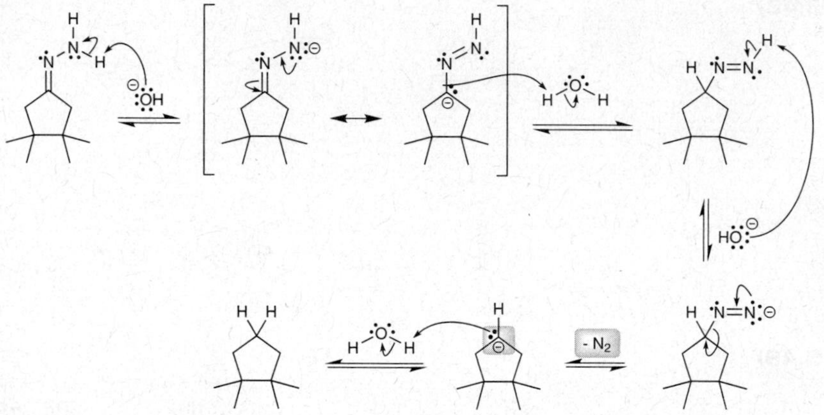

5.52)

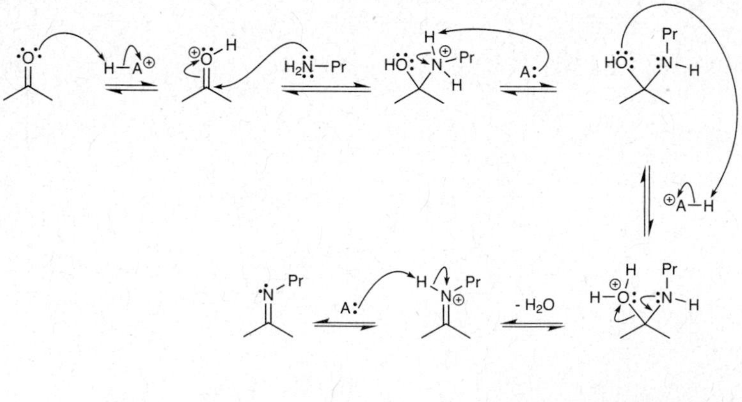

5.53)

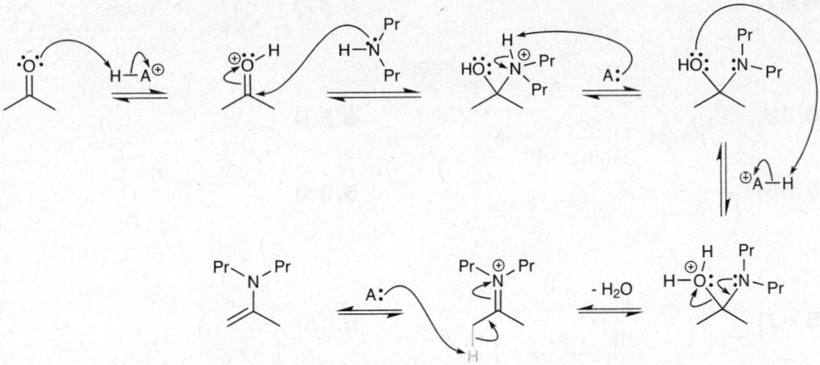

5.55)

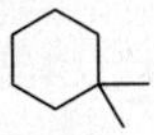

5.56)

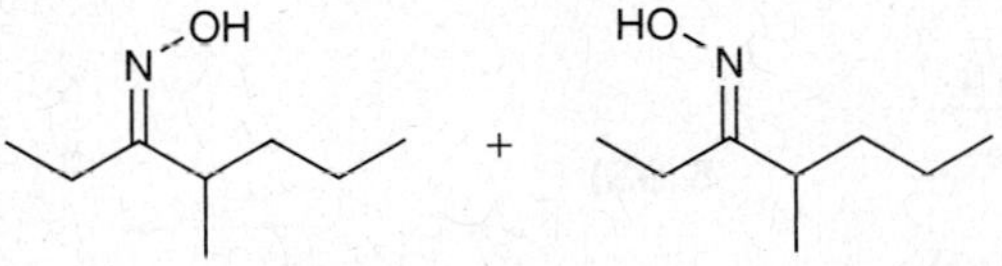

5.57) 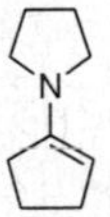 **5.58)** 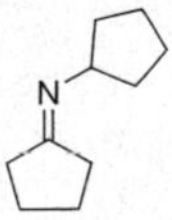 **5.59)** 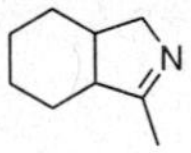 **5.60)**

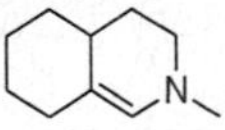

5.62) 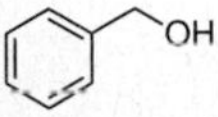 **5.63)** CH$_3$OH **5.64)** **5.65)**

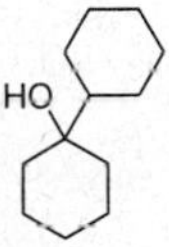

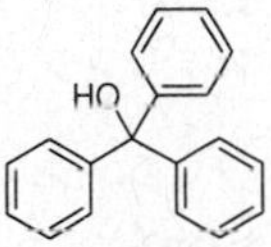

5.67) 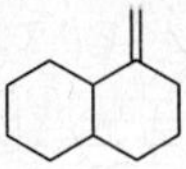 **5.68)**  **5.69)** 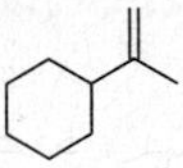 **5.71)**

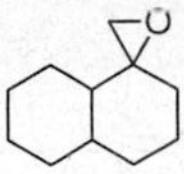

5.72) 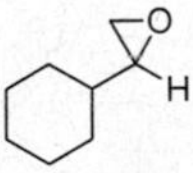 **5.73)** 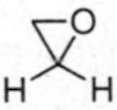 **5.75)**

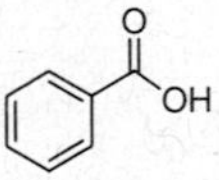

5.76)

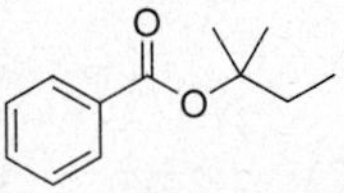

5.77)

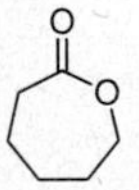

5.78)

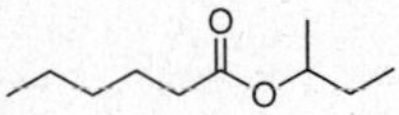

5.79)

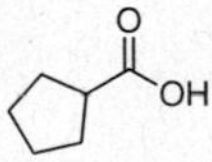

5.81)

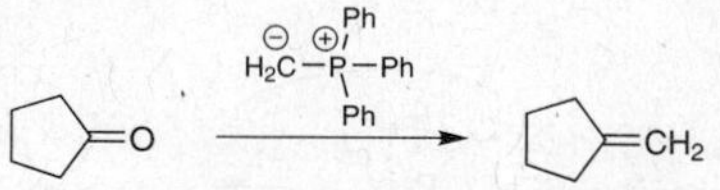
Ph
H2C–P–Ph
Ph
O
CH2

5.82)

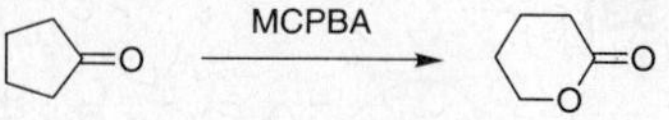
O
MCPBA
O
O

5.83)

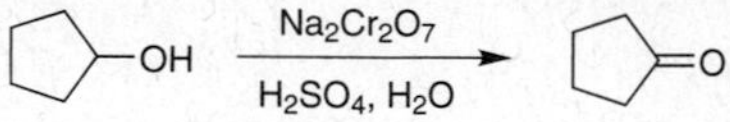
OH
Na2Cr2O7
H2SO4, H2O
O

5.84)

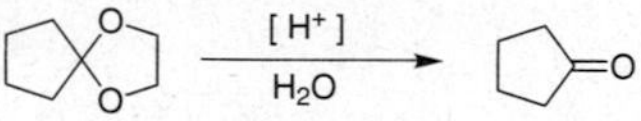
O
O
[H+]
H2O
O

5.85)

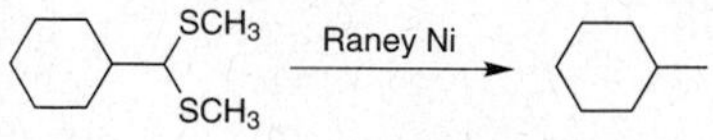
SCH3
SCH3
Raney Ni

5.86)

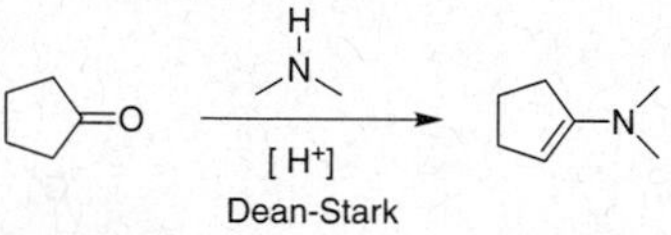
O
H
N
[H+]
Dean-Stark
N

5.87)

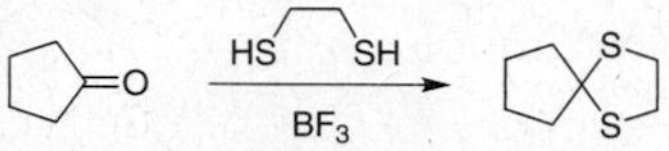
O
HS
SH
BF3
S
S

5.88)

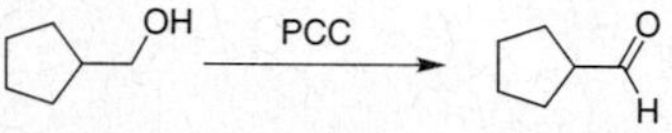
OH
PCC
O
H

5.89)

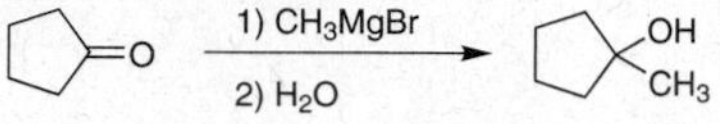
O
1) CH3MgBr
2) H2O
OH
CH3

5.90)

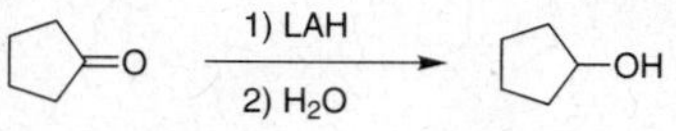
O
1) LAH
2) H2O
OH

5.91)

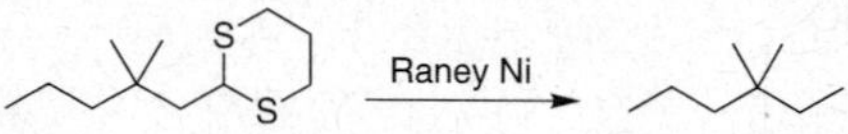
S
S
Raney Ni

5.92)

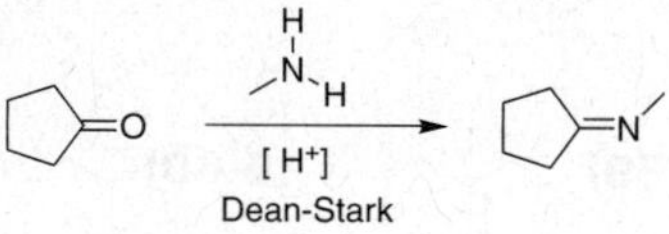
O
H
N
H
[H+]
Dean-Stark
N

5.93)

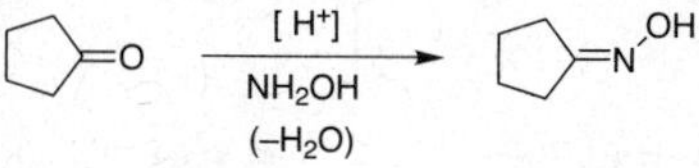
O
[H+]
NH2OH
(–H2O)
N
OH

5.94)

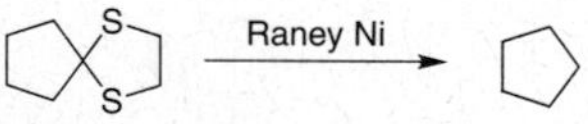
S
S
Raney Ni

5.95)

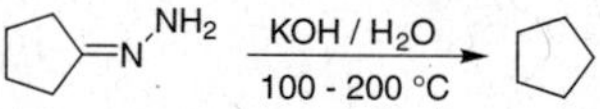
N
NH2
KOH / H2O
100 - 200 °C

5.96)

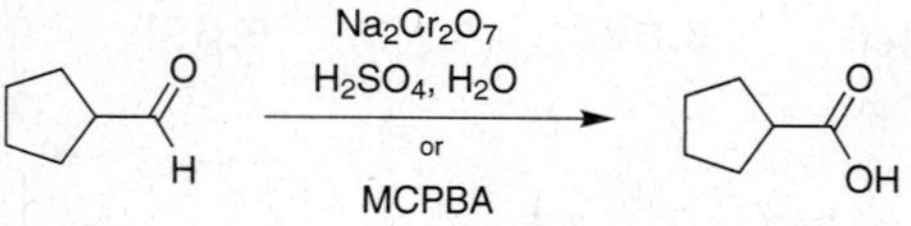
O
H
Na2Cr2O7
H2SO4, H2O
or
MCPBA
O
OH

5.97)

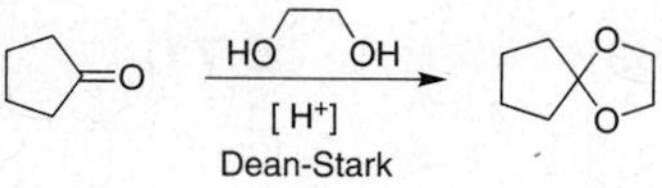
O
HO
OH
[H+]
Dean-Stark
O
O

5.98)

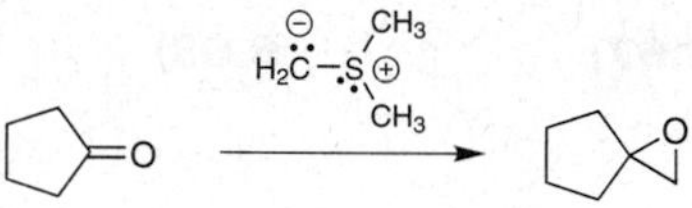
CH3
H2C–S
CH3
O
O

5.99)

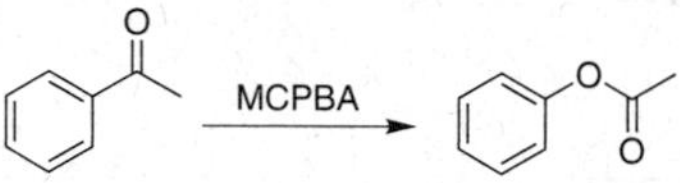
O
MCPBA
O
O

5.100)

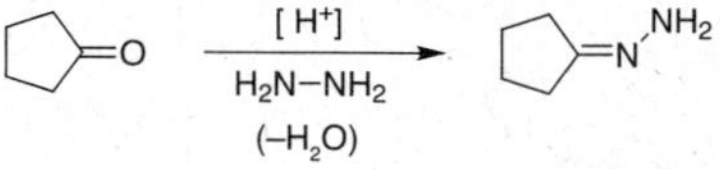
O
[H+]
H2N–NH2
(–H2O)
N
NH2

5.103)

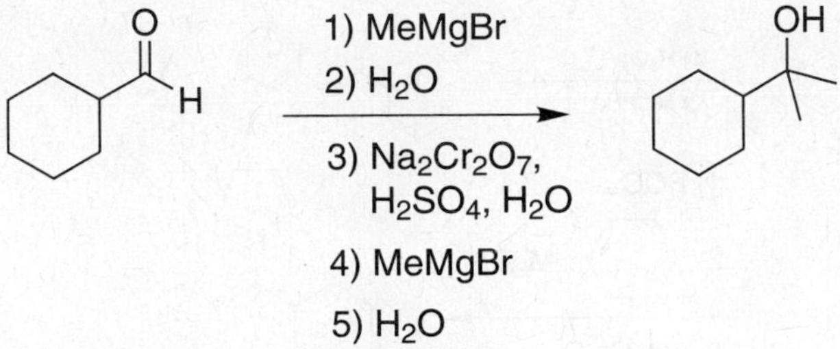

1) MeMgBr
2) H_2O
3) $Na_2Cr_2O_7$, H_2SO_4, H_2O
4) MeMgBr
5) H_2O

5.104)

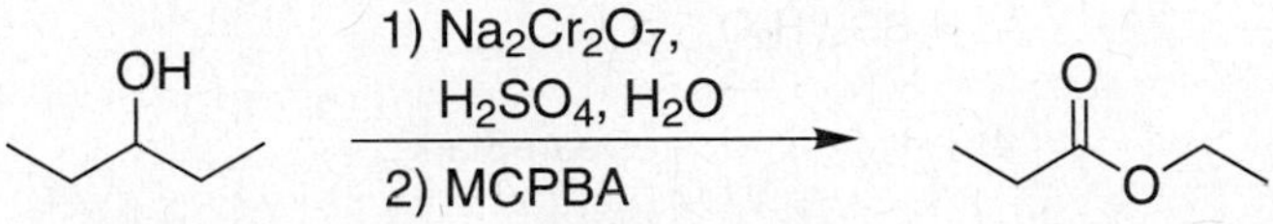

1) $Na_2Cr_2O_7$, H_2SO_4, H_2O
2) MCPBA

5.105)

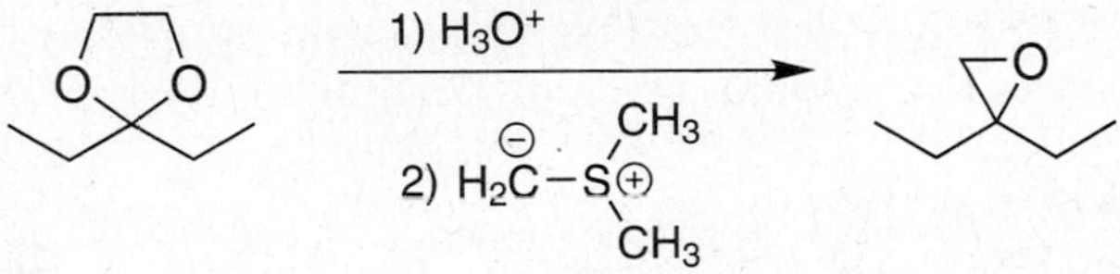

1) H_3O^+
2) $H_2\overset{\ominus}{C}-\overset{\oplus}{S}\overset{CH_3}{\underset{CH_3}{}}$

5.106)

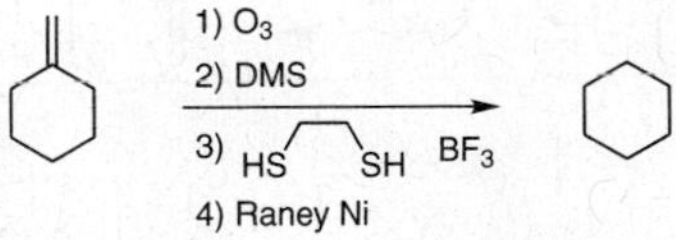

1) O_3
2) DMS
3) HS⁓SH BF_3
4) Raney Ni

5.107)

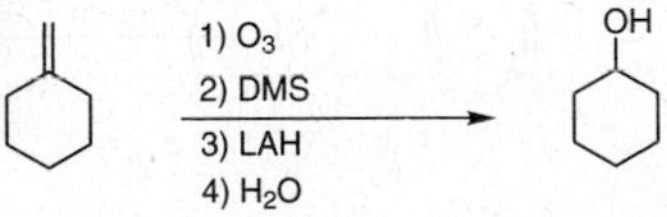

1) O_3
2) DMS
3) LAH
4) H_2O

5.108)

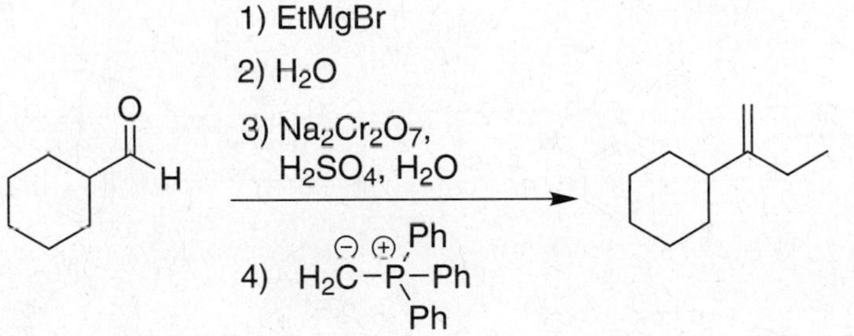

1) EtMgBr
2) H_2O
3) $Na_2Cr_2O_7$, H_2SO_4, H_2O
4) $H_2\overset{\ominus}{C}-\overset{\oplus}{P}\overset{Ph}{\underset{Ph}{}}Ph$

5.109)

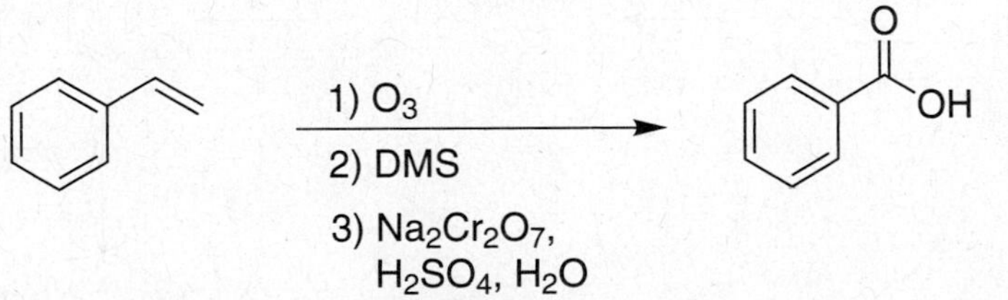

1) O_3
2) DMS
3) $Na_2Cr_2O_7$, H_2SO_4, H_2O

5.110)

1) EtMgBr
2) H_2O
3) $Na_2Cr_2O_7$, H_2SO_4, H_2O
4) MCPBA

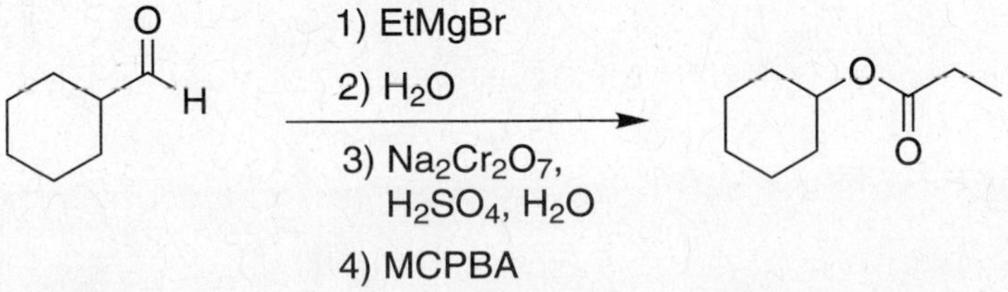

5.111)

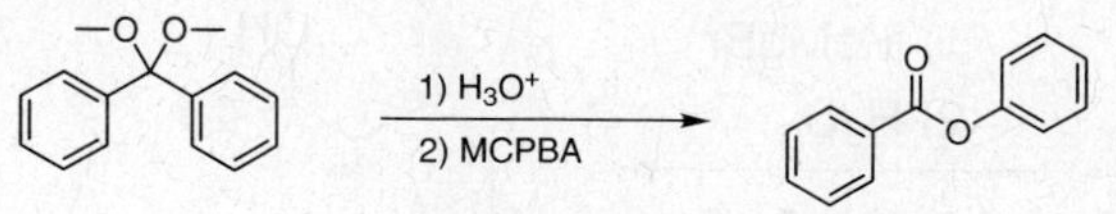

5.112)

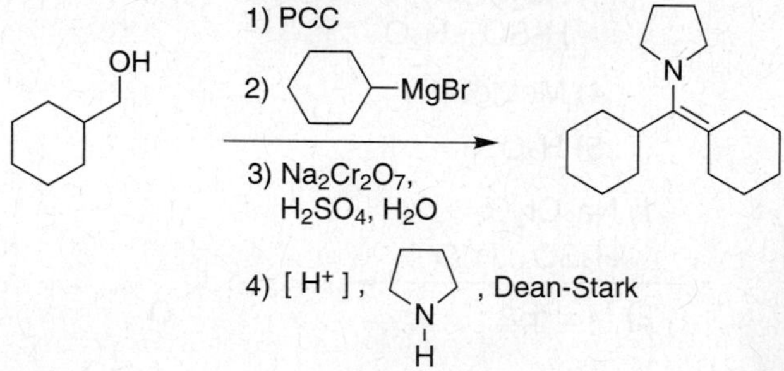

1) PCC

2) [cyclohexyl]—MgBr

3) Na₂Cr₂O₇,
   H₂SO₄, H₂O

4) [H⁺], [pyrrolidine N–H] , Dean-Stark

CHAPTER 6

6.2)

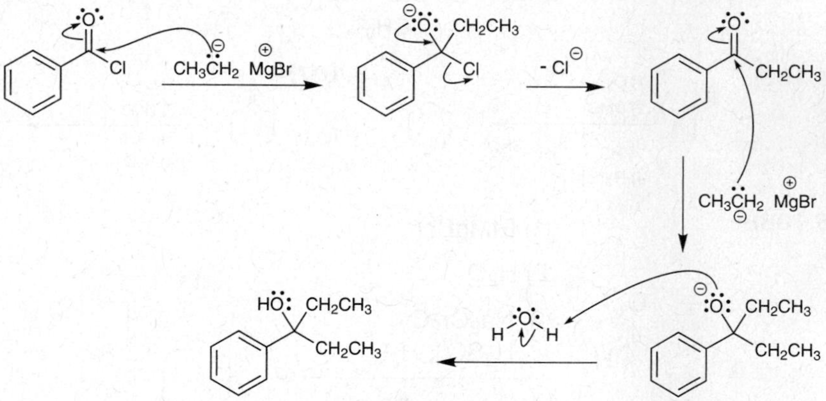

6.3)

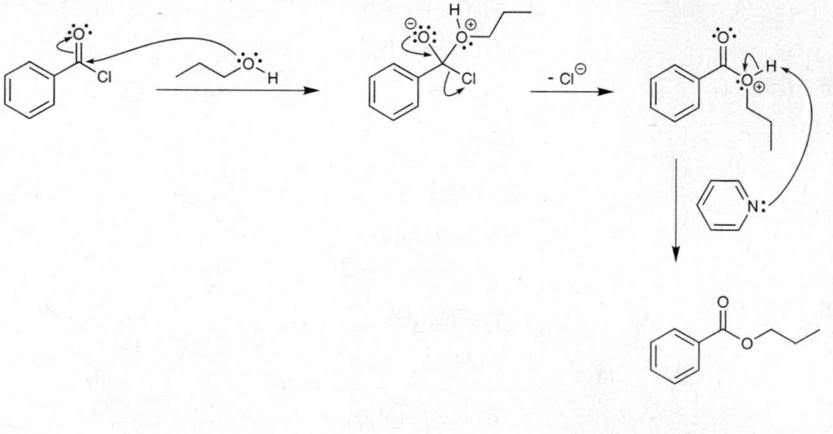

6.4)

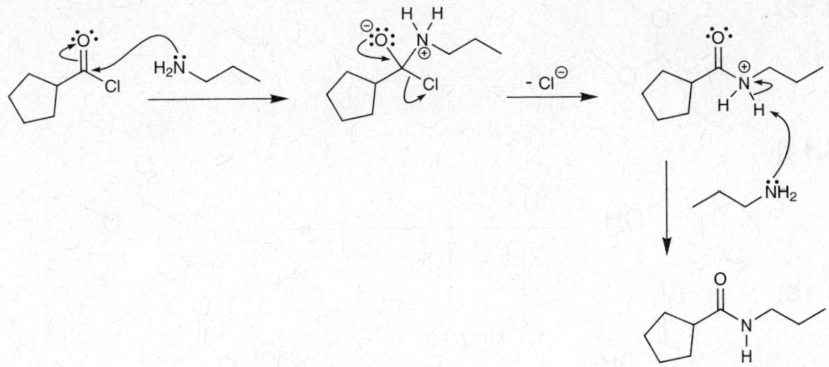

6.5)

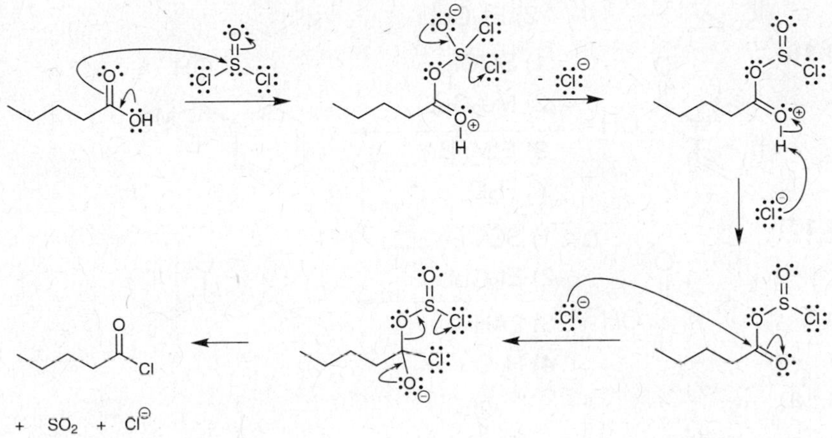

6.6)

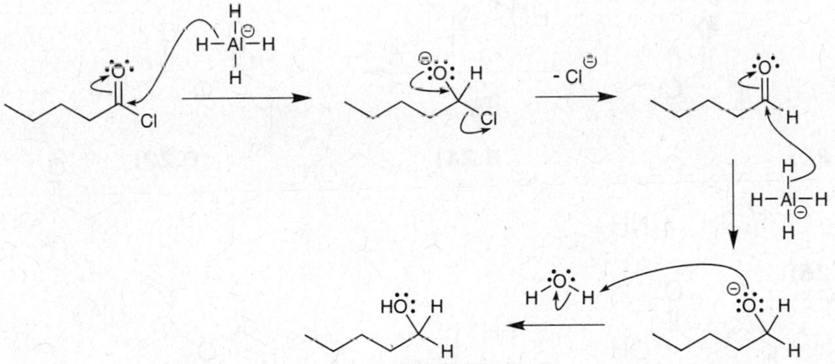

6.8)

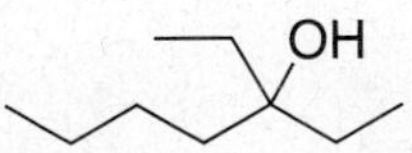

6.9)

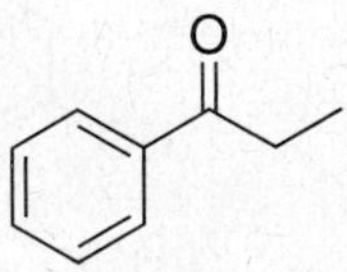

6.10)

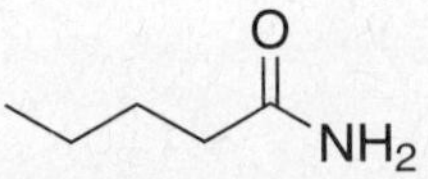

6.11)

6.12)

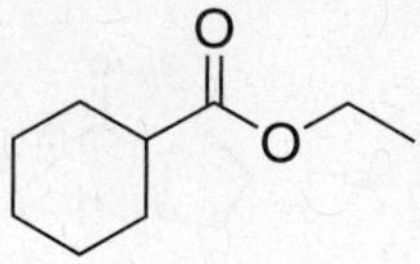

6.14)

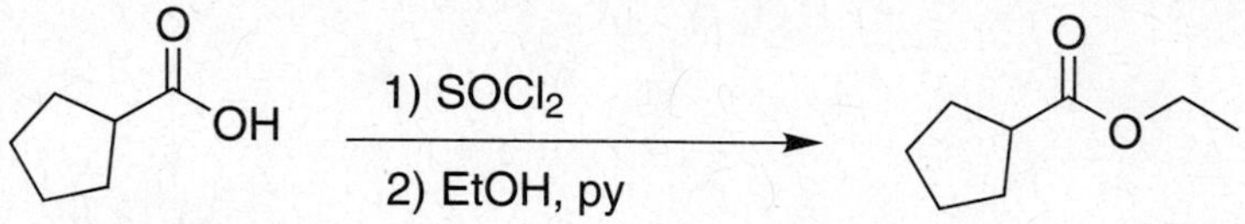

1) SOCl$_2$

2) EtOH, py

6.15)

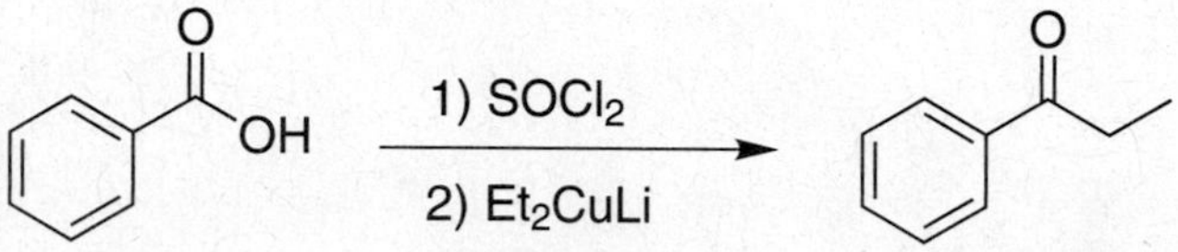

1) SOCl$_2$

2) Et$_2$CuLi

6.16)

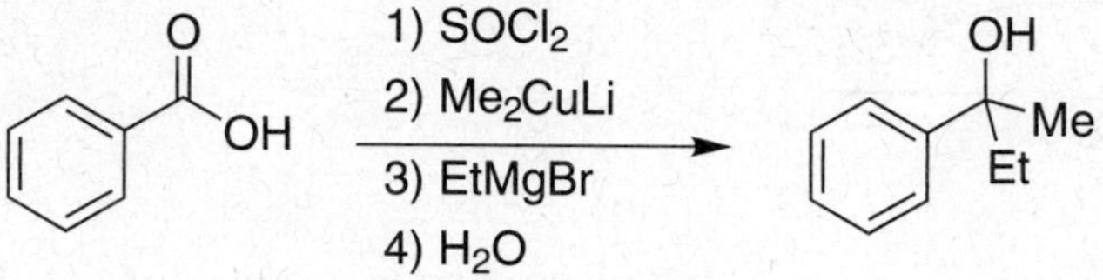

1) SOCl$_2$

2) Me$_2$CuLi

3) EtMgBr

4) H$_2$O

6.17)

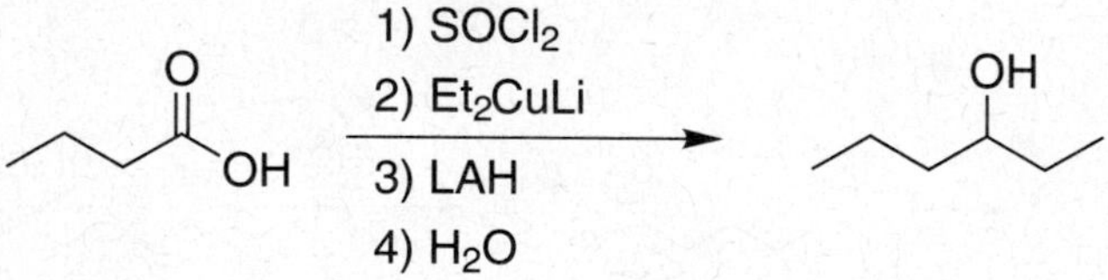

1) SOCl$_2$

2) Et$_2$CuLi

3) LAH

4) H$_2$O

6.18)

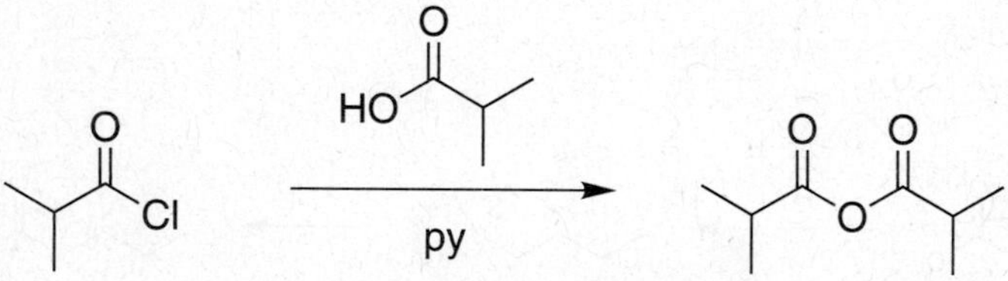

py

6.20)

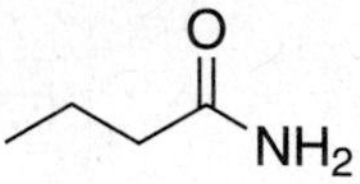

6.21)

6.22)

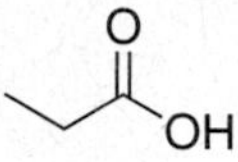

6.25)

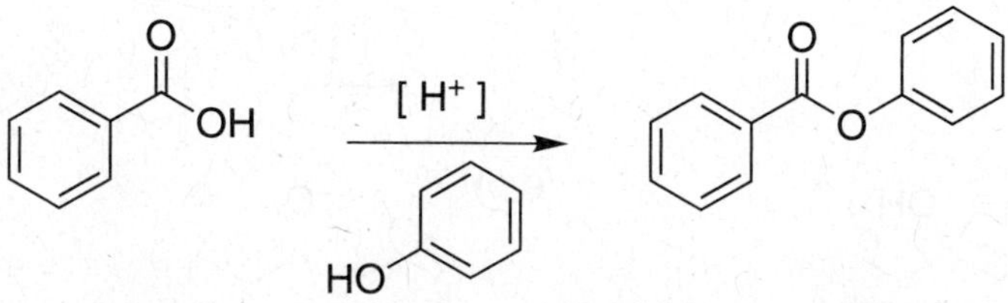

[H$^+$]

6.26)

[H$^+$]

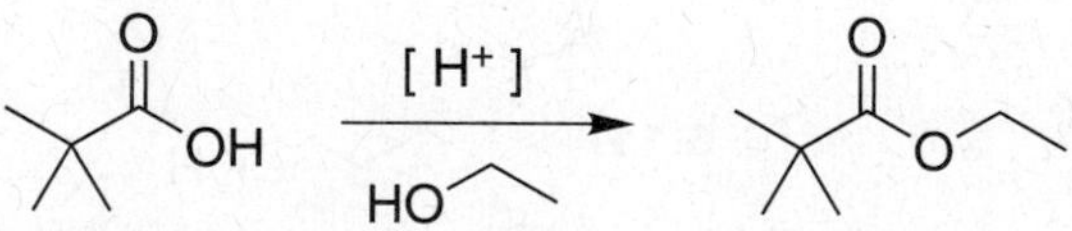

6.27)

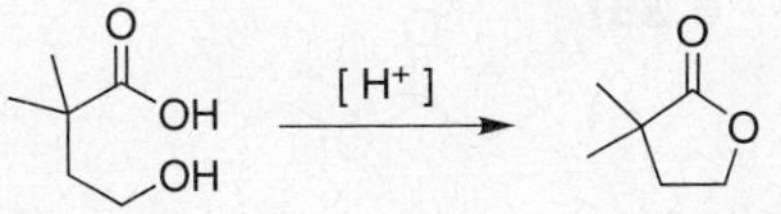

6.28)

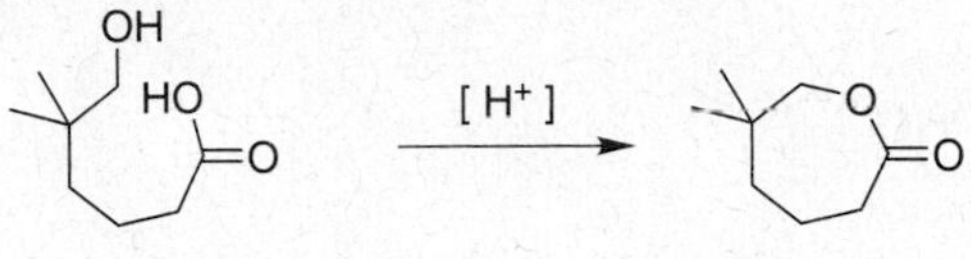

6.30)

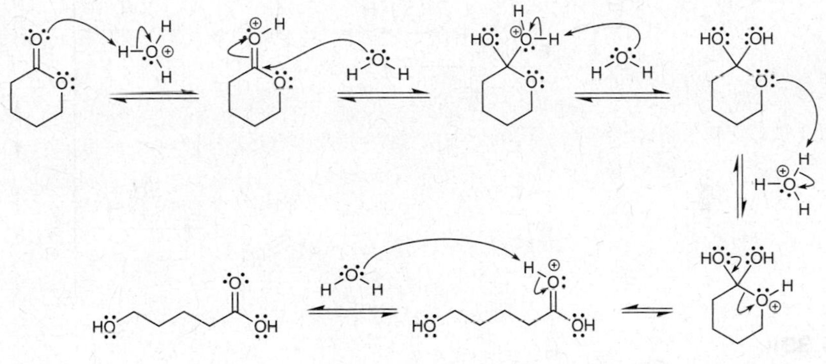

6.31)

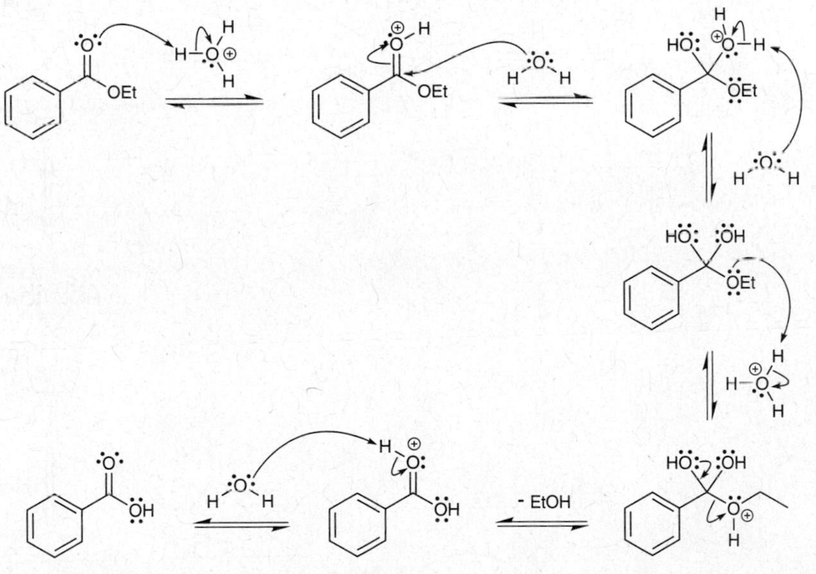

6.33)

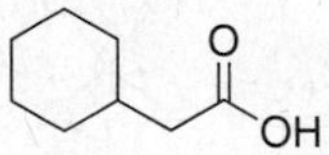

+ CH$_3$OH

6.34)

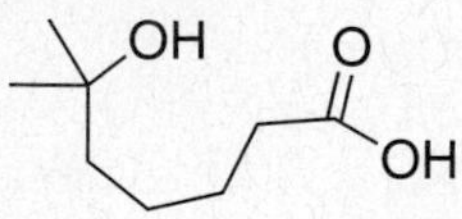

6.35)

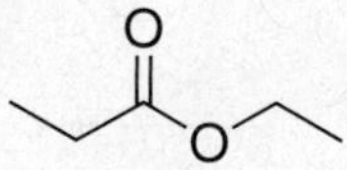

6.36)

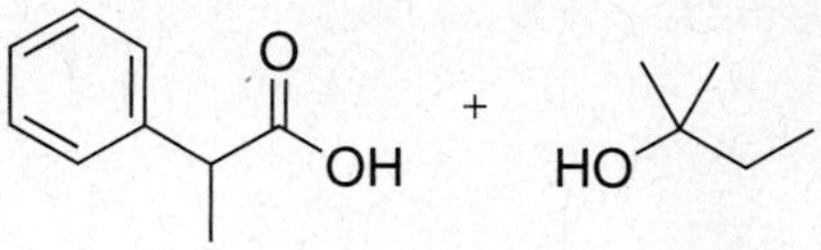

6.37)

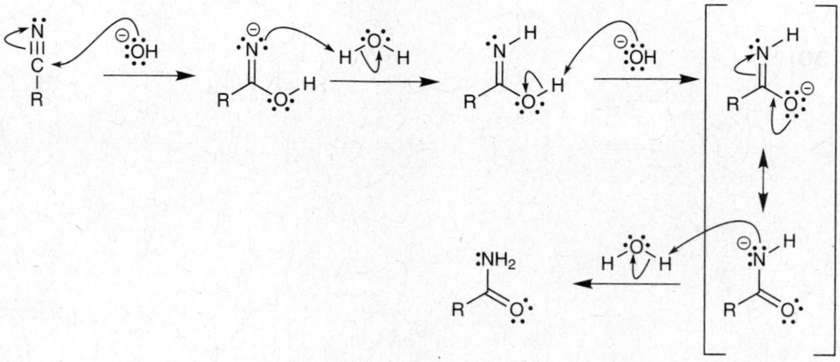

6.39)

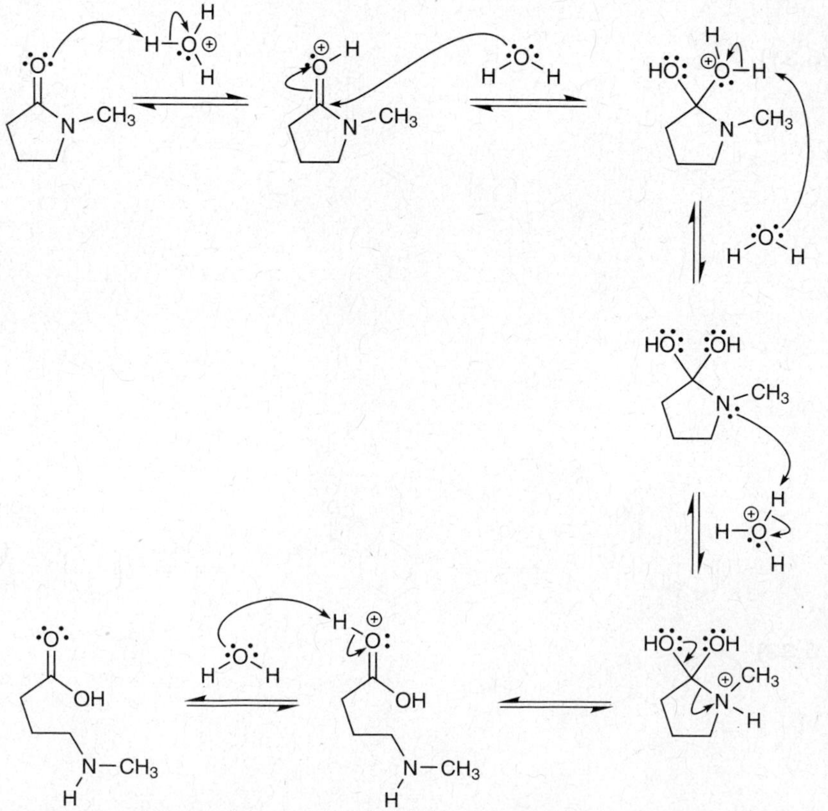

6.40)

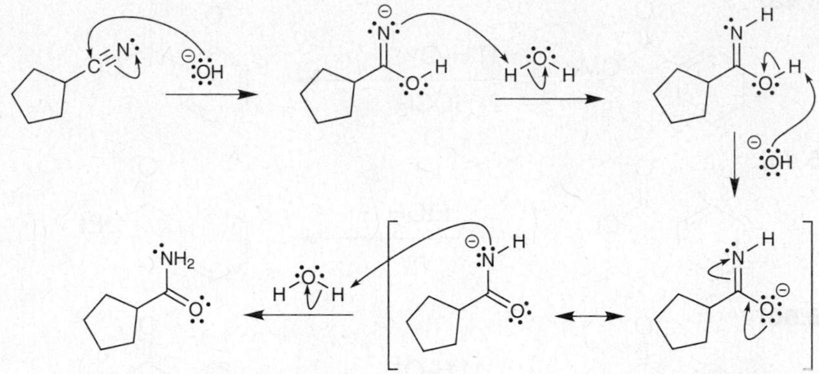

6.41)

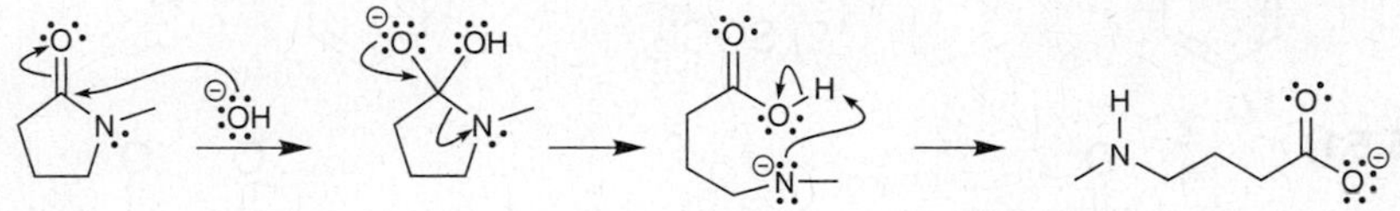

6.42)

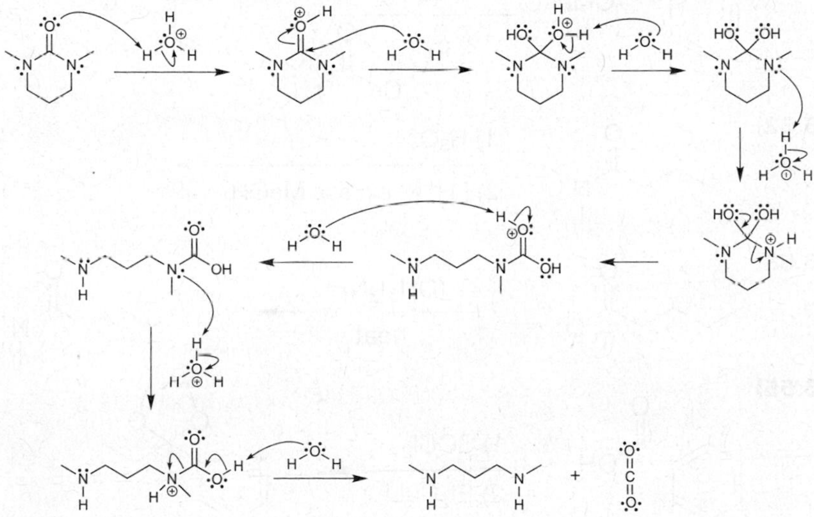

6.44)

CH_3NH_2 +

6.45)

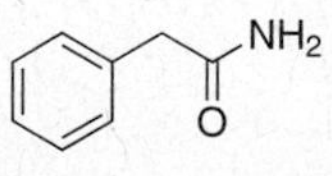

6.46)

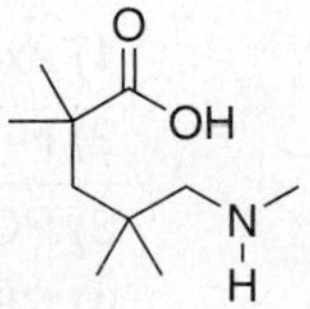

6.48)

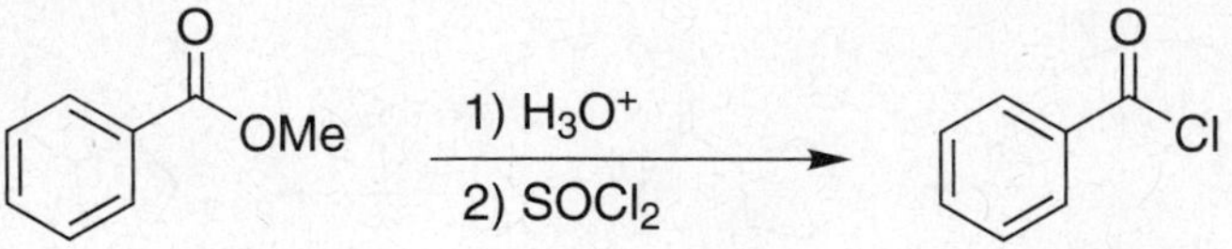
O
OMe
1) H_3O^+
2) $SOCl_2$
O
Cl

6.49)

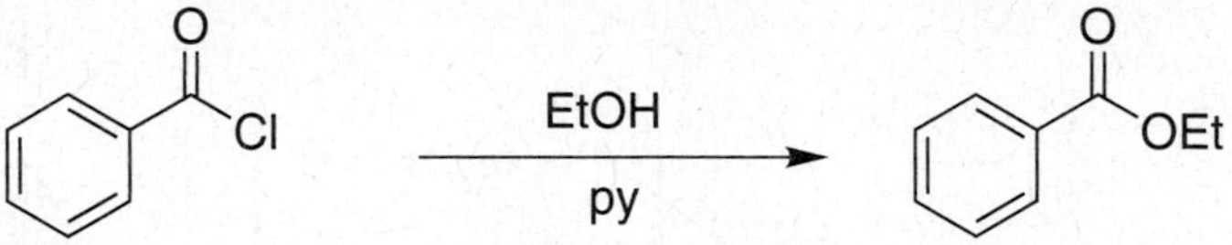
O
Cl
EtOH
py
O
OEt

6.50)

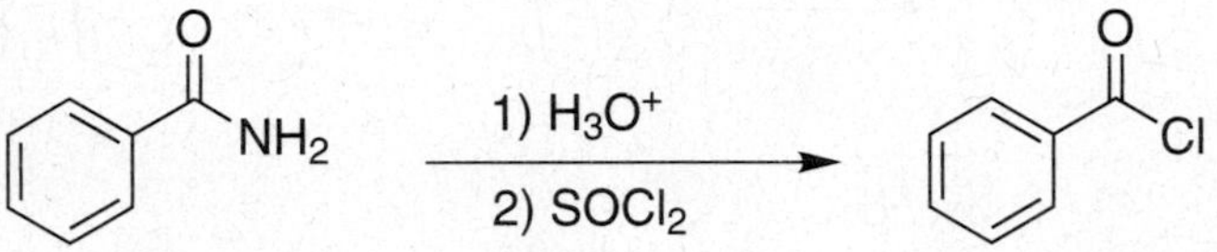
O
NH_2
1) H_3O^+
2) $SOCl_2$
O
Cl

6.51)

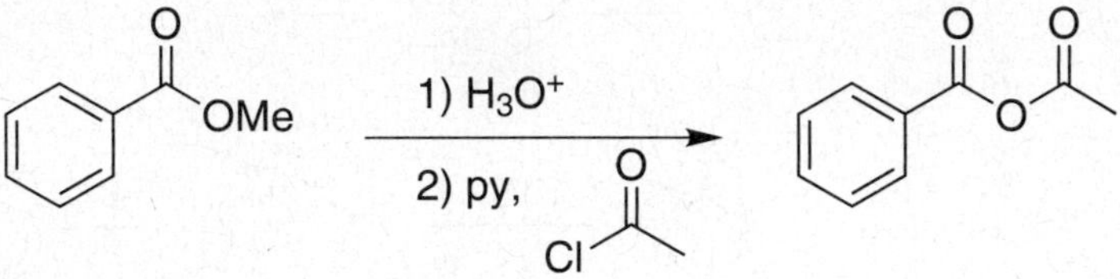
O
OMe
1) H_3O^+
2) py,
O
Cl
O O
O

6.52)

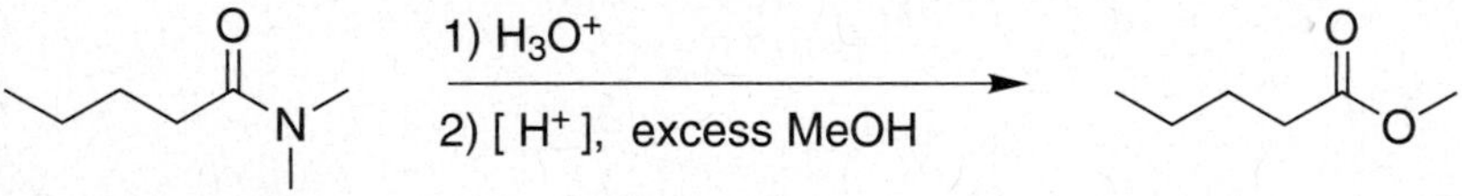
O
N
1) H_3O^+
2) [H^+], excess MeOH
O
O

6.53)

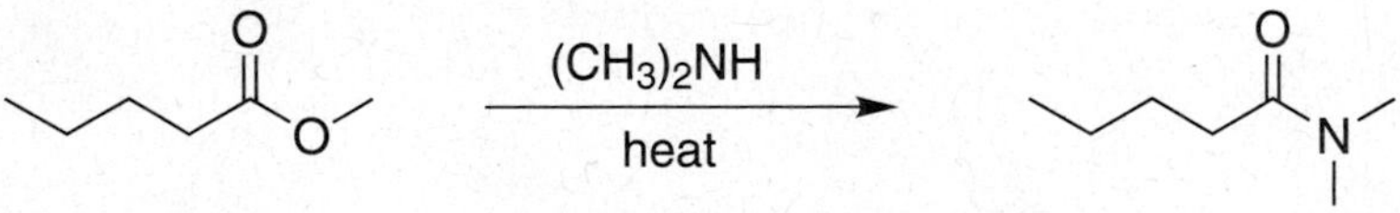
O
O
$(CH_3)_2NH$
heat
O
N

6.55)

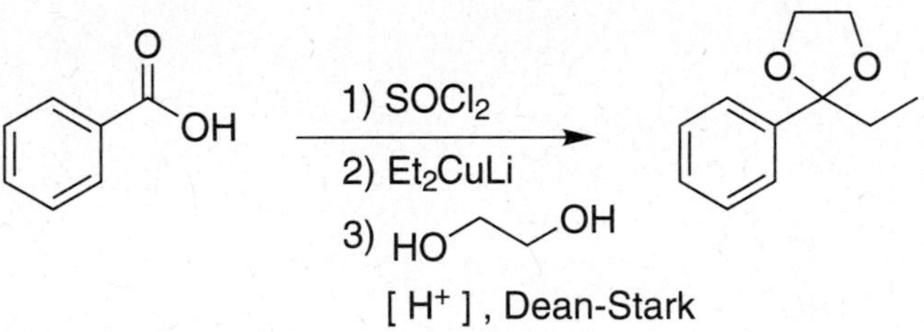
O
OH
1) $SOCl_2$
2) Et_2CuLi
3) HO OH
[H^+] , Dean-Stark
O O

6.56)

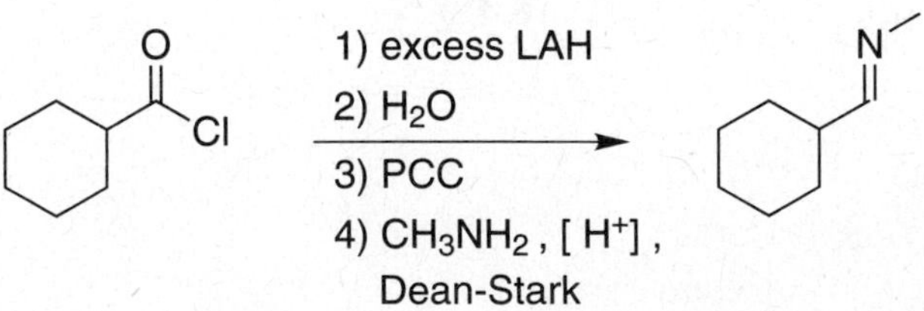
O
Cl
1) excess LAH
2) H_2O
3) PCC
4) CH_3NH_2 , [H^+] ,
Dean-Stark
N

6.57)

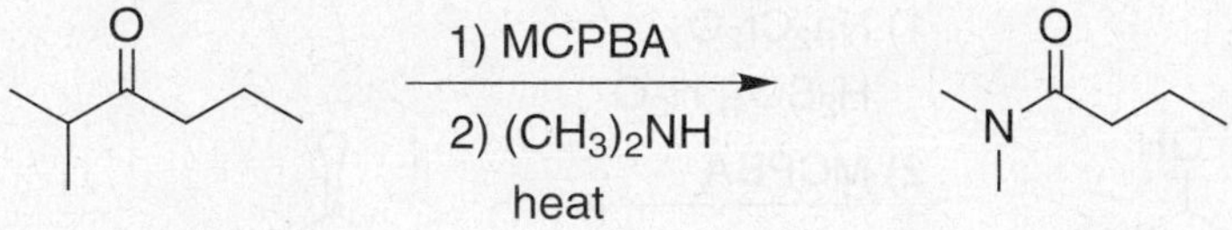

1) MCPBA

2) (CH₃)₂NH
   heat

6.58)

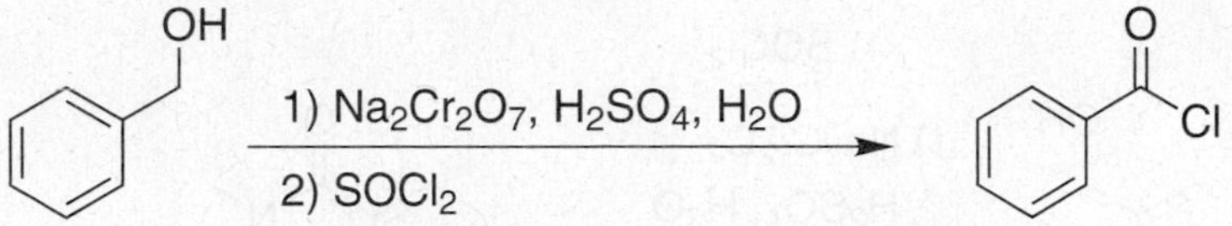

1) Na₂Cr₂O₇, H₂SO₄, H₂O

2) SOCl₂

6.59)

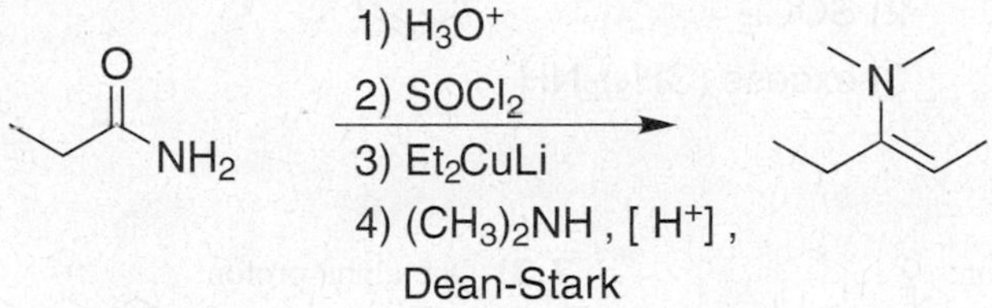

1) H₃O⁺

2) SOCl₂

3) Et₂CuLi

4) (CH₃)₂NH , [H⁺],
   Dean-Stark

6.60)

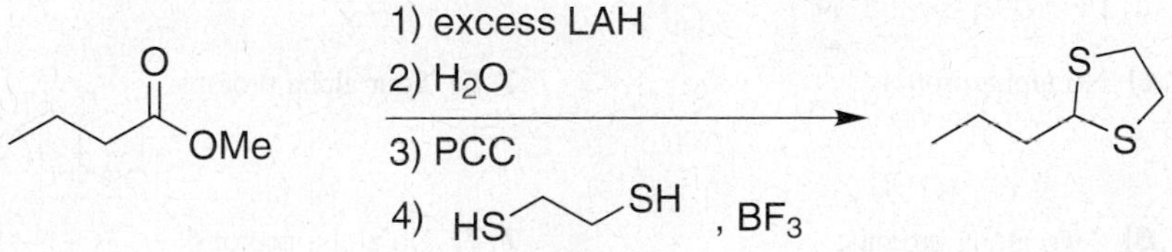

1) excess LAH

2) H₂O

3) PCC

4) HS⌒SH , BF₃

6.61)

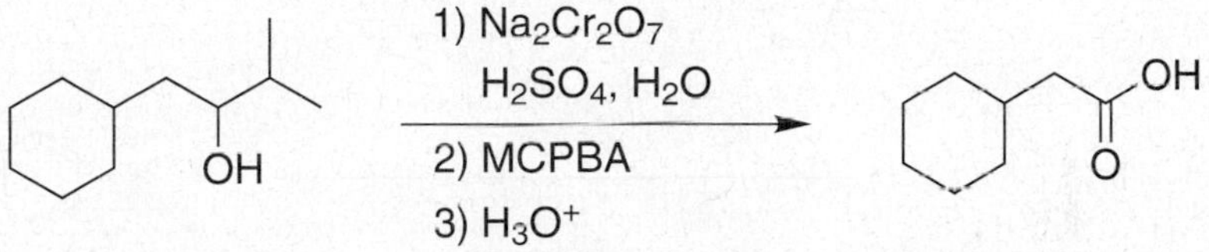

1) Na₂Cr₂O₇
   H₂SO₄, H₂O

2) MCPBA

3) H₃O⁺

6.62)

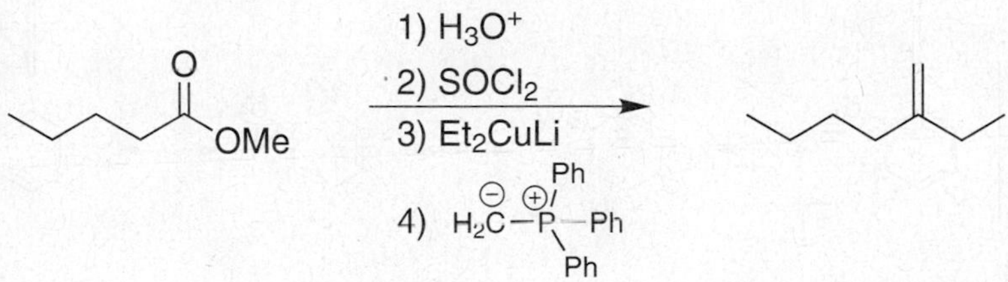

1) H₃O⁺

2) SOCl₂

3) Et₂CuLi

4) H₂C—P⁺—Ph (with Ph groups)

6.63)

1) SOCl₂

2) excess LAH

3) H₂O

4) PCC

5) HO⌒OH

[H⁺] , Dean-Stark

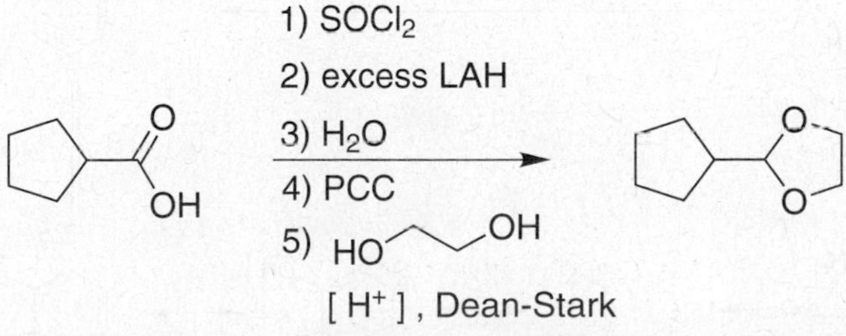

6.64)

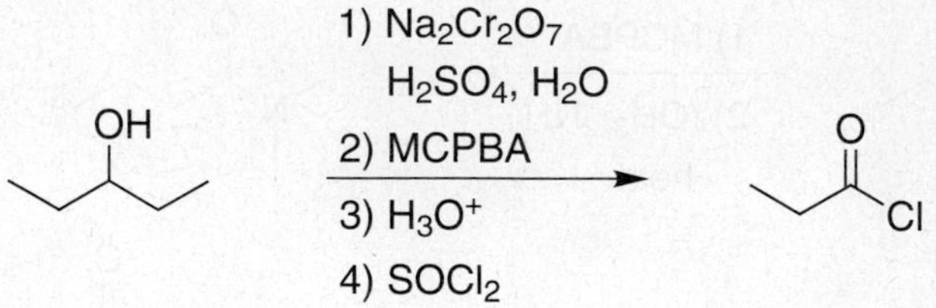

1) Na₂Cr₂O₇
   H₂SO₄, H₂O
2) MCPBA
3) H₃O⁺
4) SOCl₂

6.65)

1) Na₂Cr₂O₇
   H₂SO₄, H₂O
2) SOCl₂
3) excess (CH₃)₂NH

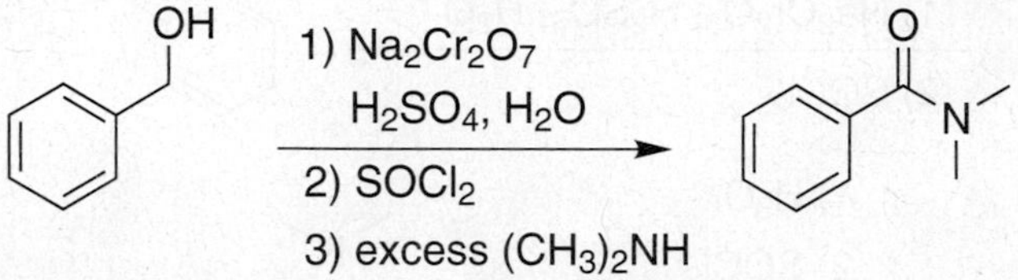

CHAPTER 7

7.2) One alpha proton:

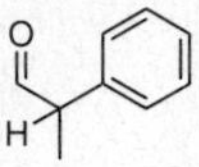

7.3) One alpha proton:

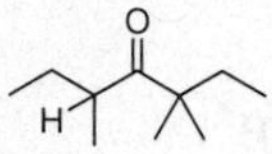

7.4) No alpha protons

7.5) Four alpha protons:

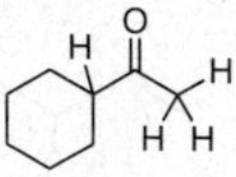

7.6) Two alpha protons:

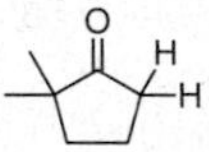

7.7) No alpha protons

7.9)

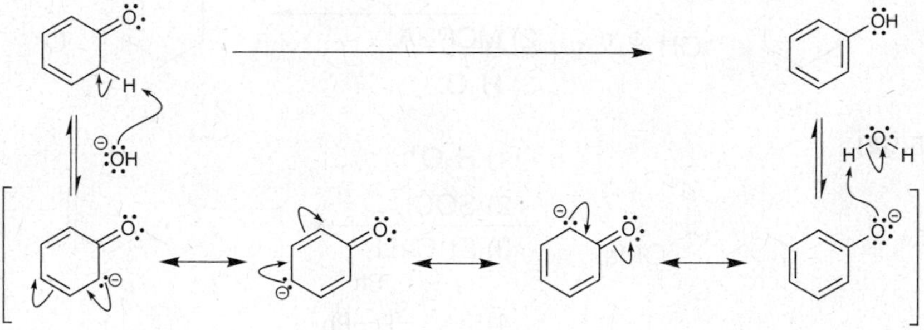

7.10)

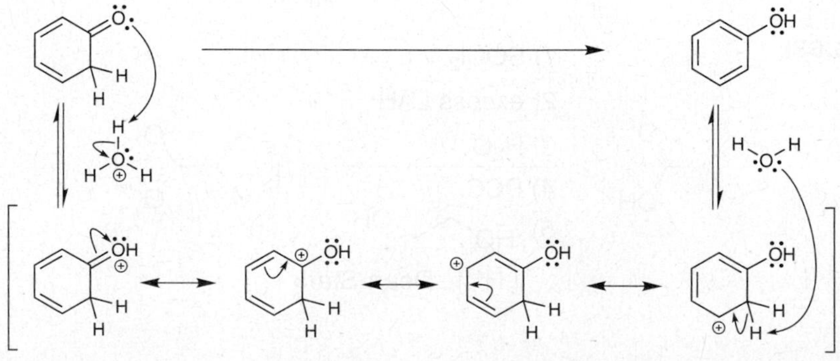

7.11)

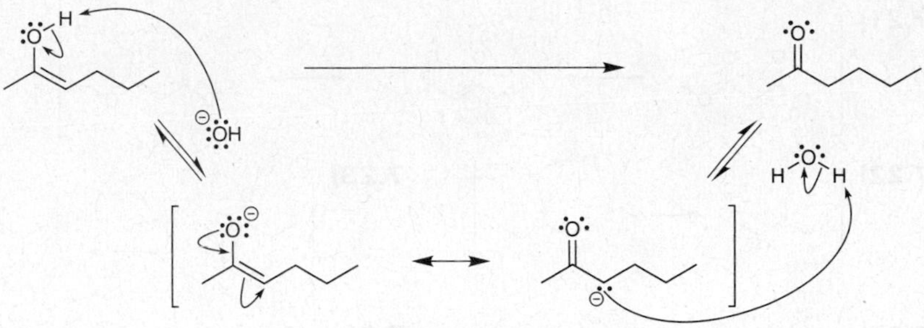

7.12)

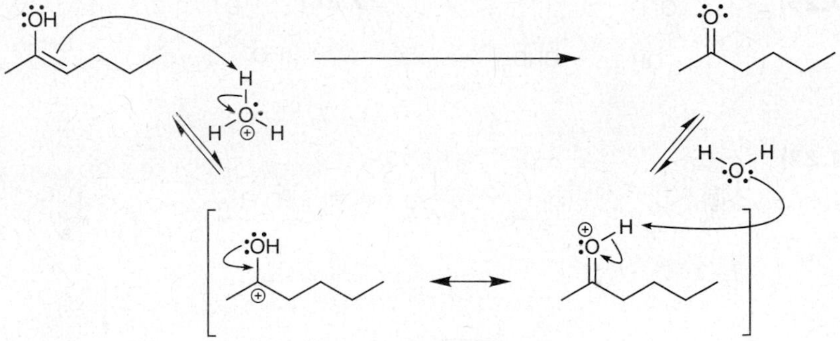

7.13)

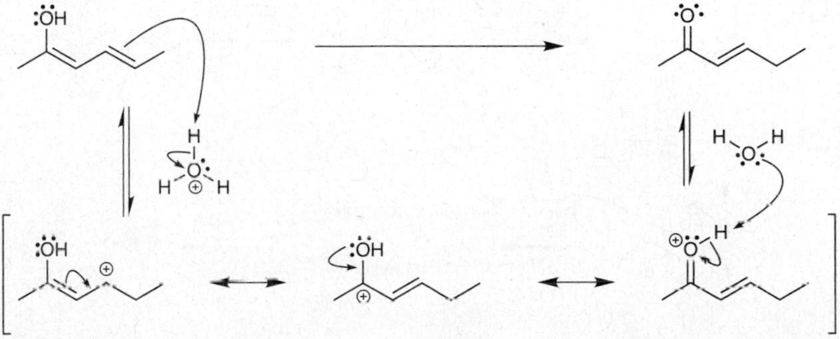

7.15)

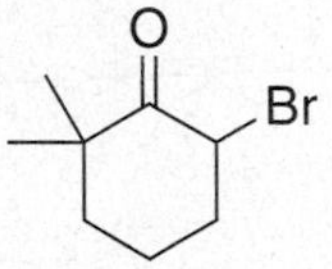

7.16)

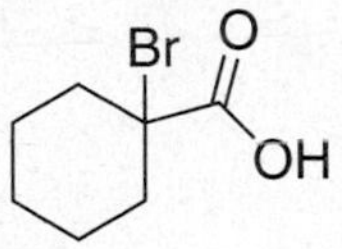

7.17)

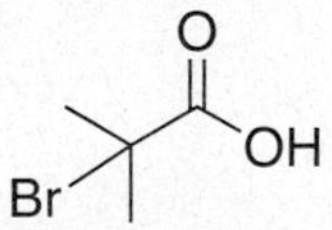

7.18)

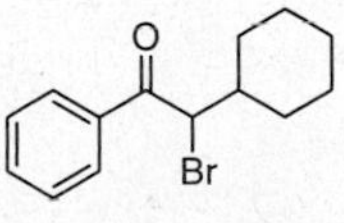

7.20)

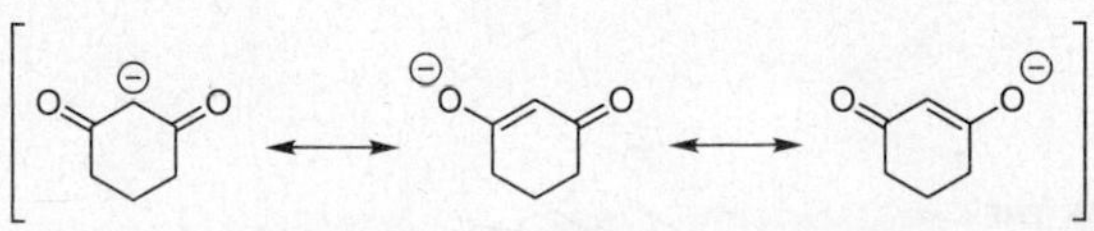

7.21)

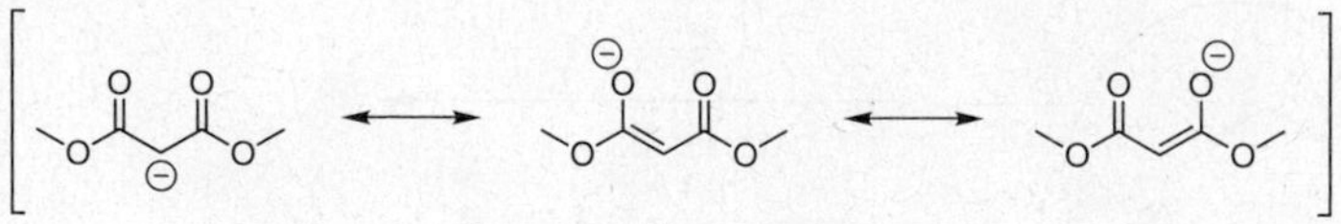

7.22)

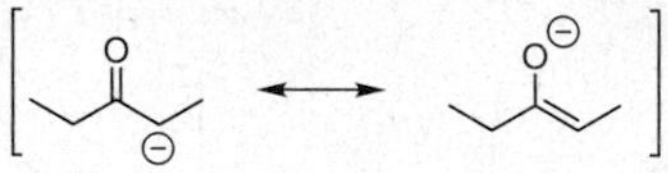

7.23)

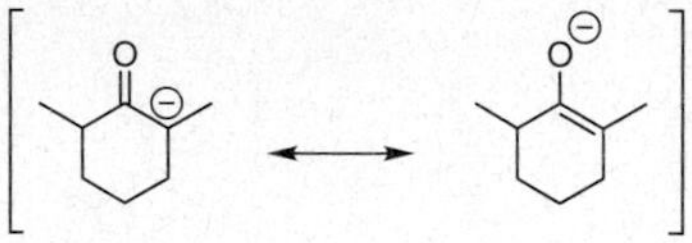

7.25)

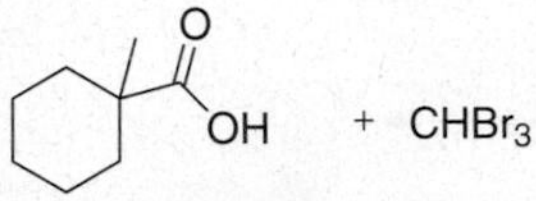

+ CHBr₃

7.26)

+ CHBr₃

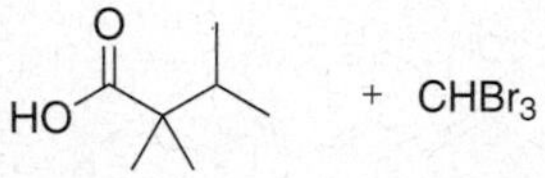

7.27)

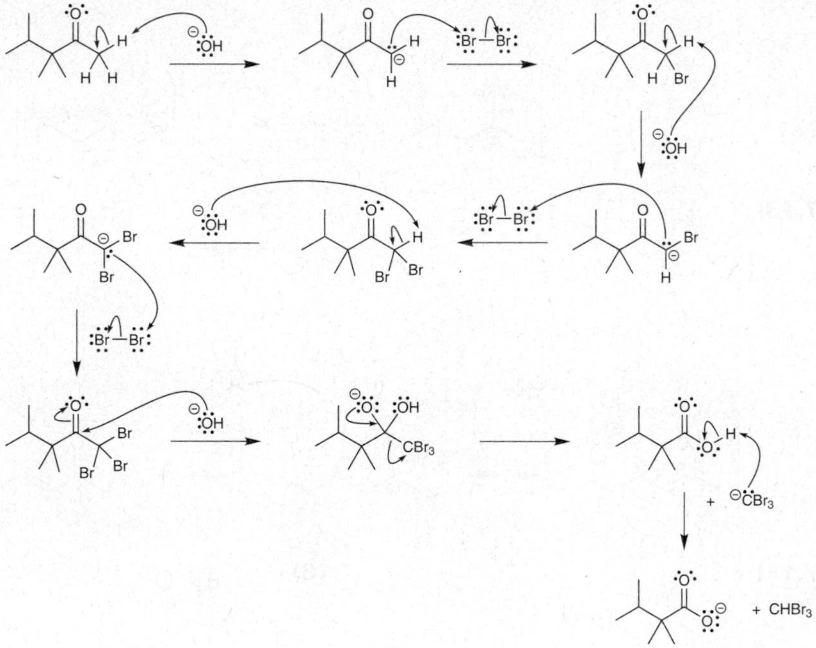

7.29)

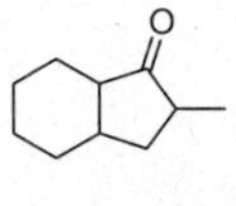

7.30)

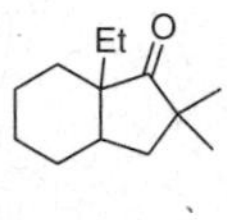

7.31)

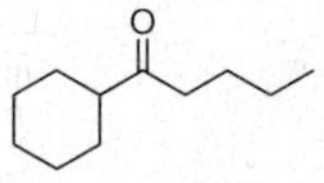

7.32)

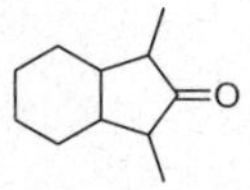

7.34)

1) LDA, THF
2) MeI

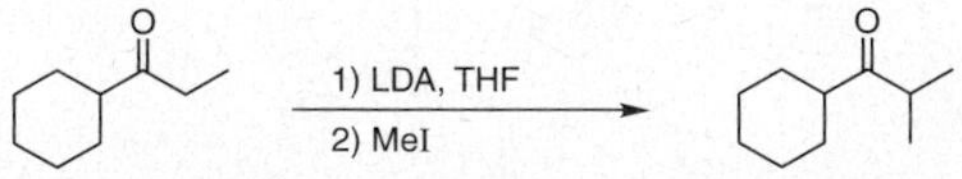

7.35)

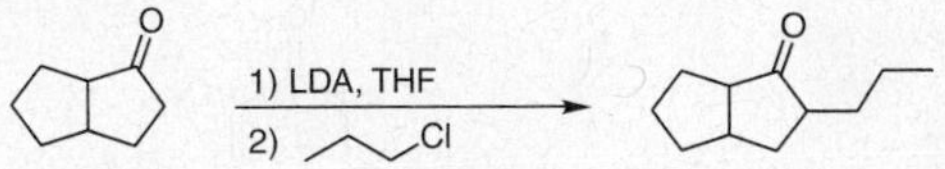

1) LDA, THF
2) [structure: CH₂=CH—CH₂—Cl]

7.36)

1) LDA, THF
2) Cl—CH₂—[cyclohexane]

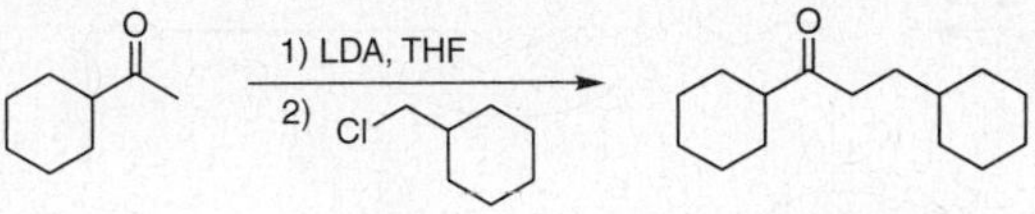

7.38)

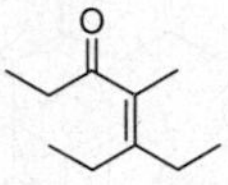

7.39)

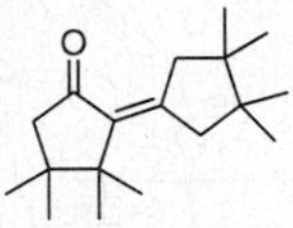

7.40)

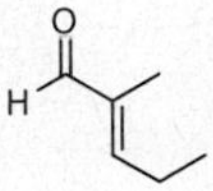

7.42)

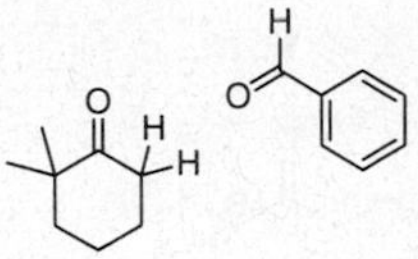

7.43)

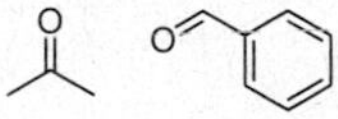

7.44)

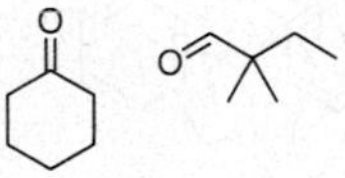

7.45)

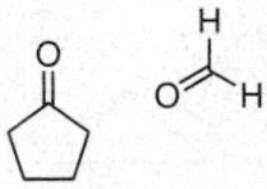

7.47)

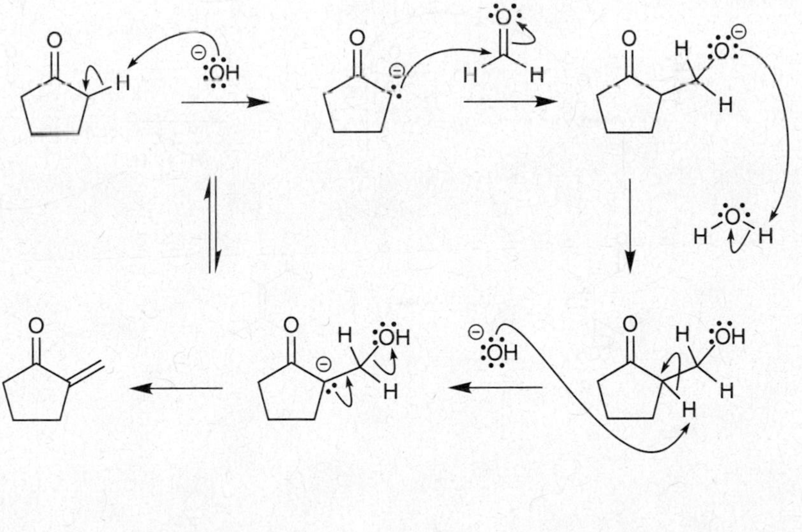

7.48)

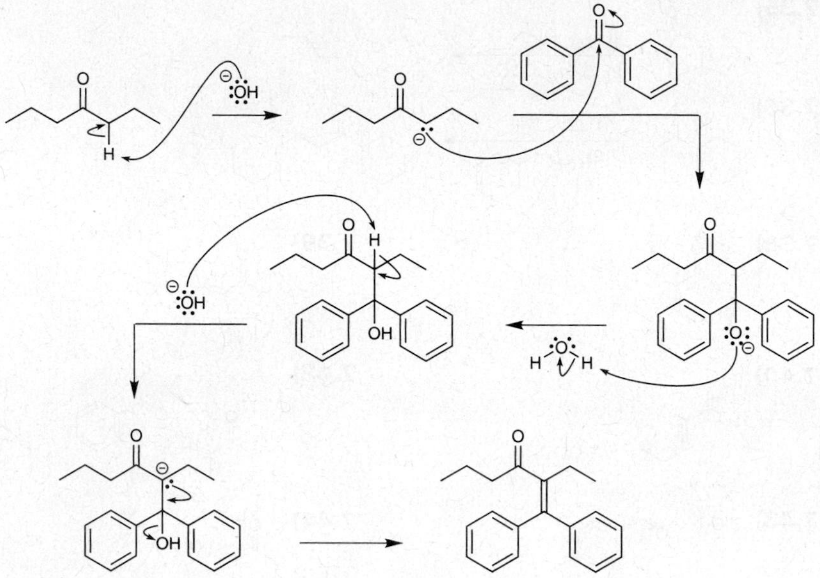

7.49)

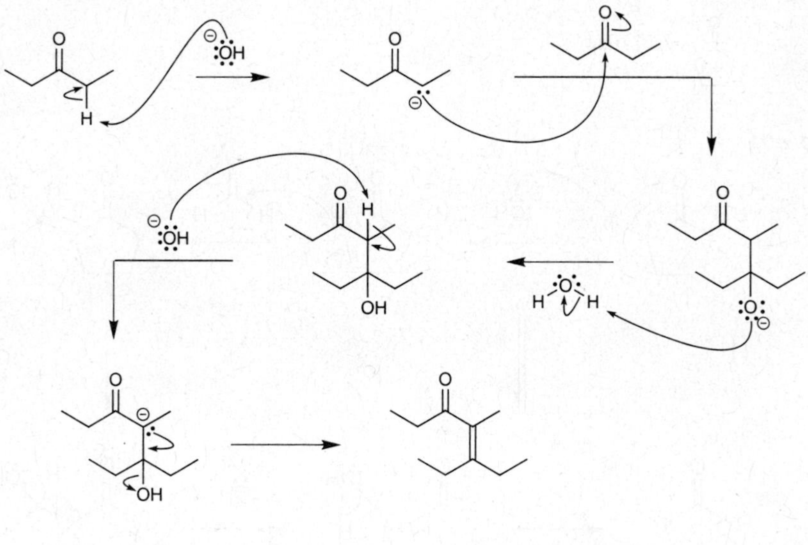

7.50)

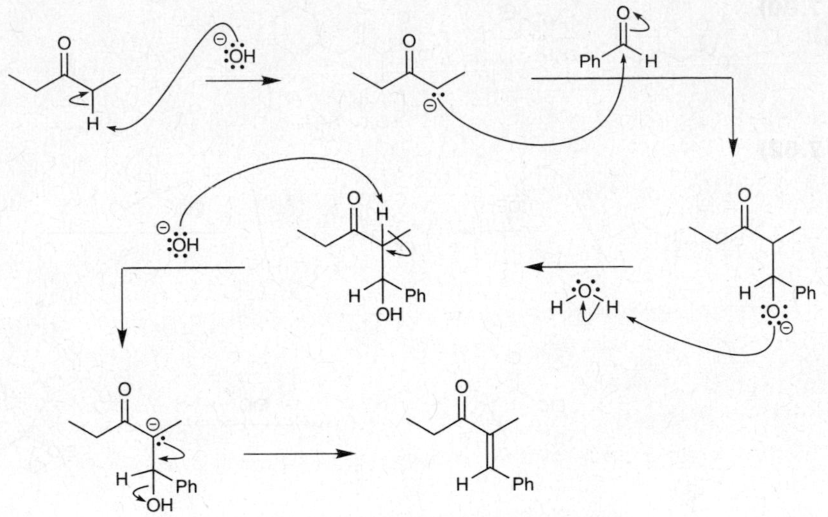

7.52)

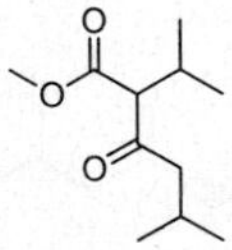

7.53)

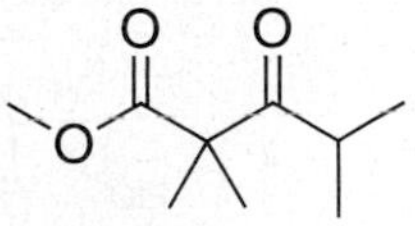

7.54)

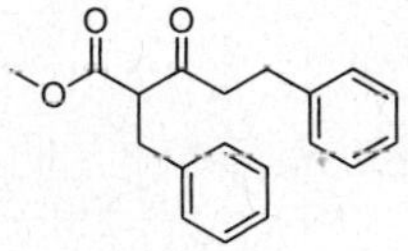

7.55)

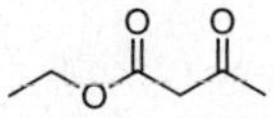

7.57)

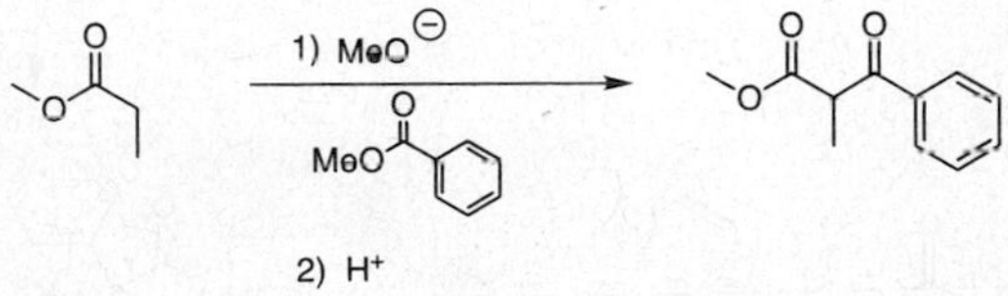

1) MeO⊖

MeO—C(=O)—Ph

2) H⁺

7.58)

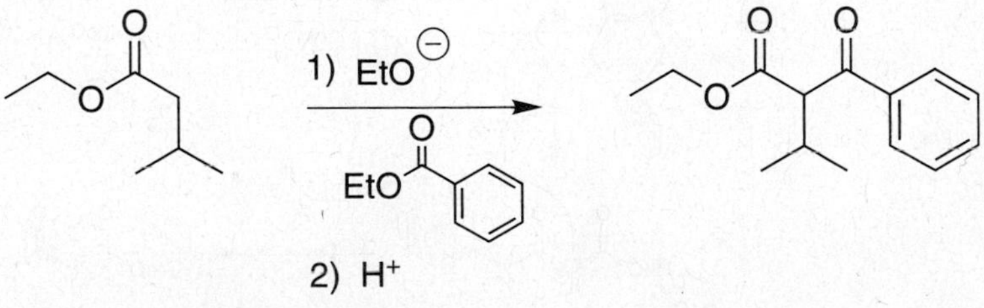

1) EtO⊖

EtO—C(=O)—Ph

2) H⁺

7.59)

1) MeO⊖

MeO—C(=O)—C(CH₃)₃

2) H⁺

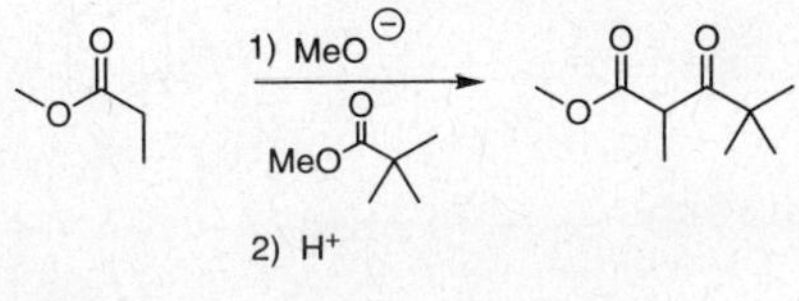

7.60)

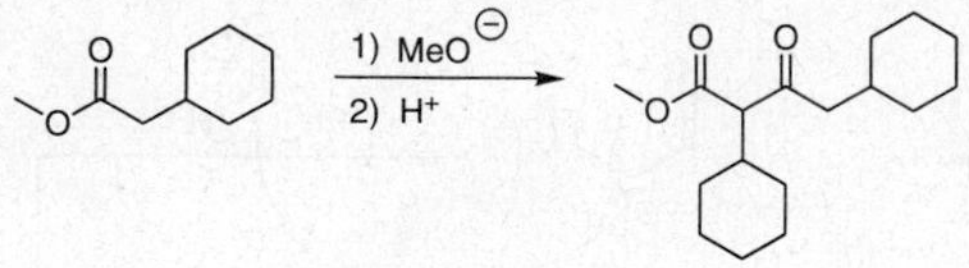
1) MeO
2) H+

7.62)

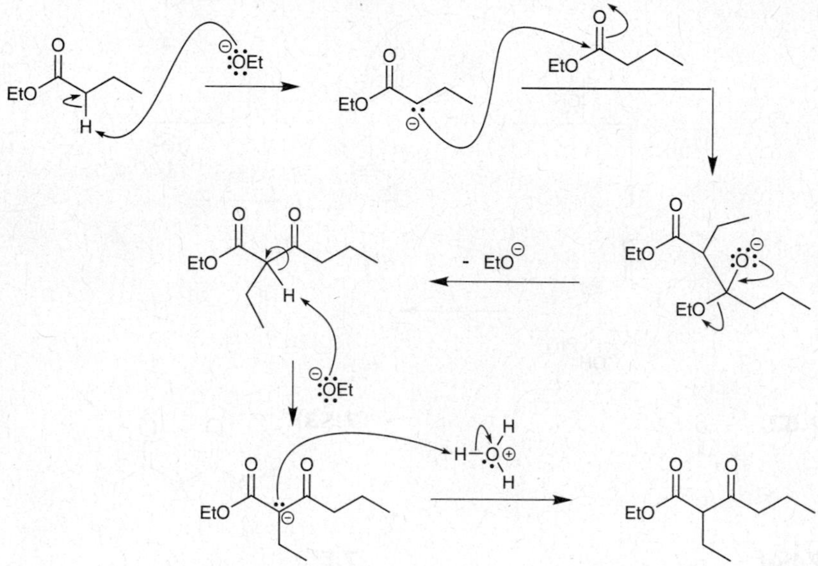
EtO
H
OEt
EtO
EtO
- EtO
EtO
EtO
H
OEt
H
H–O
H
EtO
EtO

7.63)

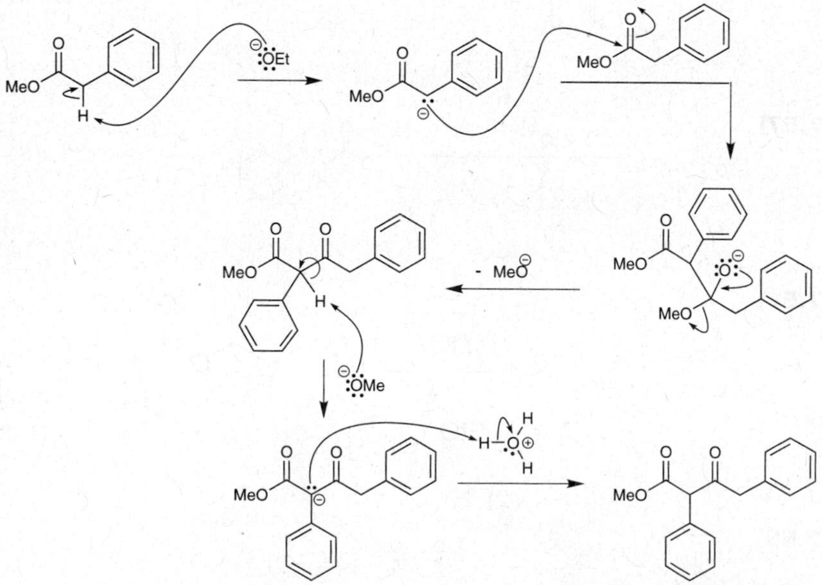
MeO
H
OEt
MeO
MeO
- MeO
MeO
MeO
H
OMe
H
H–O
H
MeO
MeO

7.64)

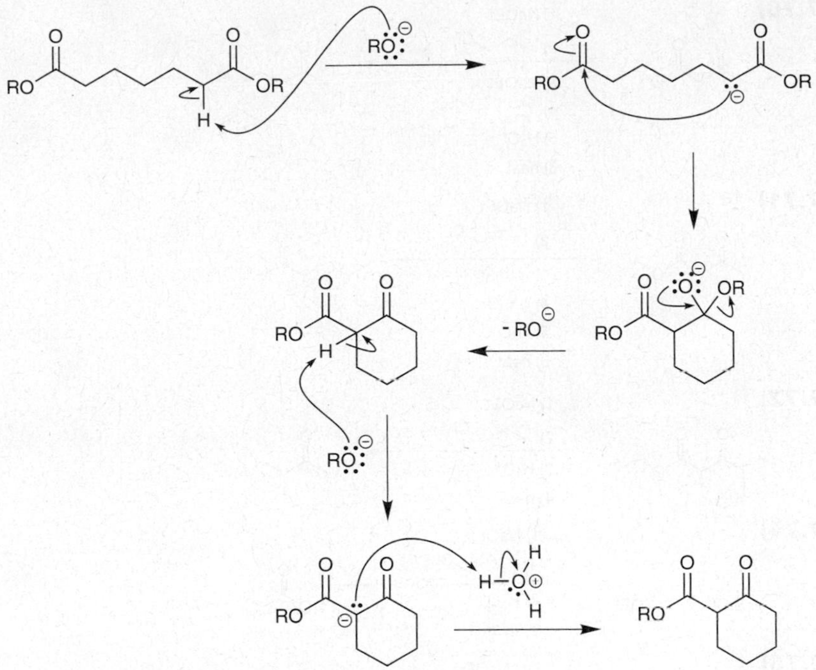
RO
OR
H
RO:
O
- RO
H–O–H
H

7.66)

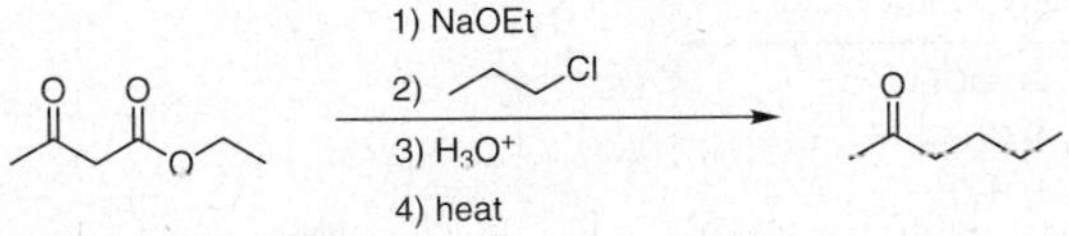
1) NaOEt
2) Cl
3) H_3O^+
4) heat

7.67)

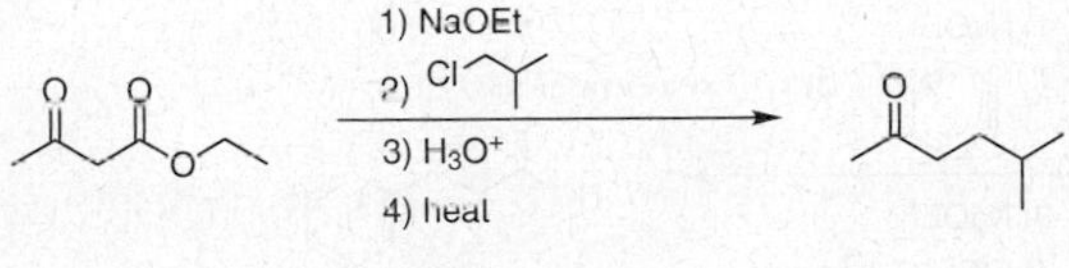
1) NaOEt
2) Cl
3) H_3O^+
4) heat

7.68)

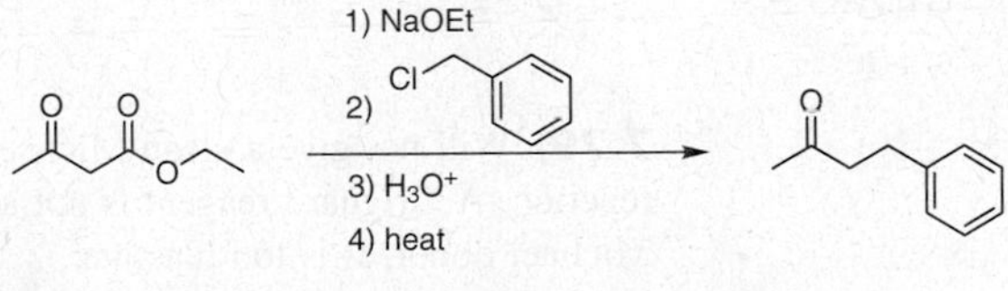
1) NaOEt
2) Cl
3) H_3O^+
4) heat

7.69) You would need to use the following halide:

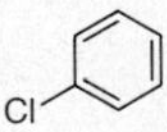
Cl

which will not undergo an S_N2 reaction (the carbon atom connected to the leaving group is sp^2 hybridized).

7.70)

1) NaOEt
2) ~~~Cl
3) NaOEt
4) Cl~~~
5) H₃O⁺
6) heat

7.71)

1) NaOEt
2) ~~~Cl
3) NaOEt
4) CH₃Cl
5) H₃O⁺
6) heat

7.72)

1) NaOEt
2) ~~~Cl
3) H₃O⁺
4) heat

7.74)

1) NaOEt
2) ~~~Cl
3) H₃O⁺
4) heat

7.75)

1) NaOEt
2) ~~~Cl
3) NaOEt
4) CH₃Cl
5) H₃O⁺
6) heat

7.76)

1) NaOEt
2) ⬡~Cl
3) NaOEt
4) EtCl
5) H₃O⁺
6) heat

7.78)

7.80)

7.79) Will not give a clean Michael reaction. A Grignard reagent is not a good Michael donor. It is too reactive.

7.82)

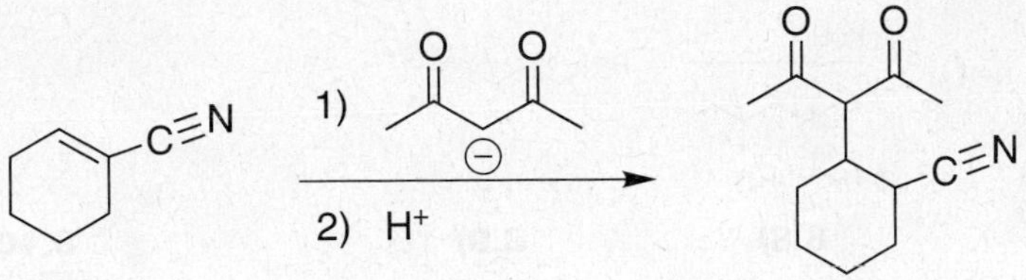

7.83)

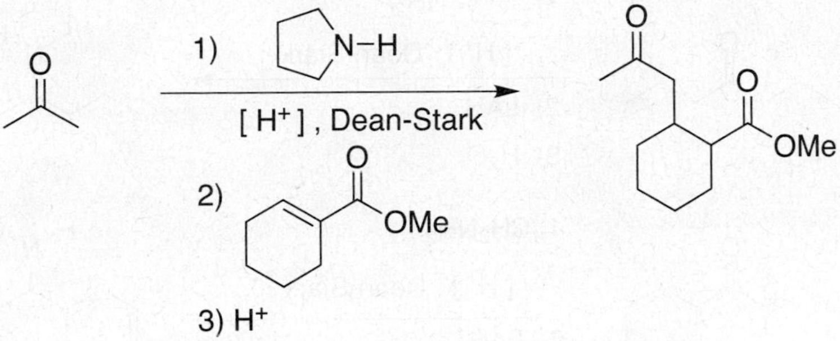

7.84)

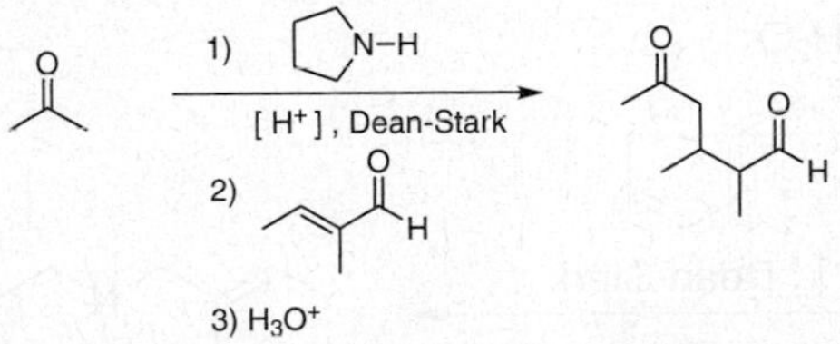

7.85)

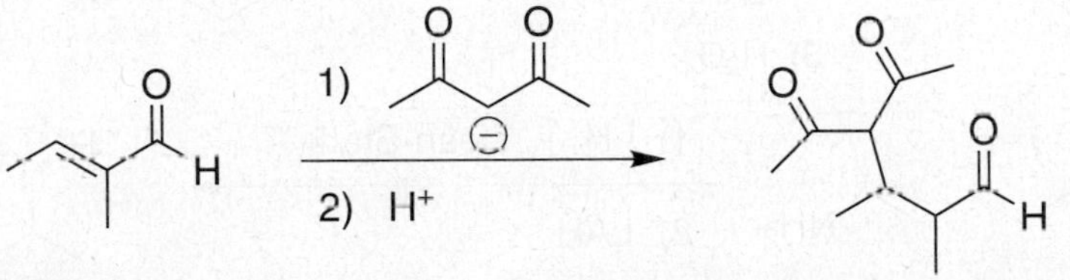

CHAPTER 8

8.2)

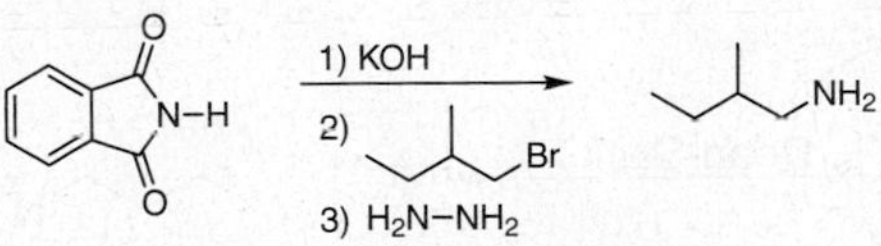

8.3)

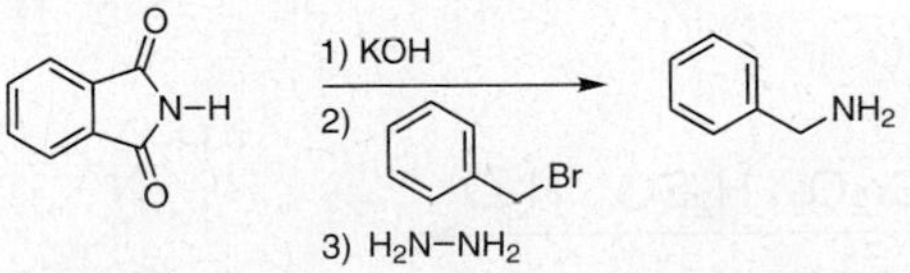

8.4)

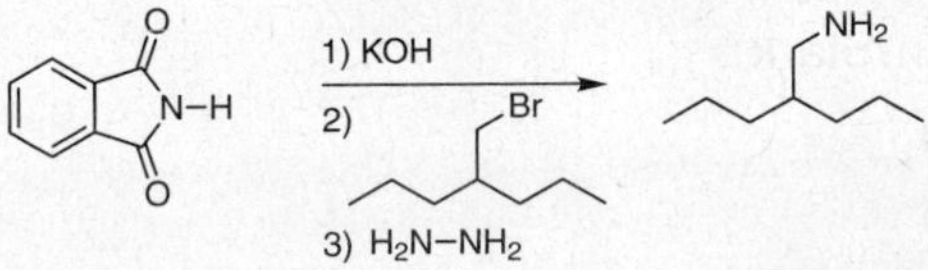

8.5)

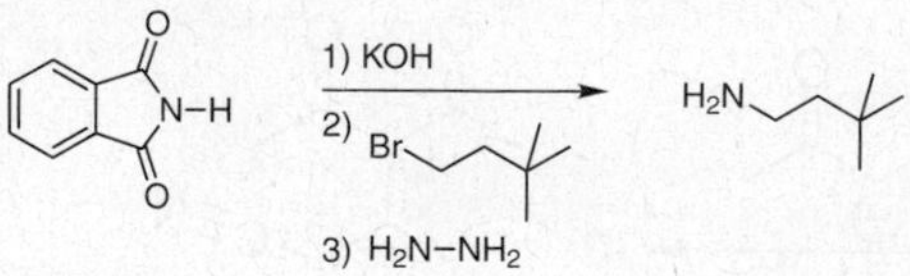

Reagents over arrow: 1) KOH 2) Br~(neopentyl bromide) 3) H$_2$N–NH$_2$

Product: H$_2$N-CH$_2$C(CH$_3$)$_3$

8.7) No   **8.8)** Yes   **8.9)** Yes   **8.10)** No

8.12)

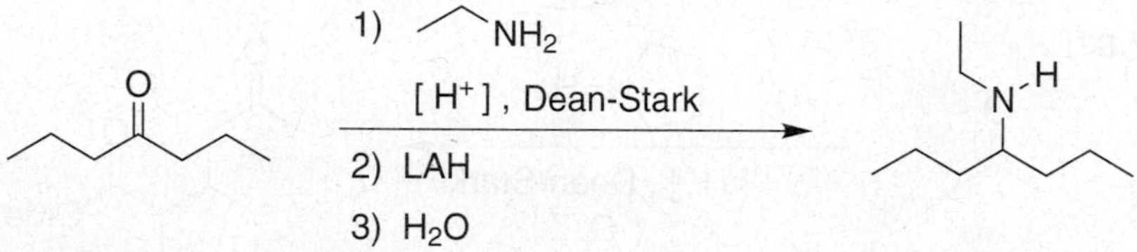

1) CH$_3$CH$_2$NH$_2$

[H$^+$], Dean-Stark

2) LAH

3) H$_2$O

8.13)

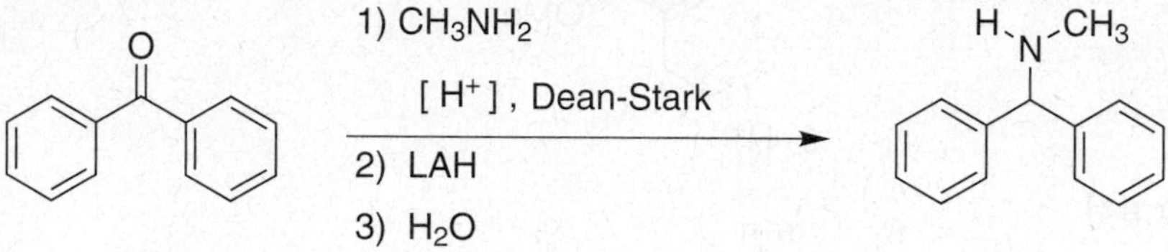

1) CH$_3$NH$_2$

[H$^+$], Dean-Stark

2) LAH

3) H$_2$O

8.14)

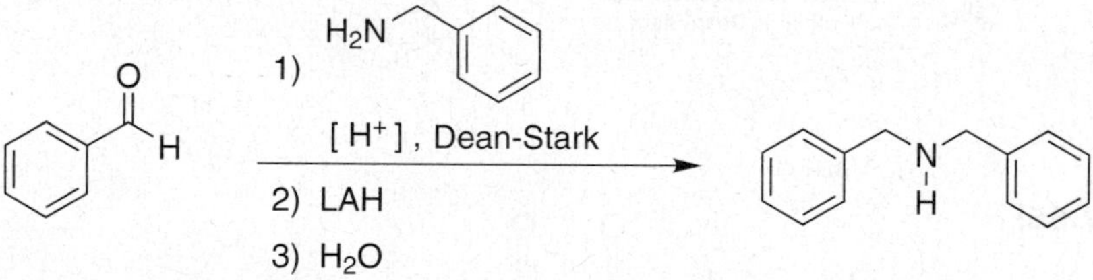

1) H$_2$N-CH$_2$-C$_6$H$_5$

[H$^+$], Dean-Stark

2) LAH

3) H$_2$O

8.15)

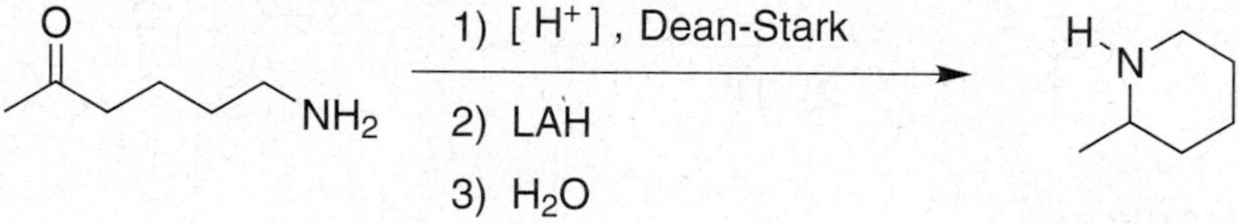

1) [H$^+$], Dean-Stark

2) LAH

3) H$_2$O

8.16)

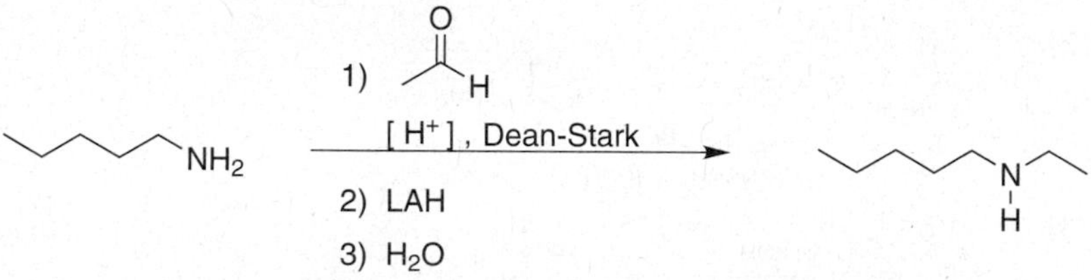

1) CH$_3$CHO

[H$^+$], Dean-Stark

2) LAH

3) H$_2$O

8.18)

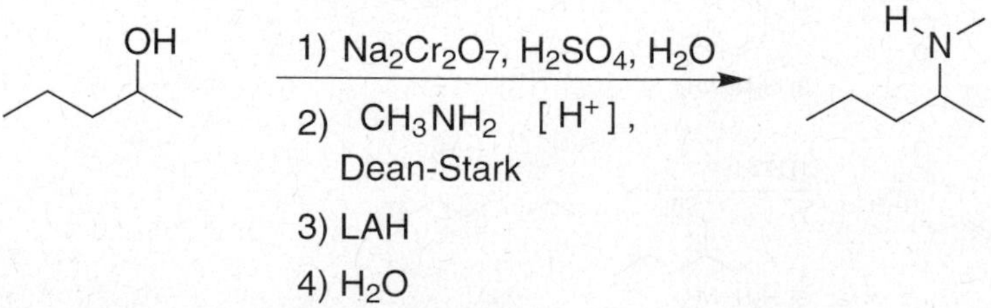

1) Na$_2$Cr$_2$O$_7$, H$_2$SO$_4$, H$_2$O

2) CH$_3$NH$_2$ [H$^+$],
Dean-Stark

3) LAH

4) H$_2$O

8.19)

1) BH₃ · THF
2) H₂O₂, NaOH
3) PCC
4) ⌃NH₂ [H⁺], Dean-Stark
5) LAH
6) H₂O

8.20)

1) H₃O⁺
2) [H⁺], CH₃CH₂CH₂NH₂
   Dean-Stark
3) LAH
4) H₂O

8.21)

1) O₃
2) DMS
3) ⌃NH₂ [H⁺],
   Dean-Stark
4) LAH
5) H₂O

8.22)

1) Et₂CuLi
2) ⌃NH₂ [H⁺],
   Dean-Stark
3) LAH
4) H₂O

8.24)

1) CH₃C(=O)Cl
2) cyclohexyl-C(=O)Cl   AlCl₃
3) H₃O⁺

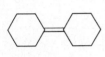

8.25)

1) CH₃C(=O)Cl
2) Cl₂, AlCl₃
3) H₃O⁺

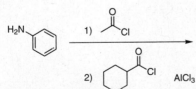

8.27)

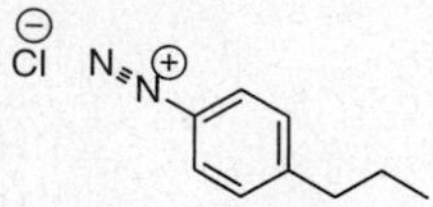

8.28)

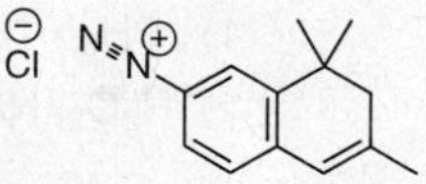

8.29)

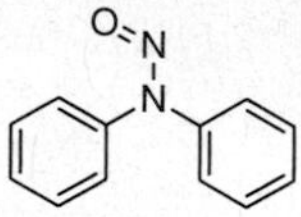

8.30)

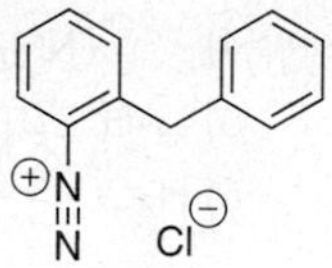

8.32)

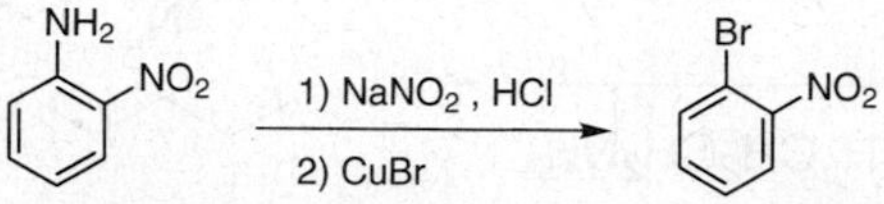

NH$_2$, NO$_2$

1) NaNO$_2$, HCl

2) CuBr

Br, NO$_2$

8.33)

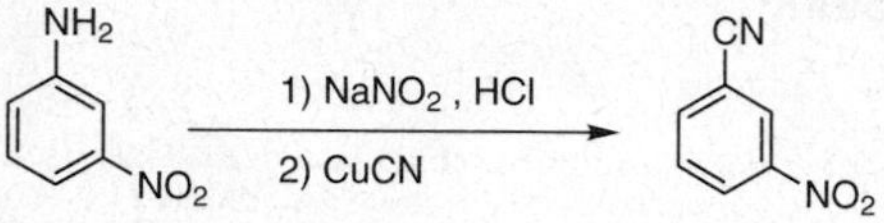

NH$_2$

1) NaNO$_2$, HCl

2) CuCN

NO$_2$

CN, NO$_2$

8.34)

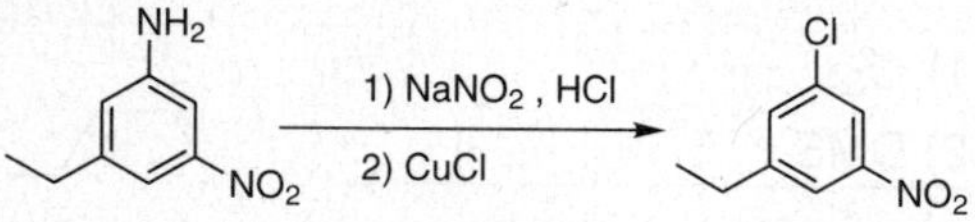

NH$_2$

1) NaNO$_2$, HCl

2) CuCl

NO$_2$

Cl, NO$_2$

8.35)

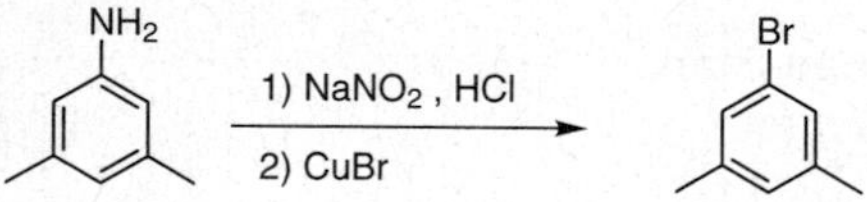

NH$_2$

1) NaNO$_2$, HCl

2) CuBr

Br

8.36)

NH$_2$

1) NaNO$_2$, HCl

2) CuCN

CN

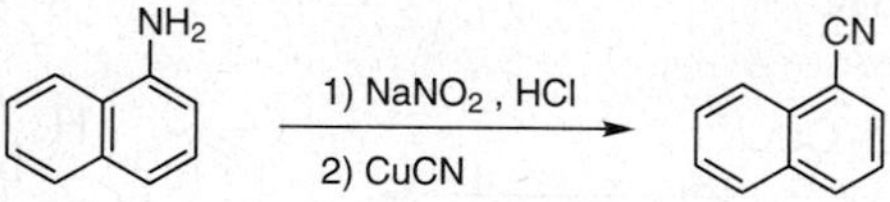

INDEX